Lecture Notes in Physics

Edited by J. Ehlers, München, K. Hepp, Zürich
R. Kippenhahn, München, H. A. Weidenmüller, Heidelberg
and J. Zittartz, Köln
Managing Editor: W. Beiglböck, Heidelberg

145

Topics in Nuclear Physics II

A Comprehensive Review of Recent Developments

Lecture Notes for the International
Winter School in Nuclear Physics
Held at Beijing (Peking),
The People's Republic of China
December 22, 1980 – January 9, 1981

Edited by T. T. S. Kuo and S. S. M. Wong

Springer-Verlag
Berlin Heidelberg New York 1981

Editors

T. T. S. Kuo
Physics Department
State University of New York at Stony Brook, NY 11794, USA

S. S. M. Wong
Physics Department
University of Toronto, Toronto, Ontario, Canada

ISBN 3-540-10853-X Springer-Verlag Berlin Heidelberg New York
ISBN 0-387-10853-X Springer-Verlag New York Heidelberg Berlin

Printing and binding: Beltz Offsetdruck, Hemsbach/Bergstr.
2153/3140-543210

TABLE OF CONTENTS (Vol. 2)

Chapter V: Microscopic Description of the Nuclear Cluster Theory (Y.C. TANG)

Chapter V

MICROSCOPIC DESCRIPTION OF THE NUCLEAR CLUSTER THEORY

Y. C. Tang
School of Physics, University of Minnesota,
Minneapolis, Minnesota 55455 USA

Abstract: The purpose of this series of lectures is to explain the foundation of the techniques used in, and the results obtained by a microscopic cluster theory (MCT). In particular, the important role played by the Pauli principle in determining nuclear characteristics will be extensively discussed.

1. Introduction

Experimental observations have shown that nuclei exhibit a variety of interesting, often perplexing, phenomena. For an explanation of these phenomena, there
have been proposed numerous types of nuclear single-particle and collective models.
Among these, one of the earliest was the compound-nucleus model suggested by Bohr
[BO 36]. As is well known, this model was specifically introduced to describe the
complexity shown by the energy spectra of many medium-heavy and heavy nuclei. It
was based on the belief that, since nucleons in nuclei interact strongly with one
another, their motion must be correlated to a very large extent. With this model,
it was indeed possible to qualitatively explain the observation of very sharp
resonance levels when heavy nuclei are bombarded by low-energy neutrons, which,
according to the uncertainty principle, indicates the existence of intermediate
compound states having long lifetimes.

For a quantitative description of the compound-nucleus behavior, Wheeler
[WH 37] proposed the method of resonating-group structure or the resonating-group
method (RGM). In this method, the main idea is that, because of the on-the-
average attractive nature of the nuclear forces, there exist in nuclei relatively
long-range correlations which manifest themselves through the formation of nucleon
clusters. The intricate phenomena exhibited by nuclear systems are, therefore,
considered to be a consequence of the dynamical interplay between various cluster
structures. During the forties and fifties, this method was extensively employed
by especially the groups at the Universities of London and Manchester [GR 60,
HE 57, LA 62, VA 59] to study the problems of nuclear scattering and reactions. The
results thus obtained agreed generally quite well with experiment. However,
because of computational difficulties, only very light systems could be investigated, namely, those systems which involve two s-shell nuclei in both the incident
and the outgoing channels.

A more static description of nuclear cluster structure has been suggested by
Margenau [MA 41]. This suggestion was subsequently extended by Bloch and Brink [BR
66], who formulated the so-called α-cluster model (sometimes referred to in the
literature as the Brink model) through the use of many-center harmonic-oscillator
shell-model wave functions. This model, which is microscopic in nature and funda-
mentally different from the older, classical α-particle model [BO 62, DE 54, HA 71,
KA 56, NO 66], has been utilized by many authors [AB 72, BA 80, BR 70, FR 71, FR 72,
GO 79, KH 71] to investigate the properties of α-particle nuclei, such as ^{16}O, ^{20}Ne,
and so on.

By adopting the Hill-Griffin-Wheeler [HI 53, GR 57] generator-coordinate procedure, the Brink model has been further extended for the purpose of providing a
dynamical description of clustering phenomena in nuclei [BR 68, DE 72, GI 73,
JA 64, TA 72, WO 70, YU 72, ZA 71]. With this extension, it became possible to

treat within this model not only nuclear bound-state structure but also scattering and reaction problems. The resultant formalism is now generally known as the generator-coordinate method (GCM) [MI 73, WO 75] and has been widely used in the past ten years to study the behavior of even relatively heavy systems (see, e.g., [BA 80a]). However, it should be mentioned that, in essence, this method is equivalent to the RGM, although it does provide an alternative viewpoint which may frequently be useful in the consideration of specific nuclear phenomena, such as rotational excitation [KE 77, PE 57, VE 63] and fission.

At a first glance, it might appear that both the RGM and the GCM are best suited only for the description of collective motions in nuclei. This is, however, not so. It has been shown that, because of the Pauli principle, these methods can be used to describe single-particle behavior equally as well. This important point concerning Pauli effects was clearly elaborated by Perring and Skyrme [PE 56] and its significance has been particularly emphasized by Wildermuth and others [AR 72, BA 58, BR 57, EL 55, TA 62, WI 58, WI 66, WI 79].

The fact that the Pauli principle reduces the differences between apparently different structures of fermion systems is crucial for a clear understanding of all aspects of nuclear structure and reactions. It is the key point to the resolution of seemingly contradictory descriptions of the nucleus by various models, all of which have had some success in predicting nuclear characteristics. In fact, the recognition of the importance of the Pauli principle in this respect was the main stimulus which motivated Wildermuth and Tang [WI 77] to propose a microscopic theory in which all non-relativistic nuclear phenomena can be considered from a unified point of view. A careful exposition of some important features concerning this unified theory, hereafter to be termed <u>Microscopic Cluster Theory</u> (MCT), will be the main purpose of this series of lectures. At the present moment, this theory is formulated mainly in the framework of the RGM; but in view of the essential equivalence between RGM and GCM, it can be formulated in terms of the GCM just as well.

Before we proceed to discuss the foundation of, the techniques used in, and the results obtained by this microscopic cluster theory, it will be useful to mention its main characteristics. These characteristics are:

(i) It is a microscopic formulation which explicitly takes cluster correlations into consideration.

(ii) It employs totally antisymmetric wave functions and, therefore, the Pauli principle is fully accounted for.

(iii) It utilizes a nucleon-nucleon potential which explains as well as possible the two-nucleon low-energy scattering data.

(iv) It treats correctly the motion of the total center of mass.

(v) It considers nuclear bound-state, scattering, and reaction problems in a unified manner.

(vi) It can be used to study cases where the particles involved in the incoming and outgoing channels are both arbitrary composite nuclei.

(vii) It is based on a variational principle; consequently, the accuracy of the result can be tested and improved by systematically expanding the basic-function space employed in the calculation.

As is unavoidable with a microscopic description possessing these features, the main difficulty is that practical calculations become frequently rather involved. In spite of this, there already existed many investigations (see, e.g., refs. [BE 69, GO 75, KA 75, MI 77, TA 78, VA 78]) which served to convincingly demonstrate the flexibility and the power of this unified theory.

In the next section, we briefly discuss the cluster representation of nuclear states. The main emphasis is to show the equivalence of cluster and shell models in the oscillator representation. Because of this equivalence, it will be possible to learn a great deal about the properties of low excited states of a nucleus without carrying out explicit calculations, but by making rather simple qualitative studies in different oscillator cluster representations   and extracting non-conflicting features. Finally, in this section, we shall also mention the way to generalize the oscillator cluster function in order to acquire further flexibility for a better description of the behavior of the system under consideration.

The formulation of the MCT is described in sect. 3. As was mentioned already, it will be based on a variational principle, with the Hilbert space spanned by a set of non-orthogonal basis wave functions. In ref. [WI 77], it has been carefully explained that, even though the use of non-orthogonal wave functions may lead to tedious calculations, it is necessary for the important purpose of introducing the incoming and outgoing channels symmetrically into the theory.

In sect. 4, we describe various aspects of the RGM. The procedure of systematically improving the calculation will be outlined. Also, we shall discuss in detail a computational technique which has been successfully used to calculate the kernel functions appearing in such calculations, namely, the complex-generator-coordinate technique (CGCT) [LE 77, TA 77, TA 78, TH 75, TH 77]. In this technique, the essential idea is to express the resonating-group wave function as a linear superposition of anti-symmetrized products of single-particle wave functions or Slater determinants. Then, by employing well-developed methods of dealing with product functions, one can usually carry out the analytical calculation of these kernel functions in a relatively straightforward manner.

The GCM will be briefly explained in sect. 5. Here the main point is to show that the physical ideas behind the GCM and the RGM are the same and, hence, either method may be adopted, depending on one's philosophical preference. The often-used contention that the GCM is computationally easier to handle (see, e.g., [BA 80a, BR 66, TA 75]) is not a valid argument. Back in the fifties and sixties, the RGM matrix elements were computed by the so-called cluster-coordinate technique

[CH 73, HA 67], where one had to perform a large number of linear transformations on the Jacobi internal coordinates [WI 77]. As a consequence, it has been frequently asserted that an application of the RGM to relatively heavy systems would be computationally infeasible [AR 72]. With the development of the CGCT mentioned in the preceding paragraph, such difficulty no longer exists because one now works directly with nucleon coordinates instead of cluster coordinates. In fact, it is my opinion that the RGM with CGCT may even be somewhat more flexible and, from a computational viewpoint, more convenient than the GCM in its present stage of development.

Results of representative bound-state, scattering, and reaction calculations are presented in sect. 6. The purpose here is not only to demonstrate the general utility of the MCT in treating nuclear problems, but also to discuss the physical information obtained from the investigations carried out until this moment.

The most important aspect of the MCT is that the Pauli principle is fully taken into account. Thus, from investigations based on this theory, one can obtain information about the effects of antisymmetrization. In sect. 7, we shall make some general discussion concerning these effects. Such discussion will obviously be very useful toward achieving a clear understanding of various nuclear phenomena and for the construction of realistic macroscopic models.

Even though there exists a substantial number of MCT-type calculations, it is somewhat surprising that the resultant many-nucleon wave functions have not been fully utilized. In sect. 8, we discuss some of the interesting results obtained by using these wave functions to determine the electromagnetic properties of nuclear systems. It is my hope that a discussion of these results will stimulate others to perform further and systematic electromagnetic and weak-interaction calculations with MCT wave functions.

Concluding remarks are given in sect. 9, where a discussion of future prospects and open problems will also be presented.

In the Appendix, we discuss the orthogonality-condition model (OCM) of Saito [FL 75, FL 76, MA 73, SA 68, SA 69, SA 77]. This semi-microscopic model was proposed for the explicit purpose of avoiding the derivation of the complicated kernel functions occurring in the RGM by taking the effects of the Pauli principle only approximately into account. It has since been extensively used to treat especially multi-cluster structure problems [HO 77, HO 78]. The results obtained in light systems seem to be fairly reasonable, suggesting that this model may be generally useful when one wishes to conduct initial, approximate studies concerning particularly the higher-excitation regions of nuclear systems.

2. Cluster Representation of Nuclear States

2.1. Oscillator cluster representations

It is well known that nuclei exhibit different kinds of behavior. Some of these are due to single-particle features, while others are connected with collective motions of the nucleons. The relative importance of the various types of behavior can change significantly from nucleus to nucleus and even from one level to the next. Thus, there exists very often a particular set of single-particle or collective coordinates, which is most appropriate for an adequate description of a given nuclear level.

The interesting phenomena of structure change have particularly been noticed in light nuclear systems. This is related to the weak nature of the intercluster interaction [IK 75, NE 69], which results from the Pauli principle and the fact that the nucleon-nucleon potential has a strong triplet-even tensor part and a large Majorana component. For instance, in the nucleus ^{7}Li, the lowest four levels have predominantly an $\alpha + t$ cluster configuration, while the level at 7.47 MeV experiences a change in structure and has instead an $n + {}^6$Li cluster configuration (see fig. 1). In the self-conjugate 4n-nuclei, the study of structure change has especially been carried out and the result can be schematically summarized by the so-called Ikeda diagram [IK 68] depicted in fig. 2 where the unlabelled smaller circles represent α clusters. From this figure, it is seen that there appear between the ground-state series dominated by shell-model character and the series formed by complete dissociation into α clusters many intermediate quasi-molecule-like series of states. Because of the weak intercluster force, the dissociation energy is rather small compared to the internal energies

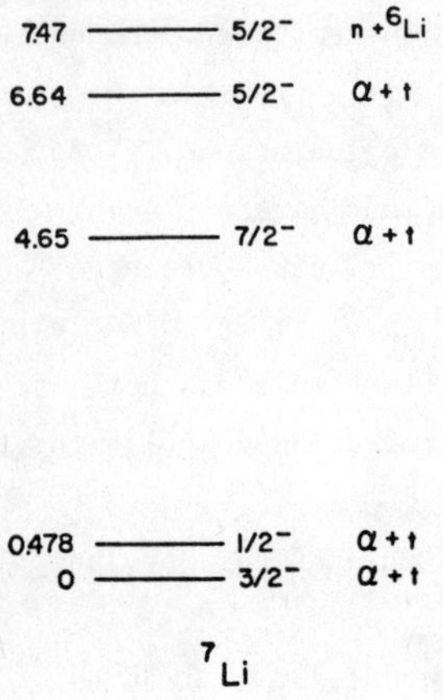

Fig. 1. Cluster structure of ^{7}Li.

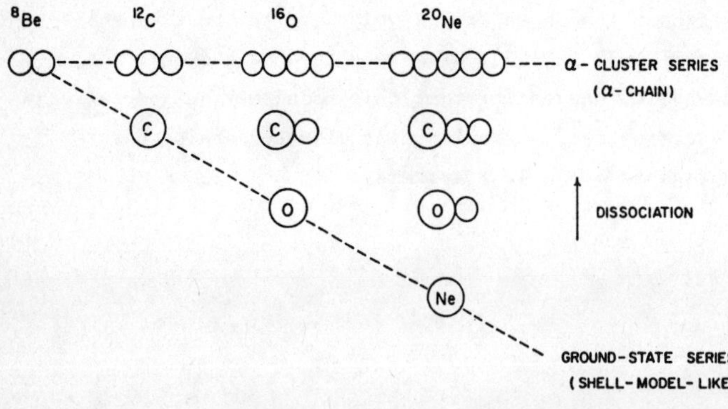

Fig. 2. Ikeda diagram of structure change in self-conjugate 4n-nuclei.

of the clusters and, consequently, structure change can already occur in the region of low excitation. As a specific example, consider the nucleus ^{16}O. The ground 0^+ state can be reasonably well described by a shell-model wave function, but the second 0^+ state has been shown by Susuki [SU 76] to have an $\alpha + {}^{12}C$ weak-coupling structure, with the ^{12}C cluster mainly in its ground state. The 4α linear-chain structure[+] [MO 56, MO 66] indicated in this figure may correspond to a rotational band with a band head which could occur at an excitation of about 16.8 MeV [CH 67]. Similar situation has also been observed in ^{20}Ne and other alpha-particle nuclei [HO 78].

For medium-heavy and heavy nuclei, the situation is somewhat different. In the interior of the nucleus where the nucleon density is large, the Pauli principle is particularly effective in reducing the differences between different cluster structures. Thus, the correlations among the nucleons are expected to be strong only in the surface region; as a consequence, it can be anticipated that the nuclear surface may have a rather granular structure.

The above discussion shows that, for the description of a chosen level or group of levels in a particular nucleus, one should first make a decision about the most appropriate coordinate set to be used. Once this decision is made, one must then find basis systems of wave functions in which the nuclear wave function employing this coordinate set can be expanded. One such system of basis functions is, for example, that generated from an oscillator potential, namely, the common harmonic-oscillator shell-model eigenfunction system (see, for instance, [DE 63]). But the shell-model system is only of limited utility in expanding trial functions, because it is restricted to single-particle coordinates. We now wish to investigate the consequences of introducing various sets of collective coordinates into the oscillator Hamiltonian in order to generate new, more practical basis systems for an approximate solution of the nuclear Schrödinger equation.

There are several reasons why we start with the oscillator potential. First, this potential reflects many broad features of the actual averaged nuclear potential, yet even upon introducing collective coordinates, the resultant Schrödinger equation is simple enough to solve exactly. Second, in order to see how antisymmetrization serves to remove the contradictions among the various collective and single-particle viewpoints, it is important that we be able to compare our new basis systems with each other and with the shell-model system; this is particularly easy if we use the oscillator potential to generate all the systems. Third, the use of the oscillator potential has the advantage that the total center-of-mass motion of the considered

---

[+]It should be noted that the term "linear-chain" is not to be literally interpreted. The 3α "linear-chain" state of ^{12}C at 7.66 MeV has an intrinsic structure where the three α clusters are loosely coupled; however, the centers of mass of these clusters do not form a straight line, but are located on the vertices of a nearly right triangle [IK 75] (see also [UE 75]).

system can always be correctly taken into account in a clearcut manner [EL 55]. Finally, the functions in these new oscillator basis systems can be generalized in a very natural way to obtain good trial functions to be used in the actual Schrödinger equation containing realistic two-nucleon interactions.

We shall first solve the N-nucleon oscillator-Hamiltonian problem using single-particle coordinates. In this case, the Schrödinger equation is

$$\frac{1}{2M}\left[\sum_{i=1}^{N}\vec{p}_i^{\,2} + \alpha^2\hbar^2\sum_{i=1}^{N}\vec{r}_i^{\,2}\right]\phi_n(\vec{r}_1,\cdots,\vec{r}_N)$$

$$= E_n\,\phi_n(\vec{r}_1,\cdots,\vec{r}_N)\,,\tag{2.1}$$

where $\vec{p}_i$ and $\vec{r}_i$ denote the momentum and position vectors of the ith nucleon, respectively. The quantity

$$\alpha = M\omega/\hbar\tag{2.2}$$

is the width parameter of the oscillator potential, with M being the nucleon mass and ω being the angular frequency of the oscillator potential.

Since the Hamiltonian is separable, the eigenfunctions $\phi_n(\vec{r}_1,\ldots,\vec{r}_N)$ of eq. (2.1) are products of single-particle harmonic-oscillator wave functions and form a complete set of orthogonal functions for the N variables $\vec{r}_i$. We can express the antisymmetrized wave function of any state of a nucleus composed of N nucleons as a superposition of anti-symmetrized combinations of these eigenfunctions, if in addition we introduce the spin and isospin functions of the nucleons. As has been mentioned already, this single-particle set has only limited utility as a basis-function set however, because many terms are required to adequately represent any collective behavior of the nucleus.

Now, collective motion occurs when a certain number of nucleons are energetically favored to move in a more or less coherent manner. In general, there may be one or more such groups within a nucleus. Therefore, to introduce more appropriate coordinates, let us divide the N nucleons into K groups, or clusters as is commonly called, with the jth cluster consisting of n_j nucleons such that

$$\sum_{j=1}^{K} n_j = N\,.\tag{2.3}$$

The set of indices $\left\{n_1, n_2, \ldots, n_k\right\}$ denotes what we call a cluster representation of the nucleus. If we now introduce center-of-mass coordinates

$$\vec{R}_j = \frac{1}{n_j}\sum_{i=1}^{n_j}\vec{r}_i\tag{2.4}$$

and center-of-mass momenta

$$\vec{P}_j = \sum_{i=1}^{n_j} \vec{p}_i \tag{2.5}$$

of these clusters, the oscillator-model Schrödinger equation (2.1) can be easily shown to become

$$\left[\sum_{j=1}^{K} \left(H_j + \frac{1}{2Mn_j} \vec{P}_j^{\,2} + \frac{\alpha^2 \hbar^2 n_j}{2M} \vec{R}_j^{\,2} \right) \right] \phi_n = E_n \phi_n , \tag{2.6}$$

where H_j is the internal Hamiltonian of the jth cluster, depending on the internal relative coordinates $(\vec{r}_i - \vec{R}_j)$ of the nucleons in this cluster and the internal relative momenta. The eigenfunctions of eq. (2.6) are, therefore, products of functions, in which each function depends either only on the internal coordinates of one cluster or only on a single center-of-mass coordinate $\vec{R}_j$.

After introducing spin and isospin functions and antisymmetrizing, all these eigenfunction systems are equivalent to the antisymmetrized single-particle wave-function system, in the sense that we can expand all the states of a nucleus in terms of any one of these eigenfunction systems. Thus, we see that to each cluster representation there belongs an antisymmetrized complete, but usually not orthogonal, oscillator eigenfunction system which we can characterize by the representation indices $\left\{ n_1, n_2, \ldots, n_K \right\}$.

We should note here one important point. This point is that, because we have only performed a mathematical transformation on our original oscillator Hamiltonian, all these eigenfunction systems must have identically the same energy spectrum and all angular frequencies appearing implicitly in eq. (2.6) must be equal to ω. This means that any eigenfunction corresponding to a given energy eigenvalue in one eigenfunction system can be expanded in any other eigenfunction system as a linear superposition of just those degenerate eigenfunctions cooresponding to the same energy eigenvalue. This makes the comparison between eigenfunctions of different oscillator eigenfunction systems especially simple.

It now depends on the nature of the nuclear force in which eigen-function system the various states of the nucleus are most simply represented, i.e., which kind of correlation among the nucleons is particularly favored in the different nuclear states. This favored representation, as we have emphasized, can change from nucleus to nucleus and from level to level. One practical criterion for what con-stitutes a simple description of a state in a certain eigenfunction system is that, in this basis system, this state is described essentially by one or by a super-position of only a small number of antisymmetrized eigenfunctions. As an example, let us consider the nucleus ^{8}Be. Because of the fact that α particles are experi-mentally found to be tightly bound, it is intuitively reasonable that, for the description of the ground and low excited states, one should choose the $\{4,4\}$ cluster representation. Indeed, an explicit calculation [PE 60] showed that each of the lowest 0^+, 2^+, and 4^+ states can be qualitatively well described by just a single term in this representation. If one adopts instead the oscillator shell-model representation, then the situation becomes rather more complicated. Here, for instance, an expansion of the $\{44\}$ ground-state cluster function will consist of a superposition of 19 shell-model terms [KA 59].

Antisymmetrized eigenfunctions of the kind discussed above are called oscillator cluster functions. It must be emphasized that the antisymmetrization very often influences the physical properties of a wave function profoundly. Therefore, one always has to investigate carefully how much of the physical contents of unanti-symmetrized cluster wave functions remain after antisymmetrization. This important point will be discussed later in more detail.

2.2. Lowest 4^+ α-cluster state of ^{8}Be

We have already mentioned that, as a result of antisymmetrization, seemingly quite different wave functions can become very similar or even equivalent to each other. This reduction in the differences between different structures by anti-symmetrization is a very general feature and not restricted to the oscillator representations discussed in the preceding subsection. It is only that, with these representations, the effects of antisymmetrization can be demonstrated in a parti-cularly clear manner.

In this subsection, the purpose is to illustrate the influence of anti-symmetrization by means of a specific example. What we shall do is to explicitly carry out the antisymmetrization in the relatively simple case of the $n = 4$, $\ell = 4$, $m_\ell = 4$, α-cluster state of ^{8}Be and show how the wave function appears when ex-pressed as a superposition of single-particle shell-model wave functions [WI 77].

The cluster wave function for this state is taken to be

$$\psi = \mathcal{A}\left[\phi_0(\tilde{\alpha}_A)\phi_0(\tilde{\alpha}_B)\chi_{444}(\vec{R})Z_0(\vec{R}_{cm})\right], \tag{2.7}$$

where $\mathcal{A}$ is an antisymmetrization operator, $\tilde{\alpha}_A$ denotes the internal spatial, spin, and isospin coordinates of all the nucleons in the α-cluster A, and similarly for $\tilde{\alpha}_B$. The relative coordinate $\vec{R}$ and the total c.m. coordinate $\vec{R}_{cm}$ are defined as

$$\vec{R} = \vec{R}_A - \vec{R}_B, \quad \vec{R}_{cm} = \frac{1}{2}(\vec{R}_A + \vec{R}_B), \tag{2.8}$$

with $\vec{R}_A$ and $\vec{R}_B$ representing the c.m. coordinates of the two α clusters, given by

$$\vec{R}_A = \frac{1}{4}\sum_{i=1}^{4}\vec{r}_i, \qquad \vec{R}_B = \frac{1}{4}\sum_{i=5}^{8}\vec{r}_i. \tag{2.9}$$

The physically irrelevant c.m. function $Z_0(\vec{R}_{cm})$ is included here in order to make the transition to the single-particle wave function.

Because in the ^{8}Be state discussed here, the α clusters have no internal excitation, their wave functions are of the form

$$\phi_0(\tilde{\alpha}_A) = exp\left(-\frac{1}{2}\alpha \sum_{i=1}^{4} \vec{P}_i^{\,2}\right) \xi_A(s_1,\cdots,t_4) \tag{2.10}$$

with

$$\xi_A(s_1,\cdots,t_4) = \alpha_1 \nu_1 \alpha_2 \pi_2 \beta_3 \nu_3 \beta_4 \pi_4 \;, \tag{2.11}$$

and a similar expression for $\phi_0(\tilde{\alpha}_B)$. In the above equation, α_i, β_i, ν_i, and π_i denote the spin-up, spin-down, isosopin-up, and isospin-down states for nucleon i, respectively. The internal spatial coordinates are defined as

$$\vec{P}_i = \vec{r}_i - \vec{R}_A \qquad \text{for } i = 1, 2, 3$$
$$= \vec{r}_i - \vec{R}_B \qquad \text{for } i = 5, 6, 7 \;. \tag{2.12}$$

The coordinates $\vec{P}_4$ and $\vec{P}_8$ are not independent coordinates; they are related to the other spatial coordinates by the relations

$$\vec{P}_4 = -(\vec{P}_1 + \vec{P}_2 + \vec{P}_3) \;, \qquad \vec{P}_8 = -(\vec{P}_5 + \vec{P}_6 + \vec{P}_7) \;. \tag{2.13}$$

Note that the function of eq. (2.11) is not an eigenfunction of the α-cluster spin and isospin operators $\vec{S}_\alpha^{\,2}$ and $\vec{T}_\alpha^{\,2}$, but will become an eigenfunction with eigenvalues $S_\alpha = 0$ and $T_\alpha = 0$ after antisymmetrization.

The relative-motion function $\chi_{444}(\vec{R})$ is a 1g oscillator wave function which has four quanta of excitation; it has the form

$$\chi_{444}(\vec{R}) = R^4 exp(-\alpha R^2) Y_{44}(\theta,\varphi) \;, \tag{2.14}$$

where $Y_{44}(\theta\varphi)$ is a spherical harmonic with the indices $\ell = 4$ and $m_\ell = 4$. Finally, $Z_0(\vec{R}_{cm})$ corresponds to a zeroth-order oscillation of the total c.m. of ^{8}Be. It is given by

$$Z_0(\vec{R}_{cm}) = exp(-4\alpha \vec{R}_{cm}^{\,2}) \;. \tag{2.15}$$

With eqs. (2.10), (2.11), (2.14) and (2.15), we obtain for the cluster function ψ the following form:

$$\psi = \mathcal{A}\left\{ \left[\exp\left(-\tfrac{1}{2}\alpha \sum_{i=1}^{4} \vec{\rho}_i^{\,2}\right) \alpha_1 \nu_1 \alpha_2 \pi_2 \beta_3 \nu_3 \beta_4 \pi_4 \right] \right.$$

$$\times \left[\exp\left(-\tfrac{1}{2}\alpha \sum_{i=5}^{8} \vec{\rho}_i^{\,2}\right) \alpha_5 \nu_5 \alpha_6 \pi_6 \beta_7 \nu_7 \beta_8 \pi_8 \right]$$

$$\left. \times \left[R^4 \exp\left(-\alpha R^2\right) Y_{44}(\theta\varphi) \right] \exp\left(-4\alpha \vec{R}_{cm}^{\,2}\right) \right\} . \qquad (2.16)$$

Next, let us express this cluster wave function in terms of nucleon coordinates $\vec{r}_i$ (i = 1 - 8) so that we may explicitly carry out the anti-symmetrization. By using eqs. (2.8), (2.9), (2.12) and (2.13), we obtain for the exponents in eq. (2.16) the following simplification:

$$\frac{\alpha}{2}\sum_{i=1}^{4} \vec{\rho}_i^{\,2} + \frac{\alpha}{2}\sum_{i=5}^{8} \vec{\rho}_i^{\,2} + \alpha \vec{R}^2 + 4\alpha \vec{R}_{cm}^{\,2} = \frac{\alpha}{2}\sum_{i=1}^{8} \vec{r}_i^{\,2} , \qquad (2.17)$$

which is a totally symmetric function of the eight spatial coordinates $\vec{r}_i$. Note that we obtain such a simple form for the exponent, because we have included in ψ the wave function for the total c.m. motion. With eq. (2.17), the cluster function becomes

$$\psi = \mathcal{A}\left\{ R^4 Y_{44}(\theta\varphi) \exp\left(-\frac{\alpha}{2}\sum_{i=1}^{8} \vec{r}_i^{\,2}\right) \alpha_1 \cdots \pi_8 \right\}$$

$$= \frac{3}{16}\sqrt{\frac{35}{2\pi}} \,\mathcal{A}\left\{ (X+iY)^4 \exp\left(-\frac{\alpha}{2}\sum_{i=1}^{8} \vec{r}_i^{\,2}\right) \alpha_1 \cdots \pi_8 \right\}$$

$$= \frac{3}{16}\sqrt{\frac{35}{2\pi}}\left(\frac{1}{4}\right)^4 \mathcal{A}\left\{ \left[(x_1+iy_1)+(x_2+iy_2)+\cdots-(x_5+iy_5) \right. \right.$$

$$\left. \left. -\cdots-(x_8+iy_8)\right]^4 \exp\left(-\frac{\alpha}{2}\sum_{i=1}^{8} \vec{r}_i^{\,2}\right) \alpha_1 \cdots \pi_8 \right\} . \qquad (2.18)$$

The term $[(x_1 + iy_1) + \cdots]^4$ in eq. (2.18) is a sum of many terms of the form

$$(x_1+iy_1)^{n_1}(x_2 + iy_2)^{n_2} \cdots (x_8 + iy_8)^{n_8} ,$$

where $n_1, \cdots, n_8$ take on integral values from 0 to 4 subject to the condition that $n_1 + n_2 + \cdots + n_8 = 4$.

In accordance with the Pauli principle, only those terms which correspond to four nucleons in the 1p-shell and four nucleons in the 1s-shell with the proper spin and isospin configuration are different from zero. Thus, even though there is a large number of terms in ψ of eq. (2.18), most of them vanish because of anti-symmetrization. In fact, it is easy to see that the terms which vanish are of the following types:

(i) Any term with one or more of the n_p larger than 1. An example of this is

$$\mathcal{A}\left\{(x_1+iy_1)^2(x_2+iy_2)(x_3+iy_3)\exp\left(-\frac{\alpha}{2}\sum_{i=1}^{8}\vec{r_i}^{\,2}\right)\alpha_1\cdots\pi_8\right\} \;,$$

which describes a system with five nucleons in the 1s-shell.

(ii) Any term with two of the non-zero n_p being (n_1, n_5), (n_2, n_6), (n_3, n_7), or (n_4, n_8). An example of such a term is

$$\mathcal{A}\left\{-(x_1+iy_1)(x_2+iy_2)(x_3+iy_3)(x_6+iy_6)\exp\left(-\frac{\alpha}{2}\sum_{i=1}^{8}\vec{r_i}^{\,2}\right)\alpha_1\cdots\pi_8\right\}$$

which describes a system with two 1s nucleons in the spin-down, isospin-down state.

Again because of antisymmetrization, the remaining non-vanishing terms are all equal to each other. Therefore, when expressed in nucleon coordinates, the function ψ can be written as

$$\psi = N_p\,\mathcal{A}\left\{(x_5+iy_5)(x_6+iy_6)(x_7+iy_7)(x_8+iy_8)\right.$$
$$\left. \times\exp\left(-\frac{\alpha}{2}\sum_{i=1}^{8}\vec{r_i}^{\,2}\right)\alpha_1\cdots\pi_8\right\} \tag{2.19}$$

with N_p being a constant factor. We see that this wave function corresponds to four 1p nucleons with parallel orbital angular momenta and four 1s nucleons. This is the way it has to be, since we have started with a state of total orbital angular momentum $\ell = 4$.

We have thus shown that the antisymmetrized cluster function of eq. (2.7) is completely identical to the antisymmetrized shell-model function of eq. (2.19). This gives us an important clue as to how one could proceed to resolve apparently contradictory physical descriptions of the same nuclear state.

At this moment, it may be interesting to briefly discuss the relationship between Brink-model and oscillator shell-model wave functions [BR 66]. For low-lying levels of ^{8}Be, the deformed intrinsic state ψ_i is constructed with two α-cluster wave functions consisting of 1s orbitals in harmonic-oscillator wells, with well centers located at points $\vec{d}$ and $-\vec{d}$. Writing

$$\varphi_A(\vec{r}) = C \exp\left[-\frac{\alpha}{2}(\vec{r}-\vec{d})^2\right] \ ,$$

$$\varphi_B(\vec{r}) = C \exp\left[-\frac{\alpha}{2}(\vec{r}+\vec{d})^2\right] \ , \tag{2.20}$$

we can express ψ_i as

$$\psi_i = \mathcal{A}\left\{\left[\prod_{i=1}^{4}\varphi_A(\vec{r}_i)\right]\left[\prod_{i=5}^{8}\varphi_B(\vec{r}_i)\right]\alpha_1\cdots\pi_8\right\}. \tag{2.21}$$

Now, suppose we choose $\vec{d}$ parallel to the z-axis and investigate the behavior of ψ_i as $d \to 0$. For this purpose, it is convenient to introduce orthogonal linear combinations of φ_A and φ_B, i.e.,

$$\varphi_+ = \varphi_A + \varphi_B \ ,$$

$$\varphi_- = \frac{1}{d}(\varphi_A - \varphi_B). \tag{2.22}$$

Then, because of the presence of the antisymmetrization operator $\mathcal{A}$ in eq. (2.21), ψ_i can be further written in the form

$$\psi_i = \left(\frac{1}{2}\right)^4 d^4 \mathcal{A}\left\{\left[\prod_{i=1}^{4}\varphi_+(\vec{r}_i)\right]\left[\prod_{i=5}^{8}\varphi_-(\vec{r}_i)\right]\right.$$

$$\left. \times \ \alpha_1\cdots\pi_8\right\}. \tag{2.23}$$

As $d \to 0$, one can easily see that

$$\varphi_+ \ \to \ 2C \exp\left(-\frac{\alpha}{2}r^2\right) \ ,$$

$$\varphi_- \ \to \ 2C\alpha z \exp\left(-\frac{\alpha}{2}r^2\right). \tag{2.24}$$

In other words, φ_+ tends toward a 1s state and φ_- tends toward a 1p state.

The limiting form of ψ_i as $d \to 0$ is, therefore, just a harmonic-oscillator shell-model function with 4 nucleons in the 1s-shell and 4 nucleons in the 1p-shell. If one further projects out from ψ_i angular-momentum eigenstates with $\ell = 0, 2, 4$, then one obtains the lowest states in the shell-model description of ^{8}Be with ℓ-s coupling.

2.3.   Qualitative discussion of α-cluster states in ^{16}O

By making qualitative studies in different oscillator cluster representations, it is frequently possible to obtain, without carrying out explicit and tedious calculations, considerable information concerning the nature of nuclear states. To illustrate this, we consider in this subsection, the low excited states of ^{16}O in some detail.

For a description of the low-lying energy levels of ^{16}O, we assume an $\alpha + {}^{12}$C cluster representation. In the oscillator cluster picture, this is completely equivalent to a representation in terms of four α clusters, if ^{12}C is considered as consisting of three α clusters without internal excitation [WI 62].

In the ground state of ^{16}O, the relative motion between the unexcited α cluster and the unexcited ^{12}C cluster has four oscillator quanta of energy and zero orbital angular momentum. This is so, because the resultant antisymmetrized oscillator cluster wave function must be mathematically equivalent to the oscillator shell-model wave function describing the configuration in which the 1s and 1p shells are completely filled.

We now consider the lowest negative-parity excited states. In these states where the α and ^{12}C clusters are not internally excited, the relative motion between the two clusters is an oscillation of fifth order with orbital angular momentum ℓ = 1, 3, or 5. Additional negative-parity states can be obtained by coupling these orbital states to the 2^{+} and 4^{+} excited states of ^{12}C which, as the ground state, are also states with no internally excited α clusters and an oscillator shell-model configuration of $(1s)^{4}(1p)^{8}$. For our present consideration, the important point to note is that, in the oscillator model, the energy eigenvalues for all these negative-parity cluster wave functions are just one oscillator quantum larger than the energy eigenvalue for the ground-state cluster wave function.

By coupling these angular-momentum values, one might think at first that there will exist many low-lying negative-parity states with J^{π} = 1^{-} to 9^{-}. However, we shall show that, after antisymmetrization, most of these negative-parity wave functions must either vanish or become identical to one another. To see this, we employ the method of considering these levels simultaneously in different representations -- the oscillator shell-model representation and the oscillator $\alpha + {}^{12}$C cluster representation. In the shell-model representation, all these negative-parity levels of ^{16}O, having one oscillator quantum of excitation energy and in which none of the four α clusters are internally excited, correspond to one-particle excitations to the 2s-1d shell. In addition, one can have only such one-particle excitations in which no spin flip and no isospin flip occur (i.e., excitation to states with T = 0 and S = 0), because otherwise an α cluster would be broken up. But this means that the total angular momenta of these states must come only from the orbital angular momenta of the nucleons. By exciting one nucleon from the 1p shell to the 2s-1d shell, the orbital angular momenta of the nucleons can couple

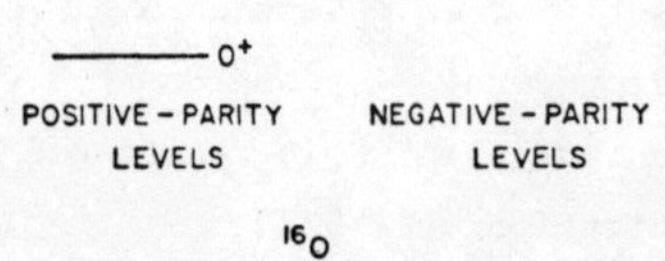

Fig. 3. Energy levels of ^{16}O.

to yield only total angular-momentum values $J = 3$, 2, and 1. Further, one sees that for $J = 3$ and 2, just one coupling possibility exists. Therefore, ^{16}O has one 3^- and one 2^- state which are one-particle-excitation states to the next higher oscillator shell and in which simultaneously no α cluster is broken up. For the $J = 1$ case, there are two coupling possibilities. But one of these couplings must describe a pure center-of-mass excitation of the ^{16}O nucleus in the space-fixed oscillator potential, which has no physical significance. The orbital angular momentum of this excited c.m. motion is $L = 1$. This is easily seen by briefly referring back to the oscillator cluster picture. There we can obtain from the ground state a negative-parity state with one additional oscillator quantum of excitation by changing the c.m. function from a 1s to a 1p oscillation. Because the c.m. coordinate is symmetric in all the nucleon spatial coordinates, this spurious state cannot vanish upon antisymmetrization. Hence, there exists also only one $J = 1$ negative-parity ^{16}O state in which a single nucleon is excited to the next higher oscillator shell and no α cluster is broken up. One sees from this that from all coupling possibilities which one expects in the $\alpha + {}^{12}$C oscillator cluster representation ($J^\pi = 1^-$, ..., 9^-), only three possibilities remain in reality. This large reduction is due to the Pauli principle.

Our considerations show that the lowest measured negative-parity levels of ^{16}O should have $J^\pi = 3^-$, 1^-, and 2^- (see fig. 3). As one might expect, the 2^- level lies above the 1^- and 3^- levels, because there the ^{12}C cluster has to be in its first excited state. The penetrating-orbit argument [WI 77], which can explain the correct level ordering in lighter nuclei such as ^{6}Li and ^{8}Be, no longer suffices to predict the ordering of the 1^- and 3^- levels, since there is large cluster overlapping in both these states and other more complicated effects connected with the Pauli principle come into play. However, realistic calculations in ^{16}O [HO 75, SU 76] and in the analogous case of ^{12}C [FU 78] did show that the 3^- state should lie slightly lower than the 1^- state, in agreement with experimental observation.

The first calculation of the negative-parity states of ^{16}O with $T = 0$ were made by Elliott and Flowers [EL 57] using one-particle excitations in the oscillator shell model. That calculation yielded the correct energy sequence of the three lowest states practically independent of the forces employed, whereas the other $T = 0$ negative-parity states do not fit in with the experimentally measured level scheme and depend rather sensitively on the choice of the nuclear force. The reason for this follows immediately from our discussion given above. The first three negative-parity states are the three α-cluster states we just considered. Due to our considerations, in the other negative-parity states of Elliott and Flowers, at least one of the α clusters must always be broken up. Therefore, their excitation energies should be around 15 MeV and higher instead of, for instance, 9.58 MeV as was found experimentally for the fourth negative-parity state of ^{16}O. It is energetically much easier to excite a higher order relative motion of the α cluster against the unexcited ^{12}C cluster than to break up an α cluster. Thus, one has to assume that the fourth negative parity state of ^{16}O is again approximately an α-cluster state, but with a relative cluster oscillation of seventh order. For the lowest α-cluster state with seven oscillator quanta of excitation, one expects a 4p oscillation between the clusters. The total angular momentum and parity of this state must therefore be 1^-, which agrees with the experimental finding (see fig. 3).

We discuss now briefly the low-excitation positive-parity states of ^{16}O. In the framework of the oscillator cluster model, the second 0^+ state at 6.06 MeV is a well-formed α-cluster state and is described by a relative oscillation of at least sixth order between an α cluster and an unexcited ^{12}C cluster. It is the band head of a $K^\pi = 0^+$ rotational band with the other members being the 2^+ state at 6.92 MeV, 4^+ state at 10.36 MeV, and 6^+ state at 16.23 MeV. In addition, it forms an inversion doublet [HO 68] together with the 9.58-MeV 1^- state, which is the band head of a $K^\pi = 0^-$ band with other members being the 3^- state at 11.63 MeV, 5^- state at 14.68 MeV, and 7^- state at 21.04 MeV. The fact that the difference in the band-head energies, being only 3.52 MeV, is not very large indicates the strong degree of $\alpha + {}^{12}$C clustering in the rotational states of these two bands [IK 75]. Also, from fig. 3, one sees that above the rotational 2^+ state, there appears a second 2^+ state at 9.84 MeV. The nature of this state is similar to that of the 6.06 - MeV, 0^+ state except that the constituent ^{12}C cluster is internally excited to its first excited state with $J^\pi = 2^+$. Such a description of the 9.84-MeV level is supported by the observation that its energy distance from the first excited 0^+ level is 3.8 MeV, which is approximately the excitation energy (4.4 MeV) of the first excited ^{12}C level. The near equality of these two energy distances is, of course, to be expected because the nuclear force is short-ranged and for the low-excited positive-parity ^{16}O states the mutual penetration of the α and ^{12}C clusters is already strongly reduced.

2.4. Generalized cluster wave functions

Our discussion above indicates that the oscillator cluster model can very often give us insight into the qualitative structure of nuclear spectra. On the other hand, for quantitative studies using one or a small number of oscillator cluster functions, one will frequently obtain rather poor results, if the system under consideration exhibits strong collective behavior. The reason is that, for a chosen oscillator cluster function, the width parameters which determine intracluster and intercluster motions are both fixed entirely by the choice of the width parameter α of the original single-particle oscillator well. Therefore, a single cluster wave function has only very limited flexibility and cannot be expected to quantitatively describe the elaborate features of almost any nuclear system.

To remedy the situation, one could, of course, employ a large number of oscillator cluster functions in calculating the energy expectation values and other relevant quantities. However, this would cause us to lose physical understanding of the structure of the system. A better way is to generalize the oscillator cluster wave functions such that only one or a few of the resultant generalized cluster wave functions [WI 77] can be used to yield satisfactory, quantitative results. To see how this generalization can be done, we shall first make a qualitative discussion of the roles played by the internal and relative-motion width parameters. As is quite evident, the internal width parameter determines the mean radius of the cluster (insofar as one can speak of the radius of a cluster in nuclear matter); as the internal width parameter becomes larger, the mean radius of the cluster becomes smaller. Similarly, the relative-motion width parameter determines, in a rough sense, the mean separation of the cluster centroids.

To make our discussion as clear as possible, let us consider a specific case where the collective feature is prominent, namely, the second 0^+ state of ^{16}O. If one calculates the expectation values of a realistic microscopic nuclear Hamiltonian for the ground and this collective state, then one will obtain, by using for each of these states a single oscillator cluster function with a width parameter $\alpha = 0.32$ fm^{-2} which yields the correct rms radius of ^{16}O, an excitation energy of about 20 MeV. To reduce this excitation energy, we mention two important effects. The first effect, the so-called radius-change effect, is associated with the fact that the width parameter (≈ 0.52 fm^{-2}) of a free α particle is appreciably larger than the width parameter (≈ 0.32 fm^{-2}) appropriate to the ^{16}O in its ground state. This means that a free α particle has a smaller mean radius than an α cluster in the ground state of ^{16}O. Since in the first excited positive-parity state at 6.06 MeV, the α and the ^{12}C clusters are bound by only 1.09 MeV, one should expect that at least some of the α clusters will spend an appreciable amount of time outside the nucleus and, therefore, behave more like free α particles. Thus, in the calculation of the energy expectation value for this state, it will be more appropriate to choose internal α-cluster width parameters larger than 0.32 fm^{-2}. The second effect, the so-called anharmonicity effect, follows from the observation that, again because of the small α-cluster separation energy, the relative motion width parameter for this excited state should be smaller than 0.32 fm^{-2}. This must be so, since in the limit case where one α cluster becomes free, this parameter must go to zero. These two effects, if properly considered, can be expected to yield a substantially better description of the nucleon correlation behavior. For instance, it is our belief that they are mainly responsible for the fact that the first excited negative-parity state of ^{16}O lies energetically higher than its first excited positive-parity state, although in the oscillator representation, the first negative-parity state needs one less oscillator quantum of excitation than the first excited positive-parity state.

The consideration given above shows that the generalized cluster wave function, obtained from the oscillator cluster wave function by letting the internal and relative-motion width parameters be independent of one another, could provide a satisfactory description of the main characteristics of a nuclear system. To proceed further, one can easily see that there is really no compelling reason in staying with internal and relative-motion functions of oscillator form. Thus, for example, a deuteron cluster, with its diffuse nature, is liable to be better represented by a superposition of Hulthén functions containing variational parameters. Similarly, in scattering and reaction problems, the relative-motion functions can be chosen more appropriately to satisfy the correct asymptotic boundary conditions.

There is also no special reason to stay within a single cluster representation. One can construct new complete sets by using a sufficient number of generalized cluster functions corresponding to different cluster structures but with the same quantum numbers. In fact, this is how the MCT is formulated in the framework of the RGM. The use of different cluster structure is particularly appropriate for an adequate description of the condition at the nuclear surface and is, therefore, especially useful to treat such nuclear problems as heavy-ion reactions and fission.

3. Formulation of the Microscopic Cluster Theory

As starting point, we rewrite the time-independent Schrödinger equation

$$(H - E_T)\,\psi = 0 \tag{3.1}$$

in the form of a projection equation

$$\langle \delta\psi \,|\, H - E_T \,|\, \psi \rangle = 0 . \tag{3.2}$$

In eqs. (3.1) and (3.2), E_T is the total energy of the system and H is a Galilean-invariant Hamiltonian operator given by

$$H = \sum_{i=1}^{N} \frac{1}{2M}\,\vec{P}_i^{\,2} + \sum_{i<j=1}^{N} V_{ij} - T_{cm} , \tag{3.3}$$

with N being the total number of nucleons, T_{cm} being the kinetic-energy operator of the total center of mass, and V_{ij} being a nucleon-nucleon potential chosen to fit the two-nucleon scattering data especially in the low-energy region.

If $\delta\psi$ represents a completely arbitrary variation in the space of all many-nucleon functions, then it is clear that eqs. (3.1) and (3.2) are entirely equivalent. As has been emphasized in ref. [WI 77], the advantage of eq. (3.2) is that this equation does allow one to conveniently treat the incoming and outgoing channels in a symmetrical way, which is an important consideration in the formulation of a flexible many-nucleon nuclear-reaction theory. In addition, of course, eq. (3.2) provides a convenient basis for approximate, variational calculations in practical problems.

To solve eq. (3.2), we make for ψ the ansatz

$$\psi(\tilde{r}_1, \cdots, \tilde{r}_N) = \sum_\tau \int \hat{\Phi}_\tau(\tilde{r}_1, \cdots, \tilde{r}_N ; \vec{\beta}_\tau) F_\tau(\vec{\beta}_\tau) \, d\vec{\beta}_\tau$$

$$+ \sum_\lambda c_\lambda \, \hat{\eta}_\lambda(\tilde{r}_1, \cdots, \tilde{r}_N) \, . \tag{3.4}$$

In the above equation, $\tilde{r}_i$ represent the space, spin, and isospin coordinates of the ith nucleon, $\vec{\beta}_\tau$ denotes a set of continuous parameter coordinates which characterize the basis functions $\hat{\Phi}_\tau(\tilde{r}_1, \ldots, \tilde{r}_N; \vec{\beta}_\tau)$, and λ enumerates the discrete set of square-integrable basis functions $\hat{\eta}_\lambda(\tilde{r}_1, \ldots, \tilde{r}_N)$. The coefficients $F_\tau(\vec{\beta}_\tau)$ and c_λ represent the continous and discrete linear variational amplitudes, respectively. The choice of the basis functions is arbitrary — they must be linearly independent, but need not be orthogonal to one another. Indeed, this latter point is an essential one, because only by choosing in general a non-orthogonal set of functions can one expect to introduce the incoming and outgoing channels symmetrically into the theory.

The basis set is selected in a way which is most suited to the problem under consideration. In the MCT, the first term on the right side of eq. (3.4) will be chosen to provide a proper description of nucleon clustering present mainly in the nuclear surface, while the second term, containing the so-called distortion functions $\hat{\eta}_\lambda$ in cluster terminology, is included to improve the wave function in the strong-interaction, compound-nucleus region.

Due to the hermiticity of H and the fact that all variational amplitudes are contained linearly in ψ , it can be easily shown [WI 77] that any two solutions ψ_m and ψ_n of eq. (3.2) are orthonormal if all degeneracies are removed, and the Hamiltonian H can be represented by a real diagonal matrix, i.e.,

$$\langle \psi_m | \psi_n \rangle = \delta(m,n) \tag{3.5}$$

and

$$\langle \psi_m | H | \psi_n \rangle = E_T^m \, \delta(m,n) \, , \tag{3.6}$$

where $\delta(m,n)$ represents the Kronecker δ-function if m is a discrete index and the

Dirac δ-function if m is a continuous index. These relations are in fact true, even when one restricts the number of linear variational parameters, or, in other words, when one works in a restricted function space. This is important, because in practical calculations one certainly cannot expect to always use a set of basis functions which span the complete function space.

By substituting eq. (3.4) into eq. (3.2) one obtains the following set of coupled equations for the determination of the variational amplitudes:

$$\sum_{\tau'} \int \langle \hat{\hat{\Phi}}_\tau \,|\, H - E_T \,|\, \hat{\hat{\Phi}}_{\tau'} \rangle \, F_{\tau'}(\vec{\beta}_{\tau'}) \, d\vec{\beta}_{\tau'}$$

$$+ \sum_{\lambda'} \langle \hat{\hat{\Phi}}_\tau \,|\, H - E_T \,|\, \hat{\eta}_{\lambda'} \rangle \, c_{\lambda'} = 0 \qquad \text{for all } \tau \,, \qquad (3.7)$$

$$\sum_{\tau'} \int \langle \hat{\eta}_\lambda \,|\, H - E_T \,|\, \hat{\hat{\Phi}}_{\tau'} \rangle \, F_{\tau'}(\vec{\beta}_{\tau'}) \, d\vec{\beta}_{\tau'}$$

$$+ \sum_{\lambda'} \langle \hat{\eta}_\lambda \,|\, H - E_T \,|\, \hat{\eta}_{\lambda'} \rangle \, c_{\lambda'} = 0 \qquad \text{for all } \lambda \,. \qquad (3.8)$$

In the above equations, the Dirac brackets indicate integration and summation over all nucleon coordinates, but not the parameter coordinates. From these equations, it is evident that, in general, the computation can become quite involved. Thus, in actual calculations, one must severely limit the number of basis functions in the trial function ψ . In this respect, it is important to realize that the Pauli principle, expressed by the antisymmetrization of the wave function, has the effect of reducing greatly the differences between apparently different non–orthogonal wave functions when the nucleons are relatively close to one another. From this it follows that, at relatively low excitation energies, the number of many–nucleon configurations which one needs for approximate calculations can usually be made so small that such calculations become quantitatively feasible.

4.   The Resonating–Group Method

4.1. Basis wave functions in the RGM

The basis wave functions $\hat{\hat{\Phi}}_\tau$ in the RGM are chosen to reflect the phenomenon of surface clustering in nuclei. The index τ will be used to indicate the type of

592

clustering and the state of internal excitation of the clusters involved. For clarity in presentation, we shall conduct the discussion by assuming that the clusters have no internal angular momentum; in actual calculations, the cluster internal angular momenta must of course be explicitly taken into consideration.

A two-cluster basis function $\hat{\Phi}^{(2)}$ with clusters A and B in any state of internal excitation, is written as

$$\hat{\Phi}^{(2)}(\tilde{r}_1, \cdots, \tilde{r}_N; \vec{R}'') = \mathcal{A}\left[\phi(A)\phi(B)\delta(\vec{R}-\vec{R}'')Z(\vec{R}_{cm})\right], \qquad (4.1)$$

where $\phi(A)$ and $\phi(B)$ are translationally invariant functions describing the internal behavior of the clusters and $Z(\vec{R}_{cm})$ is any normalizable function describing the total c.m. motion. The Jacobi relative coordinate $\vec{R}$ is a dynamical coordinate, given by

$$\vec{R} = \vec{R}_A - \vec{R}_B , \qquad (4.2)$$

where

$$\vec{R}_A = \frac{1}{N_A} \sum_{i=1}^{N_A} \vec{r}_i , \qquad \vec{R}_B = \frac{1}{N_B} \sum_{i=N_A+1}^{N_A+N_B} \vec{r}_i , \qquad (4.3)$$

with N_A and N_B being, respectively, the nucleon numbers of clusters A and B. The quantity $\vec{R}''$, on the other hand, is a parameter coordinate on which the anti-symmetrization operator $\mathcal{A}$ does not act. Also, it is important to note that the c.m. function $Z(\vec{R}_{cm})$ occurs in $\hat{\Phi}^{(2)}$ as a multiplicative factor and, hence, there will be no difficulties with spurious center-of-mass excitation.

Because of the presence of the antisymmetrization operator $\mathcal{A}$, the basis functions of eq. (4.1) with different values of $\vec{R}''$ are not orthogonal to each other. As has been pointed out previously, this does not cause any principal difficulty, because non-degenerate solutions of the projection equation (3.2) are always mutually orthogonal. On the other hand, one should be very careful in the interpretation of the basis wave function; it is only for $\vec{R}''$-values sufficiently larger than the sum of the cluster radii that one can consider $\hat{\Phi}^{(2)}$ as describing the separation of the two clusters A and B by a distance $\vec{R}''$. This is, however, sufficient for our purposes, since in nuclear reactions the measurements are always carried out under conditions where the nuclei are well separated.

A three-cluster basis function $\hat{\Phi}^{(3)}$ can be analogously defined. It is written as

$$\hat{\Phi}^{(3)}(\tilde{r}_1, \cdots, \tilde{r}_N; \vec{R}_1'', \vec{R}_2'')$$

$$= \mathcal{A}\left[\phi(A)\phi(B)\phi(C)\delta(\vec{R}_1-\vec{R}_1'')\delta(\vec{R}_2-\vec{R}_2'')Z(\vec{R}_{cm})\right], \qquad (4.4)$$

where $\vec{R}_1$ and $\vec{R}_2$ are again Jacobi relative coordinates, defined by

$$\vec{R}_1 = \vec{R}_A - \vec{R}_B \,, \tag{4.5}$$

$$\vec{R}_2 = \vec{R}_C - \frac{N_A \vec{R}_A + N_B \vec{R}_B}{N_A + N_B} \,, \tag{4.6}$$

with $\vec{R}_C$ being the dynamical c.m. coordinate of the cluster C with nucleon number N_C. In a similar way, one can proceed to define four-cluster basis function, five-cluster basis function, and so on.

The distortion functions $\hat{\eta}_\lambda$ in eq. (3.4) are included to improve the description of the behavior in the compound-nucleus region. They can be chosen as translationally-invariant shell-model functions, generalized cluster wave functions with square-integrable relative-motion functions, or any other types of normalizable functions which are regarded by intuition and experience to be the most convenient and appropriate for the problem under consideration.

In the resonating-group formulation of the MCT, the basis-function set is certainly over-complete. However, this does not pose any problem, since for practical calculations it is always necessary to use only a small number of channel and distortion functions (i.e., a small number of terms in the summations over τ and λ in eq. (3.4)) in the expansion for ψ.

Because of the microscopic nature of the RGM, the computation will in general become quite complicated if the function space is taken to be rather large. Thus, in practice, one must limit the extension of this space by using relatively simple forms for ψ, chosen according to physical intuition and energetical arguments. For instance, in the five-nucleon case, one might start by taking just one channel term in eq. (3.4), which represents a $n + \alpha$ cluster configuration with the α particle in its ground state. This results in the so-called single-channel approximation without specific distortion (see subsect. 4.2a of ref. [WI 77]). Because the α particle has a low compressibility and is, therefore, not easily distortable, one does find that this approximation can yield satisfactory results in the low-excitation region [RE 70]. If one proceeds further to consider higher energies at which the α cluster can be broken up, then one should improve the calculation by including also the d + t cluster configuration, with both the deuteron and the triton in their ground states [CH 74]. Thus one sees that, in the resonating-group approach of the MCT, one makes successive improvements until the calculation becomes computationally infeasible. When such a stage is reached, then one may have to resort to more phenomenological means, such as the introduction of imaginary potentials [BR 71] and so on.

4.2. Derivation of coupled equations

By the procedure outlined in sect. 3 and the preceding subsection, the problem of solving the Schrödinger equation is changed from one of finding $\psi(\tilde{r}_1, \ldots, \tilde{r}_N)$ to one of determining the superposition amplitudes $F_\tau(\vec{\beta}_\tau)$ and c_λ from eqs. (3.7) and (3.8). In fact, as has been mentioned by Thompson [TH 78], these amplitudes may be regarded as "new wave functions" replacing $\psi(\tilde{r}_1, \ldots, \tilde{r}_N)$ with the "new coordinates" $\vec{\beta}_\tau$ and λ replacing the nucleon coordinates $\tilde{r}_i$. The flexibility of this procedure comes, of course, from the fact that through a judicious choice of basis states, the system can be adequately represented by a relatively small number of these superposition amplitudes. Therefore, one can achieve a formulation which is amenable to quantitative studies and, perhaps more importantly, which allows for an understanding of the collective behavior of the system grounded on a microscopic description.

In this subsection, the structure of eqs. (3.7) and (3.8) will be explicitly examined in situations which typify resonating-group investigations. From this examination, we shall obtain a better feeling about the merits of the resonating-group approach and achieve some understanding concerning the connection between the MCT and existing macroscopic models.

4.2a. Single-channel calculation without specific distortion.

We start the discussion by considering the simplest case, namely, the case in which there is only one two-cluster channel function and where the specific distortion effect is neglected. In this case, the wave function ψ is

$$\psi = \int \mathcal{A}\left[\phi(A)\phi(B)\delta(\vec{R}-\vec{R}')Z(\vec{R}_{cm})\right]F(\vec{R}'')d\vec{R}'' \ , \tag{4.7}$$

which can, of course, be reduced by the trivial integration over $\vec{R}''$ to the usual form

$$\psi = \mathcal{A}\left[\phi(A)\phi(B)F(\vec{R})Z(\vec{R}_{cm})\right] \ . \tag{4.8}$$

The function space used in the calculation is specified by the variation $\delta\psi$ which is obtained by an arbitrary variation of the relative-motion function F, i.e.,

$$\delta\psi = \int \mathcal{A}\left[\phi(A)\phi(B)\delta(\vec{R}-\vec{R}')Z(\vec{R}_{cm})\right]\delta F(\vec{R}')d\vec{R}' \ . \tag{4.9}$$

Using ψ and $\delta\psi$, we obtain then from eq. (3.2) [or, equivalently, eq. (3.7)] the following equation:

$$\int \left[\mathscr{H}(\vec{R}',\vec{R}'') - E_T \mathscr{N}(\vec{R}',\vec{R}'') \right] F(\vec{R}'') d\vec{R}'' = 0 \ , \tag{4.10}$$

where

$$\mathscr{H}(\vec{R}',\vec{R}'') = \langle \phi(A)\phi(B)\delta(\vec{R}-\vec{R}')Z \mid H \mid \mathscr{A}[\phi(A)\phi(B)\delta(\vec{R}-\vec{R}'')Z] \rangle, \tag{4.11}$$

$$\mathscr{N}(\vec{R}',\vec{R}'') = \langle \phi(A)\phi(B)\delta(\vec{R}-\vec{R}')Z \mid \mathscr{A}[\phi(A)\phi(B)\delta(\vec{R}-\vec{R}'')Z] \rangle. \tag{4.12}$$

Note that, in eqs. (4.11) and (4.12), the antisymmetrization operator $\mathscr{A}$ occurs only on the ket side of the Dirac brackets; this is permissible, since this operator is a hermitian operator which commutes with the Hamiltonian operator H and satisfies the relation

$$\mathscr{A}^2 = N! \, \mathscr{A} \ . \tag{4.13}$$

To proceed, let us write

$$\mathscr{A} = \mathscr{A}' \mathscr{A}_A \mathscr{A}_B \ , \tag{4.14}$$

where $\mathscr{A}_A$ and $\mathscr{A}_B$ are, respectively, antisymmetrization operators for the nucleons in clusters A and B, and $\mathscr{A}'$ is an antisymmetrization operator which interchanges nucleons in different clusters. Then, eq. (4.12) can be written as

$$\mathscr{N}(\vec{R}',\vec{R}'') = \langle \phi(A)\phi(B)\delta(\vec{R}-\vec{R}')Z \mid \mathscr{A}'[\hat{\phi}(A)\hat{\phi}(B)\delta(\vec{R}-\vec{R}'')Z] \rangle, \tag{4.15}$$

with

$$\hat{\phi}(A) = \mathscr{A}_A \phi(A) \ , \qquad \hat{\phi}(B) = \mathscr{A}_B \phi(B) \ . \tag{4.16}$$

By defining further

$$\mathscr{A}' = 1 + \mathscr{A}'' \ , \tag{4.17}$$

we can separate $\mathscr{N}(\vec{R}',\vec{R}'')$ into two parts, i.e.,

$$\mathscr{N}(\vec{R}',\vec{R}'') = \mathscr{N}_D(\vec{R}',\vec{R}'') + \mathscr{N}_E(\vec{R}',\vec{R}'') \ , \tag{4.18}$$

where the direct part $\mathscr{N}_D$ is

$$\mathcal{N}_D(\vec{R}',\vec{R}'') = \langle \phi(A)\phi(B)\delta(\vec{R}-\vec{R}')Z \,|\, \hat{\phi}(A)\hat{\phi}(B)\delta(\vec{R}-\vec{R}'')Z \rangle \qquad (4.19)$$

and the exchange part $\mathcal{N}_E$ is

$$\mathcal{N}_E(\vec{R}',\vec{R}'') = \langle \phi(A)\phi(B)\delta(\vec{R}-\vec{R}')Z \,|\, \mathcal{A}''[\hat{\phi}(A)\hat{\phi}(B)\delta(\vec{R}-\vec{R}'')Z] \rangle. \qquad (4.20)$$

For $\mathcal{N}_D$ we now perform the integration over the relative coordinate $\vec{R}$. The result is

$$\mathcal{N}_D(\vec{R}',\vec{R}'') = \langle \phi(A)\phi(B)Z \,|\, \hat{\phi}(A)\hat{\phi}(B)Z \rangle_{\vec{R}} \; \delta(\vec{R}'-\vec{R}''), \qquad (4.21)$$

where the notation $\langle\;\rangle_{\vec{R}}$ is introduced to indicate integration over internal spatial coordinates of the clusters and the total c.m. coordinate, summation over all spin and isospin coordinates, but no integration over the relative coordinate $\vec{R}$. For convenience, we shall adopt the normalization condition

$$\langle \phi(A)\phi(B)Z \,|\, \hat{\phi}(A)\hat{\phi}(B)Z \rangle_{\vec{R}} = 1. \qquad (4.22)$$

With this particular normalization, the quantity $\mathcal{N}_D$ then takes on the simple form

$$\mathcal{N}_D(\vec{R}',\vec{R}'') = \delta(\vec{R}'-\vec{R}''). \qquad (4.23)$$

Similarly, one can separate $\mathcal{H}(R',R'')$ also into two parts, i.e.,

$$\mathcal{H}(\vec{R}',\vec{R}'') = \mathcal{H}_D(\vec{R}',\vec{R}'') + \mathcal{H}_E(\vec{R}',\vec{R}''), \qquad (4.24)$$

with

$$\mathcal{H}_D(\vec{R}',\vec{R}'') = \langle \phi(A)\phi(B)\delta(\vec{R}-\vec{R}')Z \,|\, H \,|\, \hat{\phi}(A)\hat{\phi}(B)\delta(\vec{R}-\vec{R}'')Z \rangle \qquad (4.25)$$

and

$$\mathcal{H}_E(\vec{R}',\vec{R}'') = \langle \phi(A)\phi(B)\delta(\vec{R}-\vec{R}')Z \,|\, H \,|\, \mathcal{A}''[\hat{\phi}(A)\hat{\phi}(B)\delta(\vec{R}-\vec{R}'')Z] \rangle. \qquad (4.26)$$

The expression for $\mathcal{H}_D$ can be simplified by noting that the Galilean-invariant Hamiltonian H of eq. (3.3) can be written as

$$H = H_A + H_B + H', \qquad (4.27)$$

where H_A and H_B represent, respectively, the internal Hamiltonians of the clusters

A and B, and H' is a Hamiltonian for the relative motion, given by

$$H' = -\frac{\hbar^2}{2\mu} \nabla_{\vec{R}}^2 + V' \ . \tag{4.28}$$

In the above equation,

$$V' = \sum_{i \in A} \sum_{j \in B} V_{ij} \ , \tag{4.29}$$

and

$$\mu = \mu_0 M \tag{4.30}$$

with

$$\mu_0 = N_A N_B / N \tag{4.31}$$

being the reduced nucleon number of the two clusters. The two-nucleon potential V_{ij} will in general contain all types of exchange operators; however, since totally antisymmetrized wave functions are used in resonating-group calculations, one can always replace the space-exchange operator P_{ij}^r, wherever it appears, by the operator $-P_{ij}^\sigma P_{ij}^\tau$ with P_{ij}^σ and P_{ij}^τ being the spin- and isospin-exchange operators, respectively. When this is done, we can then define a direct (local) potential $V_D(\vec{R})$ as

$$V_D(\vec{R}) = \langle \phi(A)\phi(B)Z \,|\, V' \,|\, \hat{\phi}(A)\hat{\phi}(B)Z \rangle_{\vec{R}} \ . \tag{4.32}$$

The cluster internal energies E_A and E_B are obtained by computing the expectation values of the Hamiltonians H_A and H_B, i.e.,

$$E_i = \langle \phi(A)\phi(B)Z \,|\, H_i \,|\, \hat{\phi}(A)\hat{\phi}(B)Z \rangle_{\vec{R}} \quad (i = A, B) \tag{4.33}$$

Using eqs. (4.22) and eqs. (4.27) – (4.33), we finally obtain

$$\mathscr{H}_D(\vec{R}', \vec{R}'') = \left[-\frac{\hbar^2}{2\mu} \nabla_{\vec{R}'}^2 + V_D(\vec{R}') + E_A + E_B \right] \delta(\vec{R}' - \vec{R}'') \ . \tag{4.34}$$

By utilizing this expression and the expression for $\mathscr{N}_D(\vec{R}', \vec{R}'')$ of eq. (4.23), one sees that eq. (4.10) can be written more explicitly as

$$\left[-\frac{\hbar^2}{2\mu} \nabla_{\vec{R}'}^2 + V_D(\vec{R}') - E \right] F(\vec{R}') + \int K(\vec{R}', \vec{R}'') F(\vec{R}'') d\vec{R}'' = 0 \ , \tag{4.35}$$

where E is the relative energy of the two clusters in the c.m. system, given by

598

$$E = E_T - E_A - E_B \quad , \tag{4.36}$$

and $K(\vec{R}',\vec{R}'')$ is an energy-dependent kernel function given by

$$K(\vec{R}',\vec{R}'') = \mathcal{H}_E(\vec{R}',\vec{R}'') - E_T \mathcal{N}_E(\vec{R}',\vec{R}'') \quad . \tag{4.37}$$

From this equation, it is seen that, if the two clusters are considered as structureless, then the effective interaction between them must be both nonlocal and energy-dependent.

From the above discussion, one also sees that if the antisymmetrization between nucleons in different clusters is neglected or, in other words, if the anti-symmetrization operator $\mathcal{A}'$ is set as unity, then the kernel functions $\mathcal{H}_E$ and $\mathcal{N}_E$ will be equal to zero. In this crude approximation, the effective intercluster potential will therefore just be the direct potential V_D [GR 68] (see also appendix A of ref. [TA 78]).

To obtain some feeling about the general structure of the direct potential $V_D(\vec{R}')$ and the kernel function $K(\vec{R}',\vec{R}'')$, we consider a specific example, namely, the $n+\alpha$ problem. For clarity in discussion, we choose for the α cluster wave function the simple form given by eq. (2.10) and a simple central nucleon-nucleon potential given by

$$V_{ij} = - V_0 \exp\left(-K r_{ij}^2\right)\left(w - m P_{ij}^\sigma P_{ij}^\tau + b P_{ij}^\sigma - h P_{ij}^\tau\right)$$

$$+ \frac{e^2}{4 r_{ij}}\left(1 + \tau_{iz}\right)\left(1 + \tau_{jz}\right). \tag{4.38}$$

For this problem, the derivation of $V_D(\vec{R}')$ and $K(\vec{R}',\vec{R}'')$ is elementary and is given in ref. [TH 71]. The result is

$$V_D(\vec{R}') = - V_0 \left(4w - m + 2b - 2h\right)\left(\frac{4\alpha}{4\alpha + 3K}\right)^{3/2} \exp\left(-\frac{4\alpha K}{4\alpha + 3K}\vec{R}'^2\right) , \tag{4.39}$$

$$\mathcal{N}_E(\vec{R}',\vec{R}'') = -\left(\frac{4}{5}\right)^3\left(\frac{4\alpha}{3\pi}\right)^{3/2}\exp\left[-\frac{34}{75}\alpha\left(\vec{R}'^2 + \vec{R}''^2\right) - \frac{32}{75}\alpha\vec{R}'\cdot\vec{R}''\right] , \tag{4.40}$$

$$\mathcal{H}_E(\vec{R}',\vec{R}'') = \mathcal{H}_T(\vec{R}',\vec{R}'') + \mathcal{H}_V(\vec{R}',\vec{R}'') , \tag{4.41}$$

with

$$\mathcal{H}_T(\vec{R}',\vec{R}'') = -\left(\frac{4}{5}\right)^3\left(\frac{4\alpha}{3\pi}\right)^{3/2}\frac{\hbar^2}{2M}\left[\frac{47}{5}\alpha - \frac{1216}{1125}\alpha^2(\vec{R}'^2+\vec{R}''^2)\right.$$

$$\left. - \frac{1568}{1125}\alpha^2\,\vec{R}'\cdot\vec{R}''\right]$$

$$\times\, exp\left[-\frac{34}{75}\alpha(\vec{R}'^2+\vec{R}''^2)-\frac{32}{75}\alpha\vec{R}'\cdot\vec{R}''\right], \tag{4.42}$$

$$\mathcal{H}_V(\vec{R}',\vec{R}'') = -\left(\frac{4}{5}\right)^3\left(\frac{4\alpha}{3\pi}\right)^{3/2}V_0$$

$$\times\left\{(-w+4m-2b+2h)\,exp\left[-\frac{34\alpha+48K}{75}(\vec{R}'^2+\vec{R}''^2)\right.\right.$$

$$\left.-\frac{32\alpha-96K}{75}\vec{R}'\cdot\vec{R}''\right] + (-3w-3m)\left(\frac{3\alpha}{3\alpha+2K}\right)^{3/2}$$

$$\times\left[exp\left(-\frac{34\alpha^2+28\alpha K}{75\alpha+50K}\vec{R}'^2 - \frac{34\alpha^2+108\alpha K}{75\alpha+50K}\vec{R}''^2\right.\right.$$

$$\left. - \frac{32\alpha^2+64\alpha K}{75\alpha+50K}\vec{R}'\cdot\vec{R}''\right)$$

$$+ exp\left(-\frac{34\alpha^2+108\alpha K}{75\alpha+50K}\vec{R}'^2 - \frac{34\alpha^2+28\alpha K}{75\alpha+50K}\vec{R}''^2\right.$$

$$\left.\left. - \frac{32\alpha^2+64\alpha K}{75\alpha+50K}\vec{R}'\cdot\vec{R}''\right)\right]$$

$$+ (-3w-3m)\left(\frac{\alpha}{\alpha+2K}\right)^{3/2}exp\left[-\frac{34}{75}\alpha(\vec{R}'^2+\vec{R}''^2)-\frac{32}{75}\alpha\vec{R}'\cdot\vec{R}''\right]\right\}$$

$$-e^2\left(\frac{4}{5}\right)^3\left(\frac{4\alpha}{3\pi}\right)^{3/2}\left(\frac{2\alpha}{\pi}\right)^{1/2}exp\left[-\frac{34}{75}\alpha(\vec{R}'^2+\vec{R}''^2)-\frac{32}{75}\alpha\vec{R}'\cdot\vec{R}''\right].$$

$$\text{....}(4.43)$$

From these expressions one sees that, because of the translational and rotational
invariance of the chosen Hamiltonian, $V_D(\vec{R}')$ and $K(\vec{R}',\vec{R}'')$ depends, as they should,
only on the invariant quantities $\vec{R}'^2$, $\vec{R}''^2$, and $\vec{R}'\cdot\vec{R}''$. In addition, we should point
out the important characteristic that the coefficients of $\vec{R}'\cdot\vec{R}''$ in the exponents
of the kernel function take on both positive and negative values (note that, for a
realistic nucleon-nucleon potential, $K\approx\alpha$). As will be discussed in a later
section, this is essential for an understanding of the features of the scattering
differential cross section in the entire angular region. In this respect, it is
interesting to mention that the phenomenological kernel proposed by Frahn and
Lemmer [FR 57], which has been extensively used to analyze experimental data and to

study general properties of nonlocal interaction (see, e.g., refs. [AU 65, FI 67, HO 80, PE 62]), lacks this important feature and, therefore, cannot be expected to faithfully represent the main character of the microscopically derived effective internuclear potential.

Next, we consider very briefly the case where the trial function is represented by a single three-cluster term [SC 80], i.e.,

$$\psi = \int \mathcal{A}\left[\phi(A)\phi(B)\phi(C)\,\delta(\vec{R}_1 - \vec{R}_1'')\,\delta(\vec{R}_2 - \vec{R}_2'')\,Z(\vec{R}_{cm})\right]$$
$$\times\, F(\vec{R}_1'', \vec{R}_2'')\, d\vec{R}_1''\, d\vec{R}_2'' . \tag{4.44}$$

Concerning this wave function, there is one important point which should be mentioned; that is, this trial function can also describe the behavior of some two-cluster configurations. Consider the six-nucleon system as an example. Here the three-cluster $n + p + \alpha$ channel includes the two-cluster $d + \alpha$ channel. This is so, because the function $F(\vec{R}_1'', \vec{R}_2'')$ can assume a product form consisting of the deuteron-cluster bound-state function and the $d + \alpha$ relative-motion function. Similarly, one can easily see that the $n + {}^5\mathrm{Li}$ and $p + {}^5\mathrm{He}$ channels are also included. On the other hand, the $t + {}^3\mathrm{He}$ channel cannot be properly described by the wave function of eq. (4.44). This means that, if one wishes to consider the breakup reaction ${}^3\mathrm{He}\,(t,np)\alpha$, then one must, in addition, introduce a two-cluster $t + {}^3\mathrm{He}$ term into the resonating-group formulation.

From eq. (3.7), one obtains in this case

$$\int \left[\mathcal{H}(\vec{R}_1', \vec{R}_2', \vec{R}_1'', \vec{R}_2'') - E_T \mathcal{N}(\vec{R}_1', \vec{R}_2', \vec{R}_1'', \vec{R}_2'')\right]$$
$$\times\, F(\vec{R}_1'', \vec{R}_2'')\, d\vec{R}_1''\, d\vec{R}_2'' = 0 , \tag{4.45}$$

where

$$\mathcal{H}(\vec{R}_1', \vec{R}_2', \vec{R}_1'', \vec{R}_2'') = \langle\, \phi(A)\phi(B)\phi(C)\,\delta(\vec{R}_1 - \vec{R}_1')\,\delta(\vec{R}_2 - \vec{R}_2')\,Z$$
$$|H|\,\mathcal{A}\left[\phi(A)\phi(B)\phi(C)\,\delta(\vec{R}_1 - \vec{R}_1'')\,\delta(\vec{R}_2 - \vec{R}_2'')\,Z\right]\,\rangle , \tag{4.46}$$

$$\mathcal{N}(\vec{R}_1', \vec{R}_2', \vec{R}_1'', \vec{R}_2'') = \langle\, \phi(A)\phi(B)\phi(C)\,\delta(\vec{R}_1 - \vec{R}_1')\,\delta(\vec{R}_2 - \vec{R}_2')\,Z$$
$$|\,\mathcal{A}\left[\phi(A)\phi(B)\phi(C)\,\delta(\vec{R}_1 - \vec{R}_1'')\,\delta(\vec{R}_2 - \vec{R}_2'')\,Z\right]\,\rangle . \tag{4.47}$$

Proceeding now in exactly the same manner as described above in the two-cluster case, we can again write eq. (4.45) in the following more explicit form:

$$\left[-\frac{\hbar^2}{2\mu_1} \nabla_{\vec{R}_1'}^2 - \frac{\hbar^2}{2\mu_2} \nabla_{\vec{R}_2'}^2 + V_D(\vec{R}_1', \vec{R}_2') - E \right] F(\vec{R}_1', \vec{R}_2')$$

$$+ \int K(\vec{R}_1', \vec{R}_2', \vec{R}_1'', \vec{R}_2'') \, F(\vec{R}_1'', \vec{R}_2'') \, d\vec{R}_1'' d\vec{R}_2'' = 0 \ , \tag{4.48}$$

where

$$\mu_1 = \frac{N_A N_B}{N_A + N_B} M \quad , \qquad \mu_2 = \frac{N_C(N_A + N_B)}{N} M \ , \tag{4.49}$$

and E is the relative energy of the clusters given by

$$E = E_T - E_A - E_B - E_C \ , \tag{4.50}$$

with E_A, E_B, and E_C being cluster internal energies.

If one is concerned with bound and sharp resonance states, then eq. (4.48) may
be approximately solved by using the Ritz variational procedure employing trial wave
functions which vanish in the asymptotic regions ([FU 78, KH 61, TA 62a] and refs.
[89-97] quoted in ref. [TA 78]). For the consideration of three-body breakup
reactions, some possibilities to utilize the derived direct potential $V_D(\vec{R}_1', \vec{R}_2')$
and the energy-dependent kernel functions $K(\vec{R}_1', \vec{R}_2', \vec{R}_1'', \vec{R}_2'')$ have been recently
discussed by Schmid [SC 80]. It should be mentioned, however, that these functions
have, in general, quite complicated structures. Thus, together with the fact that
three-body boundary conditions are also complicated, we are of the opinion that for
future progress to be realized in solving the breakup problem, more advanced
mathematical techniques and computational innovations must first be made.

4.2b. <u>Single-channel calculation with specific distortion.</u>

For some systems, the specific distortion effect may play an important role.
This occurs in the $d + \alpha$ case [JA 69, KA 80, TH 73], for example, where the weak
binding of the deuteron cluster means that this cluster may be strongly distorted
when it overlaps appreciably with the α cluster. Therefore, for a proper consideration of such systems, it will be necessary to introduce square-integrable distortion
functions into the formulation such that the behavior in the region of strong
interaction may be adequately described.

Let us assume, for simplicity in discussion, that only one distortion
function is introduced. The trial function is then given by

$$\psi = \psi_0 + c_1 \hat{\eta}_1 \ , \tag{4.51}$$

where ψ_0 is a channel function of the form given by eq. (4.7) and $\hat{\eta}_l$ is taken as

$$\hat{\eta}_l = \mathcal{A}\left[\,\mathcal{S}_l\, z\,(\vec{R}_{cm})\right] \tag{4.52}$$

to assure translational invariance. By performing independent variations $\delta\psi_0$ and $\hat{\eta}_l\,\delta c_l$, the projection equation (3.2) becomes

$$\langle\,\delta\psi_0\,|\,H-E_T\,|\,\psi_0\,\rangle + c_l\,\langle\,\delta\psi_0\,|\,H-E_T\,|\,\hat{\mathcal{S}}_l\,z\,\rangle = 0\,, \tag{4.53}$$

$$\langle\,\hat{\mathcal{S}}_l\,z\,|\,H-E_T\,|\,\psi_0\,\rangle + c_l\,\langle\,\hat{\mathcal{S}}_l\,z\,|\,H-E_T\,|\,\hat{\mathcal{S}}_l\,z\,\rangle = 0\,, \tag{4.54}$$

with

$$\hat{\mathcal{S}}_l = \mathcal{A}\,\mathcal{S}_l\,. \tag{4.55}$$

Solving eq. (4.54) for c_l and substituting the resultant expression into eq. (4.53) yields

$$\langle\,\delta\psi_0\,|\,H+\tilde{V}-E_T\,|\,\psi_0\,\rangle = 0\,, \tag{4.56}$$

where

$$\tilde{V} = -\frac{(H-E_T)\,|\,\hat{\mathcal{S}}_l\,z\,\rangle\langle\,\hat{\mathcal{S}}_l\,z\,|\,(H-E_T)}{\langle\,\hat{\mathcal{S}}_l\,z\,|\,H-E_T\,|\,\hat{\mathcal{S}}_l\,z\,\rangle}\,. \tag{4.57}$$

By using eq. (4.22) and the normalization condition

$$\langle\,\mathcal{S}_l\,z\,|\,\hat{\mathcal{S}}_l\,z\,\rangle = 1\,, \tag{4.58}$$

we can write eq. (4.56) more explicitly as follows:

$$\left[-\frac{\hbar^2}{2\mu}\nabla^2_{\vec{R}'} + V_D(\vec{R}') - E\right] F(\vec{R}')$$

$$+ \int\left[K(\vec{R}',\vec{R}'') + K_l(\vec{R}',\vec{R}'')\right] F(\vec{R}'')\,d\vec{R}'' = 0\,, \tag{4.59}$$

where

$$K_l(\vec{R}',\vec{R}'') = -\frac{\lambda(\vec{R}')\lambda^*(\vec{R}'')}{\hat{E}_l - E_T}\,, \tag{4.60}$$

with

$$\lambda(\vec{R}') = \langle \phi(A)\phi(B)\delta(\vec{R}-\vec{R}')z \mid H-E_T \mid \hat{\mathcal{S}}_1 z \rangle \tag{4.61}$$

representing the coupling between the channel and the distortion terms and $\hat{E}_1$ being the expectation value of H with respect to the distortion function, given by

$$\hat{E}_1 = \langle \hat{\mathcal{S}}_1 z \mid H \mid \hat{\mathcal{S}}_1 z \rangle . \tag{4.62}$$

Thus, one sees that the addition of a distortion function introduces an extra nonlocal, energy-dependent, separable term into the effective internuclear potential.

It is a simple procedure to generalize to the case where many distortion functions are added to ψ_0 . The only complication arises because the basis functions $\hat{\eta}_\lambda$ are not orthogonal to each other. But this complication can be easily overcome by transforming to a basis set of orthonormal functions which diagonalize the Hamiltonian H in the subspace spanned by the original distortion functions $\hat{\eta}_\lambda$.

Because of the separable nature of $K_1(\vec{R}',\vec{R}'')$, it has been shown in chapter 9 of ref. [WI 77] that the introduction of the distortion-function term gives rise to resonance structure in the wave function and, consequently, the scattering cross section. Thus, one can interpret the wave function $\psi(\tilde{r}_1, \ldots, \tilde{r}_N)$ of eq. (3.4) in a somewhat different way. If the first term (i.e., the term with $\hat{\Phi}_\tau$) in this equation is omitted, the orthonormalization of the non-orthogonal functions $\hat{\eta}_\lambda$ in the manner described above yields a set of internal functions $\tilde{\eta}_\lambda$ with eigen-energies $\hat{E}_\lambda$ which constitute the energy spectrum of this system, consisting of levels with zero width. This is the type of bound-state calculations which have been frequently performed by many authors, adopting especially the shell-model viewpoint. By adding open-channel terms $\hat{\Phi}_\tau$ into the formulation, one is then able to probe the internal structure of the nucleus by scattering and reaction processes. At the same time, of course, one introduces energy shifts and energy widths into these levels and, thereby, enriches the comparison between theory and experiment.

As a consequence of non-orthogonality, caution must be exercised when many $\hat{\Phi}_\tau$ and $\hat{\eta}_\lambda$ terms are employed. It could happen that some of these terms are linearly dependent of one another. For example, it may turn out that the bound-state structures can be replaced by a linear superposition of open-channel terms. To overcome the difficulty resulting from this problem, one has to project out the bound structures from the open-channel functions. This projection procedure can be easily carried out and is carefully described in ref. [WI 77]; hence, it will not be further elaborated here.

4.2c. <u>Coupled-channel calculation.</u>

We discuss in this subsection the resonating-group formulation of a two-channel problem. Here the trial function ψ has the form

$$\psi = \psi_f + \psi_g \ , \tag{4.63}$$

where

$$\psi_f = \int \mathcal{A} \left[\phi(A)\phi(B)\delta(\vec{R}_f - \vec{R}_f'') Z(\vec{R}_{cm})\right] F(\vec{R}_f'')\,d\vec{R}_f'' \ , \tag{4.64}$$

$$\psi_g = \int \mathcal{A} \left[\phi(C)\phi(D)\delta(\vec{R}_g - \vec{R}_g'') Z(\vec{R}_{cm})\right] G(\vec{R}_g'')\,d\vec{R}_g'' . \tag{4.65}$$

In the above equations, the $\phi's$ are cluster internal functions, chosen to satisfy the normalization condition (see p. 114 of ref. [WI 77])

$$\left\langle \phi(A)\phi(B)Z(\vec{R}_{cm}) \,\middle|\, \hat{\phi}(A)\hat{\phi}(B)Z(\vec{R}_{cm}) \right\rangle_{\vec{R}_f}$$

$$= \left\langle \phi(C)\phi(D)Z(\vec{R}_{cm}) \,\middle|\, \hat{\phi}(C)\hat{\phi}(D)Z(\vec{R}_{cm}) \right\rangle_{\vec{R}_g} = 1 \tag{4.66}$$

in order to insure the unitarity of the S-matrix.

From eq. (3.7), one obtain the following coupled equations satisfied by $F(\vec{R}_f'')$ and $G(\vec{R}_g'')$:

$$\int \hat{K}_{ff}(\vec{R}_f', \vec{R}_f'') F(\vec{R}_f'')\,d\vec{R}_f'' + \int K_{fg}(\vec{R}_f', \vec{R}_g'') G(\vec{R}_g'')\,d\vec{R}_g'' = 0 \ , \tag{4.67}$$

$$\int \hat{K}_{gg}(\vec{R}_g', \vec{R}_g'') G(\vec{R}_g'')\,d\vec{R}_g'' + \int K_{gf}(\vec{R}_g', \vec{R}_f'') F(\vec{R}_f'')\,d\vec{R}_f'' = 0 \ , \tag{4.68}$$

where

$$\hat{K}_{ff}(\vec{R}_f', \vec{R}_f'') = \left\langle \phi(A)\phi(B)\delta(\vec{R}_f - \vec{R}_f')Z \,\middle|\, H - E_T \,\middle|\, \mathcal{A}[\phi(A)\phi(B)\delta(\vec{R}_f - \vec{R}_f'')Z]\right\rangle, \tag{4.69}$$

$$\hat{K}_{gg}(\vec{R}_g', \vec{R}_g'') = \left\langle \phi(C)\phi(D)\delta(\vec{R}_g - \vec{R}_g')Z \,\middle|\, H - E_T \,\middle|\, \mathcal{A}[\phi(C)\phi(D)\delta(\vec{R}_g - \vec{R}_g'')Z]\right\rangle, \tag{4.70}$$

$$K_{fg}(\vec{R}_f', \vec{R}_g'') = \left\langle \phi(A)\phi(B)\delta(\vec{R}_f - \vec{R}_f')Z \,\middle|\, H - E_T \,\middle|\, \mathcal{A}[\phi(C)\phi(D)\delta(\vec{R}_g - \vec{R}_g'')Z]\right\rangle, \tag{4.71}$$

$$K_{gf}(\vec{R}_g', \vec{R}_f'') = \left\langle \phi(C)\phi(D)\delta(\vec{R}_g - \vec{R}_g')Z \,\middle|\, H - E_T \,\middle|\, \mathcal{A}[\phi(A)\phi(B)\delta(\vec{R}_f - \vec{R}_f'')Z]\right\rangle. \tag{4.72}$$

If we now use the procedure described in subsect. 4.2a, then it is easily seen that these coupled equations may be reduced to the following more familiar form:

$$\left[-\frac{\hbar^2}{2\mu_f} \nabla^2_{\vec{R}'_f} + V_{Df}(\vec{R}'_f) - E_f \right] F(\vec{R}'_f) + \int K_{ff}(\vec{R}'_f, \vec{R}''_f) F(\vec{R}''_f) d\vec{R}''_f$$

$$+ \int K_{fg}(\vec{R}'_f, \vec{R}''_g) G(\vec{R}''_g) d\vec{R}''_g = 0 , \qquad (4.73)$$

$$\left[-\frac{\hbar^2}{2\mu_g} \nabla^2_{\vec{R}'_g} + V_{Dg}(\vec{R}'_g) - E_g \right] G(\vec{R}'_g) + \int K_{gg}(\vec{R}'_g, \vec{R}''_g) G(\vec{R}''_g) d\vec{R}''_g$$

$$+ \int K_{gf}(\vec{R}'_g, \vec{R}''_f) F(\vec{R}''_f) d\vec{R}''_f = 0 , \qquad (4.74)$$

where μ_f and μ_g are, respectively, the reduced masses of the clusters in channels f and g, defined as

$$\mu_f = \frac{N_A N_B}{N} M \quad , \qquad \mu_g = \frac{N_C N_D}{N} M , \qquad (4.75)$$

V_{Df} and V_{Dg} are direct potentials, E_f and E_g are relative energies given by

$$E_f = E_T - E_A - E_B , \quad E_g = E_T - E_C - E_D , \qquad (4.76)$$

and K_{ff} and K_{gg} are energy-dependent kernel functions given by

$$K_{ff}(\vec{R}'_f, \vec{R}''_f) = \langle \phi(A)\phi(B)\delta(\vec{R}_f - \vec{R}'_f)z \mid H - E_T \mid \mathcal{A}''[\hat{\phi}(A)\hat{\phi}(B)\delta(\vec{R}_f - \vec{R}''_f)z] \rangle, \quad (4.77)$$

$$K_{gg}(\vec{R}'_g, \vec{R}''_g) = \langle \phi(C)\phi(D)\delta(\vec{R}_g - \vec{R}'_g)z \mid H - E_T \mid \mathcal{A}''[\hat{\phi}(C)\hat{\phi}(D)\delta(\vec{R}_g - \vec{R}''_g)z] \rangle. \quad (4.78)$$

From the expressions for these kernel functions, one sees that

$$K_{ff}(\vec{R}'_f, \vec{R}''_f) = K_{ff}^*(\vec{R}''_f, \vec{R}'_f) , \qquad (4.79)$$

$$K_{gg}(\vec{R}'_g, \vec{R}''_g) = K_{gg}^*(\vec{R}''_g, \vec{R}'_g) , \qquad (4.80)$$

$$K_{fg}(\vec{R}'_f, \vec{R}''_g) = K_{gf}^*(\vec{R}''_g, \vec{R}'_f) . \qquad (4.81)$$

These relations follow, of course, from the hermiticity of the Hamiltonian operator.

By solving eqs. (4.73) and (4.74) subject to appropriate boundary conditions, one obtains information concerning various scattering and reaction processes. For example, if the channels f and g denote, respectively, the d + t channel and the n + α channel, then performing a two-channel calculation in the manner described here will yield results for the scattering processes t(d,d)t and α(n,n)α, and the reaction process $\alpha(n,d)t$.

The above-described two-channel formulation can be improved by the addition of one or more distortion functions. This will, of course, result in a more complicated set of coupled equations. The procedure to derive these equations is, however, essentially the same as that described in subsect. 4.2b; therefore, we shall not further go into it here.

4.3. Complex-generator-coordinate technique

In resonating-group calculations, the major task is to compute the kernel functions given, for example, by eqs. (4.69) – (4.72). For these computations, one needs to carry out a substantial number of multi-dimensional spatial integrations. Because of the complicated nature of the resonating-group trial wave function caused by the requirement of total antisymmetrization and correct treatment of the c.m. motion, these integrations are generally tedious to perform, especially when the number of nucleons involved in the system is large. Indeed, this was the main reason which led a number of people to comment in the past that the RGM is practicable only for very light nuclear systems.

The integration technique, which was extensively used about ten years ago, was the so-called cluster-coordinate technique [CH 73]. In this technique, one introduces as spatial variables of integration the internal and relative coordinates of the clusters and the total c.m. coordinate. This seems to be a natural set of integration variables to use, because the resonating-group trial wave function is explicitly expressed in terms of these coordinates. Then, by choosing functions of Gaussian dependence for both the spatial part of the wave function and the spatial part of the nucleon-nucleon potential, multi-dimensional integrals can be evaluated analytically.

The evaluation of matrix elements by the cluster-coordinate technique is, however, usually a quite tedious procedure. To compute the multidimensional integrals which involve integrands of Gaussian functions multiplied by polynomials of spatial coordinates, one must first diagonalize the quadratic terms in the exponents by applying linear coordinate transformations. For systems containing a relatively small number of nucleons, this is not too difficult. However, when one deals with systems which contain a rather large number of nucleons, this diagonalization procedure can become very tedious due to the antisymmetrization of the wave function, because for every permutation of nucleons in different clusters, one has to introduce a different coordinate transformation. In addition, it is clear that the cluster-coordinate technique is also not particularly suited for reaction calculations. This is so, since the natural choices for internal and relative coordinates in the incoming and outgoing channels are obviously not the same.

Because of these difficulties mentioned above, the cluster-coordinate technique has now been replaced by another technique, the complex-generator-coordinate technique (CGCT), which uses as integration variables the nucleon coordinates themselves and which is especially useful for reaction calculations and for systems involving a relatively small number of large clusters containing many nucleons. This technique has been developed [HO 72, LE 77, SU 72, SU 76a, TA 77, TA 78, TH 75, TH 77] and employed for practical calculations [LE 76, LE 79, LE 80, ST 78, SU 75, SU 79, TH 76] within the last six or seven years. Its flexibility and, consequently, extensive use has resulted in a great expansion of the scope of resonating-group investigations. As was mentioned in the Introduction, the main idea is to express the trial wave function as an integral of antisymmetrized products of single-particle functions and then make use of techniques similar to those used in shell-model calculations to carry out an analytical evaluation of the required matrix elements.

In subsect. 4.3a, we give a general description of the CGCT. Then, in subsect. 4.3b, we consider a special case where the internal behavior of each cluster is described by a harmonic-oscillator shell-model wave function or a linear superposition of such wave functions. This special case is of great significance, because with such internal functions the CGCT becomes particularly convenient.

4.3a. Description of the CGCT.

Let us consider a two-cluster function of the form

$$\psi = A\left[\phi(A)\phi(B)F(\vec{R}_A - \vec{R}_B)Z(\vec{R}_{cm})\right], \tag{4.82}$$

where

$$\phi(K) = \phi_\alpha(K)\xi_K(s_K, t_K) \tag{4.83}$$

with K = A or B and ξ_K being an appropriate spin-isospin function. The functions $\phi_\alpha(A)$ and $\phi_\alpha(B)$ are spatial parts of the cluster internal functions; they are assumed to be products of functions, with each function describing the motion of a nucleon with respect to the c.m. of the corresponding cluster. That is, we choose $\phi_\alpha(A)$ and $\phi_\alpha(B)$ to have the forms

$$\phi_\alpha(A) = \prod_{j=1}^{N_A} \varphi_j(\vec{r}_j - \vec{R}_A), \tag{4.84}$$

$$\phi_\alpha(B) = \prod_{k=N_A+1}^{N_A+N_B} \varphi_k(\vec{r}_k - \vec{R}_B), \tag{4.85}$$

with $\vec{R}_A$ and $\vec{R}_B$ defined by eq. (4.3). The functions φ_j with j = 1 to N_A and φ_k with k = N_A + 1 to N_A + N_B are chosen such that ϕ_Δ(A) and ϕ_Δ(B) describe properly the spatial behavior of the clusters; they do not have to be generated from an oscillator well, or even a Woods-Saxon well, but can be generated from any well which is appropriate.

Because of the presence of the cluster c.m. coordinates $\vec{R}_A$ and $\vec{R}_B$, the trial function ψ is not in the form of an antisymmetrized product of single-particle wave functions. However, by introducing an integral representation for ψ, we can show that the integrand can be represented in such a product form. To show this, we rewrite ψ in the form

$$\psi = \int \mathcal{A} \left[\phi_\Delta (A; \vec{R}_A'') \phi_\Delta (B; \vec{R}_B'') \delta (\vec{R}_A - \vec{R}_A'') \delta (\vec{R}_B - \vec{R}_B'') \xi_A \xi_B \right]$$

$$\times \; F(\vec{R}_A'' - \vec{R}_B'') Z \left(\frac{N_A \vec{R}_A'' + N_B \vec{R}_B''}{N} \right) d\vec{R}_A'' d\vec{R}_B'' \tag{4.86}$$

by introducing parameter coordinates $\vec{R}_A''$ and $\vec{R}_B''$. In the above equation, it is noted that these parameter coordinates are included in the arguments of the internal wave functions. This is purposedly done, in order to call attention to the fact that in $\phi_\Delta(A;\vec{R}_A'')$ and $\phi_\Delta(B;\vec{R}_B'')$ the parameter coordinates $\vec{R}_A''$ and $\vec{R}_B''$, instead of the dynamical cluster coordinates $\vec{R}_A$ and $\vec{R}_B$, are used [see eqs. (4.84) and (4.85)].

Consider now the term $\phi_\Delta(K;\vec{R}_K'') \delta(\vec{R}_K-\vec{R}_K'')$ in eq. (4.86). If we use the integral representation for $\delta(\vec{R}_K - \vec{R}_K'')$, i.e.,

$$\delta(\vec{R}_K - \vec{R}_K'') = \left(\frac{1}{2\pi} \right)^3 \int exp \left[i \vec{S}_K'' \cdot (\vec{R}_K - \vec{R}_K'') \right] d\vec{S}_K'' \; , \tag{4.87}$$

then it can be easily shown that

$$\phi_\Delta (K; \vec{R}_K'') \delta(\vec{R}_K - \vec{R}_K'') = \left(\frac{1}{2\pi} \right)^3 \int \left[\prod_i \varphi_i (\vec{r}_i - \vec{R}_K'') exp \left(i \frac{1}{N_K} \vec{S}_K'' \cdot \vec{r}_i \right) \right]$$

$$\times \; exp \left(-i \vec{S}_K'' \cdot \vec{R}_K'' \right) d\vec{S}_K'' \tag{4.88}$$

represents an integral over a product of single-particle wave functions with respect to the nucleon coordinates $\vec{r}_i$. Using eq. (4.88), we obtain therefore for ψ the following expression:

$$\psi = \left(\tfrac{1}{2\pi}\right)^6 \int \mathcal{A}\left\{\left[\xi_A \prod_{j=1}^{N_A} \varphi_j\left(\vec{r}_j - \vec{R}_A''\right)\exp\left(i\,\tfrac{1}{N_A}\,\vec{S}_A''\cdot\vec{r}_j\right)\right]\right.$$

$$\times \left[\xi_B \prod_{k=N_A+1}^{N} \varphi_k\left(\vec{r}_k - \vec{R}_B''\right)\exp\left(i\,\tfrac{1}{N_B}\,\vec{S}_B''\cdot\vec{r}_k\right)\right]\right\}$$

$$\times \exp\left(-i\,\vec{S}_A''\cdot\vec{R}_A'' - i\,\vec{S}_B''\cdot\vec{R}_B''\right) F\left(\vec{R}_A'' - \vec{R}_B''\right)$$

$$\times Z\left(\frac{N_A\vec{R}_A'' + N_B\vec{R}_B''}{N}\right) d\vec{S}_A''\,d\vec{S}_B''\,d\vec{R}_A''\,d\vec{R}_B'' \,. \tag{4.89}$$

This shows that, by the introduction of the parameter coordinates $\vec{R}_A''$, $\vec{R}_B''$ and the generator coordinates $\vec{S}_A''$, $\vec{S}_B''$, the wave function ψ can be represented as an integral over antisymmetrized products of single-particle wave functions or Slater determinants.

By making the transformations

$$\vec{R}'' = \vec{R}_A'' - \vec{R}_B''\,, \tag{4.90}$$

$$\vec{R}_{cm}'' = \frac{1}{N}\left(N_A\vec{R}_A'' + N_B\vec{R}_B''\right), \tag{4.91}$$

and by expressing $\delta\psi$ in a similar form involving parameter coordinates $\vec{R}_A'$, $\vec{R}_B'$ and generator coordinates $\vec{S}_A'$, $\vec{S}_B'$, we obtain explicit expressions for the kernel functions $\mathcal{H}(\vec{R}',\vec{R}'')$ and $\mathcal{N}(\vec{R}',\vec{R}'')$. Consider the kernel function $\mathcal{H}(\vec{R}',\vec{R}'')$, for example. Because the operator H contains a sum of one- and two-particle operators, this kernel function has the form of an integral over the generator coordinates $\vec{S}_A'$, $\vec{S}_B'$, $\vec{S}_A''$, $\vec{S}_B''$ and the parameter coordinates $\vec{R}_{cm}'$, $\vec{R}_{cm}''$ ($\vec{R}'$ and $\vec{R}''$ are not integrated over), with its integrand consisting of a sum of products of matrix elements

$$\langle \hat{\varphi}_i | \hat{\varphi}_j \rangle\,, \qquad \langle \hat{\varphi}_i | O_p | \hat{\varphi}_j \rangle\,, \qquad \langle \hat{\varphi}_i \hat{\varphi}_j | O_{pq} | \hat{\varphi}_\ell \hat{\varphi}_m \rangle\,,$$

where the $\hat{\varphi}_i$'s are one-particle functions depending on the generator and parameter coordinates. The orthogonality relations between the $\hat{\varphi}_i$'s in the same cluster will usually reduce very drastically the number of terms which need to be evaluated.

Even though the discussion given above was based on a two-cluster wave function, it is clear that one can easily generalize it to a wave function which describes the behavior of a system composed of a large number of clusters.

4.3b. <u>Special case – cluster internal functions in harmonic-oscillator shell-model representation</u>.

Even though the technique described above is useful when the cluster internal functions have the general form given by eqs. (4.84) and (4.85), it becomes especially convenient when the internal structure of each cluster is described by a translationally-invariant harmonic-oscillator shell-model function of the lowest configuration or a sum of such functions, with each of them specified by a different value for the oscillator width parameter.

In this subsection, we shall consider first the simpler case where one such oscillator internal function is used, i.e.,

$$\phi_{\mathcal{A}}(K; \vec{R}_K) = \prod_i h_i(\vec{r}_i - \vec{R}_K)\, exp\left[-\tfrac{1}{2}\alpha_K(\vec{r}_i - \vec{R}_K)^2\right], \tag{4.92}$$

with α_K being the oscillator width parameter for cluster K (K = A or B) and the functions h_i being polynomials in single-particle spatial coordinates. With this choice and upon performing the transformation

$$\vec{Q}_K'' = \frac{1}{N_K\alpha_K}\vec{S}_K'' - i\vec{R}_K'' , \tag{4.93}$$

the wave function ψ of eq. (4.89) can then be written in the following form:

$$\psi = \left(\frac{N_A\alpha_A}{2\pi}\right)^3 \left(\frac{N_B\alpha_B}{2\pi}\right)^3$$

$$\times \int \mathcal{A}'\left\{\mathcal{A}_A\left[\xi_A\prod_{j=1}^{N_A} h_j(\vec{r}_j - \vec{R}_A'')exp\left[-\tfrac{1}{2}\alpha_A(\vec{r}_j - i\vec{Q}_A'')^2\right]\right]\right.$$

$$\times \mathcal{A}_B\left[\xi_B\prod_{k=N_A+1}^{N} h_k(\vec{r}_k - \vec{R}_B'')exp\left[-\tfrac{1}{2}\alpha_B(\vec{r}_k - i\vec{Q}_B'')^2\right]\right]\right\}$$

$$\times F(\vec{R}_A'' - \vec{R}_B'')exp\left[\tfrac{1}{2}N_A\alpha_A(\vec{R}_A'' - i\vec{Q}_A'')^2 + \tfrac{1}{2}N_B\alpha_B(\vec{R}_B'' - i\vec{Q}_B'')^2\right]$$

$$\times Z\left(\frac{N_A\vec{R}_A'' + N_B\vec{R}_B''}{N}\right) d\vec{Q}_A''\, d\vec{Q}_B''\, d\vec{R}_A''\, d\vec{R}_B'' . \tag{4.94}$$

The above equation for ψ can be reduced by noting that, because the internal function is chosen to have the lowest configuration in a harmonic-oscillator well, the arguments $(\vec{r}_j - \vec{R}_A'')$ of h_j and $(\vec{r}_k - \vec{R}_B'')$ of h_k can be replaced, respectively, by the arguments $(\vec{r}_j - \vec{C}_A'')$ and $(\vec{r}_k - \vec{C}_B'')$ with $\vec{C}_A''$ and $\vec{C}_B''$ being any constant vectors independent of the nucleon spatial coordinates.[+] Thus, by using eqs. (4.90) and (4.91) and the transformations

$$\vec{S}'' = \alpha_A \vec{Q}_A'' - \alpha_B \vec{Q}_B'' \quad , \tag{4.95}$$

$$\vec{S}_{cm}'' = \frac{1}{N}\left(N_A \alpha_A \vec{Q}_A'' + N_B \alpha_B \vec{Q}_B''\right), \tag{4.96}$$

and by choosing

$$Z(\vec{R}_{cm}) = exp\left[-\frac{1}{2}(N_A \alpha_A + N_B \alpha_B)\vec{R}_{cm}''^2\right], \tag{4.97}$$

we can easily perform the integration over the variables $\vec{R}''_{cm}$ and $\vec{S}''_{cm}$, and obtain the following simplified expression:

$$\psi = \left(\frac{N_A N_B}{2\pi N}\right)^3 \int \mathscr{A}'\left[\hat{\phi}'(A;\vec{S}'',\vec{R}'')\hat{\phi}'(B;\vec{S}'',\vec{R}'')\right]$$
$$\times \Gamma(\vec{S}'',\vec{R}'')F(\vec{R}'')d\vec{S}''d\vec{R}'', \tag{4.98}$$

where

$$\Gamma(\vec{S}'',\vec{R}'') = exp\left[-\frac{1}{2}\frac{N_A N_B(N_A \alpha_A + N_B \alpha_B)}{N^2 \alpha_A \alpha_B}\left(\vec{S}'' + i\frac{N_B \alpha_A + N_A \alpha_B}{N}\vec{R}''\right)^2\right], \tag{4.99}$$

and

$$\hat{\phi}'(A;\vec{S}'',\vec{R}'') = \mathscr{A}_A\left\{\xi_A \prod_{j=1}^{N_A} h_j \; exp\left[-\frac{1}{2}\alpha_A(\vec{r}_j - \vec{\eta}_A'')^2\right]\right\}, \tag{4.100}$$

$$\hat{\phi}'(B;\vec{S}'',\vec{R}'') = \mathscr{A}_B\left\{\xi_B \prod_{k=N_A+1}^{N} h_k \; exp\left[-\frac{1}{2}\alpha_B(\vec{r}_k - \vec{\eta}_B'')^2\right]\right\}, \tag{4.101}$$

with

$$\vec{\eta}_A'' = \frac{iN_B}{N\alpha_A}\left[\vec{S}'' + i\frac{N_A}{N}(\alpha_B - \alpha_A)\vec{R}''\right], \tag{4.102}$$

$$\vec{\eta}_B'' = -\frac{iN_A}{N\alpha_B}\left[\vec{S}'' - i\frac{N_B}{N}(\alpha_B - \alpha_A)\vec{R}''\right]. \tag{4.103}$$

The meaning of the function $\hat{\phi}'(K;\vec{S}'',\vec{R}'')$, with K = A or B, is clear; it is a shell-model function of the lowest configuration in a harmonic-oscillator well of width

---

[+] Even more general choices can be made. For example, $\vec{C}_A''$ and $\vec{C}_B''$ may be chosen as any symmetric functions of the nucleon spatial coordinates.

parameter α_K , with the center of the well located at the point $\vec{\eta}_K''$. Since it is seen from eqs. (4.102) and (4.103) that $\vec{\eta}_K''$ is a complex quantity, we have therefore chosen to call the technique described here a <u>complex-generator-coordinate technique</u>.

The wave function ψ has now acquired the form of eq. (3.4), with the continuous parameter coordinates being $\vec{S}''$ and $\vec{R}''$. In the Hill-Wheeler terminology [HI 53, GR 57], the part $\mathcal{A}'[\hat{\phi}'(A;\vec{S}'',\vec{R}'')\hat{\phi}'(B;\vec{S}'',\vec{R}'')]$ is the intrinsic or generating function, while the part $\Gamma(\vec{S}'', \vec{R}'')F(\vec{R}'')$ is the weight function. Thus, through the procedure described in this subsection, we have created a connection between the resonating-group theory and the Hill-Griffin-Wheeler generator-coordinate formalism.

If one makes the assumption that α_A is equal to α_B , then the expressions for $\vec{\eta}_A''$ and $\vec{\eta}_B''$ are simpler and the computation of the required matrix elements becomes also somewhat simplified. The power of the CGCT lies, however, in the fact that the computation does not become much more complicated even when the width parameters α_A and α_B are chosen to have different values. This is in contrast to the generator-coordinate method used by many other investigators (see sect. 5 and refs. [29-45] given in ref. [TA 78]). In this latter method, a real generator coordinate (i.e., the distance vector between the centers of the harmonic-oscillator potential wells) is used and the calculation of the matrix elements becomes straightforward only when α_A and α_B are assumed as equal, which is of course a frequently unrealistic assumption.

By carrying out the same procedure for the variation $\delta\psi$ which will contain a generator coordinate $\vec{S}'$, a parameter coordinate $\vec{R}'$, and two arbitrary constant vectors $\vec{C}_A'$ and $\vec{C}_B'$, one obtains then for the kernel function $\hat{K}(\vec{R}',\vec{R}'')$ [see eq. (4.10)] the expression

$$\hat{K}(\vec{R}',\vec{R}'') = \mathcal{H}(\vec{R}',\vec{R}'') - E_T \mathcal{N}(\vec{R}',\vec{R}'')$$

$$= \int \mathcal{M}(\vec{S}',\vec{R}';\vec{S}'',\vec{R}'')\,\Gamma^*(\vec{S}',\vec{R}')\,\Gamma(\vec{S}'',\vec{R}'')\,d\vec{S}'\,d\vec{S}'' , \tag{4.104}$$

where

$$\mathcal{M}(\vec{S}',\vec{R}';\vec{S}'',\vec{R}'') = \left(\frac{N_A N_B}{2\pi N}\right)^6$$

$$\times \left\langle \left[\xi_A \prod_{j=1}^{N_A} h_j(\vec{r}_j - \vec{C}_A')\,exp\left[-\tfrac{1}{2}\alpha_A(\vec{r}_j - \vec{\eta}_A')^2\right]\right] \right.$$

$$\times \left[\xi_B \prod_{k=N_A+1}^{N} h_k(\vec{r}_k - \vec{C}_B')\,exp\left[-\tfrac{1}{2}\alpha_B(\vec{r}_k - \vec{\eta}_B')^2\right]\right]$$

$$|H - E_T| \mathcal{A}'\left\{ \mathcal{A}_A\left[\xi_A \prod_{j=1}^{N_A} h_j(\vec{r}_j - \vec{C}_A'')\,exp\left[-\tfrac{1}{2}\alpha_A(\vec{r}_j - \vec{\eta}_A'')^2\right]\right] \right.$$

$$\left. \times \mathcal{A}_B\left[\xi_B \prod_{k=N_A+1}^{N} h_k(\vec{r}_k - \vec{C}_B'')\,exp\left[-\tfrac{1}{2}\alpha_B(\vec{r}_k - \vec{\eta}_B'')^2\right]\right]\right\} \right\rangle \tag{4.105}$$

with

$$\vec{\eta}_A' = \frac{iN_B}{N\alpha_A}\left[\vec{S}' + i\,\frac{N_A}{N}(\alpha_B - \alpha_A)\vec{R}'\right] , \qquad (4.106)$$

$$\vec{\eta}_B' = -\frac{iN_A}{N\alpha_B}\left[\vec{S}' - i\,\frac{N_B}{N}(\alpha_B - \alpha_A)\vec{R}'\right] . \qquad (4.107)$$

From eq. (4.105) it is noted that in the expression for $\mathcal{M}(\vec{S}',\ \vec{R}';\ \vec{S}'',\ \vec{R}'')$ there appear only products of complex functions, each depending on a single-particle (space, spin, and isospin) coordinate. Thus, the antisymmetrization procedure causes no great difficulty and the computation of $\mathcal{M}(\vec{S}',\ \vec{R}';\ \vec{S}'',\ \vec{R}'')$ can be performed in a rather straightforward manner (for more details, see refs. [LE 77, TH 75]). In addition, we should point out that the integration over the nucleon coordinates may be further facilitated by making use of the freedom associated with the choice of the quantities $\vec{C}_A'$, $\vec{C}_B'$, $\vec{C}_A''$, and $\vec{C}_B''$ appearing in the arguments of h_j and h_k. In most cases, it is appropriate to choose

$$\vec{C}_A'^* = \vec{C}_A'' = \frac{1}{2}(\vec{\eta}_A'^* + \vec{\eta}_A'') ,$$

$$\vec{C}_B'^* = \vec{C}_B'' = \frac{1}{2}(\vec{\eta}_B'^* + \vec{\eta}_B'') ; \qquad (4.108)$$

although when $N_A \approx N_B$ the choice

$$\vec{C}_A'^* = \vec{C}_B'' = \frac{\alpha_A}{\alpha_A + \alpha_B}\,\vec{\eta}_A'^* + \frac{\alpha_B}{\alpha_A + \alpha_B}\,\vec{\eta}_B'' ,$$

$$\vec{C}_B'^* = \vec{C}_A'' = \frac{\alpha_B}{\alpha_A + \alpha_B}\,\vec{\eta}_B'^* + \frac{\alpha_A}{\alpha_A + \alpha_B}\,\vec{\eta}_A'' \qquad (4.109)$$

may be somewhat more convenient. These particular choices are made, in order to take advantage of the orthonormality relations satisfied by the single-particle wave functions in a harmonic-oscillator potential well.

It should be mentioned that, in all investigations carried out so far with the CGCT, the matrix element $\mathcal{M}(\vec{S}',\ \vec{R}';\ \vec{S}'',\ \vec{R}'')$ has been analytically evaluated. If one adopts the assumption $\alpha_A = \alpha_B$, then $\mathcal{M}$ will no longer depend on $\vec{R}'$ and $\vec{R}''$, but depend only on $\vec{S}'$ and $\vec{S}''$. In this simplified situation, it may turn out that a numerical evaluation of $\mathcal{M}$ is a more convenient procedure.

Next, we discuss the more general case [TH 77] where the cluster internal function is given by a sum of translationally-invariant harmonic-oscillator shell-model functions of the lowest configuration. That is, the internal functions are now chosen to be

$$\phi(A) = \sum_{p=1}^{n_p} \phi_p(A) \quad , \qquad \phi(B) = \sum_{q=1}^{n_q} \phi_q(B) \quad , \tag{4.110}$$

where

$$\phi_p(A) = C_{Ap}\, \mathcal{A}_A \left\{ \xi_A \prod_{j=1}^{N_A} h_j(\vec{r}_j - \vec{R}_A)\exp\left[-\tfrac{1}{2}\alpha_{Ap}(\vec{r}_j - \vec{R}_A)^2\right]\right\} , \tag{4.111}$$

$$\phi_q(B) = C_{Bq}\, \mathcal{A}_B \left\{ \xi_B \prod_{k=N_A+1}^{N} h_k(\vec{r}_k - \vec{R}_B)\exp\left[-\tfrac{1}{2}\alpha_{Bq}(\vec{r}_k - \vec{R}_B)^2\right]\right\} , \tag{4.112}$$

with α_{Ap} (p = 1 to n_p) and α_{Bq} (q = 1 to n_q) being a set of appropriately chosen width parameters, and C_{Ap} and C_{Bq} being appropriate amplitudes. We should mention that, even though the internal functions given above are expressed explicitly in terms of oscillator functions, there is no severe restriction regarding the flexibility required in properly describing the internal behavior of the clusters, because any cluster internal function may be well approximated by the linear superposition of a sufficiently large number of oscillator cluster functions with different width parameters.

With $\phi(A)$ and $\phi(B)$ given by eqs. (4.110) – (4.112), eq. (4.10) then becomes

$$\sum_{p,r=1}^{n_p} \sum_{q,t=1}^{n_q} \int \widehat{K}_{pq,rt}(\vec{R}',\vec{R}'')F(\vec{R}'')\,d\vec{R}'' = 0 \quad , \tag{4.113}$$

where

$$\widehat{K}_{pq,rt}(\vec{R}',\vec{R}'') = \langle \phi_p(A)\phi_q(B)\,\delta(\vec{R}-\vec{R}')\,Z(\vec{R}_{cm})\,|\,H-E_T\,|$$

$$\mathcal{A}\left[\phi_r(A)\phi_t(B)\,\delta(\vec{R}-\vec{R}'')\,Z(\vec{R}_{cm})\right]\rangle . \tag{4.114}$$

As discussed above, a particular form for the function $Z(\vec{R}_{cm})$ must be chosen in order to express $\widehat{K}(\vec{R}',\vec{R}'')$ in the form of eq. (4.104). Since this choice involves the oscillator width parameters of the cluster wave functions [see eq. (4.97)], it is not clear in the present case how this choice should be made because different sets of width parameters appear in the bra and ket sides of eq. (4.114). However, by using the fact that the operator $(H - E_T)$ is Galilean-invariant, one can write eq. (4.114) as

615

$$\hat{K}_{pq,rt}(\vec{R}',\vec{R}'') = \frac{\langle Z(\vec{R}_{cm}) \mid Z(\vec{R}_{cm})\rangle}{\langle \chi_{pq}(\vec{R}_{cm}) \mid \chi_{rt}(\vec{R}_{cm})\rangle}$$

$$\times \ \langle \phi_p(A)\phi_q(B)\delta(\vec{R}-\vec{R}')\chi_{pq}(\vec{R}_{cm}) \mid H-E_T \mid$$

$$A\left[\phi_r(A)\phi_t(B)\delta(\vec{R}-\vec{R}'')\chi_{rt}(\vec{R}_{cm})\right]\rangle \ . \tag{4.115}$$

The arbitrary function $\chi_{pq}(\vec{R}_{cm})$ can now be chosen, in accordance with eq. (4.97), to be

$$\chi_{pq}(\vec{R}_{cm}) = exp\left[-\tfrac{1}{2}(N_A\alpha_{Ap} + N_B\alpha_{Bq})\vec{R}_{cm}^{\,2}\right] \ . \tag{4.116}$$

With this choice, the matrix elements $\hat{K}_{pq,rt}(\vec{R}',\vec{R}'')$ can be written in the form of eq. (4.104), and its evaluation can be accomplished just as in the single-width-parameter case ($n_p = n_q = 1$) discussed above.

4.4. Redundant solutions

Even with the adoption of the CGCT, it is still rather tedious to derive explicit expressions for the required kernel functions. In this subsection, we discuss a redundant-solution test [HO 54, TA 63] which can be used to check the correctness of these functions, in the case where each of the clusters is described by a single harmonic-oscillator shell-model function in the lowest configuration and a common width parameter is used to characterize all the clusters involved.

For simplicity in discussion, let us consider a single-channel two-cluster case [+] where the trial function is given by eq. (4.82). Under conditions mentioned above for the cluster internal functions, there exist then intercluster relative-motion functions $F_{red}(\vec{R})$, called redundant solutions, for which

$$\psi_{red} = A\left[\phi(A)\phi(B)F_{red}(\vec{R})Z(\vec{R}_{cm})\right] = 0 \ . \tag{4.117}$$

In light systems, these redundant solutions can be readily determined by using the exact correspondence between oscillator cluster and oscillator shell models (see sect. 2 and also ref. [HO 77b]). For example, in the $^3H + \alpha$ system, it can be easily shown that, to comply with the Pauli principle, the relative motion between

---

[+] A more general discussion for a multi-cluster system is given in ref. [HO 77a].

the clusters must have at least three oscillator quanta of excitation. Thus, the redundant solutions are those which correspond to lower-order oscillations with number of quanta less than three, i.e., 1s, 1p, 1d, and 2s oscillator functions.

Because ψ_{red} is identically equal to zero, one obtains also

$$\langle \phi(A)\phi(B)\delta(\vec{R}-\vec{R}')Z(\vec{R}_{cm})\,|\,\mathcal{A}[\phi(A)\phi(B)F_{red}(\vec{R})Z(\vec{R}_{cm})]\rangle = 0 \,, \tag{4.118}$$

$$\langle \phi(A)\phi(B)\delta(\vec{R}-\vec{R}')Z(\vec{R}_{cm})\,|\,H\,|\,\mathcal{A}[\phi(A)\phi(B)F_{red}(\vec{R})Z(\vec{R}_{cm})]\rangle = 0 \,. \tag{4.119}$$

By using eqs. (4.11) and (4.12), these equations can be further written as

$$\int \mathcal{N}(\vec{R}',\vec{R}'')F_{red}(\vec{R}'')\,d\vec{R}'' = 0 \,, \tag{4.120}$$

$$\int \mathcal{H}(\vec{R}',\vec{R}'')F_{red}(\vec{R}'')\,d\vec{R}'' = 0 \,. \tag{4.121}$$

That is, the redundant solutions are the eigenfunctions of both the norm or overlap kernel $\mathcal{N}(\vec{R}',\vec{R}'')$ and the energy or Hamiltonian kernel $\mathcal{H}(\vec{R}',\vec{R}'')$, with eigenvalues equal to zero. Thus, the correctness of these two kernel functions can be ascertained by simply examining their eigenvalue spectra and counting the number of times when eigenvalues equal to zero occur.

As an illustration, let us consider the $^3H+\alpha$ and $\alpha+{}^6Li$ systems. In table 1, we list the eigenvalues $\mu_{n\ell}$ of the norm kernel, with ℓ being the quantum number representing the orbital-angular-momentum between the clusters and n being an ordering index. For our present consideration where $\alpha_A = \alpha_B = \alpha$, $\mu_{n\ell}$ is

Table 1

Norm eigenvalues $\mu_{n\ell}$ for $^3H+\alpha$ and $\alpha+{}^6Li$ systems in the case where $\alpha_A = \alpha_B$.

| n | $H^3+\alpha$ | | $\alpha+{}^6Li$ | | | |
	μ_{no}	μ_{n1}	μ_{no}	μ_{n1}	μ_{n2}	μ_{n3}
0	0		0			
1		0		0		
2	0		0		0	
3		1.1910		0		0
4	0.5955		0.6430		0.5224	
5		1.1592		0.2813		0.4320
6	0.8064		1.0658		0.9997	
7		1.1269		0.5376		0.6126
8	0.8918		1.1685		1.1382	
9		1.0740		0.7092		0.7430
10	0.9432		1.1533		1.1398	
11		1.0420		0.8212		0.8362
12	0.9683		1.1106		1.1046	
13		1.0236		0.8930		0.8997
14	0.9824		1.0713		1.0686	
15		1.0130		0.9383		0.9412

Table 2

Norm eigenvalues $\mu_{n\ell}$ for $^3H+\alpha$ and $\alpha+{}^6Li$ systems in the case where $\alpha_A \neq \alpha_B$.

| n | $^3H+\alpha$ | | $\alpha+{}^6Li$ | | | |
	μ_{n0}	μ_{n1}	μ_{n0}	μ_{n1}	μ_{n2}	μ_{n3}
0	0.0000		0.0000			
1		0.0009		0.0006		
2	0.0313		0.0074		0.0066	
3		1.1434		0.0600		0.0539
4	0.6135		0.5961		0.5004	
5		1.2104		0.3613		0.4753
6	0.8163		0.9648		0.9193	
7		1.1129		0.6086		0.6686
8	0.9031		1.1577		1.1331	
9		1.0681		0.7604		0.7878
10	0.9467		1.0998		1.0849	
11		1.0390		0.8547		0.8668
12	0.9704		1.0639		1.0544	
13		1.0220		0.9140		0.9193
14	0.9836		1.0421		1.0368	
15		1.0121		0.9505		0.9528

independent of α [HO 77b][+] and n can be considered as the number of oscillation quanta in the relative motion. From this table, one finds that there do exist expected numbers of zero eigenvalues, indicating that the norm kernels for these two systems are correctly derived at least in the equal-width-parameter case. In addition, it is noted that many eigenvalues are larger than 1; this is connected with the fact that, in our calculation of the norm kernels, antisymmetric basis functions possessing correct c.m. behavior have been used [FL 77].

One may also demonstrate the existence of redundant solutions in another interesting way. Consider, as an example, the seven-nucleon ${}^3\mathrm{He} + \alpha$ case [KO 74, TA 63]. If one takes eq. (4.35), makes the partial-wave expansion

$$F(\vec{R}') = \sum_{\ell} \frac{1}{R'} f_{\ell}(R') P_{\ell}(\cos\theta') , \tag{4.122}$$

$$K(\vec{R}',\vec{R}'') = \frac{1}{R'R''} \sum_{\ell,m} k_{\ell}(R',R'') Y_{\ell m}(\theta',\phi') Y_{\ell m}^{*}(\theta'',\phi'') , \tag{4.123}$$

and solves numerically the resultant integrodifferential equation

$$\left\{ \frac{\hbar^2}{2\mu}\left[\frac{d^2}{dR'^2} - \frac{\ell(\ell+1)}{R'^2} \right] + E - V_D(R') \right\} f_{\ell}(R')$$

$$= \int_0^{\infty} k_{\ell}(R',R'') f_{\ell}(R'') dR'' , \tag{4.124}$$

then the redundant solutions may be exhibited at any value of E by plotting $f_{\ell}(R')$ obtained using different values for the integration step size $\tilde{h}_i$ in the numerical procedure. If the norm and Hamiltonian kernel functions are correctly derived, then the functions $f_{\ell}(R')$ for $\ell \le 2$ should differ at small values of R' but remain the same in the asymptotic region. On the other hand, for $\ell > 2$, $f_{\ell}(R')$ should be independent of $\tilde{h}_i$ for all values of R'. In fig. 4, we illustrate the $\ell = 0$ and 3 solutions for ${}^3\mathrm{He} + \alpha$ scattering at 20 MeV, obtained by using $\tilde{h}_i$ equal to 0.24 and 0.28 fm. As is seen, these curves do behave in exactly the way as discussed above, thus demonstrating that the kernel functions given in the appendices of refs. [KO 74, TA 63] have indeed been correctly derived.

When α_A is not equal to α_B , redundant solutions do not exist. However, the eigenvalue spectra of $\mathcal{N}(\vec{R}',\vec{R}'')$ and $\mathcal{H}(\vec{R}', \vec{R}'')$ are still expected to be similar to those in the equal-width-parameter case, which follows from the fact that the Pauli principle tends to reduce greatly the differences between apparently different structures. This is demonstrated in table 2 for $\mathcal{N}(\vec{R}',\vec{R}'')$ in the ${}^3\mathrm{H} + \alpha$ and $\alpha + {}^6\mathrm{Li}$ systems. Here one sees that, even with different width parameters equal

---

[+]In the equal-width-parameter ${}^3\mathrm{H} + \alpha$ case, $\mu_{n\ell}$ depends also only on n, but not on ℓ [HO 77b]. This is no longer so, when α_A is chosen to be different from α_B .

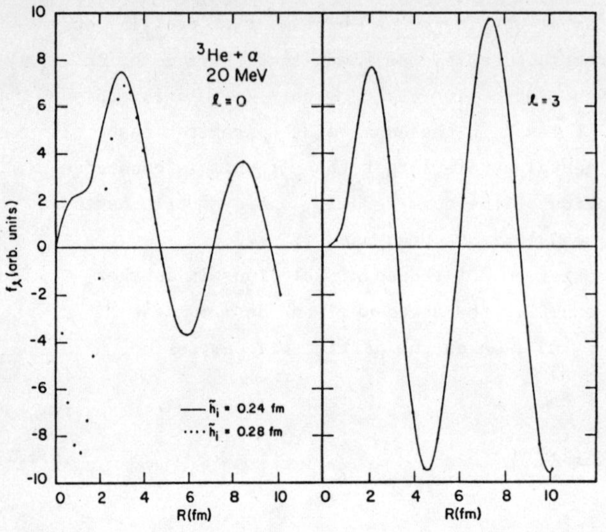

Fig. 4. Demonstration, in the ^{3}He + α system, of the existence and non-existence of reduntant solutions in $\ell = 0$ and 3 states, respectively.

to 0.378, 0.514, and 0.305 fm^{-2} for the ^{3}H, α , and ^{6}Li clusters, respectively, the norm eigenvalues $\mu_{n\ell}$ are still rather close to the corresponding values given in table 1 for the equal-width-parameter case. Thus, although there is, strictly speaking, no redundant-solution test when $\alpha_A \neq \alpha_B$, one can still examine the eigenvalue spectra to gain some confidence about the correctness of the derived kernel functions.

In the $\alpha_A \neq \alpha_B$ case, the absence of redundant solutions is replaced by the presence of spurious resonances which occur with certain definite characteristic energies [CH 72, MU 76, SA 73, WE 77]. Consider again the ^{3}H + α system as an

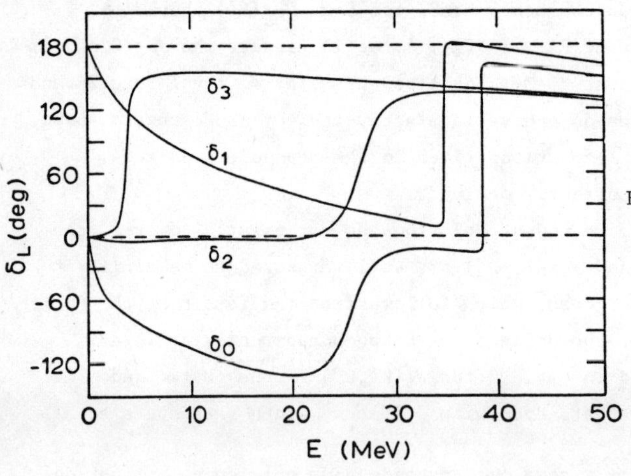

Fig. 5. Demonstration of the presence of spurious resonances in the ^{3}H + α system. (Adapted from ref.[CL 75])

example. Now, instead of redundant solutions, there appear relatively sharp spurious
resonances at E > 20 MeV in the ℓ = 0, 1, and 2 phase-shift curves [CL 75] (see fig.5).
These spurious resonances do not represent the existence of observable resonance
states in the compound system, and their presence can be easily suppressed either
by the adoption of a least-square variational procedure [SC 73, SC 73a] or by the
introduction of a phenomenological imaginary potential into the resonating-group
formulation [CH 73a].

5. The Generator-Coordinate Method

5.1. Basis wave functions in the GCM

For clarity in discussion, we consider the simplest case where there are two
spin-zero clusters A and B. In the GCM formulation, the basis function is chosen as
a Brink-type wave function [BR 66], in which each cluster is assumed to be described
by a single shell-model function of the lowest configuration in a harmonic-oscillator
well of width parameter α_K (K = A or B), with the center of the potential well
located at a point $\vec{X}_K''$. That is, we write the basis function $\hat{\tilde{\Phi}}$, with real generator
or parameter coordinates $\vec{X}_A''$ and $\vec{X}_B''$ in the form

$$\hat{\tilde{\Phi}} (\vec{r}_1,\cdots,\vec{r}_N ; \vec{X}_A'',\vec{X}_B'') = \mathcal{A}' \left[\tilde{\phi}(A;\vec{X}_A'') \tilde{\phi}(B;\vec{X}_B'') \right] , \tag{5.1}$$

where

$$\tilde{\phi}(K;\vec{X}_K'') = \mathcal{A}_K \left[\phi_\alpha(K;\vec{X}_K'')\xi_K \right] \tag{5.2}$$

with

$$\phi_\alpha(A;\vec{X}_A'') = \prod_{j=1}^{N_A} h_j(\vec{r}_j - \vec{X}_A'') \exp\left[-\tfrac{1}{2}\alpha_A(\vec{r}_j - \vec{X}_A'')^2\right] , \tag{5.3}$$

$$\phi_\alpha(B;\vec{X}_B'') = \prod_{k=N_A+1}^{N} h_k(\vec{r}_k - \vec{X}_B'') \exp\left[-\tfrac{1}{2}\alpha_B(\vec{r}_k - \vec{X}_B'')^2\right] . \tag{5.4}$$

By using the relation

$$\sum (\vec{r}_i - \vec{X}_K'')^2 = \sum (\vec{r}_i - \vec{R}_K)^2 + N_K(\vec{R}_K - \vec{X}_K'')^2 \tag{5.5}$$

and the fact that, because of antisymmetrization, the arguments $(\vec{r}_j - \vec{X}_A'')$ of h_j and
$(\vec{r}_k - \vec{X}_B'')$ of h_k in eqs. (5.3) and (5.4) can be replaced, respectively, by the

arguments $(\vec{r}_j - \vec{R}_A)$ and $(\vec{r}_k - \vec{R}_B)$, one can further write $\hat{\underline{\Phi}}$ of eq. (5.1) as

$$\hat{\underline{\Phi}}(\tilde{r}_1,\cdots,\tilde{r}_N\,;\,\vec{X}_A'',\vec{X}_B'') = \mathcal{A}'\{\,\hat{\phi}(A)\hat{\phi}(B)$$

$$\times \exp\left[-\frac{1}{2}N_A\alpha_A\left(\vec{R}_A-\vec{X}_A''\right)^2 - \frac{1}{2}N_B\alpha_B\left(\vec{R}_B-\vec{X}_B''\right)^2\right]\}$$

$$= \mathcal{A}'\{\,\hat{\phi}(A)\hat{\phi}(B)$$

$$\times \exp\left[-\tilde{\alpha}\left(\vec{R}_{cm}-\vec{X}_{cm}''\right)^2 - \tilde{\beta}\left(\vec{R}_{cm}-\vec{X}_{cm}''\right)\cdot\left(\vec{R}-\vec{X}''\right) - \tilde{\gamma}\left(\vec{R}-\vec{X}''\right)^2\right]\}\,, \tag{5.6}$$

where $\hat{\phi}(A)$ and $\hat{\phi}(B)$ represent translationally-invariant antisymmetrized cluster internal functions [see eq. (4.16)] and

$$\vec{X}'' = \vec{X}_A'' - \vec{X}_B''\,, \tag{5.7}$$

$$\vec{X}_{cm}'' = \frac{1}{N}\left(N_A\vec{X}_A'' + N_B\vec{X}_B''\right), \tag{5.8}$$

$$\tilde{\alpha} = \frac{1}{2}\left(N_A\alpha_A + N_B\alpha_B\right), \tag{5.9}$$

$$\tilde{\beta} = \mu_0\left(\alpha_A - \alpha_B\right), \tag{5.10}$$

$$\tilde{\gamma} = \frac{1}{2}\frac{\mu_0}{N}\left(N_B\alpha_A + N_A\alpha_B\right). \tag{5.11}$$

In the general case where α_A is not equal to α_B, it is seen from eq. (5.6) that, in contrast to the RGM wave function of eq. (4.8), the c.m. function does not appear as a multiplicative factor and, hence, the basis or intrinsic function $\hat{\underline{\Phi}}$ contains necessarily spurious excitations for the total c.m. motion.

For the special case of equal width parameters, i.e., $\alpha_A=\alpha_B=\alpha$, the situation is rather different. Here it is convenient to make the further simplification of setting $\vec{X}_{cm}''$ equal to zero and write the intrinsic function as

$$\hat{\underline{\Phi}}_e(\tilde{r}_1,\cdots,\tilde{r}_N\,;\,\vec{X}'') = \mathcal{A}'\left[\tilde{\phi}\left(A\,;\,\frac{N_B}{N}\vec{X}''\right)\tilde{\phi}\left(B\,;\,-\frac{N_A}{N}\vec{X}''\right)\right]$$

$$= \mathcal{A}'\{\,\hat{\phi}(A)\hat{\phi}(B)\,\tilde{\Gamma}_e\left(\vec{R},\vec{X}''\right)Z\left(\vec{R}_{cm}\right)\}\,, \tag{5.12}$$

where the subscripts e are used to emphasize the fact that we are specifically considering the equal-width-parameter case. In the above equation,

$$\tilde{\Gamma}_e\left(\vec{R},\vec{X}''\right) = \exp\left[-\tilde{\gamma}_e\left(\vec{R}-\vec{X}''\right)^2\right] \tag{5.13}$$

with

$$\tilde{\gamma}_e = \frac{1}{2}\mu_0\alpha \tag{5.14}$$

and $Z(\vec{R}_{cm})$ is given by eq. (4.97), which now appears appropriately as a multiplicative factor. Equation (5.12) is very important, because it establishes, in the equal-width-parameter case, a definite connection between GCM and RGM basis functions.

From the intrinsic function $\hat{\tilde{\Phi}}_e$, one can project out states of definite angular momentum [BR 66, WI 77]. In our present consideration where the clusters involved have no internal spin, the angular-momentum projection K on the symmetry axis is equal to zero and the total angular momentum J is equal to the relative orbital angular momentum ℓ between the clusters. In addition, there exists the simplifying feature that positive-parity states have even angular-momentum values, while negative-parity states have odd angular-momentum values. Thus, the projection can be simply achieved by operating on $\hat{\tilde{\Phi}}_e$ with the operator [KR 73, MI 77a]

$$C_{MO}^{J} = \frac{2J+1}{8\pi^2} \int d\Omega \, D_{MO}^{J}{}^{*}(\Omega) R(\Omega) \, , \tag{5.15}$$

where D_{MO}^{J} is the Wigner D-function with K = 0 and $R(\Omega)$ is the rotation operator [ED 60]. Following the procedure outlined in Chapter 14 of ref. [WI 77], one can then obtain the so-called static energy curves (or static energy surfaces) by the expression

$$\tilde{E}_{J}(x'') = \frac{\tilde{H}_{J}(x'')}{\tilde{N}_{J}(x'')} \, , \tag{5.16}$$

where

$$\left\{ \begin{array}{c} \tilde{H}_{J}(x'') \\ \tilde{N}_{J}(x'') \end{array} \right\} = \frac{2J+1}{2} \int_{0}^{\pi} d\beta \, \sin\beta \, d_{oo}^{J}(\beta) \langle \hat{\tilde{\Phi}}_e | \left\{ \begin{array}{c} H \\ 1 \end{array} \right\} e^{i\beta J_y/\hbar} | \hat{\tilde{\Phi}}_e \rangle \tag{5.17}$$

with

$$d_{oo}^{J}(\beta) = P_{J}(\cos\beta) \tag{5.18}$$

and $\langle \hat{\tilde{\Phi}}_e | H e^{i\beta J_y/\hbar} | \hat{\tilde{\Phi}}_e \rangle$ and $\langle \hat{\tilde{\Phi}}_e | e^{i\beta J_y/\hbar} | \hat{\tilde{\Phi}}_e \rangle$ representing, respectively, the overlap integrals for the Hamiltonian and unity operators between two intrinsic functions whose symmetry axes are oriented at an angle β with respect to each other. From the viewpoint of the Ritz variational principle, the minima of these energy curves can be interpreted as being the approximate energy values of the bound and resonance states in the corresponding compound nuclear system having predominantly A + B cluster structure [BA 79, FR 75].

It has been further proposed [RE 72] that for an approximate but somewhat more dynamic study of the A + B scattering problem, one may consider the generator coordinate X" as a quantal variable of scattering and treat the function

$$\tilde{V}_J(x'') = \tilde{\tilde{E}}_J(x'') - \tilde{\tilde{E}}_J(\infty) \tag{5.19}$$

as an effective local optical potential including Coulomb and centrifugal terms. In other words, one solves the equation

$$\left[-\frac{\hbar^2}{2\mu} \frac{1}{X''} \frac{d^2}{dx''^2} X'' + \tilde{V}_J(x'') \right] \tilde{\varphi}_J(x'') = E \tilde{\varphi}_J(x'') \tag{5.20}$$

to obtain bound-state energies and scattering phase shifts. Although valid objections have been raised concerning such an utilization of the energy curves [FR 75a], a series of investigations by Baye and his collaborators [BA 77, BA 79, BA 79a] did show that for collisions between sufficiently heavy ions [BA 80a], this approximate procedure does yield reasonable results especially in energy regions where rather sharp resonances exist and, therefore, can be adopted to provide qualitative or even semi-quantitative information about the resonance behavior of the compound system.

5.2. Hill-Wheeler equation and equivalence between RGM and GCM

In this subsection, we consider exclusively the equal-width-parameter case and use the basis function $\hat{\tilde{\Phi}}_e(\tilde{r}_1, \cdots, \tilde{r}_N ; \vec{X}'')$ or simply $\hat{\tilde{\Phi}}_e(\vec{X}'')$ of eq. (5.12). The trial wave function Ψ in the GCM formulation is given by

$$\Psi = \int \hat{\tilde{\Phi}}_e(\vec{X}'') \tilde{F}(\vec{X}'') d\vec{X}'' . \tag{5.21}$$

In the above equation, $\tilde{F}(\vec{X}'')$ is a weight function which, according to eq. (3.7), satisfies the Hill-Wheeler equation [GR 57, HI 53]

$$\int \left[\tilde{\mathscr{H}}(\vec{X}', \vec{X}'') - E_T \tilde{\mathscr{N}}(\vec{X}', \vec{X}'') \right] \tilde{F}(\vec{X}'') d\vec{X}'' = 0 , \tag{5.22}$$

where

$$\left\{ \begin{matrix} \tilde{\mathscr{H}}(\vec{X}', \vec{X}'') \\ \tilde{\mathscr{N}}(\vec{X}', \vec{X}'') \end{matrix} \right\} = \left\langle \Phi_e(\vec{X}') \left| \left\{ \begin{matrix} H \\ 1 \end{matrix} \right\} \right| \hat{\Phi}_e(\vec{X}'') \right\rangle \tag{5.23}$$

with Φ_e being the unantisymmetrized part of $\hat{\tilde{\Phi}}_e$, i.e.,

$$\hat{\tilde{\Phi}}_e = \mathscr{A} \Phi_e . \tag{5.24}$$

By substituting eq. (5.12) into eq. (5.21) and by comparing the resultant expression with eq. (4.7), one obtains the following relationship between the RGM relative-motion function $F(\vec{R}'')$ and the GCM weight function $\tilde{F}(\vec{X}'')$:

$$F(\vec{R}'') = \int \tilde{\Gamma}_e(\vec{R}'',\vec{X}'')\tilde{F}(\vec{X}'')\,d\vec{X}'' \tag{5.25}$$

or, written more simply,

$$F = \tilde{\Gamma}_e\,\tilde{F}\;. \tag{5.26}$$

Similarly, it is easy to find the connection between RGM and GCM kernels. By utilizing again eq. (5.12), the result is

$$\left\{ \begin{matrix} \tilde{\mathscr{H}}(\vec{X}',\vec{X}'') \\ \tilde{\mathscr{N}}(\vec{X}',\vec{X}'') \end{matrix} \right\} = \left\langle \phi(A)\phi(B)\tilde{\Gamma}_e(\vec{R},\vec{X}')z(\vec{R}_{cm}) \left| \left\{ \begin{matrix} H \\ 1 \end{matrix} \right\} \right| \right.$$

$$\left. \mathscr{A}'\left[\hat{\phi}(A)\hat{\phi}(B)\tilde{\Gamma}_e(\vec{R},\vec{X}'')z(\vec{R}_{cm}) \right] \right\rangle$$

$$= \int \tilde{\Gamma}_e(\vec{R}',\vec{X}')\left\{ \begin{matrix} \mathscr{H}(\vec{R}',\vec{R}'') \\ \mathscr{N}(\vec{R}',\vec{R}'') \end{matrix} \right\} \tilde{\Gamma}_e(\vec{R}'',\vec{X}'')\,d\vec{R}'d\vec{R}'' \tag{5.27}$$

or

$$\left\{ \begin{matrix} \tilde{\mathscr{H}} \\ \tilde{\mathscr{N}} \end{matrix} \right\} = \tilde{\Gamma}_e \left\{ \begin{matrix} \mathscr{H} \\ \mathscr{N} \end{matrix} \right\} \tilde{\Gamma}_e\;. \tag{5.28}$$

From the above equation, one notes that, once the GCM kernels have been derived, the RGM kernels can be obtained by using the inverse operator $\tilde{\Gamma}_e^{-1}$ of the folding operator $\tilde{\Gamma}_e$, i.e.,

$$\left\{ \begin{matrix} \mathscr{H} \\ \mathscr{N} \end{matrix} \right\} = \tilde{\Gamma}_e^{-1} \left\{ \begin{matrix} \tilde{\mathscr{H}} \\ \tilde{\mathscr{N}} \end{matrix} \right\} \tilde{\Gamma}_e^{-1}\;. \tag{5.29}$$

Based on this discussion, one can therefore conclude that in the equal-width-parameter case and under the condition mentioned in subsect. 5.1 concerning the choice of cluster internal functions, the GCM and the RGM are entirely and rather trivially equivalent and either of these methods may be adopted as a quantitative formulation of the MCT, discussed qualitatively in sect. 2.

The above discussion shows that the connection between RGM and GCM is through the folding kernel $\tilde{\Gamma}_e$ and its inverse $\tilde{\Gamma}_e^{-1}$. To obtain a useful expression for $\tilde{\Gamma}_e^{-1}$, one proceeds in the following manner. From the form of $\tilde{\Gamma}_e(\vec{R}',\vec{X}')$ given by eq. (5.13), it is clear that the eigenfunctions of this integral kernel are simply plane waves, i.e.,

$$\int \tilde{\Gamma}_e(\vec{R}',\vec{X}')\,\tilde{\varphi}_{\vec{k}}(\vec{X}')\,d\vec{X}' = \left(\frac{\pi}{\tilde{\gamma}_e}\right)^{3/2} exp\left(-\frac{k^2}{4\tilde{\gamma}_e}\right)\tilde{\varphi}_{\vec{k}}(\vec{R}') \quad , \tag{5.30}$$

with

$$\tilde{\varphi}_{\vec{k}}(\vec{R}') = \left(\frac{1}{2\pi}\right)^{3/2} exp\left(i\vec{k}\cdot\vec{R}'\right) . \tag{5.31}$$

Thus, the spectral representations of $\tilde{\Gamma}_e$ and $\tilde{\Gamma}_e^{-1}$ are [HO 77b]

$$\tilde{\Gamma}_e(\vec{R}',\vec{X}') = \int \tilde{\varphi}_{\vec{k}}(\vec{R}')\left(\frac{\pi}{\tilde{\gamma}_e}\right)^{3/2} exp\left(-\frac{k^2}{4\tilde{\gamma}_e}\right)\tilde{\varphi}_{\vec{k}}^{*}(\vec{X}')\,d\vec{k} \quad , \tag{5.32}$$

$$\tilde{\Gamma}_e^{-1}(\vec{X}',\vec{R}') = \int \tilde{\varphi}_{\vec{k}}(\vec{X}')\left(\frac{\tilde{\gamma}_e}{\pi}\right)^{3/2} exp\left(\frac{k^2}{4\tilde{\gamma}_e}\right)\tilde{\varphi}_{\vec{k}}^{*}(\vec{R}')\,d\vec{k} \quad . \tag{5.33}$$

Equation (5.33) shows that $\tilde{\Gamma}_e^{-1}$ is a singular kernel [GI 73, GI 73a, GI 75, YU 72], which is a well-known fact associated with the so-called high-frequency catastrophe of the GCM with real generator coordinates. This means that for certain RGM relative-motion wave functions $F(\vec{R}'')$, there exist no corresponding GCM weight functions $\tilde{F}(\vec{X}'')$ which are non-singular. On the other hand, it is noted that, in the Hill-Wheeler equation (5.22), highly regular GCM kernels also appear; consequently, no fundamental difficulties are expected when one solves this integral equation to obtain results for physical quantities [DE 72].

In terms of the spectral representation of $\tilde{\Gamma}_e^{-1}$, one can write eq. (5.29) more explicitly as

$$\left\{\begin{matrix} \mathcal{H}(\vec{R}',\vec{R}'') \\ \mathcal{N}(\vec{R}',\vec{R}'') \end{matrix}\right\} = \left(\frac{1}{2\pi}\right)^{6}\left(\frac{\tilde{\gamma}_e}{\pi}\right)^{3}$$

$$\times \int exp\left[i\vec{k}'\cdot(\vec{R}'-\vec{X}') - i\vec{k}''\cdot(\vec{R}''-\vec{X}'') + \frac{1}{4\tilde{\gamma}_e}(k'^2+k''^2)\right]$$

$$\times \left\{\begin{matrix} \tilde{\mathcal{H}}(\vec{X}',\vec{X}'') \\ \tilde{\mathcal{N}}(\vec{X}',\vec{X}'') \end{matrix}\right\} d\vec{k}'d\vec{k}''d\vec{X}'d\vec{X}'' \quad . \tag{5.34}$$

This particular procedure [GI 71, TO 75, TO 77] of transforming the GCM kernel into the RGM kernel is known as the double Fourier transform (DFT). In comparing with the CGCT discussed in subsect. 4.3b, one notes that for this transformation one needs to perform the integration over four variables $\vec{k}'$, $\vec{k}''$, $\vec{X}'$, and $\vec{X}''$, while in the CGCT, only the integration over two variables $\vec{S}'$ and $\vec{S}''$ [see eq. (4.104)] is required. Thus, generally speaking, the transformation procedure used in the CGCT is somewhat more convenient than that used in the DFT.

There is another analytical procedure which can be used to transform the GCM kernel into the RGM kernel. This procedure, to be called the harmonic-oscillator expansion procedure (HOE), was proposed by Kamimura and Matsuse [KA 74]. Here the idea is to expand the RGM kernel in the form

$$\left\{ \begin{matrix} \mathcal{H}(\vec{R}',\vec{R}'') \\ \mathcal{N}(\vec{R}',\vec{R}'') \end{matrix} \right\} = \sum_{ij} \left\{ \begin{matrix} h_{ij} \\ n_{ij} \end{matrix} \right\} u_i(\vec{R}') u_j^*(\vec{R}'') \ , \tag{5.35}$$

where the u_i's represent harmonic-oscillator wave functions containing an exponential factor $\exp\left(-\gamma_H R'^2\right)$. By using the orthonormality property of these functions, it is easily shown that the expansion coefficients h_{ij} and n_{ij} are given by

$$\left\{ \begin{matrix} h_{ij} \\ n_{ij} \end{matrix} \right\} = \int u_i^*(\vec{R}') \left\{ \begin{matrix} \mathcal{H}(\vec{R}',\vec{R}'') \\ \mathcal{N}(\vec{R}',\vec{R}'') \end{matrix} \right\} u_j(\vec{R}'')\, d\vec{R}'\, d\vec{R}'' \ . \tag{5.36}$$

For the actual calculation of these coefficients, one utilizes the GCM kernels. What one does is to simply find the function $\tilde{u}_i$ which is the Gauss-transform of the function u_i , i.e.,

$$\tilde{u}_i = \tilde{\Gamma}_e^{-1} u_i \ . \tag{5.37}$$

The explicit form of $\tilde{u}_i$ was determined by Kamimura and Matsuse. The important point to note is that, because of the singular nature of $\tilde{\Gamma}_e^{-1}$, it is necessary to impose the condition that $\gamma_H < \tilde{\gamma}_e$.[+] Using eq. (5.37), one obtains then

$$\left\{ \begin{matrix} h_{ij} \\ n_{ij} \end{matrix} \right\} = \int \tilde{u}_i^*(\vec{X}')\, \tilde{\Gamma}_e(\vec{R}',\vec{X}') \left\{ \begin{matrix} \mathcal{H}(\vec{R}',\vec{R}'') \\ \mathcal{N}(\vec{R}',\vec{R}'') \end{matrix} \right\} \tilde{\Gamma}_e(\vec{R}'',\vec{X}'')\, \tilde{u}_j(\vec{X}'')\, d\vec{R}'\, d\vec{R}''\, d\vec{X}'\, d\vec{X}''$$

$$= \int \tilde{u}_i^*(\vec{X}') \left\{ \begin{matrix} \tilde{\mathcal{H}}(\vec{X}',\vec{X}'') \\ \tilde{\mathcal{N}}(\vec{X}',\vec{X}'') \end{matrix} \right\} \tilde{u}_j(\vec{X}'')\, d\vec{X}'\, d\vec{X}'' \ . \tag{5.38}$$

Thus, once the GCM kernels are derived, one can find the expansion coefficients and, consequently, the corresponding RGM kernels.

Since the intrinsic function $\hat{\Phi}_e$ of eq. (5.12) is a simple two-center harmonic-oscillator shell-model wave function, the computation of the GCM kernels $\tilde{\mathcal{H}}(\vec{X}',\vec{X}'')$ and $\tilde{\mathcal{N}}(\vec{X}',\vec{X}'')$ by eq. (5.23) is straightforward. We shall not describe the computational procedure here, because it has already been extensively discussed by a number of authors [BR 66, HO 77b, TO 77]. The only point we wish to mention is that

---

[+] In the actual computation, it was found convenient to take the limit $\gamma_H \rightarrow \tilde{\gamma}_e$ after the integrations over $\vec{X}'$ and $\vec{X}''$ are performed.

the use of symbolic algebra to analytically derive these kernels on the computer, described by Tohsaki-Suzuki [TO 77], seems particularly interesting and should be seriously considered in any future attempt to study scattering and reaction problems by the MCT approach.

5.3. General case of unequal width parameters

In the general case where α_A is not equal to α_B , the situation is more complicated. However, even within the restriction of adopting real generator coordinates, there are still methods [GI 75a, KA 74, TO 78] by which one can take advantage of simplifications resulting from the use of two-center harmonic-oscillator shell-model wave functions. In this subsection, we discuss two of these methods, namely, the method of double Fourier transform and the method of generalized intrinsic functions.

5.3a. Method of double Fourier transform

In the method of the double Fourier transform [HO 77b, TO 77, TO 78], one uses the intrinsic function of eq. (5.1) but with $\vec{X}''_{cm}$ set as zero, i.e.,

$$\hat{\Phi}(\tilde{r}_1, \cdots, \tilde{r}_N ; \vec{X}'') = \mathcal{A}\, \Phi(\tilde{r}_1, \cdots, \tilde{r}_N ; \vec{X}'')$$

$$= \mathcal{A}'\left[\tilde{\phi}(A ; \frac{N_B}{N}\vec{X}'')\tilde{\phi}(B ; -\frac{N_A}{N}\vec{X}'')\right]$$

$$= \mathcal{A}'\left\{\hat{\phi}(A)\hat{\phi}(B)\exp\left[-\tilde{\alpha}\vec{R}_{cm}^2 - \tilde{\beta}\vec{R}_{cm}\cdot(\vec{R}-\vec{X}'') - \tilde{\gamma}(\vec{R}-\vec{X}'')^2\right]\right\} \tag{5.39}$$

Because of the fact that $\hat{\Phi}$ contains spurious c.m. excitations, the GCM kernels, given by

$$\left\{\begin{array}{l}\tilde{\mathcal{H}}(\vec{X}',\vec{X}'')\\ \tilde{\mathcal{N}}(\vec{X}',\vec{X}'')\end{array}\right\} = \left\langle \Phi(\tilde{r}_1,\cdots,\tilde{r}_N ; \vec{X}')\left|\left\{\begin{array}{c}H\\1\end{array}\right\}\right|\hat{\Phi}(\tilde{r}_1,\cdots,\tilde{r}_N ; \vec{X}'')\right\rangle , \tag{5.40}$$

are not directly useful. However, they can be used to construct RGM kernels which have no problem with spurious c.m. motion. For this latter purpose, one notes first that

$$\int \exp(i\vec{k}''\cdot\vec{X}'')\,\hat{\tilde{\Phi}}(\tilde{r}_1,\cdots,\tilde{r}_N;\vec{X}'')\,d\vec{X}''$$

$$= \left(\frac{\pi}{\tilde{\gamma}}\right)^{3/2}\exp\left(-\frac{k''^2}{4\tilde{\gamma}}\right)\mathcal{A}'\left[\hat{\phi}(A)\hat{\phi}(B)\exp(i\vec{k}''\cdot\vec{R})\right.$$

$$\left.\times\;\exp\left(-\tilde{\omega}\vec{R}_{cm}^2+\frac{i\tilde{\beta}}{2\tilde{\gamma}}\vec{k}''\cdot\vec{R}_{cm}\right)\right] \tag{5.41}$$

with

$$\tilde{\omega} = \tilde{\alpha} - \frac{1}{4\tilde{\gamma}}\tilde{\beta}^2 = \frac{N_A N_B\,\alpha_A\,\alpha_B}{4\,\tilde{\gamma}} \tag{5.42}$$

Using eq. (5.41), one finds then

$$\int \exp(-i\vec{k}'\cdot\vec{X}')\left\{\begin{matrix}\hat{\tilde{\mathcal{H}}}(\vec{X}',\vec{X}'')\\[4pt]\tilde{\mathcal{N}}(\vec{X}',\vec{X}'')\end{matrix}\right\}\exp(i\vec{k}''\cdot\vec{X}'')\,d\vec{X}'d\vec{X}''$$

$$= \left(\frac{\pi}{\tilde{\gamma}}\right)^{3}\exp\left(-\frac{k'^2+k''^2}{4\tilde{\gamma}}\right)\left\{\begin{matrix}I_H(\vec{k}',\vec{k}'')\\[4pt]I_N(\vec{k}',\vec{k}'')\end{matrix}\right\} \tag{5.43}$$

where

$$\left\{\begin{matrix}I_H(\vec{k}',\vec{k}'')\\[4pt]I_N(\vec{k}',\vec{k}'')\end{matrix}\right\} = \left\langle \phi(A)\phi(B)\exp(i\vec{k}'\cdot\vec{R})\exp\left(-\tilde{\omega}\vec{R}_{cm}^2+\frac{i\tilde{\beta}}{2\tilde{\gamma}}\vec{k}'\cdot\vec{R}_{cm}\right)\right.$$

$$\left|\left\{\begin{matrix}H\\I\end{matrix}\right\}\right|\mathcal{A}'\left[\hat{\phi}(A)\hat{\phi}(B)\exp(i\vec{k}''\cdot\vec{R})\exp\left(-\tilde{\omega}\vec{R}_{cm}^2+\frac{i\tilde{\beta}}{2\tilde{\gamma}}\vec{k}''\cdot\vec{R}_{cm}\right)\right]\right\rangle$$

$$= \frac{1}{\langle Z(\vec{R}_{cm})|Z(\vec{R}_{cm})\rangle}\left\langle \exp\left(-\tilde{\omega}\vec{R}_{cm}^2+\frac{i\tilde{\beta}}{2\tilde{\gamma}}\vec{k}'\cdot\vec{R}_{cm}\right)\right|\exp\left(-\tilde{\omega}\vec{R}_{cm}^2+\frac{i\tilde{\beta}}{2\tilde{\gamma}}\vec{k}''\cdot\vec{R}_{cm}\right)\right\rangle$$

$$\times\;\left\langle \phi(A)\phi(B)\exp(i\vec{k}'\cdot\vec{R})Z(\vec{R}_{cm})\left|\left\{\begin{matrix}H\\I\end{matrix}\right\}\right|\mathcal{A}'\left[\hat{\phi}(A)\hat{\phi}(B)\exp(i\vec{k}''\cdot\vec{R})Z(\vec{R}_{cm})\right]\right\rangle$$

$$= \left(\frac{\tilde{\alpha}}{\tilde{\omega}}\right)^{3/2}\exp\left[-\frac{\tilde{\beta}^2}{32\,\tilde{\gamma}^2\tilde{\omega}}(\vec{k}'-\vec{k}'')^2\right]$$

$$\times\;\left\langle \phi(A)\phi(B)\exp(i\vec{k}'\cdot\vec{R})Z(\vec{R}_{cm})\left|\left\{\begin{matrix}H\\I\end{matrix}\right\}\right|\mathcal{A}'\left[\hat{\phi}(A)\hat{\phi}(B)\exp(i\vec{k}''\cdot\vec{R})Z(\vec{R}_{cm})\right]\right\rangle , \tag{5.44}$$

with $Z(\vec{R}_{cm})$ being the c.m. function of eq. (4.97). By recognizing that

$$\left\langle \phi(A)\phi(B)\exp(i\vec{k}'\cdot\vec{R})z \left| \left\{ {H \atop 1} \right\} \right| A'\left[\hat{\phi}(A)\hat{\phi}(B)\exp(i\vec{k}''\cdot\vec{R})z \right] \right\rangle$$

$$= \int \exp(-i\vec{k}'\cdot\vec{Q}') \left\{ {\mathcal{H}(\vec{Q}',\vec{Q}'') \atop \mathcal{N}(\vec{Q}',\vec{Q}'')} \right\} \exp(i\vec{k}''\cdot\vec{Q}'') \, d\vec{Q}'d\vec{Q}'' \quad , \tag{5.45}$$

where $\mathcal{H}(\vec{Q}',\vec{Q}'')$ and $\mathcal{N}(\vec{Q}',\vec{Q}'')$ are RGM kernels given by eqs. (4.11) and (4.12), one obtains finally, by performing inverse Fourier transformations, the relation

$$\left\{ {\mathcal{H}(\vec{R}',\vec{R}'') \atop \mathcal{N}(\vec{R}',\vec{R}'')} \right\} = \left(\frac{1}{2\pi}\right)^6 \left(\frac{\tilde{\gamma}}{\pi}\right)^3 \left(\frac{\tilde{\omega}}{\tilde{\alpha}}\right)^{3/2}$$

$$\times \int \exp\left[i\vec{k}'\cdot(\vec{R}'-\vec{x}') - i\vec{k}''\cdot(\vec{R}''-\vec{x}'') + \frac{k'^2+k''^2}{4\tilde{\gamma}} \right.$$

$$\left. + \frac{\tilde{\beta}^2}{32\tilde{\gamma}^2\tilde{\omega}}(\vec{k}'-\vec{k}'')^2 \right] \left\{ {\tilde{\mathcal{H}}(\vec{x}',\vec{x}'') \atop \tilde{\mathcal{N}}(\vec{x}',\vec{x}'')} \right\} d\vec{k}'d\vec{k}''d\vec{x}'d\vec{x}'' \quad . \tag{5.46}$$

In the special case of equal width parameters, $\tilde{\beta} = 0$ and this equation reduces to the simpler expression given by eq. (5.34).

Tohsaki-Suzuki [TO 78] has further proposed the derivation of new GCM kernels from the RGM kernels of eq. (5.46) by using a folding procedure similar to that given by eq. (5.27), i.e.,

$$\left\{ {\tilde{\mathcal{H}}(\vec{S}',\vec{S}'') \atop \tilde{\mathcal{N}}(\vec{S}',\vec{S}'')} \right\} = \int \tilde{\Gamma}(\vec{R}',\vec{S}') \left\{ {\mathcal{H}(\vec{R}',\vec{R}'') \atop \mathcal{N}(\vec{R}',\vec{R}'')} \right\} \tilde{\Gamma}(\vec{R}'',\vec{S}'') \, d\vec{R}'d\vec{R}'' \tag{5.47}$$

with[+]

$$\tilde{\Gamma}(\vec{R}',\vec{S}') = \exp\left[-\tilde{\gamma}(\vec{R}'-\vec{S}')^2 \right]. \tag{5.48}$$

These new kernels are free of spurious components and can be used to set up a Hill-Wheeler equation for the weight function. However, in view of the fact that solving the GCM Hill-Wheeler equation is generally not any more convenient than solving the RGM integrodifferential equation, it is not clear that, since the RGM kernels are already known, there is much advantage in deriving these new kernel functions.

[+] The choice of the width parameter in the Gaussian folding function $\tilde{\Gamma}$ of eq. (5.48) is arbitrary; the adoption of the particular value $\tilde{\gamma}$ given by eq. (5.11) is merely for convenience.

5.3b. Method of generalized intrinsic functions.

To avoid the difficulty arising from the spurious c.m. motion, we define the generalized intrinsic function $\widehat{\Phi}_G(\tilde{r}_1, \ldots, \tilde{r}_N; \vec{X}'')$ or simply $\widehat{\Phi}_G(\vec{X}'')$ as

$$
\begin{aligned}
\widehat{\Phi}_G(\tilde{r}_1, \cdots, \tilde{r}_N; \vec{X}'') &= \mathcal{A}\,\Phi_G(\tilde{r}_1, \cdots, \tilde{r}_N; \vec{X}'') \\[2mm]
&= \frac{1}{V^{1/2}}\left(\frac{\tilde{\alpha}}{2\pi}\right)^{3/4} \int \widehat{\Phi}(\tilde{r}_1, \cdots, \tilde{r}_N; \vec{X}'', \vec{X}''_{cm})\, d\vec{X}''_{cm} \;,
\end{aligned}
\tag{5.49}
$$

where $\widehat{\Phi}(\tilde{r}_1, \ldots, \tilde{r}_N; \vec{X}'', \vec{X}''_{cm})$ or $\widehat{\Phi}(\vec{X}'', \vec{X}''_{cm})$ is given by eq. (5.1) and the arbitrarily large volume V is the domain of integration for the variable $\vec{X}''_{cm}$. By using eq. (5.6), the integration over $\vec{X}''_{cm}$ can be explicitly carried out. The result is

$$
\widehat{\Phi}_G(\vec{X}'') = \frac{1}{V^{1/2}}\left(\frac{\pi}{2\tilde{\alpha}}\right)^{3/4} \mathcal{A}'\left[\hat{\phi}(A)\hat{\phi}(B)\tilde{\Gamma}_G(\vec{R}, \vec{X}'')\right] \;,
\tag{5.50}
$$

where

$$
\tilde{\Gamma}_G(\vec{R}, \vec{X}'') = \exp\left[-\tilde{\gamma}_G(\vec{R} - \vec{X}'')^2\right]
\tag{5.51}
$$

with

$$
\tilde{\gamma}_G = \tilde{\gamma} - \frac{1}{4\tilde{\alpha}}\tilde{\beta}^2 = \frac{N_A N_B \alpha_A \alpha_B}{4\tilde{\alpha}} \;.
\tag{5.52}
$$

Therefore, one sees that, by integrating over $\vec{X}''_{cm}$, the generalized intrinsic function $\widehat{\Phi}_G$ is now free of spurious c.m. excitations [HO 77b, KA 74].

The GCM trial wave function is written as

$$
\psi = \int \widehat{\Phi}_G(\vec{X}'')\tilde{F}_G(\vec{X}'')\,d\vec{X}'' \;,
\tag{5.53}
$$

where $\tilde{F}_G(\vec{X}'')$ is a generalized weight function. The corresponding Hill-Wheeler equation has the form

$$
\int\left[\tilde{\mathcal{H}}_G(\vec{X}', \vec{X}'') - E_T\tilde{\mathcal{N}}_G(\vec{X}', \vec{X}'')\right]\tilde{F}_G(\vec{X}'')\,d\vec{X}'' = 0 \;,
\tag{5.54}
$$

where

$$\left\{ \begin{array}{c} \widetilde{\mathscr{H}}_G\,(\vec{X}',\vec{X}'') \\ \widetilde{\mathscr{N}}_G\,(\vec{X}',\vec{X}'') \end{array} \right\} = \left\langle \Phi_G(\vec{X}')\left|\left\{\begin{array}{c} H \\ 1 \end{array}\right\}\right|\widehat{\Phi}_G(\vec{X}'') \right\rangle$$

$$= \frac{1}{V}\left(\frac{\tilde{\alpha}}{2\pi}\right)^{3/2} \int \left\{ \begin{array}{c} H_G\,(\vec{X}',\vec{X}'',\vec{X}'_{cm},\vec{X}''_{cm}) \\ N_G\,(\vec{X}',\vec{X}'',\vec{X}'_{cm},\vec{X}''_{cm}) \end{array} \right\} d\vec{X}'_{cm}\,d\vec{X}''_{cm} \quad , \tag{5.55}$$

with

$$\left\{ \begin{array}{c} H_G\,(\vec{X}',\vec{X}'',\vec{X}'_{cm},\vec{X}''_{cm}) \\ N_G\,(\vec{X}',\vec{X}'',\vec{X}'_{cm},\vec{X}''_{cm}) \end{array} \right\} = \left\langle \Phi\,(\vec{X}',\vec{X}'_{cm})\left|\left\{\begin{array}{c} H \\ 1 \end{array}\right\}\right|\widehat{\Phi}\,(\vec{X}'',\vec{X}''_{cm}) \right\rangle . \tag{5.56}$$

This indicates that, in the unequal-width-parameter case, the computation becomes
more complicated because now the GCM kernels H_G and N_G depend on four variables $\vec{X}'$,
$\vec{X}''$, $\vec{X}'_{cm}$, and $\vec{X}''_{cm}$. It should be pointed out, however, that there is one simplifying
feature in this problem. Since the operator H is Galilean-invariant, one can easily
show that H_G and N_G can depend only on the difference $(\vec{X}'_{cm} - \vec{X}''_{cm})$. Thus, by defining

$$\vec{Y} = \vec{X}'_{cm} - \vec{X}''_{cm} \quad , \qquad \vec{Y}_{cm} = \frac{1}{2}\left(\vec{X}'_{cm} + \vec{X}''_{cm}\right) , \tag{5.57}$$

one can immediately integrate over $\vec{Y}_{cm}$ and obtain

$$\left\{ \begin{array}{c} \widetilde{\mathscr{H}}_G\,(\vec{X}',\vec{X}'') \\ \widetilde{\mathscr{N}}_G\,(\vec{X}',\vec{X}'') \end{array} \right\} = \left(\frac{\tilde{\alpha}}{2\pi}\right)^{3/2} \int \left\{ \begin{array}{c} H_G(\vec{X}',\vec{X}'',\vec{Y},0) \\ N_G\,(\vec{X}',\vec{X}'',\vec{Y},0) \end{array} \right\} d\vec{Y} \quad , \tag{5.58}$$

indicating that only the simpler kernels H_G $(\vec{X}', \vec{X}'', \vec{Y}, 0)$ and $N_G(\vec{X}', \vec{X}'', \vec{Y}, 0)$ need
to be evaluated.

The relationship between the GCM kernels $\widetilde{\mathscr{H}}_G(\vec{X}',\vec{X}'')$, $\widetilde{\mathscr{N}}_G(\vec{X}',\vec{X}'')$ and the RGM
kernels $\mathscr{H}\,(\vec{R}',\vec{R}'')$, $\mathscr{N}\,(\vec{R}',\vec{R}'')$ can be obtained in the following way. By using eq.
(5.50), it is found that

$$\left\{ \begin{array}{c} \widetilde{\mathscr{H}}_G\,(\vec{X}',\vec{X}'') \\ \widetilde{\mathscr{N}}_G\,(\vec{X}',\vec{X}'') \end{array} \right\} = \frac{1}{V}\left(\frac{\pi}{2\tilde{\alpha}}\right)^{3/2} \left\langle \phi(A)\phi(B)\widetilde{\Gamma}_G(\vec{R},\vec{X}')\left|\left\{\begin{array}{c} H \\ 1 \end{array}\right\}\right.\right.$$

$$\left.\left. \mathscr{A}'\left[\widehat{\phi}(A)\widehat{\phi}(B)\widetilde{\Gamma}_G(\vec{R},\vec{X}'')\right]\right\rangle$$

$$= \left(\frac{\pi}{2\tilde{\alpha}}\right)^{3/2} \left\langle \phi(A)\phi(B)\widetilde{\Gamma}_G(\vec{R},\vec{X}')\left|\left\{\begin{array}{c} H \\ 1 \end{array}\right\}\right| \mathscr{A}'\left[\widehat{\phi}(A)\widehat{\phi}(B)\widetilde{\Gamma}_G(\vec{R},\vec{X}'')\right]\right\rangle_{\vec{R}_{cm}} \quad , \tag{5.59}$$

where the notation $\langle \ \rangle_{\vec{R}_{cm}}$ means that the integration is over all internal and relative coordinates, but not over the c.m. coordinate $\vec{R}_{cm}$. By again introducing the c.m. function $Z(\vec{R}_{cm})$ of eq. (4.97), one then obtains the equation

$$\left\{ \begin{array}{c} \widetilde{\mathscr{H}}_G(\vec{X}',\vec{X}'') \\ \widetilde{\mathscr{N}}_G(\vec{X}',\vec{X}'') \end{array} \right\} = \left(\frac{\pi}{2\widetilde{\alpha}}\right)^{3/2} \frac{1}{\langle Z|Z\rangle}$$

$$\times \int \widetilde{\Gamma}_G(\vec{R}',\vec{X}') \langle \phi(A)\phi(B)\delta(\vec{R}-\vec{R}')Z \left| \left\{ \begin{array}{c} H \\ 1 \end{array} \right\} \right| \mathscr{A}'[\hat{\phi}(A)\hat{\phi}(B)\delta(\vec{R}-\vec{R}'')Z] \rangle$$

$$\times \widetilde{\Gamma}_G(\vec{R}'',\vec{X}'')\, d\vec{R}'\, d\vec{R}''$$

$$= \int \widetilde{\Gamma}_G(\vec{R}',\vec{X}') \left\{ \begin{array}{c} \mathscr{H}(\vec{R}',\vec{R}'') \\ \mathscr{N}(\vec{R}',\vec{R}'') \end{array} \right\} \widetilde{\Gamma}_G(\vec{R}'',\vec{X}'')\, d\vec{R}'\, d\vec{R}'' \ , \tag{5.60}$$

which has the form of eq. (5.27). Thus, one can proceed to perform the GCM to RGM conversion in exactly the same way as described in subsect. 5.2.

5.4. Procedures for solving the equation of motion

For computational purposes, it will be useful to perform partial-wave expansion for the Hill-Wheeler equation (5.22). This can be easily accomplished by defining

$$\widetilde{F}(\vec{X}'') = \sum_{\ell} \frac{1}{X''} \widetilde{f}_\ell(X'') P_\ell(\cos\theta_X'') \tag{5.61}$$

and

$$\left\{ \begin{array}{c} \widetilde{\mathscr{H}}(\vec{X}',\vec{X}'') \\ \widetilde{\mathscr{N}}(\vec{X}',\vec{X}'') \end{array} \right\} = \frac{1}{X'X''} \sum_{\ell m} \left\{ \begin{array}{c} \widetilde{\mathscr{H}}_\ell(X',X'') \\ \widetilde{\mathscr{N}}_\ell(X',X'') \end{array} \right\} Y_{\ell m}(\theta_X',\phi_X') Y_{\ell m}^*(\theta_X'',\phi_X'') \ . \tag{5.62}$$

Performing some simple algebra then yields

$$\int [\widetilde{\mathscr{H}}_\ell(X',X'') - E_T \widetilde{\mathscr{N}}_\ell(X',X'')] \widetilde{f}_\ell(X'')\, dX'' = 0 \ . \tag{5.63}$$

Similarly, one can express the GCM-RGM transformation formulas of subsect. 5.2 in partial-wave forms. Writing

$$\widetilde{\Gamma}_e(\vec{R}'',\vec{X}'') = \frac{1}{R''X''} \sum_{\ell m} \widetilde{\Gamma}_{e\ell}(R'',X'') Y_{\ell m}(\theta'',\phi'') Y_{\ell m}^*(\theta_X'',\phi_X'') \ , \tag{5.64}$$

$$\left\{\begin{matrix} \mathcal{H}(\vec{R}',\vec{R}'') \\ \mathcal{N}(\vec{R}',\vec{R}'') \end{matrix}\right\} = \frac{1}{R'R''} \sum_{\ell m} \left\{\begin{matrix} \mathcal{H}_\ell(R',R'') \\ \mathcal{N}_\ell(R',R'') \end{matrix}\right\} Y_{\ell m}(\theta',\phi') Y_{\ell m}^*(\theta'',\phi'') \,, \tag{5.65}$$

and using eq. (4.122), one obtains

$$f_\ell(R'') = \int \tilde{\Gamma}_{e\ell}(R'',X'') \tilde{f}_\ell(X'') \, dX'' \,, \tag{5.66}$$

and

$$\left\{\begin{matrix} \tilde{\mathcal{H}}_\ell(X',X'') \\ \tilde{\mathcal{N}}_\ell(X',X'') \end{matrix}\right\} = \int \tilde{\Gamma}_{e\ell}(R',X') \left\{\begin{matrix} \mathcal{H}_\ell(R',R'') \\ \mathcal{N}_\ell(R',R'') \end{matrix}\right\} \tilde{\Gamma}_{e\ell}(R'',X'') \, dR' dR'' \,, \tag{5.67}$$

where

$$\tilde{\Gamma}_{e\ell}(R'',X'') = 4\pi R''X'' \, i_\ell(2\tilde{\gamma}_e R''X'') \exp\left[-\tilde{\gamma}_e(R''^2 + X''^2)\right] \tag{5.68}$$

With i_ℓ being a modified spherical Bessel function of the first kind [AB 72a].

After the GCM kernels are derived, one must then solve the resultant equation of motion to obtain results for physical quantities. This particular step of the MCT calculational chain has recently been carefully discussed and summarized by Kamimura [KA 77] and Baye and Heenen [BA 77a]. Therefore, we shall only describe it very briefly here.

There are two general types of calculational procedure. These are as follows:
(i) Direct use of GCM kernels.
This is indicated by the block diagram labelled (a1) in fig. 6. In one version

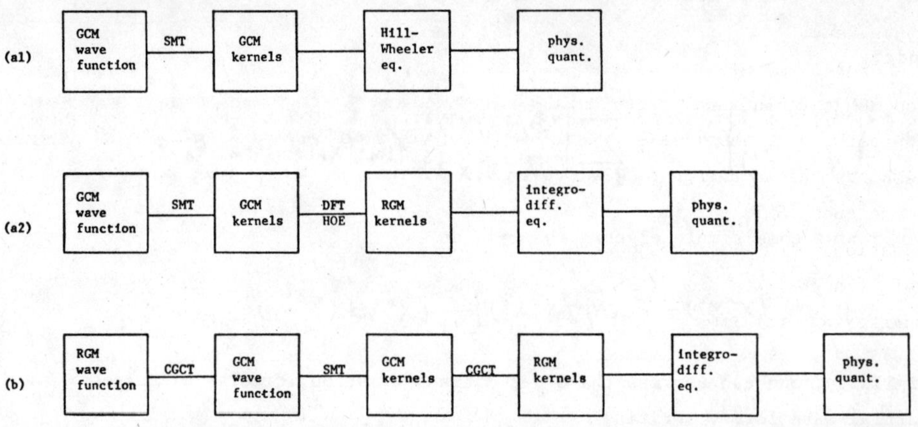

Fig. 6. MCT formulations by GCM and RGM.

of this procedure, one solves the Hill-Wheeler equation in the whole generator-coordinate space. The asymptotic boundary condition, which can be obtained by transforming the RGM Coulomb wave functions through the use of the inverse folding operator $\tilde{\Gamma}_{e\ell}^{-1}$, is then applied to the weight function $\tilde{f}_{\ell}$ itself. In the literature, the methods of Beck et al. [BE 75], Canto and Brink no. 1 [CA 77], and many other authors [DE 72, FR 74, MI 76, RE 73, TA 73, TA 74] are of this type. As has been mentioned by Kamimura [KA 77], this particular procedure tends to become laborious when the mass numbers of the colliding nuclei become large.

In another version, the Hill-Wheeler equation is solved only in the nuclear interaction region for the GCM weight function and the asymptotic boundary condition is applied to the corresponding RGM relative-motion wave function. Calculational procedures of this kind are the microscopic R-matrix method (MRM) of Baye and Heenen [BA 74, HE 76] (see also refs. [HO 70, LA 69]) and those described by Mito and Kamimura [KA 77a, MI 76a], Canto and Brink no. 2 [CA 77], and Nagata and Yamamoto [NA 77]. As was pointed out by Baye and Heenen [BA 77a] and Kamimura [KA 77], the methods of refs. [NA 77] and [CA 77] are equivalent to each other, and to the method discussed in refs. [BA 74, HE 76]. An extension of the MRM to coupled-channel calculations has recently been described by Baye, Heenen, and Libert-Heinemann [BA 77b].

(ii) Indirect use of GCM kernels by transforming into RGM kernels.

This procedure is indicated in fig. 6 by the block diagram labelled (a2). The GCM kernels are first transformed into RGM kernels as described in subsections 5.2 and 5.3. The resultant RGM integrodifferential equation is then solved by the matrix-inversion method of Robertson [HO 56] or by the variational method described by Hackenbroich [HA 69, HE 70]. Many calculations (see, e.g., refs. [FR 75, MA 75, TO 75]) have been performed in this way and the procedure was found to be rather convenient.

For completeness, we also show by the block diagram labelled (b) in fig. 6 the RGM calculational sequence for the MCT formulation. As is seen, this particular approach, with the flexible CGCT providing a convenient link between RGM and GCM, combines the advantage of the GCM in computing kernel functions with shell-model techniques (SMT) and the advantage of the RGM in solving the integro-differential equation. It has been particularly adopted by the Minnesota group (see, e.g., ref. [TA 78]) and the Tübingen group (see, e.g., ref. [SU 76a]) to solve many scattering problems.

6. Examples of MCT calculations

In this section, we show the usefulness of the MCT in treating nuclear many-body problems by discussing the results of some bound-state, scattering, and reaction calculations. These calculations are selected mainly for illustrative purposes; however, they do serve to indicate the importance of including various features, such as specific distortion effects, reaction channels, etc., in a practical calculation.

6.1. Bound-state calculation

As an example of a bound-state calculation, we describe the investigation of Fukushima and Kamimura, in which the low-energy T = 0 natural-parity states of ^{12}C are considered in the 3α cluster representation [FU 78, HO 78]. The trial function adopted has the form of eq. (4.44), with the three α clusters assumed to have the same spatial distribution given by a $(1s)^4$ configuration in a harmonic-oscillator well. The resultant low-energy spectrum is shown in fig. 7, together with the spectrum experimentally determined [AJ 80, MO 66]. To facilitate a comparison between these two spectra, Fukushima and Kamimura have slightly shifted the energy origin of the calculated spectrum such that the calculated and experimental 3α threshold energies occur at the same vertical position.

The major success of this calculation is that the second 0^+ state occurs with nearly the correct excitation energy. This is a significant achievement, because from the shell-model viewpoint this state must be described by a complicated superposition of many-particle, many-hole configurations. In the MCT, this latter feature may be considered as arising from the existence of a strong degree of surface α-clustering. Indeed, the present calculation shows that this state has predominantly an $\alpha + {}^8$Be (0^+) cluster configuration, as can be seen from the behavior

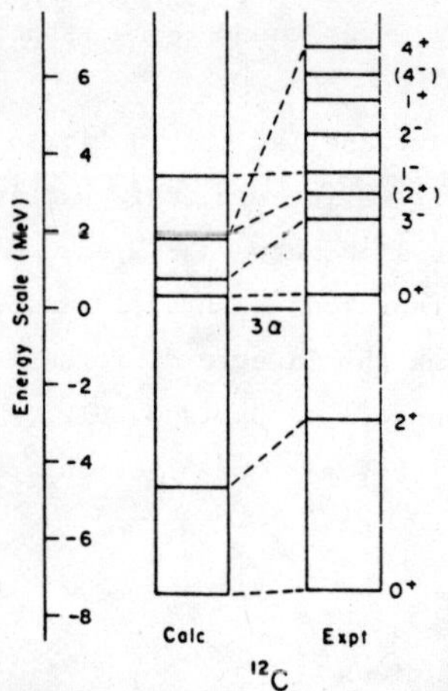

Fig. 7. Comparison of calculated and experimental energy spectra of ^{12}C. Only the lowest natural-parity level of each spin and the 0_2^+ and 2_2^+ levels are shown. (Adapted from ref.[FU 78])

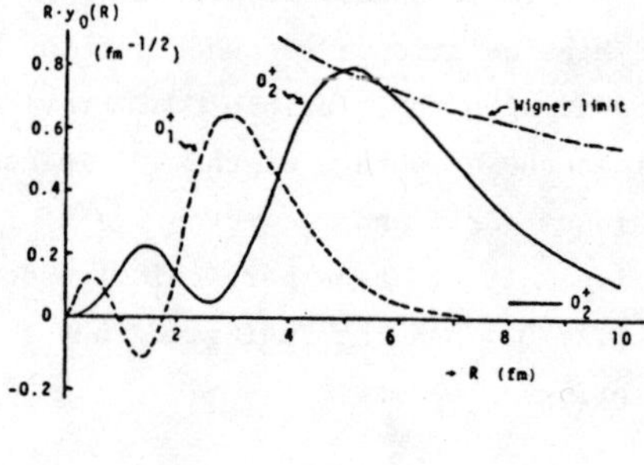

Fig. 8. Reduced-width amplitudes of the $\alpha + {}^8$Be(0^+) breakup for the 0_1^+ and 0_2^+ states. (Adapted from ref.[FU 78])

of the reduced-width amplitudes for the ground and 0_2^+ states (fig. 8). From this figure, one finds that the α-clustering in the 0_2^+ state is almost complete and much stronger than that in the ground state, a feature which has been pointed out by many authors (see fig. 4 of ref. [HO 78]).

On the other hand, it should be mentioned that this calculation is somewhat oversimplified and, consequently, does yield some undesirable results. For instance, the states in the calculated ground-state band are too closely spaced, which is a basic defect of the 3α model with each α cluster in its lowest configuration. As has been emphasized by Takigawa and Arima [TA 71], this defect can be remedied by taking into account the effects caused by the spin-orbit component in the nucleon-nucleon potential. In addition, it will be useful to improve the calculation by adopting different width parameters for the three α clusters, in order that the radius-change effect, mentioned in subsect. 2.4, can be taken into consideration.

6.2. Scattering calculations

6.2a. <u>Scattering of protons by ^{3}He and α with noncentral nucleon-nucleon potentials</u>

Because of the predominance of cluster formation in light systems, MCT calculations are particularly successful in describing phenomena of light-ion scattering by light nuclei. To show this, we discuss in this subsection the scattering calculations of $p + {}^3$He and $p + \alpha$ with noncentral nucleon-nucleon potentials containing central, spin-orbit, and tensor components.

The $p + {}^3$He calculation described here was performed by Furutani <u>et al</u>. [FU 79b] using the GCM. They investigated the level structure of ^{4}Li and analyzed the experimental scattering data in the energy region below about 15 MeV where the reaction cross section is small [SO 76] and, hence, the omission of reaction channels is a reasonable approximation. For simplicity, the coupling of various states with different L and S values due to the tensor force is omitted, but the coupling of singlet and triplet states due to the spin-orbit force is fully taken into account.

A comparison between calculated and experimental results for the differential cross section, proton polarization, and ^{3}He polarization is shown in fig. 9. From this figure, one sees that the agreement is quite good. In addition, it should be mentioned that the data on spin correlations and polarization transfers can be explained equally as well. This shows that the compound system ^{4}Li does contain the feature of strong clustering in the low-excitation region and, consequently, even a single-channel calculation without specific distortion can lead to impressive consequences.

The $p + \alpha$ calculation was that of Kanada <u>et al</u>. [KA 79a] with a single-channel RGM formulation. In this calculation, the α-cluster internal wave function used

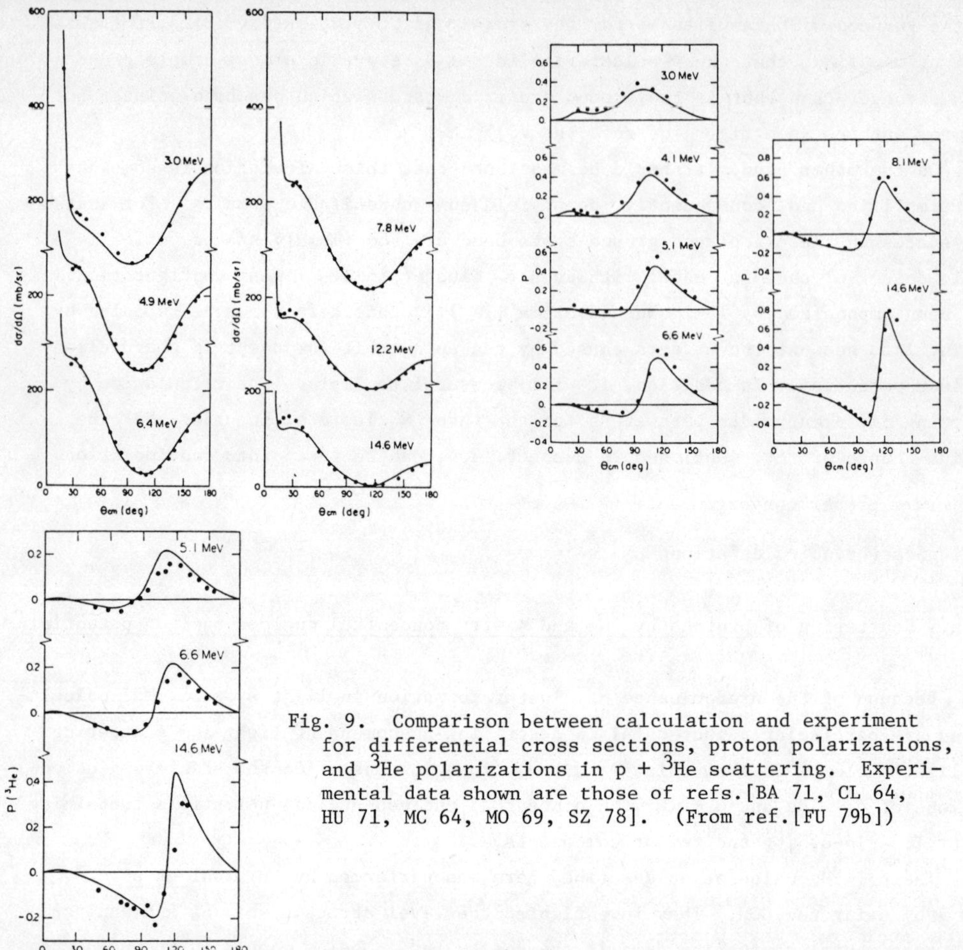

Fig. 9. Comparison between calculation and experiment
for differential cross sections, proton polarizations,
and ^{3}He polarizations in p + ^{3}He scattering. Experimental data shown are those of refs.[BA 71, CL 64,
HU 71, MC 64, MO 69, SZ 78]. (From ref.[FU 79b])

contains both S and principal D states, with the parameters chosen to yield

reasonable values for the rms radius, the binding energy, and the D-state probability.

The results for the differential cross sections and polarizations are shown in

fig. 10, together with the experimental values. Here again, one notes that the

agreement between calculation and experiment is quite satisfactory, thus further

supporting our assertion made at the beginning of this subsection.

6.2b. $\alpha + \alpha$ scattering with specific distortion effect.

The formulation of a scattering problem with specific distortion effect taken

into consideration is described in subsect. 4.2b. In this subsection, we illustrate

this by discussing the $\alpha + \alpha$ case studied by Thompson et al. [TH 77a].

The calculation was performed with a sufficient number of distortion functions

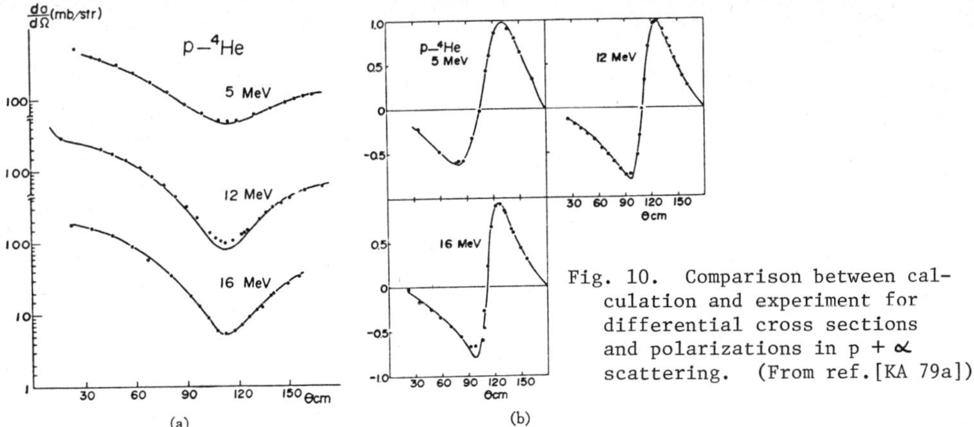

Fig. 10. Comparison between calculation and experiment for differential cross sections and polarizations in p + α scattering. (From ref.[KA 79a])

to insure proper convergence of the phase-shift values. The results are given in fig. 11, where ℓ = 0, 2, and 4, $\alpha + \alpha$ phase shifts at c.m. energies from 0 to 20 MeV are shown. In this figure, the solid curves represent the result obtained with specific distortion, while the dashed curves represent the no-distortion result. The empirical data points are those of refs. [CH 74a, HE 56, NI 58, TO 63, WE 65]. As is seen, there is a good over-all agreement between the calculated result with distortion and the empirical result. In all angular-momentum states, the effect of including specific distortion is not too large in this case, which is related to the fact that the α clusters have a rather low compressibility.

By studying the results of distortion calculations in different systems [BR 68a, LE 75, TH 73], one can make the following general comments: (i) Specific distortion effects of a nucleus are more important the higher its compressibility, (ii) a nucleus is more distorted the larger the number of nucleons in the other nucleus which causes the distortion, (iii) except near energies where resonances occur, specific distortion effects decrease with increasing energy, (iv) except for resonance effects and odd-even effects, the influence of distortion decreases with increasing orbital angular momentum, and (v) distortion appears to be stronger in "Pauli-favored" states [LE 75]. Based on these comments, one anticipates then that, for the scattering of nucleons by s-shell nuclei ^{3}He and ^{4}He, and doubly-closed-shell nuclei ^{16}O and ^{40}Ca, the effects of specific distortion should be relatively unimportant. That this is indeed so has been demonstrated in all these cases (see subsect. 6.2a and ref. [TH 77a]), where one finds that the results calculated without specific distortion do compare very favorably with experimental results.

6.2c. <u>Scattering of α particles by ^{3}He and ^{6}Li — odd-even ℓ-dependence.</u>

In light-ion scattering where the nucleon-number difference of the interacting

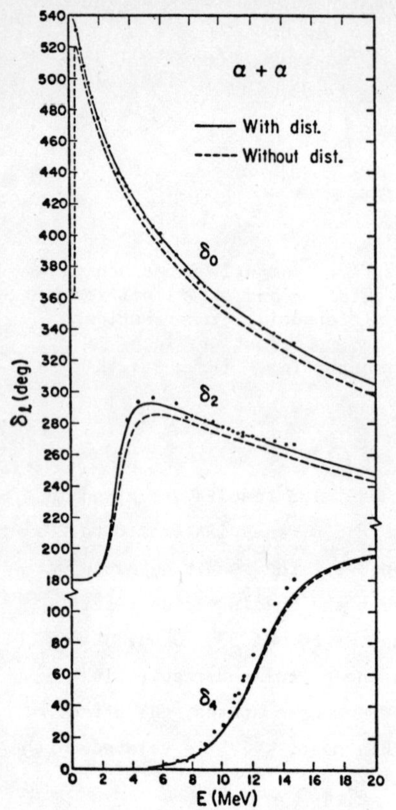

Fig. 11. $\alpha + \alpha$ scattering with and without specific distortion effect. (From ref. [TH77a])

nuclei is small, the experimental angular-distribution curve for the differential cross section shows a well-formed V-shape at a relatively high energy larger than about 20 MeV/nucleon. That is, this curve has a rapidly decreasing behavior in the forward angular region and a rapidly increasing behavior in the backward angular region.[+] As will be discussed in sect. 7, these distinct angular-distribution characteristics can be attributed to the fact that, under the conditions mentioned above, core-exchange effects [LE 79a] have an important and predominant influence on the cross-section values at large angles. For a description of such experimental data, it is expected that MCT calculations should be particularly appropriate. This is so, since in these calculations exchange effects are fully taken into consideration through the use of totally antisymmetric wave functions.

To demonstrate the features mentioned above, we show in fig. 12 the ^{3}He $+ \alpha$ experimental data [RO 80] at 60.2 MeV (i.e., 35.1 MeV/nucleon), together with the result of a resonating-group calculation in which reaction effects are approximately taken into account by the introduction of a phenomenological imaginary potential into the formulation [FU 76]. Here one sees that the calculation explains the experimental result quite well; in particular, the feature at back angles is correctly reproduced. The only discrepancy is that the angle at which the calculated cross section has a minimum value is too large. This is somewhat puzzling, since at neighboring energies, such a discrepancy does not seem to occur [KO 74].

A detailed study of the resonating-group kernel function (see sect. 7) has indicated that, if a local potential is adopted to represent the effective inter-nuclear interaction, the real-central part of this potential must generally contain a Majorana or odd-even ℓ-dependent component. In addition, it was shown that,

---

[+] For heavy-ion scattering on a target of similar nucleon number such as ^{16}O $+ ^{17}$O scattering, it is expected that the angular-distribution curve will also show such a behavior. However, there exist at present no experimental data at high enough energies to confirm this prediction.

especially in scattering systems involving nuclei of nearly equal mass, this odd–even component has significant effects and its inclusion in the internuclear potential is imperative for a satisfactory explanation of the cross-section behavior at backward angles when the scattering energy is relatively high. In the present case of ^{3}He $+\alpha$ scattering. This finding is indeed fully borne out. The optical-model analysis discussed in ref. [RO 80] does show that, if such a component is not included, the calculated differential cross sections at large angles will be a few orders of magnitude smaller than the experimental values.

A similar situation occurs also in the $\alpha + {}^6$Li case. In fig. 13, we show a comparison between resonating-group [SU 79] and experimental [BA 72, HA 69a] results for the cross-section ratio $\sigma(\theta)/\sigma_c(\theta)$. Here again, it is found that there is a significant rise in the back-angle cross section, which is reasonably explained by the resonating-group calculation. The rise is, however, not as dramatic as that in the ^{3}He $+\alpha$ case. This is partially related to the existence of a blocking effect [ST 78] which arises from the presence of two nucleons in the nonclosed 1p-shell of the ^{6}Li nucleus and which affects appreciably the amount of exchange contributions from antisymmetrization.

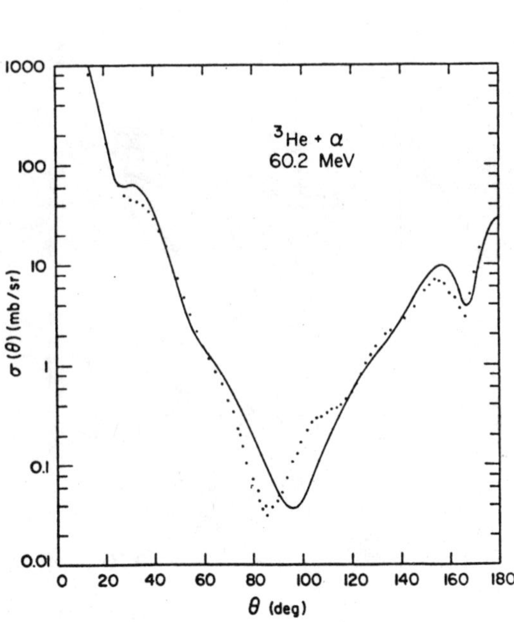

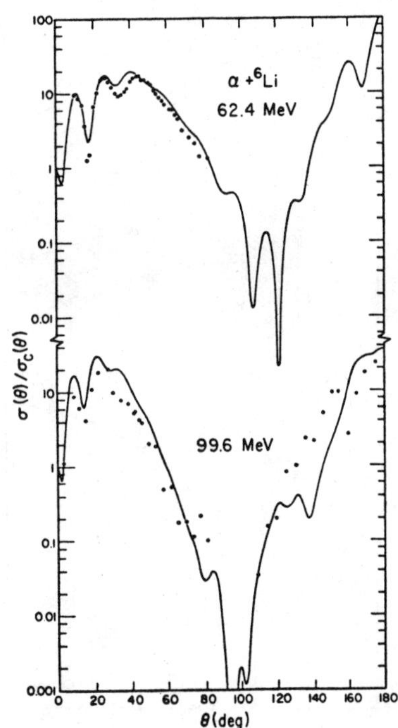

Fig. 12. Comparison of calculated and experimental ^{3}He $+\alpha$ differential cross sections at 60.2 MeV.

Fig. 13. Comparison of calculated and experimental values for the ratio $\sigma(\theta)/\sigma_c(\theta)$ in $\alpha + {}^6$Li scattering. (From ref. [SU 79])

6.2d. <u>Scattering of α particles by Ca-isotopes:phenomenon of ALAS.</u>

In the scattering of α particles by the Ca-isotopes (^{40}Ca, ^{42}Ca, ^{44}Ca, and ^{48}Ca), the phenomenon of anomalous large-angle scattering (ALAS) was observed [EC 75, GR 66, GU 81]. At angles near 180°, it was experimentally found that the elastic cross sections for $\alpha + {}^{40}$Ca scattering are orders of magnitude larger than those for $\alpha + {}^{44}$Ca scattering. For a macroscopic analysis of these data, it was additionally discovered that the use of conventional local potentials of the Woods-Saxon shape cannot adequately explain the cross-section behavior in the intermediate- and large-angle regions [MI 76b].

From careful investigations of this phenomenon, it now appears that the main difference between "normal" (as in $\alpha + {}^{44}$Ca) and "anomalous" (as in $\alpha + {}^{40}$Ca) scattering results from the reduced absorption (i.e., reduced strength of the imaginary part of the optical potential) in the latter case [BR 77, BR 78, DE 78, EB 79]. As a consequence of this comparatively low absorption, cross sections are enhanced at large angles and the angular distribution in the backward angular region becomes more sensitive to the detailed nature of the real part of the optical potential

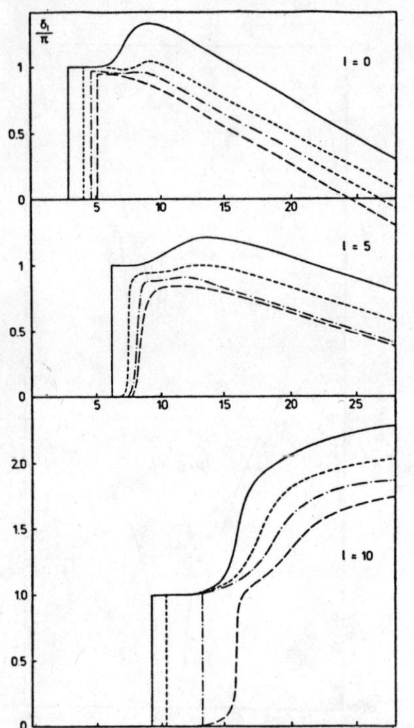

Fig. 14. Phase shifts for the scattering of α particles by ^{40}Ca($— — —$), ^{42}Ca($—\cdot—$), ^{44}Ca($-----$), and ^{48}Ca ($————$). As typical examples, only ℓ = 0, 5, and 10 phases are shown. (From ref.[LA 80])

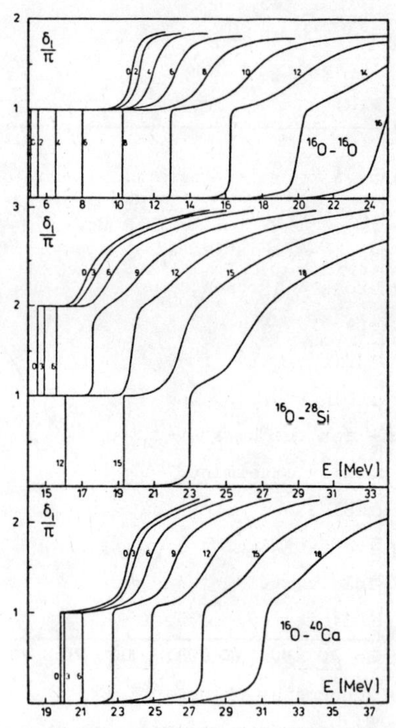

Fig. 15. Calculated phase shifts for the scattering of ^{16}O by ^{16}O, ^{28}Si, and ^{40}Ca. (From ref.[LA 80])

at smaller separation distances. On a microscopic level, such details depend appreciably on the contributions from intercluster nucleon exchanges (see sect. 7 and ref. [AN 75]) and, apparently, these could not be adequately approximated by a local potential having a standard Woods-Saxon form factor.

To give further credence to the above assertion, it will be important to obtain convincing evidence that the real parts of the internuclear interaction for the $\alpha + {}^{40}\text{Ca}$, $\alpha + {}^{42}\text{Ca}$, $\alpha + {}^{44}\text{Ca}$, and $\alpha + {}^{48}\text{Ca}$ systems do not show any abrupt change of property in going from one system to a neighboring system. This has, in fact, been accomplished recently by Langanke [LA 80], who performed single-channel GCM calculations for these systems. In fig. 14, typical phase shifts in some ℓ-states are shown. From this figure, one notes that these phase shifts do change smoothly with the mass number of the target nucleus, indicating that there is no anomalous behavior insofar as the real part of the internuclear interaction is concerned.

6.2e. <u>Heavy-ion scattering of ${}^{16}\text{O}$ by ${}^{16}\text{O}$, ${}^{28}\text{Si}$, and ${}^{40}\text{Ca}$.</u>

As examples of heavy-ion scattering, we use the single-channel GCM calculations of Langanke [LA 80] in the ${}^{16}\text{O} + {}^{16}\text{O}$, ${}^{16}\text{O} + {}^{28}\text{Si}$, and ${}^{16}\text{O} + {}^{40}\text{Ca}$ cases. In these calculations, the cluster internal wave functions are chosen to be as simple as possible.   For ${}^{16}\text{O}$ and ${}^{40}\text{Ca}$, doubly-closed-shell wave functions are adopted, while for ${}^{28}\text{Si}$ the outer nucleons are placed entirely in the $d_{5/2}$-shell. The results for some typical phase shifts are shown in fig. 15, where one sees that there are several resonances in each partial wave. The low-lying ones are narrow and can be classified as quasi-molecular states. The higher-lying resonances are very broad and strongly overlapping; they are located near the tops of the potential barriers and, hence, can be appropriately labelled as "barrier resonances".

Additionally, it was found that these resonances form well developed rotational bands. In table 3, the values of the band-head energy E_0 and the rotational constant $\hbar^2/2\mathcal{J}$ for the barrier-resonance bands are listed [LA 80]. For comparison, the empirically determined values (E_0 - values are extrapolated) are also shown. From this table, one notes that the agreement between calculated and empirical results is quite satisfactory, which is a strong indication that even simple MCT calculations can yield useful results also in heavy-ion studies.

A microscopic investigation of the ${}^{16}\text{O} + {}^{16}\text{O}$ problem has also been carried out by Ando <u>et al</u>. [AN 79]. In this calculation, absorption effects have been taken into account by phenomenologically reducing the magnitudes of the calculated S-matrix elements according to a smooth cutoff model. The results for the angular distributions in the c.m. energy range of 12.5 to 31.5 MeV are shown in fig. 16. Here it is seen that the gross features of the experimental data are reasonably reproduced, thus further demonstrating the utility of the MCT calculation in

Table 3

Values of E_0 and $\hbar^2/2\mathcal{J}$ for the barrier-resonance bands in the $^{16}O + ^{16}O$, $^{16}O + ^{28}Si$, and $^{16}O + ^{40}Ca$ systems.

System	E_0 (MeV)		$\hbar^2/2\mathcal{J}$ (keV)	
	GCM	Expt	GCM	Expt
$^{16}O + ^{16}O$	10.6	9.8	58	51
$^{16}O + ^{28}Si$	17.8	18.6	33.5	34.5
$^{16}O + ^{40}Ca$	23.5	23.0	29	28

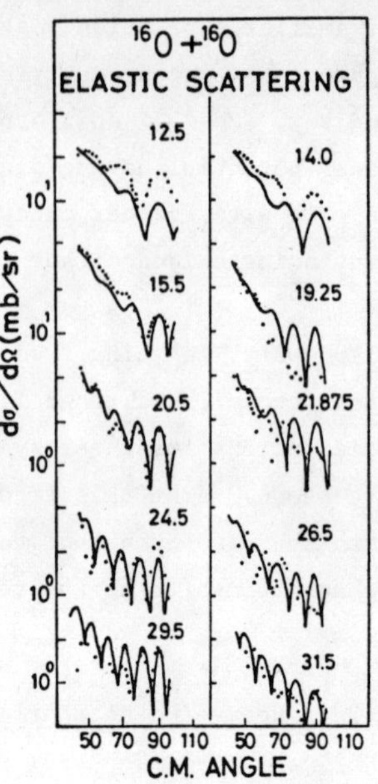

Fig. 16. Comparison of calculated and experimental $^{16}O + ^{16}O$ angular distributions at various c.m. energies. (From ref.[AN 79])

explaining the general properties of a heavy-ion system.

6.3. Reaction calculations

As in the scattering cases, reaction calculations using the MCT have also yielded useful results. In this subsection, we illustrate this by discussing a calculation in the five-nucleon system where both the $d + ^3He$ (or $d + ^3H$) channel and the $p + \alpha$ (or $n + \alpha$) channel are included [CH 74].

The formulation of a coupled-channel problem is given in subsect. 4.2c. For the description of the cluster internal structures, single-Gaussian wave functions are used for the three-nucleon and the α clusters, but a flexible three-Gaussian wave function is used for the deuteron cluster. Reaction effects, arising from the presence of other open channels, are approximately taken into account by the use of phenomenological imaginary potentials.

In fig. 17, we show a comparison between calculated coupled-channel (solid curve) and single-channel (dashed curve) differential scattering cross sections for $d + ^3H$ scattering at 2.02 MeV. The experimental data shown are those of ref. [IV 68]. Here one sees that the coupled-channel result is not only considerably improved over the single-channel result [CH 73b], but also in excellent agreement with experiment. This shows that a systematic improvement of the MCT calculation by expanding the trial-function space can indeed quickly lead to a satisfactory explanation of observed phenomena.

The calculated $\alpha(p,d){}^{3}He$ differential reaction cross sections at a c.m. energy of 68 MeV in the $p + \alpha$ channel are shown by the solid curve in fig. 18. Considering the fact that there are no adjustable parameters in computing these cross sections, one may conclude that the agreement with experimental data [VO 74] is fairly satisfactory. The presence of deep minima and the slight underestimate in the calculated cross section are likely due to the fact that in the nucleon-nucleon potential employed there are no noncentral components.

From fig. 18, it is also noted that the differential reaction cross section has a decreasing trend in the forward angular region, but begins to increase when θ passes about 90°. As will be discussed in sect. 7, the reason for this is that, at a relatively high energy, the reaction proceeds mainly through different mechanisms in the forward and backward angular regions. Thus, in the forward angular region it proceeds mainly through a one-nucleon pickup process, while in the backward angular region it proceeds mainly through a two-nucleon pickup process. Both of these processes are automatically included in the coupled-channel calculation described here, because in this calculation a totally antisymmetrized wave function is employed.

Other light-ion reaction calculations using the MCT approach are reported in refs. [94, 141-146, 148, 155, 156] quoted in ref. [TA 78] (see also ref. [LI 80]). In addition, there exist calculations of this type involving heavy ions. These are the ${}^{12}C + {}^{16}O$ inelastic-scattering calculation of Baye et al. [BA 78a] and the ${}^{28}Si({}^{16}O,\alpha){}^{40}Ca$ rearrangement-collision calculation of Langanke [LA 80].

6.4. Summary

Until this moment, many RGM and GCM scattering and reaction calculations in both light and relatively heavy systems have been performed. The results obtained have been generally encouraging and supported the assertion that, because of the use of energetically favorable cluster-type trial wave functions, even rather simplified MCT investigations can lead to useful conclusions concerning the general properties of nuclear systems.

There are, however, a number of problems which need to be carefully examined. At relatively high energies, the existence of many open multi-cluster channels has necessitated the introduction of a phenomenological imaginary potential into the formulation. Presently, the imaginary potentials adopted [BR 71, LA 78] have rather simple forms and lack microscopic justification. It is clearly desirable to approach this problem from a more basic viewpoint. In this respect, we note that, based on the considerations of Saloner et al. [SA 77a, SA 77b], Weiguny has recently made some interesting discussion in this direction [WE 77]. By pursuing this discussion further, there is the possibility that one may achieve a better

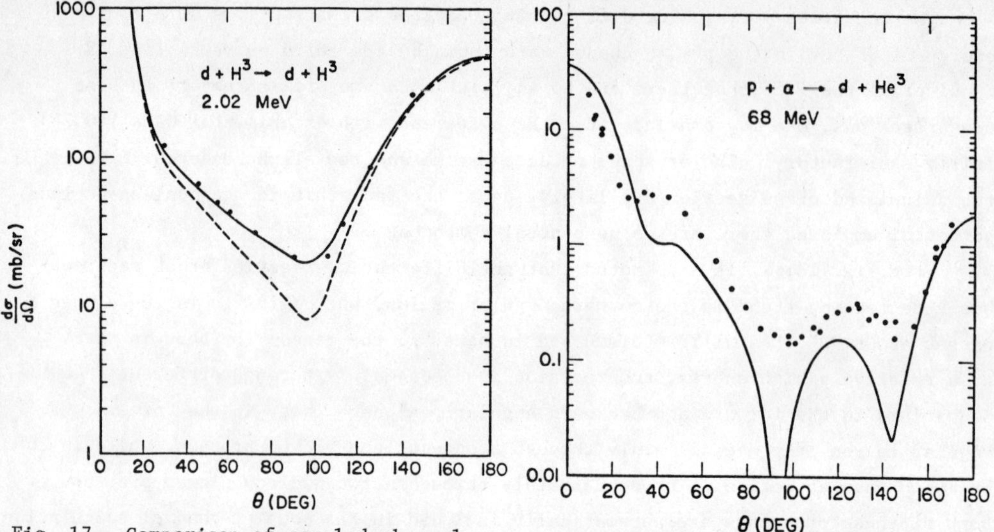

Fig. 17. Comparison of coupled-channel
(solid curve) and single-channel
(dashed curve) differential cross
sections for d + ^{3}H scattering at
2.02 MeV with experimental data
(solid dots). (From ref.[CH 74])

Fig. 18. Comparison of calculated and
experimental differential reaction
cross sections for the process
α (p,d)^{3}He. (From ref.[CH 74])

understanding concerning a more proper way of taking absorption effects approximately
into account in a MCT calculation.

For direct-reaction calculations, there is additionally the threshold problem
which has been particularly pointed out by Baye [BA 80a]. Because of the use of
relatively simple nucleon-nucleon potentials and cluster internal functions, the
threshold energies for various reaction channels are frequently not accurately
calculated and this may yield rather serious consequences especially in the low-
energy region. To remedy this situation, one must obviously adopt more refined
nucleon-nucleon potentials and use more flexible cluster wave functions. However,
in view of the complicated nature of MCT calculations, such a remedy may not be
easily accomplished and simpler procedures of alleviating the threshold problem
may have to be sought.

7. Antisymmetrization effects in nuclear systems

As has been frequently mentioned, one of the essential features of the MCT
is the use of totally antisymmetric wave functions. Thus, from an examination of
the MCT formulation, one should be able to obtain information concerning the
importance of nucleon-exchange effects arising from antisymmetrization. In this
section, we shall show that, by studying the structure of the RGM exchange kernels,
one can in fact learn much about the main characteristics of nuclear scattering and
direct-reaction problems.

7.1. Effects of antisymmetrization on the effective internuclear potential

For the discussion of nucleon-exchange effects, we consider the case where a nucleus A containing N_A nucleons is scattered by another nucleus B containing $N_B (N_B < N_A)$ nucleons. For simplicity in presentation, we shall assume that the spin of these nuclei are both equal to zero and the charge of the proton is infinitesimally small. In addition, the internal wave functions for the two nuclei will be chosen to be described by translationally-invariant shell-model functions in harmonic-oscillator wells have a common width parameter α.

From eq. (4.35), it is clear that the information we seek is contained in the function $K(\vec{R}',\vec{R}'')$, given by eq. (4.37), which consists of the exchange-normalization kernel $\mathcal{N}_E$ and the exchange-Hamiltonian kernel $\mathcal{H}_E$[see eqs. (4.20) and (4.26)].[+] To extract this information, we must, therefore, study the structure of these kernel functions [LE 79a, LE 79b]. However, as has been shown in refs. [LE 79b], a study of $\mathcal{H}_E$ alone should be sufficient for our present purpose, since all the features in $\mathcal{N}_E$ are already contained in the kernel function $\mathcal{H}_E$.

7.1a. <u>Study of the exchange-Hamiltonian kernel function.</u>

The summations and integrations in eq. (4.26) can be facilitated by using the CGCT discussed in subsect. 4.3. The result for the exchange-Hamiltonian kernel $\mathcal{H}_E$ is

$$\mathcal{H}_E (\vec{R}',\vec{R}'') = \sum_{x} \sum_{q} \mathcal{H}_{Eq}^{x} (\vec{R}',\vec{R}'') \quad , \quad (q = a,b,c,d,e) \tag{7.1}$$

where $x(N_B \geq x \geq 1)$ is the number of nucleons interchanged between the colliding nuclei and

$$\mathcal{H}_{Eq}^{x} (\vec{R}',\vec{R}'') = P_{xq} \exp\left(-A_{xq}\vec{R}'^{2} - C_{xq}\vec{R}'\cdot\vec{R}'' - B_{xq}\vec{R}''^{2}\right) + h.c. \tag{7.2}$$

with P_{xq} (q = a,b,c,d,e) being a polynomial in $\vec{R}'^2$, $\vec{R}' \cdot \vec{R}''$, and $\vec{R}''^2$. From eqs. (7.1) and (7.2) one sees that the expression for $\mathcal{H}_E$ is quite complicated; for each value of x, there appear five distinct types of exponential factors. In the special case of equal-width-parameter considered in the present study, the situation is somewhat simplified. Here the exponential factors in type-d and type-e terms are the same and, therefore, only four types of terms i.e., types a,b,c, and d, need to be examined. Also, it should be noted that, in the core-exchange case where $x = N_B$, there are only three types of contributing terms, namely, types a, c, and d.

[+]It is important to note that in our discussion the term nucleon-exchange is defined through the kernel functions, while in the discussion of Baldock <u>et al</u>. [BA 81] it is defined through the wave function. Although the definition of the latter authors has certain mathematical merits, our way of defining the nucleon-exchange has the advantage that the resultant finding can be used to simplify practical calculations.

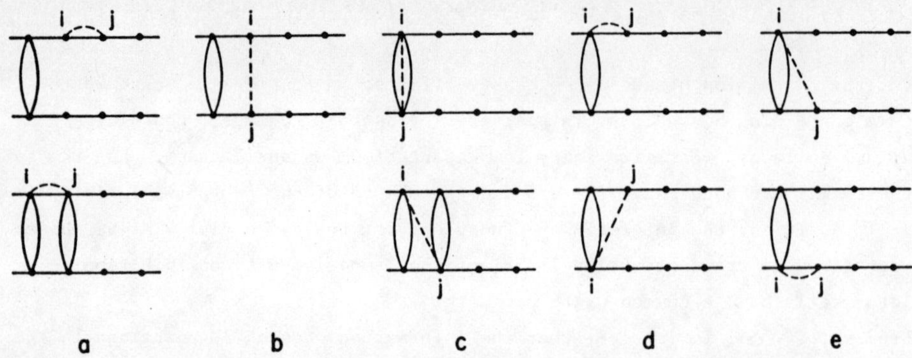

Fig. 19. Diagrammatical representation of nucleon-exchange terms.

To explain briefly the origin of these five types of exchange terms, we make use
of the diagrammatic representation introduced by LeMere et al.[LE 77, LE 79b] and
assume, for simplicity in presentation, that the nucleon-nucleon potential has a
purely Wigner character. This is shown in fig. 19, where representative diagrams
for each type are depicted. In this diagram, each dot on the upper or lower line
represents a set of all occupied single-nucleon spatial states with the same spin
and isospin in cluster A or B. The dashed line is used to call attention to the
fact that nucleons i and j are interacting, while the solid-line exchange loop is
used to indicate the nucleons which are interchanged between the clusters. Also,
we have adopted the convention that diagrams shown in fig. 19 are considered to
represent implicitly all other diagrams which contain additional exchange loops not
involving the interacting nucleons.

The type-a exchange term arises from both the kinetic-energy and the potential-
energy operators in the Hamiltonian H, while exchange terms of other types come only
from the potential-energy operator. Also, it should be noted that the exchange-
normalization kernel $\mathcal{N}_E$ has the same exponential factor as the type-a term. This
indicates therefore that, for each value of x, this particular exchange type and the
exchange-normalization kernel have rather similar structure.

General expressions for the coefficients A_{xg}, B_{xg}, and C_{xg} can be derived by
usig the CGCT. The results are

$$A_{xg} = \frac{1}{2}\mu_0 \propto \frac{\Gamma_n + \lambda \tilde{A}_{ng}}{\Gamma_d + \lambda \tilde{A}_{dg}} \, , \tag{7.3}$$

$$B_{xg} = \frac{1}{2}\mu_0 \propto \frac{\Gamma_n + \lambda \tilde{B}_{ng}}{\Gamma_d + \lambda \tilde{B}_{dg}} \, , \tag{7.4}$$

$$C_{xg} = 2\mu_0^2 \propto \frac{\Delta + \lambda \tilde{C}_{ng}}{\Gamma_d + \lambda \tilde{C}_{dg}} \, , \tag{7.5}$$

Table 4

Expressions for $\tilde{A}_{nq}$, $\tilde{A}_{dq}$, $\tilde{B}_{nq}$, $\tilde{B}_{dq}$, $\tilde{C}_{nq}$, and $\tilde{C}_{dq}$.

Exch. type q	$\tilde{A}_{nq}$	$\tilde{A}_{dq}$	$\tilde{B}_{nq}$	$\tilde{B}_{dq}$	$\tilde{C}_{nq}$	$\tilde{C}_{dq}$
a	0	0	0	0	0	0
b	$2(x-\mu_0)$	$-2x$	$2(x-\mu_0)$	$-2x$	1	$-2x$
c	$-2(x-\mu_0)$	$2(x-2\mu_0)$	$-2(x-\mu_0)$	$2(x-2\mu_0)$	-1	$2(x-2\mu_0)$
d	μ_0	$-\mu_0$	$-\mu_0$	$-\mu_0$	0	$-\mu_0$

where

$$\Gamma_n = x^2 + 2\mu_0(\mu_0 - x) , \tag{7.6}$$

$$\Gamma_d = x(2\mu_0 - x) , \tag{7.7}$$

$$\Delta = x - \mu_0 , \tag{7.8}$$

and the constants $\tilde{A}_{nq}$, $\tilde{A}_{dq}$, $\tilde{B}_{nq}$, $\tilde{B}_{dq}$, $\tilde{C}_{nq}$, and $\tilde{C}_{dq}$ are tabulated in table 4. The quantity λ is defined as

$$\lambda = K/(\alpha + 2K) , \tag{7.9}$$

with K being the range parameter of the nucleon-nucleon potential [see eq. (4.38)]. The range of this quantity is between 0 and 1/2. In a realistic situation where K is approximately equal to α, λ has a value around 1/3.

7.1b. **Effective local potentials.**

The structure of the exchange-Hamiltonian kernel (EHK) is rather complicated. Thus, to determine, at relatively high energies, the relative importance of the various nucleon-exchange terms, we adopt the following simplifying procedure. We construct effective local energy-dependent "exchange" potentials which yield, in the Born approximation, the same scattering amplitudes as the nucleon-exchange terms, and then examine the energy and spatial dependence of these exchange potentials.

For a kernel function

$$K_{ex}(\vec{R}',\vec{R}'') = P_{ex} \exp(-a\vec{R}'^2 - c\vec{R}'\cdot\vec{R}'' - b\vec{R}''^2) \tag{7.10}$$

Table 5

Expressions for x_q, J_q, and K_q.

Exchange type q	x_q	J_q	K_q
a	μ_0	μ_0^2	0
b	$\mu_0-\lambda$	$(\mu_0+\lambda)^2$	$-\lambda^2$
c	$\mu_0+\lambda$	$(\mu_0+\lambda)^2$	$-\lambda^2$
d	μ_0	$\mu_0(\mu_0+\lambda)$	0

with P_{ex} being a polynomial in $\vec{R}'^2$, $\vec{R}'\cdot\vec{R}''$, and $\vec{R}''^2$, the effective local potential $\tilde{V}_{ex}$ which yields the same Born scattering amplitude can be easily shown to have the following expressions:

(i) $c < 0$. In this case, the Born scattering amplitude is forward-peaked and the effective local potential is given by

$$\tilde{V}_{ex}(\vec{R}) = \tilde{P}_{ex}\, exp\left[-(R/R_{ex})^2\right]$$
$$\times\, exp\left(-E/E_{ex}\right) ,$$

$$\cdots(7.11)$$

where $\tilde{P}_{ex}$ is a polynomial in R^2 and E. This effective potential is characterized by a characteristic range R_{ex} and a characteristic energy E_{ex}, which have the forms

$$R_{ex} = \left(\frac{2|c|}{4ab-c^2}\right)^{1/2} ,$$

$$(7.12)$$

$$E_{ex} = \frac{\hbar^2}{2M\mu_0}\, \frac{4ab-c^2}{a+b-|c|} ,$$

$$(7.13)$$

with M being the nucleon mass.

(ii) $c > 0$. In this case, the Born scattering amplitude is backward-peaked and the effective local potential is given by

$$\tilde{V}_{ex}(\vec{R}) = \tilde{P}_{ex}\, exp\left[-(R/R_{ex})^2\right] exp\left(-E/E_{ex}\right) P^R ,$$

$$(7.14)$$

where the characteristic quantities are again given by eqs. (7.12) and (7.13), and P^R is a Majorana space-exchange operator interchanging the position coordinates of the clusters which are now treated as structureless point particles.

For the study of the EHK, one determines, for each exchange type, first the value of x, to be called x_q, for which C_{xq} is equal to zero. Then, depending upon whether x is smaller or larger than x_q, the resultant effective exchange potential $\tilde{V}_{xq}$ is a Wigner-type or Majorana-type potential containing a polynomial factor $\tilde{P}_{xq}$ and characterized by a characteristic range R_{xq} and a characteristic energy E_{xq}.

By using eqs. (7.12) and (7.13) we find, in a straightforward manner, the general expressions for R_{xq} and E_{xq}. These expressions are

$$R_{xq} = \left[\frac{4}{\alpha} \frac{|x - x_q|}{J_q - (x - x_q)^2} \right]^{1/2} , \tag{7.15}$$

$$E_{xq} = \frac{\hbar^2 \alpha}{2M} \frac{J_q - (x - x_q)^2}{[\mu_0 - |x - x_q|]^2 + K_q} . \tag{7.16}$$

In table 5, we list the constants x_q, J_q, and K_q. From this table, one notes that x_q differs from μ_0 only for the exchange types b and c. In fact, even for these two types, the difference, being equal to λ, is relatively insignificant. The reason for this is that, in a realistic situation, the value of λ is only about 1/3 which, except in the case of nucleon-nucleus scattering, is substantially smaller than the value of μ_0.

By examining the general properties of the characteristic quantities R_{xq} and E_{xq}, one finds that (i) for $x < x_q$, R_{xq} and E_{xq} decrease monotonically with increasing x and have largest values when $x = 1$, and (ii) for $x > x_q$, R_{xq} and E_{xq} increase monotonically with increasing x and have largest values when $x = N_B$ ($q = a,c,d$). Since in the expression for $\tilde{V}_{xq}$ the exponential factor $\exp (-E/E_{xq})$ appears, it is reasonable to expect that, at relatively high energies, the effective potentials with large characteristic energies should make dominant contributions. Therefore, the EHK study indicates that, among all exchange terms, the one-exchange term ($x = 1$) has the largest influence for $x < x_q$ and the core-exchange term ($x = N_B$) has the largest influence for $x > x_q$.

The above assertion concerning the importance of one-exchange and core-exchange terms is further strengthened when one considers situations where large absorptions are present and, hence, grazing collisions are dominant. In such situations, it is of course evident that longer-ranged effective potentials will have larger influence. Therefore, in cases such as heavy-ion scattering, α-scattering by medium- and heavy-weight nuclei, and so on, one anticipates that approximate calculations in which all exchange effects except one-exchange and core-exchange effects are omitted will usually yield quite satisfactory results.

7.1c. <u>One-exchange and core-exchange potentials.</u>

Because the one-exchange and core-exchange terms are found to be particularly important among all exchange terms, we concentrate in this subsection in studying the properties of these terms. For the one-exchange case, a close examination shows that, for all exchange types, the characteristic range and the characteristic energy of the type-c term are the largest [LE 79b]. This indicates, therefore, that the type-c term is the dominant one-exchange term. The situation is not so clearcut in the core-exchange case. Here one finds that both type-a and type-d terms are important, but the type-c term may make less contributions.

In the following discussion, we shall regard the type-c and type-d terms as representing the major influence of one-exchange and core-exchange processes, respectively. With this viewpoint, the one-exchange and core-exchange characteristic ranges and characteristic energies are, therefore, given by

$$R_1 = R_{1c} = \left[\frac{4(\mu_0 + \lambda - 1)}{2\mu_0 + 2\lambda - 1} \frac{1}{\alpha} \right]^{1/2} , \qquad (7.17)$$

$$E_1 = E_{1c} = \frac{\hbar^2}{2M} \frac{2\mu_0 + 2\lambda - 1}{1 - 2\lambda} \alpha , \qquad (7.18)$$

$$R_c = R_{cd} = \left[\frac{4}{(N_A - N_B) + (N_A/N_B)\lambda} \frac{1}{\alpha} \right]^{1/2} , \qquad (7.19)$$

$$E_c = E_{cd} = \frac{\hbar^2}{2M} \frac{(N_A - N_B) + (N_A/N_B)\lambda}{(N_A - N_B)^2} (N_A + N_B)\alpha . \qquad (7.20)$$

Because of the small value of λ ($0 < \lambda < 0.5$), it is interesting to note that all these characteristic quantities depend only weakly on λ . Physically, this is easily understandable. Since nucleon-exchange processes occur predominantly when the colliding nuclei are in close proximity, one may plausibly expect that the range of the nucleon-nucleon potential will not qualitatively influence antisymmetrization effects to a large extent.

We now use eqs. (7.17) and (7.19) to study the spatial dependence of the one-exchange effective potential $\tilde{V}_{1c}$ and the core-exchange effective potential $\tilde{V}_{cd}$ (in the following discussion, these potentials will be referred to simply as $\tilde{V}_1$ and $\tilde{V}_c$, and the polynomial factors contained in them will be written as $\tilde{P}_1$ and $\tilde{P}_c$). For this purpose, we need to know the range of the direct potential V_D which has the expression

$$V_D(\vec{R}) = \tilde{P}_D \exp\left[-(R/R_D)^2\right], \qquad (7.21)$$

with $\tilde{P}_D$ being a polynomial in R^2. By examining the diagrammatical representations shown in fig. 19, it can be easily seen that R_D must be equal to the characteristic range of the type-b term with x = 0; that is,

$$R_D = \left(\frac{\mu_0 - \lambda}{\mu_0 \lambda} \frac{1}{\alpha} \right)^{1/2} . \qquad (7.22)$$

In addition, it is important to note that the highest powers of R^2 in the polynomial factors $\tilde{P}_1$ and $\tilde{P}_D$ are quite similar and for the interesting case where N_A and N_B are nearly equal (see the discussion below), these are also approximately the same as the highest power appearing in the polynomial factor $\tilde{P}_c$ [BA 77c]. Therefore, since the polynomial factors in $\tilde{V}_1$, $\tilde{V}_c$, and V_D have similar values for their highest powers in R^2, it is appropriate to simply examine the exponential factors in order to determine the situations under which the effective potentials $\tilde{V}_1$ and $\tilde{V}_c$ make important contributions.

By comparing the values of R_1 and R_c with the value of R_D, one can make the following general remarks:

(i) The ratio R_1/R_D is given by

$$R_1/R_D = \left[\frac{4\mu_0\lambda}{\mu_0-\lambda} \cdot \frac{\mu_0 - (1-\lambda)}{2\mu_0 - (1-2\lambda)} \right]^{1/2} , \tag{7.23}$$

which is smaller than but close to 1. For example, in the realistic case where λ is around 1/3 and μ_0 is appreciably larger than 1, the value of R_1/R_D is approximately equal to 0.8. This indicates, therefore, that the one-exchange contribution may be generally important, which agrees with the results obtained from a number of previous investigations [LE 77a, KO 74]. In these investigations, the purpose was to find whether the resonating-group phase-shift values (calculated with central nucleon-nucleon potential only) can be apprximately reproduced by using a simple potential model in which one solves the equation

$$\left[-\frac{\hbar^2}{2\mu} \nabla^2 + \tilde{V}(\vec{R}) - E \right] F(\vec{R}) = 0 \tag{7.24}$$

with $\tilde{V}(\vec{R})$ having the form

$$\tilde{V}(\vec{R}) = V_D(R) + V_W(R) + V_M(R)P^R . \tag{7.25}$$

Indeed, it has invariably been found that the V_W term in $\tilde{V}(\vec{R})$, which arises mainly from an one-exchange process, must have a non-negligible magnitude in comparing with the V_D term obtained by a double-folding procedure.

(ii) The value of R_c becomes substantially smaller than that of R_D when the nucleon-number difference

$$\delta = N_A - N_B \tag{7.26}$$

is large. For instance, in the $\alpha + {}^{16}O$ case where δ has a rather large value equal to 12, the value of R_c/R_D obtained with $\lambda = 1/3$ is only about 0.33. This means that one expects the core-exchange effect to become less important as δ increases.

Table 6

Values of $\bar{E}_1$ and $\bar{E}_c$ in various systems.

System	δ	$\alpha(\mathrm{fm}^{-2})$	$\bar{E}_1$ (MeV/nucleon)	$\bar{E}_c$ (MeV/nucleon)
$^3\mathrm{He} + \alpha$	1	0.46	47	55
$\alpha + {}^6\mathrm{Li}$	2	0.40	46	22
$\alpha + {}^{16}\mathrm{O}$	12	0.36	46	4
$^{16}\mathrm{O} + {}^{17}\mathrm{O}$	1	0.32	46	37
$^{16}\mathrm{O} + {}^{20}\mathrm{Ne}$	4	0.30	45	7

Indeed, we have reached a similar conclusion based on the results of many resonating-group calculations. There it was found that the degree of odd-even ℓ-dependence, exhibited by the calculated phase shift, turns out to be quite strong in scattering systems involving two s-shell nuclei where δ is small, and weak in systems such as $\alpha + {}^{16}\mathrm{O}$ and $n + {}^{40}\mathrm{Ca}$ where δ takes on much larger values. In addition, of course, the finding that core-exchange effects are important in $\alpha + {}^6\mathrm{Li}$, ${}^{12}\mathrm{C} + {}^{13}\mathrm{C}$, ${}^{12}\mathrm{C} + {}^{16}\mathrm{O}$, and ${}^{16}\mathrm{O} + {}^{19}\mathrm{F}$ scattering [BA 72, BA 76, FU 75, VO 70] supports the assertion reached by the present analysis.

Next, we discuss the energy dependence of exchange effects. For this, we examine the expressions of the one-exchange characteristic energy per nucleon $\bar{E}_1 = E_1/\mu_0$ and the core-exchange characteristic energy per nucleon $\bar{E}_c = E_c/\mu_0$. By using eqs. (7.18) and (7.20), one easily sees that $\bar{E}_1$ takes on rather large values in all scattering systems (remember that λ is around 1/3), indicating that the one-exchange term has generally an important influence over a wide range of energies. On the other hand, because of the factor $(N_A - N_B)^2$ occuring in the denominator of eq. (7.20), $\bar{E}_c$ is large only when δ is relatively small.

In table 6, we list the values of $\bar{E}_1$ and $\bar{E}_c$ for various systems, calculated with $\kappa = 0.4\ \mathrm{fm}^{-2}$. These values may be used to obtain a semi-quantitative estimate of the energy range in which exchange effects may be significant. For example, consider the $^3\mathrm{He} + \alpha$ system in which the core-exchange effect is known to be important at relatively low energies [KO 74]. Since the depth of the core-exchange effective potential decreases with energy according to the factor $\exp[-(E/\mu_0)/\bar{E}_c]$, one notes that even when E/μ_0 is 2 or 3 times larger than the value of 55 MeV listed in table 6, the core-exchange effect may still have an appreciable influence. Thus, one expects that for this particular system the scattering angular distribution may exhibit a noticeable backward rise even at a c.m. energy as high as about 300 MeV.

7.1d. **Explicit study of ^{3}He $+ \alpha$ and $\alpha + {}^{16}$O systems.**

The considerations given in the above subsection are made in the Born approximation; hence, one might expect at first that the results obtained should have only semi-quantitative significance at relatively high energies.[+] In this subsection, however, we shall show by an explicit study of the ^{3}He $+ \alpha$ and $\alpha + {}^{16}$O systems that these results may in fact have general utility at energies less than about 50 MeV/nucleon where experimental nuclear-structure and nuclear-reaction studies are commonly carried out and, therefore, may be used to make interesting predictions even in the low-energy region.

In the ^{3}He $+ \alpha$ study, we choose $\alpha = 0.46$ fm^{-2} and use the nucleon-nucleon potential of eq. (4.38) with $\kappa = 0.46$ fm^{-2} and a Serber exchange mixture. Also, as was mentioned previously, all charge effects are omitted by letting the charge of the proton to be infinitesimally small.

The procedure we use to study the importance of various nucleon-exchange terms ($x = 1,2,3$) is as follows. We compare the binding energies, phase shifts, and differential cross sections obtained by solving the integrodifferential equation (4.35) with the full kernel (to be referred to as resonating-group calculation) and with different nucleon-exchange terms turned off. For the $\ell = 1$ ground state, for example, we obtain a resonating-group binding energy of 3.55 MeV, which should be compared with the values of 0.29, 3.10, and 2.64 MeV obtained by omitting,

---

[+]It is interesting to note that a WKB study by Horiuchi [HO 80] has yielded similar conclusions.

Table 7

Characteristic range and characteristic

energy in the ^{3}He $+\alpha$ case.

x	Exchange type q	R_{xq} (fm)	E_{xq} (MeV)	Exchange nature
	a	1.60	23	W
1	b	0.91	23	W
	c	1.71	89	W
	d	1.44	29	W
	a	0.93	13	M
2	b	1.19	33	M
	c	0.31	15	W
	d	0.85	16	M
	a	2.95	67	M
3	c	1.60	67	M
	d	2.45	97	M

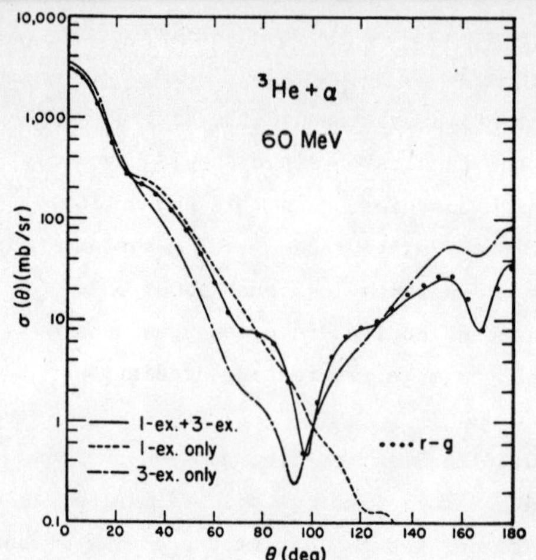

Fig. 20. Comparison of ^{3}He + α differential cross sections at 60 MeV calculated with the full resonating-group kernel and with various nucleon-exchange terms omitted. (From ref.[LE 79b])

respectively, one-exchange, two-exchange, and three-exchange terms. From these values, one already sees that the three-exchange terms have a larger influence than the two-exchange terms (note that $\mu_0 = 12/7$), in agreement with the discussion given in subsect. 7.1c.

Various values of R_{xq} and E_{xq} are listed in table 7, where the symbols W and M indicate the exchange nature of the effective potential being Wigner and Majorana, respectively. Here one sees that, as discussed in subsect. 7.1c, the one-exchange and core-exchange terms have rather large characteristic ranges and energies (the range R_D of the direct potential is 2.29 fm). On the other hand, the characteristic quantities of the two-exchange terms are significantly smaller, indicating that, at relatively high energies, these exchange terms will have much less influence [LE 79b].

In fig. 20, we show a cross-section comparison at 60 MeV (i.e., 35 MeV/nucleon), an energy which is significantly higher than the characteristic energies of the two-exchange terms. From this figure, it is seen that the differential cross sections obtained with the resonating-group calculation (solid circles) and with two-exchange terms turned off (solid curve) are nearly the same at all angles, thus demonstrating that the two-exchange terms are not important at this energy. Also, in this figure, the importance of one-exchange and core-exchange terms is clearly shown. Here one sees that if one-exchange terms alone are included (dashed curve; the direct

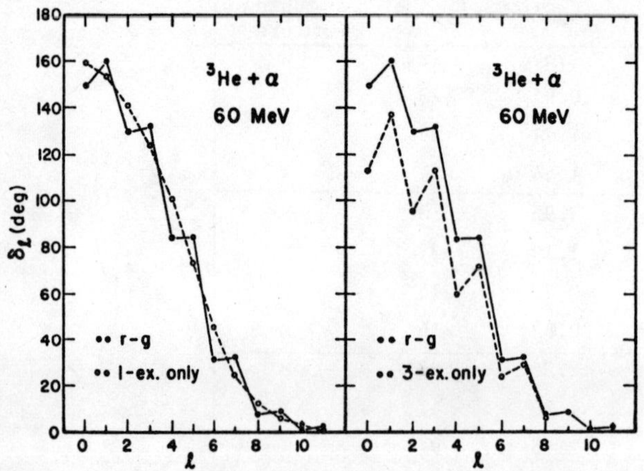

Fig. 21. Comparison of ^{3}He + α phase shifts at 60 MeV calculated with the full resonating group kernel and with various nucleon-exchange terms omitted. (From ref. [LE 79b])

potential V_D is understood to be always included in the calculation), the cross-section behavior in the forward angular region can be reasonably reproduced but the strong rise in the backward direction does not at all materialize. In a different way one can also see from this figure that one-exchange terms must necessarily be considered in the calculation. If one includes only core-exchange terms (dot-dashed curve), then one finds that the calculation does yield a back-angle rise in the cross section; however, the cross sections at forward angles are severely underestimated and the over-all agreement with the resonating-group result is quite poor.

Individual contributions of one-exchange and core-exchange terms at 60 MeV are shown in fig. 21 (remember that two-exchange terms have little influence at this energy). With one-exchange terms alone, it is seen from the left side of this figure that the calculated phase-shift points (open circles) lie on a smooth curve which cuts right across the zigzag line joining the resonating-group phase-shift points (solid circles). This demonstrates that the core-exchange effective potential is essentially a Majorana potential. The situation is quite different when only core-exchange terms are considered. The right side of fig. 21 shows that the dashed line joining the calculated phase-shift points has now a zigzag behavior, but in all partial waves the phase-shift values are smaller than those from the resonating-group calculation. This is an indication that, as discussed above, the one-exchange effective potential is essentially an attractive Wigner potential.

A similar procedure can be employed to study the importance of different nucleon-exchange terms (x = 1, 2, 3, 4) in the $\alpha + ^{16}O$ system. In this case, we use $\alpha = 0.32$ fm^{-2} and adopt the nucleon-nucleon potential of eq. (4.38) with w = 0.334, m = 0.481, b = 0.076, and h = 0.109. Again, for clarity in discussion, all charge effects will be omitted from the calculation.

As in the ^{3}He + α case, the analysis will be guided by a critical study of the characteristic ranges and energies. Here the calculation shows that the one-exchange and two-exchange characteristic energies have rather large values; for the type-c terms, for example, these are equal to 158 and 27 MeV, respectively. On the other hand, the characteristic energies for the three-exchange and four-exchange terms are quite small, being equal to only about 10 MeV. Thus, if a study is performed at an energy around 20 MeV, one would anticipate that the resonating-group result can be well reproduced by a simpler calculation

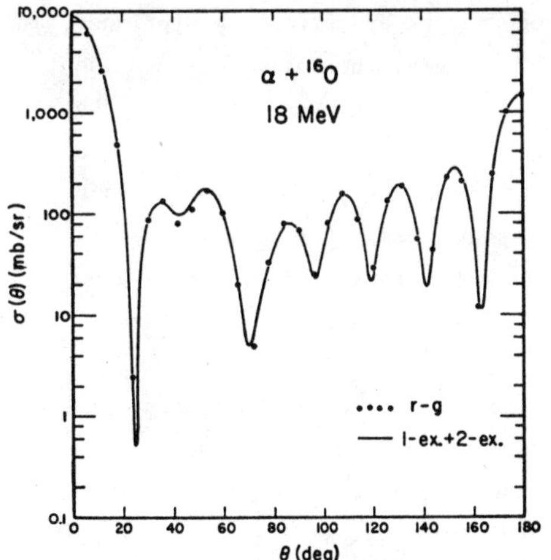

Fig. 22. Comparison of $\alpha + ^{16}O$ differential cross sections at 18 MeV calculated with the full resonating-group kernel and with the three-exchange and four-exchange terms omitted. (From ref. [LE79b])

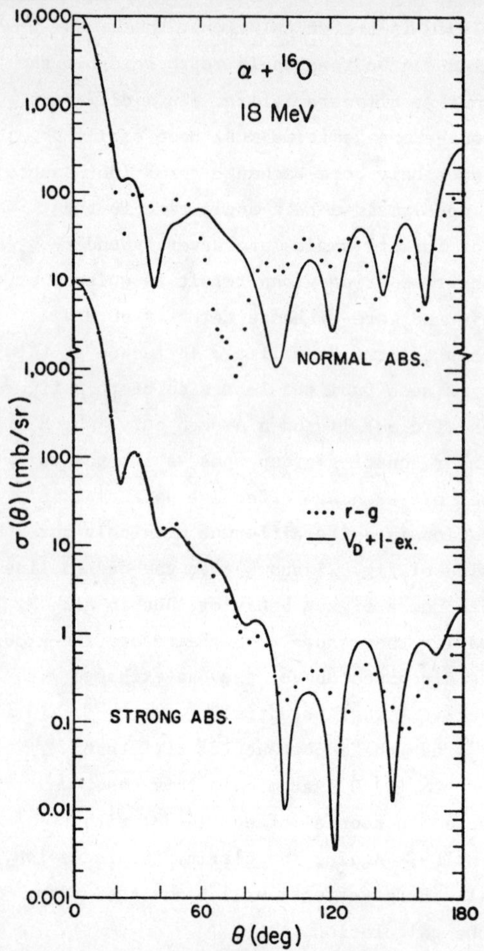

Fig. 23. Comparison of $\alpha + {}^{16}O$ differential cross sections at 18 MeV calculated
with the full resonating-group kernel
and with only one-exchange terms taken
into account.

in which three-exchange and four-exchange terms are omitted.

The above assertion is indeed borne out by explicit calculations. In fig. 22, we compare at 18 MeV differential cross sections calculated with the full resonating-group kernel and with only one-exchange and two-exchange terms included. Here one sees that there is good agreement at all angles, thus fully demonstrating that three- and four-exchange terms are not important at this energy.

The two-exchange characteristic ranges are significantly shorter than the characteristic range of the direct potential. This means that, if there is strong absorption present, the contribution from two-exchange terms will be greatly reduced. To show this, we have performed calculations by including an imaginary potential in the formulation. The results at 18 MeV are shown in fig. 23 for $\alpha + {}^{16}O$ scattering with both normal and strong absorptions. In this figure, the solid dots represent resonating-group results, while the solid curves represent results obtained with two-, three-, and four-exchange terms omitted. For both of these comparisons, the absorptive strengths are adjusted such that the resonating-group and the V_D + one-exchange calculations yield the same values for the reaction cross section.

The normal-absorption case refers to the situation where the reflection coefficient exp $(-2\delta_\ell^I)$, with δ_ℓ^I being the imaginary part of the phase shift, acquires values around 0.5 for $\ell \lesssim 7$. In this case, the comparison shown in the top part of fig. 23 indicates that it is not sufficient to take just one-exchange terms into account. As the absorption becomes stronger, the comparison between the resonating-group and the V_D + one-exchange results does become more favorable. In the strong-absorption case where the reflection coefficient takes on very small values around 0.03 for $\ell \lesssim 7$, it is seen from the lower part of fig. 23 that the

V_D + one-exchange calculation does yield all the essential features of the resonating-group calculation.

The above finding concerning the influence of absorption on the importance of exchange terms is very interesting. As is well known, MCT calculations for relatively heavy systems are rather tedious to perform even with the CGCT and the other techniques described in sect. 5. Thus, the possibility of omitting many exchange terms, especially in heavy-ion problems where absorptions are strong, will represent considerable saving in computational effort and is certainly very important from a practical viewpoint.

7.2. Effects of antisymmetrization in nuclear direct reactions

Consider a direct-reaction process A(a,b)B, which involves in the incident channel (channel f) two clusters a and A having nucleon numbers N_a and N_A, and in the outgoing channel (channel g) two clusters b and B having nucleon numbers N_b and N_B. For definiteness, we adopt the convention that $N_a < N_A$, $N_a < N_b$, and $N_b \leq N_B$. Also, for simplicity in discussion, we shall assume that all these clusters have spin zero and are described by translationally-invariant shell-model functions in harmonic-oscillator wells having the same width parameter α .

To examine antisymmetrization effects in direct reactions [LE 79c, LE 80a], one proceeds in a way which is very similar to that discussed in subsect. 7.1 for the scattering problem. What one does in this case is to study the properties of the coupling-Hamiltonian kernel (CHK) (see subsect. 4.2c), which has the form

$$\mathcal{H}_c(\vec{R}_g', \vec{R}_f'')$$

$$= \left\langle \hat{\phi}_b(1, \cdots, N_a\,;\, N_a+1, \cdots, N_a+N_t)\,\hat{\phi}_B(N_a+N_t+1, \cdots, N_a+N_t+N_B)\,\delta(\vec{R}_g-\vec{R}_g')z \right.$$

$$\left. |H|\mathcal{A}'\left[\hat{\phi}_a(1, \cdots, N_a)\,\hat{\phi}_A(N_a+1, \cdots, N_a+N_t\,;\, N_a+N_t+1, \cdots, N_a+N_A)\,\delta(\vec{R}_f-\vec{R}_f'')z\right] \right\rangle , \tag{7.27}$$

where $\hat{\phi}_a$, $\hat{\phi}_A$, $\hat{\phi}_b$, and $\hat{\phi}_B$ denote antisymmetrized cluster internal functions, and N_t is the number of nucleons in the transferred cluster t, given by

$$N_t = N_b - N_a . \tag{7.28}$$

Also, it is noted that, in eq. (7.27), the arguments of the internal functions of the clusters indicate which nucleons are involved in these clusters; such a labelling is permissible, because the wave functions are totally antisymmetrized.

Similar to the EHK, the CHK can also be expressed as

$$\mathcal{H}_C(\vec{R}_g', \vec{R}_f'') = \sum_x \sum_q \mathcal{H}_{Cq}^x(\vec{R}_g', \vec{R}_f''),\qquad(7.29)$$

where q denotes the coupling type (see ref. [LE 80a]). The quantity x denotes the number of nucleons interchanged between the group of nucleons labelled by $(1,\ldots,N_a)$ in cluster a and the group of nucleons labelled by $(N_a + N_t + 1, \ldots, N_a + N_A)$ in cluster A [see eq. (7.27)]; its minimum and maximum values are equal to 0 and N_a, respectively.

The coupling kernel $\mathcal{H}_{Cq}^x$ consists of a polynomial function multiplied by an exponential factor in $\vec{R}_f''^2$, $\vec{R}_f''\cdot\vec{R}_g'$, and $\vec{R}_g'^2$. By using the CGCT, one can derive a general expression for the exponential factor from which the information about the relative importance of the various nucleon-exchange terms can be gleaned. The procedure to do this in the plane-wave Born approximation (PWBA) is carefully explained in ref. [LE 80a]. Here we shall only quote the results:

(i) The critical quantity in the analysis is x_q which is, in almost all cases of interest, quite close to the quantity μ_t defined as

$$\mu_t = N_a N_B / N.\qquad(7.30)$$

For $x < x_q$, the PWBA reaction amplitude is forward-peaked, while for $x > x_q$, the PWBA reaction amplitude is backward-peaked.

(ii) In the PWBA, the coupling kernels can be represented by equivalent coupling potentials, with each of these potentials characterized by a characteristic range and a characteristic energy. For $x < x_q$, it can be shown that, for all coupling types, both the characteristic range and the characteristic energy decrease monotonically with increasing x and have largest values when x = 0. Since direct reactions occur predominantly in the peripheral region, it is clear that longer-ranged equivalent potentials will have larger influence. Therefore, this indicates that the x = 0 term (no-exchange term) is the most important one among all exchange terms with $x < x_q$.

(iii) For $x > x_q$, both the characteristic range and the characteristic energy increase monotonically with increasing x and have largest values when $x = N_a$. This indicates that the $x = N_a$ term (maximum-exchange term) is the dominant one among all exchange terms with $x > x_q$.

(iv) The no-exchange and maximum-exchange terms correspond to light-particle and heavy-particle pickup processes commonly employed in phenomenological analyses [CH 77, HU 72]. In a macroscopic description, these processes can be accounted for by the use of three-body models. Take the reaction $^{20}\mathrm{Ne}(\alpha,\ ^6\mathrm{Li})^{18}\mathrm{F}$ as an example. For the light-particle pickup process, the nucleus $^{20}\mathrm{Ne}$ is considered to have a

$d + {}^{18}F$ cluster structure and the incident α-particle picks up the light d-cluster to form ${}^{6}Li$, while for the heavy-particle pickup process, the nucleus ${}^{20}Ne$ is considered to have a ${}^{6}Li + {}^{14}N$ cluster structure and the incident α-particle picks up the heavy ${}^{14}N$ cluster to emerge as a ${}^{18}F$ nucleus.

In phenomenological studies, the adoption of such three-body simplifications has frequently been shown to yield fairly reasonable results [GR 74]. Thus the finding here that the no-exchange and maximum-exchange terms yield the dominant contributions does provide some justification for such a simplified treatment.

(v) By studying the properties of the no-exchange term, it can be shown that there is a general tendency in favor of the pickup of a light cluster containing a relatively small number of nucleons. We should mention, however, that this conclusion was reached without taking into account the effect arising from the formation of nucleon clusters, such as α-clusters, in the various nuclei involved. A proper consideration of this latter effect will definitely modify the above statement to a certain extent.

(vi) From a careful investigation of the maximum-exchange term, one obtains the interesting information that back-angle reaction cross sections will have large values only if N_a and N_A do not greatly differ. In a systematic experiment of (p,α) reaction on light nuclei at $E_p = 38$ MeV [GA 69], it was indeed found that the ratio of backward-angle to forward-angle cross section does become progressively smaller as the target nucleus becomes heavier.

The above findings must, of course, be checked against the systematics of experimental direct-reaction results. In this respect, it is interesting to note that there appeared recently a review of such results by Teplov et al. [TE 77]. In comparison, we find that there is general agreement between the conclusions reached here and the systematic trends reported in this reference. We should mention, however, that these authors have performed their analyses mainly at relatively low energies where our present consideration is expected to have only qualitative significance. For our purpose, it would be much more interesting to check against experimental measurements at higher energies of about 20 to 50 MeV/nucleon.

7.3. Concluding remarks

By examining the structures of the exponential factors appearing in the resonating-group kernel functions, some general understanding concerning the effects of antisymmetrization has been achieved. In scattering problems, it was found that the one-exchange and core-exchange terms are especially important. For direct reactions, we have obtained interesting information about the general behavior of such processes and some justification for the use of three-body models in phenomenological analyses.

For further progress, it would be very useful to obtain a better understanding of the core-exchange effect in scattering problems. To achieve this, one needs to carry out both theoretical and experimental investigations in systems where the nuclei involved have similar mass. In addition, it is important that experimental measurements should be performed at energies of about 20 to 50 MeV/nucleon where resonance effects are small and cover as large an angular region as feasible. Cases of particular interest would be the scattering of ^{16}O by ^{17}O, ^{18}O, ^{19}F, and ^{20}Ne. For these cases, systematic resonating-group studies have not yet been carried out, but are certainly feasible with the computational techniques presently available.

Finally, it should be mentioned that our present effort represents just an initial attempt to study the effects of antisymmetrization in scattering and reaction problems. Many interesting aspects have yet to be examined. For example, it will certainly be important to investigate the influence which comes from the possibility of the formation of nucleon clusters in the various nuclei involved. Within the present formulation, one can achieve this by studying the general properties of the polynomial factors occurring in the kernel functions. Quite obviously, this will be a major and difficult task, but is definitely worth carrying out for the purpose of achieving a deeper understanding of the important role played by the Pauli principle in nuclear problems.

8. Electromagnetic calculations with MCT wave functions

The examples discussed in sect. 6 have given strong indications that, especially in lighter systems, MCT calculations can yield very satisfactory results for scattering and reaction cross sections. In this section, we shall further show that the many-nucleon wave function obtained can also give a good description of the behavior of the system in the compound-nucleus region. To show this, we illustrate with two examples; namely, the study of the electromagnetic properties of the seven-nucleon system, and the calculations of ^{12}C and ^{20}Ne elastic and inelastic form factors.

8.1. Electromagnetic properties of the seven-nucleon system

8.1a. Charge form factor of ^{7}Li

To calculate the charge form factor of ^{7}Li [KA 80a], we take both proton and neutron distributions into consideration. For this we first compute, in a definite magnetic substate, the bare form factors for proton and neutron distributions in the Born approximation given by

$$F_M^p = \frac{1}{N_p} \frac{\langle \bar{\Psi}_M | \sum_{i=1}^{N} exp[i\vec{q}\cdot(\vec{r}_i - \vec{R}_{cm})]\frac{1}{2}(1+\tau_{iz})|\Psi_M\rangle}{\langle \bar{\Psi}_M | \Psi_M \rangle} \tag{8.1}$$

$$F_M^n = \frac{1}{N_n} \frac{\langle \bar{\Psi}_M | \sum_{i=1}^{N} exp[i\vec{q}\cdot(\vec{r}_i - \vec{R}_{cm})]\frac{1}{2}(1-\tau_{iz})|\Psi_M\rangle}{\langle \bar{\Psi}_M | \Psi_M \rangle} \tag{8.2}$$

In the above equations, $\vec{q}$ is the momentum transfer divided by $\hbar$, and N_p and N_n denote the proton and neutron numbers in ^{7}Li, respectively. The function Ψ_M, with $\bar{\Psi}_M$ being its unantisymmetrized part, represents the ^{7}Li ground-state wave function; it has the form

$$\Psi_M = A\,\bar{\Psi}_M = A\left[\phi_\alpha \phi_t \frac{1}{R} f_{JL}(R)\, \mathcal{Y}_{JLS}^M\, Z(\vec{R}_{cm})\right], \tag{8.3}$$

where the function $\mathcal{Y}_{JLS}^M$ is a spin-isospin-angle function appropriate for $T = 1/2$, $S = 1/2$, and required values of orbital angular momentum $\ell = 1$ and total angular momentum $J = 3/2$ with z-component M. In defining the magnetic substates, we have chosen the quantization or z-axis to be in the direction of the momentum transfer $\vec{q}$.

Upon averaging over initial magnetic substates and summing over final magnetic substates, one obtains for the square of the charge form factor the expression

$$F_{ch}^2 = \frac{1}{2}\left(F_{3/2}^2 + F_{1/2}^2\right), \tag{8.4}$$

where

$$F_{3/2} = F_{3/2}^p F_p + \frac{N_n}{N_p} F_{3/2}^n F_n, \tag{8.5}$$

$$F_{1/2} = F_{1/2}^p F_p + \frac{N_n}{N_p} F_{1/2}^n F_n, \tag{8.6}$$

with F_p and F_n being the charge form factors of the proton and the neutron, respectively [JA 66].

From $F_{3/2}$ and $F_{1/2}$, one obtains the C0 charge form factor F_{C0} defined as

$$F_{C0} = \frac{1}{2}\left(F_{3/2} + F_{1/2}\right), \tag{8.7}$$

and the C2 charge form factor F_{C2} defined as

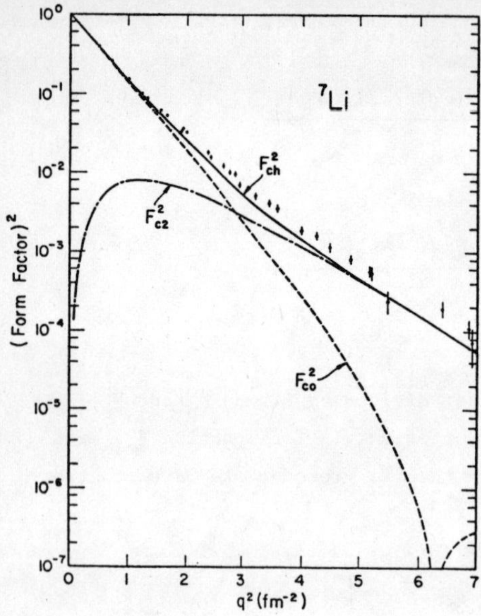

Fig. 24. Comparison of calculated and empirical results for F_{ch}^2. Contributions from C0 and C2 interactions are also shown. (From ref.[KA 80a])

$$F_{C2} = \frac{1}{2}(F_{3/2} - F_{1/2}) \tag{8.8}$$

In terms of these form factors, one can further write F_{ch}^2 as

$$F_{ch}^2 = F_{C0}^2 + F_{C2}^2 \tag{8.9}$$

The mean-square charge radius R_{ch}^2 and the spectroscopic quadrupole moment Q can then be determined by examining the low-q^2 behavior of F_{C0} and F_{C2}; these are given by

$$R_{ch}^2 = -6\left(\frac{dF_{C0}}{dq^2}\right)_{q^2=0} \tag{8.10}$$

and

$$Q = -18\left(\frac{dF_{C2}}{dq^2}\right)_{q^2=0} \tag{8.11}$$

From the above discussion, it is clear that the quantities one needs to compute are the bare form factors F_M^p and F_M^n, with M = 3/2 and 1/2. By using the ground-state radial wave function obtained from a resonating-group calculation [FU 76, KA 80a] which explains quite well the ^{7}Li and ^{7}Be bound-state data, and the experimental data for ^{3}He + α scattering in the low-energy region, one can easily compute these quantities by employing the CGCT discussed in subsect. 4.3.

In fig. 24, we show by the solid curve the calculated F_{ch}^2 values for q^2 up to 7 fm^{-2}. The empirical data shown are those of Suelzle et al. [SU 67]. From this figure, one sees that there is a rather satisfactory agreement over the whole range of q^2 studied. Considering the fact that in the present investigation the charge form factor is calculated straightforwardly from a resonating-group wave function obtained in a nuclear-structure study and no adjustments are made, one may view this as yet another strong evidence in establishing the soundness of the resonating-group approach to solve problems at least in light nuclear systems.

Contributions from C0 and C2 interactions are also depicted separately in fig. 24. Here one finds that the C0 contribution is dominant for low-q^2 values smaller than about 2 fm^{-2}. At high momentum transfers, the contribution comes instead mainly from the C2 term, which indicates that the ^{7}Li nucleus has a large intrinsic quadrupole deformation in its ground state. This latter feature is correctly contained in the wave function used in our calculation, because it is one of the main advantages of a resonating-group or cluster-model calculation that the collective behavior in the form of nucleon clustering is generally taken into account in an adequate manner.

Using eqs. (8.10) and (8.11), and the low-q^2 behavior of the form factors, one determines the rms charge radius R_{ch} and the spectroscopic quadrupole moment Q. The results are

$$R_{ch} = 2.44 \text{ fm} \quad , \qquad (8.12)$$

$$Q = - 3.70 \text{ fm}^2 . \qquad (8.13)$$

Comparing with the empirical values for R_{ch} of 2.35 ± 0.10 fm obtained by a model-independent analysis [BU 72] of experimental data at low-momentum transfers less than about 1 fm^{-1} and 2.39 ± 0.03 fm obtained by an oscillator shell-model analysis [SU 67] of higher-q^2 data, one can conclude that the result obtained here is quite good. As for the spectroscopic quadrupole moment, the experimentally determined values are $-4.1 \pm 0.6 \text{ fm}^2$ [OR 75] and $-3.4 \pm 0.6 \text{ fm}^2$ [EG 80], which agree well with the result given by eq. (8.13)

8.1b. Radiative-capture reaction of ^{3}He by α.

In this subsection, we describe another interesting electromagnetic process involving the seven-nucleon system, namely, the electric dipole (E1) radiative-capture reaction of ^{3}He by α , leading to the ground and first excited states of ^{7}Be. This is a timely and useful investigation, because a careful examination of this parti-cular reaction at very-low-energies is important for a thorough understanding of the solar-neutrino problem [BA 71a, BA 80b], which is receiving extensive theoretical and experimental attention at the present moment.

In contrast to other considerations [KI 81, TO 63a] of this capture problem, the investigation described here [LI 81] is entirely microscopic in nature. That is, the seven-nucleon initial-state continuum wave function and the final-state bound ^{7}Be wave function will be taken directly from a resonating-group calculation and there will be no adjustments in the computation of the capture cross sections.

In the low-energy region, the long-wavelength limit of the E1 operator is a valid approximation. Thus, the cross section for E1 capture from an initial state with channel spin $S(= 1/2)$ to a final state with total angular momentum J_f, accompanied by the emission of a γ-ray of energy $E_\gamma = \hbar\omega$, is

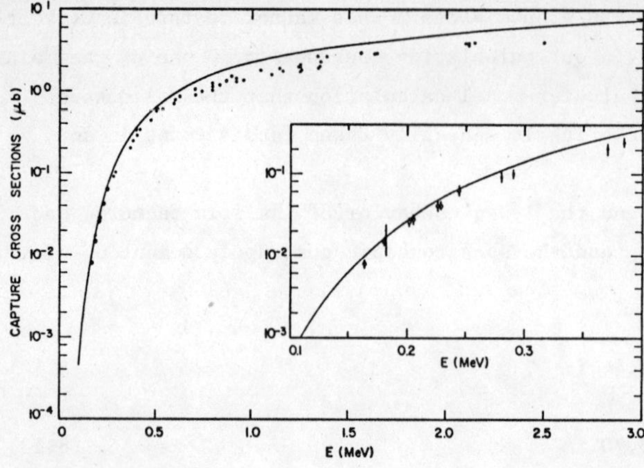

Fig. 25. Comparison of calculated and experimental
total capture cross sections.
(From ref.[LI 81])

$$\sigma_{J_f}(E) = \frac{16\pi}{9} \frac{1}{\hbar v} \left(\frac{\omega}{c}\right)^3 B(E1) \, , \tag{8.14}$$

where v is the relative velocity of the ^{3}He and α nuclei at infinite separation, and B(E1) is the reduced E1 transition probability having the form

$$B(E1) = \frac{e^2}{2S+1} \sum_{M_f M M_S} |\tilde{Q}_{M_f M M_S}|^2 \, , \tag{8.15}$$

with e being the proton charge. In the above equation, $\tilde{Q}_{M_f M M_S}$ represents the transition matrix element and is given by

$$\tilde{Q}_{M_f M M_S} = \left\langle \psi_{J_f M_f} \left| \sum_{i=1}^{N} \rho_i Y_1^{M*}(\hat{\rho}_i) \frac{1+\tau_{iz}}{2} \right| \psi_{SM_S}^{(+)} \right\rangle \, , \tag{8.16}$$

where the nucleon spatial coordinate is measured relative to the coordinate of the total c.m., i.e.,

$$\vec{\rho}_i = \vec{r}_i - \vec{R}_{cm} \, . \tag{8.17}$$

The computation of the partial cross sections $\sigma_{3/2}(E)$ and $\sigma_{1/2}(E)$ can be easily carried out by using the CGCT. The details are given in ref. [LI 81] and will not be further described here. After these partial cross sections are computed, one obtains then, at a given c.m. relative energy E between the two nuclei, the total capture cross section σ_t given by

$$\sigma_t(E) = \sigma_{3/2}(E) + \sigma_{1/2}(E) \, . \tag{8.18}$$

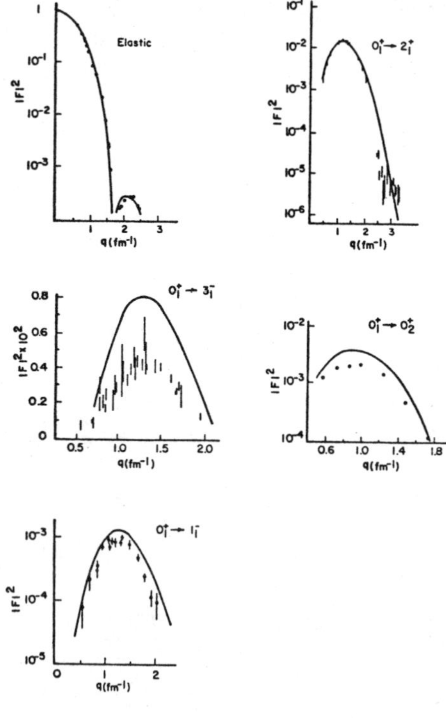

Fig. 26. Calculated and experimental electron-scattering elastic and inelastic form factors of ^{12}C. (From ref.[FU 78])

Calculated results for the total capture cross section is depicted in fig. 25, where for clarity an expanded version at very low energies is also given in the insert. In this figure, the data points shown are those of Parker and Kavanagh [PA 63] and Nagatani et al. [NA 69]. Here one sees that there is an over-all satisfactory agreement between theory and experiment; in particular, the calculated cross section does increase with the same trend as the experimental result. However, the calculated values are somewhat too large. The main reason for this discrepancy is likely that the resonating-group wave function used in this investigation has a single $^{3}He + \alpha$ cluster configuration. For simplicity, other cluster configurations are not considered. Although it is anticipated that, at low energies, the $^{3}He + \alpha$ cluster structure should play a major role, the omission of other cluster structures will certainly affect the calculated result to a significant extent.

The calculated value for the branching ratio $\tilde{\rho}$, defined as

$$\tilde{\rho} = \sigma_{1/2} / \sigma_{3/2} , \tag{8.19}$$

is nearly constant in the low-energy region. It ranges from 0.417 at 0.1 MeV to 0.443 at 2.0 MeV. These values agree very well with the measured results obtained recently by the Münster group [KR 79, RO 81].

8.2. Electron scattering elastic and inelastic form factors of ^{12}C and ^{20}Ne

For heavier systems, resonating-group wave functions seem also capable of providing reliable nuclear-structure information. Here we illustrate this by discussing the electron-scattering elastic and inelastic form factors of the nuclei ^{12}C and ^{20}Ne.

RGM calculations of the T = 0, natural-parity, 3α -clustering states of ^{12}C are discussed in subsect. 6.1. In fig. 26, we show the elastic and various

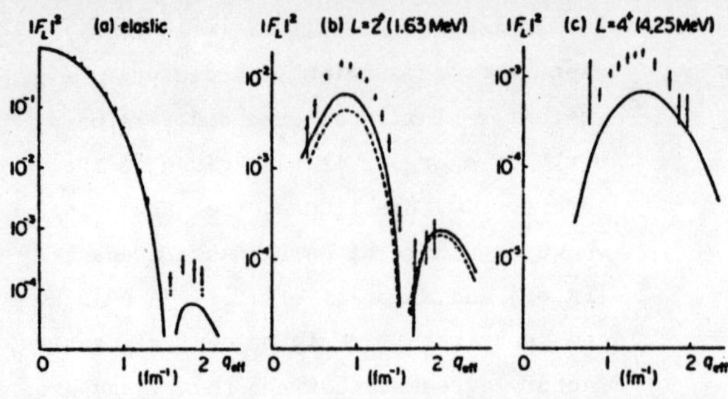

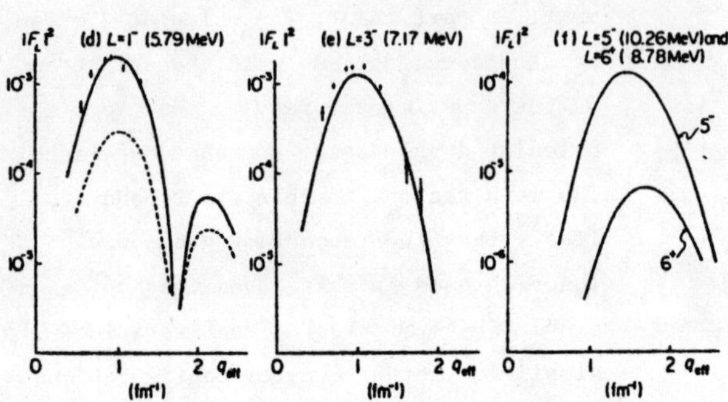

Fig. 27. Elastic and inelastic Coulomb form factors of ^{20}Ne. (from ref.[FU 76a])

inelastic form factors calculated by Fukushima and Kamimura [FU 78]. From this figure, one sees that the experimental data are fairly well reproduced by the RGM investigation. This is of course an important indication that the microscopic 3α model does provide us with a good description of the static and transition charge densities in ^{12}C.

The situation is also rather favorable in the ^{20}Ne case. Here it has been shown by many authors [LE 76, MA 75] that single-channel $\alpha + {}^{16}$O RGM calculations can predict not only the existence of $K^{\pi} = 0^{+}$ and 0^{-} rotational bands in ^{20}Ne, but also nearly correct excitation energies for the levels in these bands. In addition, these calculations further show that the states in the excited $K^{\pi} = 0^{-}$ band have a strong $\alpha + {}^{16}$O molecule-like character, while the states in the ground $K^{\pi}=0^{+}$ band have rather shell-model like character with large $(8,0)$ SU_3 components.

A comparison of calculated and measured elastic and inelastic Coulomb form factors is shown in fig. 27 [FU 76a]. In this figure, the solid curve represents the results obtained with resonating-group wave functions, while the dashed curves are obtained when the 2^{+} and 1^{-} states are considered to be pure $(8,0)$ and $(9,0)$ SU_3 states, respectively. As is seen, the agreement between the resonating-group and the experimental results is especially satisfactory for transitions from the ground state to the strong α-clustering states in the $K^{\pi}= 0^{-}$ band. For transitions within the ground $K^{\pi}=0^{+}$ band, the calculated strengths are somewhat too small. The reason for this is that the $\alpha + {}^{16}$O cluster-model space is not large enough for a detailed description of the shell-model like states in this band. To improve the form-factor results, one will need to extend the calculation by including also the $(sd)^4$ shell-model space, as has been suggested by Tomoda and Arima [TO 75a].

8.3. Concluding remarks

The examples discussed in this section show quite convincingly that even rather simplified resonating-group wave functions can give a good description of the behavior of nuclear states which exhibit a large degree of nucleon clustering. Therefore, one should make use of existing MCT wave functions for further electro-magnetic and weak-interaction investigations, such as the M1 and M3 form-factor study in ^{7}Li, the electron capture of ^{7}Be, the ^{3}He $+ \alpha$ and $\alpha + \alpha$ bremsstrahlungs, the p + ^{7}Be radiative-capture reaction, and so on. All these problems are rather easy to perform, but are certainly interesting because the solution of which will undoubtedly lead to a better understanding of the detailed structures of nuclear systems.

9. Conclusion and Outlook

In this series of lectures, we have discussed some aspects of a microscopic cluster theory (MCT) which can describe all low-energy nuclear phenomena from a unified point of view. It is variational in nature and based on the idea that relatively long-ranged correlations, manifested through the formation of nucleon clusters, are important in determining the properties of nuclear systems.

The MCT has been formulated either by the resonating-group method (RGM) or by the generator-coordinate method (GCM). These methods may appear to be not the same, but are actually entirely equivalent. They are founded on exactly the same physical viewpoint, with the only difference being the choice of the sets of non-orthogonal basis wave functions. In fact, the direction is such that in practical calculations one now tends to adopt the computational convenience embodied in both approaches and, consequently, there is no longer any need to make distinctions between these two methods.

In light systems where clustering features are prominent, MCT calculations have been found to be particularly successful. Especially at low energies, one can systematically improve the result by enlarging the trial-function space employed. However, even here it seems that, because of tediousness in computation, the large body of detailed investigations has been concentrated in scattering problems. It would certainly be very useful to perform more, carefully selected reaction studies. For example, one such study could be a three-channel investigation involving

$p + {}^6\text{Li}$, $p + {}^6\text{Li}^*(3^+)$, and ${}^3\text{He} + \alpha$ cluster configurations. In my opinion, this will be a very interesting calculation, because from the results obtained one can learn not only about the influence of target clustering and open channels on antisymmetrization effects but also about the significance of two-step processes in direct reactions.

For heavier systems, the present emphasis seems to be mainly on the study of quasi-molecular resonances. Because of the presence of many multi-cluster open channels, detailed comparisons with experiment on scattering and reaction cross sections will be quite difficult. In fact, even for the modest purpose of explaining the gross features of measured angular distributions, one has been compelled to introduce phenomenological imaginary potentials into the formulation. This is clearly a rather undesirable procedure, since the concept of an imaginary potential is certainly not within the realm of the MCT framework. However, there does not appear to be any attractive alternative at this moment and, therefore, future effort will have to be expended to learn at least the essential features which these imaginary potentials must possess.

One of the major achievements of MCT investigations was to attain an understanding of the roles played by the Pauli principle in nuclear scattering and reactions. For example, it was shown that in scattering problems the essential result of anti-symmetrization is to introduce an odd-even ℓ-dependent component into the effective internuclear potential, the importance of which seems to decrease with increasing value of the difference between the nucleon numbers of the incident and target nuclei. This means that one would expect this odd-even component to be generally important in those cases where the nuclei involved have similar mass, but to have a comparatively minor influence when the nucleon-number difference is large. Indeed, it is precisely this latter feature which enables the conventional optical model to yield a reasonable description, over a large angular region, of light-ion scattering by medium- and heavy-weight nuclei.

For future developments, it would clearly be necessary to perform many more detailed calculations, preferably with realistic nucleon-nucleon potentials containing repulsive-core, spin-orbit, and tensor components. The resultant wave functions should then be employed to compute electromagnetic and weak-interaction transition rates in order to firmly establish the qualities of these wave functions. In addition, one should also use the MCT to study more pragmatic types of problems. For instance, one promising direction would be a detailed investigation to understand the connection between microscopic and phenomenological descriptions of many-nucleon systems. The achievement of such an understanding will certainly be important, because in complicated situations such as deep-inelastic scattering in heavy-ion problems, one will likely be forced to adopt semi-classical or essentially phenomenological viewpoints.

Within the next decade, it is quite probable that one will be able to analytically derive the complicated MCT kernels by performing symbolic algebra on the computer.

If this should materialize into a realistic endeavor, then one will witness an explosive growth in the number of MCT calculations. This will of course be a much heralded occurrence, because one will then be able to microscopically explain many of those exotic phenomena which can only be crudely considered at the present moment.

In conclusion, the MCT is a sound and practical theory built on firm physical foundations. With further mathematical developments and computational innovations, there is no doubt in my mind that it will eventually lead us into achieving a satisfactory understanding of the intricate behavior exhibited by nuclear systems.

Acknowledgements

I would like to take this opportunity to express my sincere gratitude to all my colleagues and friends who have rendered valuable help and offered constructive criticisms over the years. In particular, I wish to thank D. R. Thompson and M. LeMere for their dedication to the group effort in nuclear-theory research carried out at the University of Minnesota.

This work was supported in part by the U. S. Department of Energy under contract No. DOE/DE-AC02-79 ER 10364.

Appendix A

The Orthogonality Condition Model

A.1. Description of the model

The orthogonality condition model (OCM) was proposed by Saito [SA 69] more than ten years ago as an approximation to the RGM equation of motion. Since then, it has been extensively applied to treat especially multi-cluster structure problems [HO 78], where a straightforward application of the RGM approach would be quite difficult. In this Appendix, we present a rather detailed description of this model, in regard to its basic ideas and its limitations (see, also, refs. [BA 78, BU 77, HO 77b, SA 77]). Again, for the sake of clarity, we shall consider the simple case consisting of two spin-zero clusters A and B, which are described by translationally-invariant shell-model wave functions of lowest configurations in harmonic-oscillator wells with width parameters α_A and α_B , respectively.

The central quantity in this model is the norm operator $\mathcal{N}$ or the norm kernel $\mathcal{N}(\vec{R}',\vec{R}'')$ [see eq. (4.12)]. Thus, we begin by discussing some properties of this operator. By solving the equation

$$\mathcal{N}\chi_i = \mu_i \chi_i \ , \tag{A.1}$$

we obtain a set of eigenvalues μ_i and a set of normalized eigenstates χ_i . In the equal-width-parameter case where $\alpha_A = \alpha_B = \alpha$, some of the eigenvalues are equal to zero (see table 1) and the corresponding eigenstates will then be referred to as redundant or forbidden states (FS). When α_A is not equal to α_B , no forbidden states will exist; however, there appear now almost forbidden states (AFS) which are characterized by norm eigenvalues nearly equal to zero (see table 2) and which approach the forbidden states as α_A approaches α_B .

In this discussion,[+] we shall consider the general case of unequal width parameters, i.e., $\alpha_A \neq \alpha_B$. By means of the eigenvalues μ_i and the eigenvectors χ_i , one can express the norm operator $\mathcal{N}$, its square root $\mathcal{N}^{1/2}$ and the inverse $\mathcal{N}^{-1/2}$ as

$$\mathcal{N} = 1 - \sum_i (1-\mu_i)|\chi_i\rangle\langle\chi_i| , \tag{A.2}$$

$$\mathcal{N}^{1/2} = 1 - \sum_i (1-\sqrt{\mu_i})|\chi_i\rangle\langle\chi_i| , \tag{A.3}$$

$$\mathcal{N}^{-1/2} = 1 - \sum_i (1 - 1/\sqrt{\mu_i})|\chi_i\rangle\langle\chi_i| . \tag{A.4}$$

For these operators, it is easy to show that

$$\mathcal{N}^{1/2}\mathcal{N}^{-1/2} = \mathcal{N}^{-1/2}\mathcal{N}^{1/2} = 1 , \tag{A.5}$$

$$\mathcal{N}^{1/2}\mathcal{N}^{1/2} = \mathcal{N}, \tag{A.6}$$

$$\mathcal{N}^{-1/2}\mathcal{N} = \mathcal{N}\mathcal{N}^{-1/2} = \mathcal{N}^{1/2} . \tag{A.7}$$

The basic idea of the OCM is to divide the Hilbert space into two parts, the almost forbidden space and the physical space. For this division, one constructs a set of orthonormal states Ψ_i' given by

$$\Psi_i' = \frac{1}{\sqrt{N!\,\mu_i}} \, \mathcal{A}\left[\phi(A)\phi(B)\chi_i z\right] , \tag{A.8}$$

where χ_i represents the AFS, and another set of orthonormal states Ψ_i which is defined in the same manner except that χ_i is not the AFS. The almost forbidden and physical spaces are then considered as those which are spanned by Ψ_i' and Ψ_i , respectively.[++] In the OCM, the main purpose is to determine an effective

---

[+] A significant part of the discussion in this subsection is based on lectures given in 1978 by H. Horiuchi at the University of Minnesota (see, also, ref. [SA 73]).

[++] The symbol prime on Ψ_i' is used only to emphasize the fact that it belongs to the almost forbidden space. It is evident, of course, that Ψ_i' and Ψ_i are mutually orthogonal.

intercluster potential which is defined only in the physical space and which is simple enough but capable of reproducing the essential results of the RGM calculation.

The RGM equation (4.10) can be written as

$$E \mathcal{N} F = \hat{\mathcal{H}} F \, , \tag{A.9}$$

where

$$\hat{\mathcal{H}} = \mathcal{H} - (E_A + E_B)\mathcal{N} \, . \tag{A.10}$$

Operating on eq. (A.9) with $\mathcal{N}^{-1/2}$ then yields

$$E \mathcal{N}^{1/2} F = \mathcal{N}^{-1/2} \hat{\mathcal{H}} \mathcal{N}^{-1/2} \mathcal{N}^{1/2} F \, . \tag{A.11}$$

By defining

$$\hat{F} = \mathcal{N}^{1/2} F \, , \tag{A.12}$$

one obtains

$$E \hat{F} = \mathcal{N}^{-1/2} \hat{\mathcal{H}} \mathcal{N}^{-1/2} \hat{F} \, . \tag{A.13}$$

Because of the fact that μ_i approaches 1 as i becomes large, $\hat{F}$ and F have the same asymptotic forms and, thus, yield the same intercluster binding energies and scattering phase shifts.

Since the operation on χ_{AFS} by $\mathcal{N}^{1/2}$ and $\mathcal{N}^{-1/2}$ does not yield null states, it is useful to define two other operators $\tilde{\mathcal{N}}^{1/2}$ and $\tilde{\mathcal{N}}^{-1/2}$ as

$$\tilde{\mathcal{N}}^{1/2} = \tilde{\Lambda} - \sum_{i \neq AFS} (1 - \sqrt{\mu_i}) |\chi_i\rangle\langle\chi_i| \, , \tag{A.14}$$

$$\tilde{\mathcal{N}}^{-1/2} = \tilde{\Lambda} - \sum_{i \neq AFS} (1 - 1/\sqrt{\mu_i}) |\chi_i\rangle\langle\chi_i| \, , \tag{A.15}$$

where $\tilde{\Lambda}$ is a projection operator given by

$$\tilde{\Lambda} = 1 - \sum_{i = AFS} |\chi_i\rangle\langle\chi_i| \, . \tag{A.16}$$

For these operators, one sees easily that

$$\tilde{\mathcal{N}}^{1/2} \chi_{AFS} = \tilde{\mathcal{N}}^{-1/2} \chi_{AFS} = \tilde{\Lambda} \chi_{AFS} = 0 \, . \tag{A.17}$$

Now, by using the relation

$$\mathcal{N}^{-1/2} = \tilde{\mathcal{N}}^{-1/2} + \sum_{i=AFS} \frac{1}{\sqrt{\mu_i}} |\chi_i\rangle\langle\chi_i| \quad , \tag{A.18}$$

one can express eq. (A.13) in the form

$$E\hat{F} = \tilde{\mathcal{N}}^{-1/2} \hat{\mathcal{H}} \tilde{\mathcal{N}}^{-1/2} \hat{F}$$

$$+ \left[\tilde{\mathcal{N}}^{-1/2} \hat{\mathcal{H}} \sum_{i=AFS} \frac{1}{\sqrt{\mu_i}} |\chi_i\rangle\langle\chi_i| + \sum_{i=AFS} \frac{1}{\sqrt{\mu_i}} |\chi_i\rangle\langle\chi_i| \hat{\mathcal{H}} \tilde{\mathcal{N}}^{-1/2} \right.$$

$$\left. + \sum_{i,j=AFS} \frac{1}{\sqrt{\mu_i}} |\chi_i\rangle\langle\chi_i| \hat{\mathcal{H}} |\chi_j\rangle\langle\chi_j| \frac{1}{\sqrt{\mu_j}} \right] \hat{F} \quad . \tag{A.19}$$

The above equation can be simplified by defining a state W_i as

$$W_i = \frac{1}{\sqrt{\mu_i}} \hat{\mathcal{H}} \chi_i \quad , \tag{A.20}$$

which has the spatial representation

$$W_i(\vec{R}') = \frac{1}{\sqrt{\mu_i}} \langle \phi(A)\phi(B)\delta(\vec{R}-\vec{R}')z \mid H - (E_A + E_B) \mid$$

$$\mathcal{A}\left[\phi(A)\phi(B)\chi_i(\vec{R})z \right] \rangle$$

$$= \sqrt{N!} \langle \phi(A)\phi(B)\delta(\vec{R}-\vec{R}')z \mid H - (E_A + E_B) \mid \Psi_i' \rangle \tag{A.21}$$

Using W_i , one can then rewrite eq. (A.19) as follows:

$$E\hat{F} = \tilde{\mathcal{N}}^{-1/2} \hat{\mathcal{H}} \tilde{\mathcal{N}}^{-1/2} \hat{F}$$

$$+ \sum_{i=AFS} \left[\tilde{\mathcal{N}}^{-1/2} |W_i\rangle\langle\chi_i|\hat{F}\rangle + |\chi_i\rangle\langle W_i| \tilde{\mathcal{N}}^{-1/2}\hat{F}\rangle \right]$$

$$+ \sum_{i,j=AFS} |\chi_i\rangle\langle \Psi_i' | H - (E_A + E_B) | \Psi_j' \rangle\langle\chi_j|\hat{F}\rangle \quad . \tag{A.22}$$

Next, one operates on the above equation with $\tilde{\Lambda}$ and uses the fact that

$$\tilde{\Lambda}\tilde{\mathcal{N}}^{-1/2} = \tilde{\mathcal{N}}^{-1/2} \quad . \tag{A.23}$$

The result is

$$E \tilde{\Lambda} \hat{F} = \tilde{N}^{-1/2} \hat{H} \tilde{N}^{-1/2} \hat{F} + \sum_{i=AFS} \tilde{N}^{-1/2} |W_i\rangle\langle\chi_i|\hat{F}\rangle \ . \tag{A.24}$$

Also, one can obtain, by taking the scalar product of χ_i (i = AFS) with eq. (A.22), the following equation:

$$E\langle\chi_i|\hat{F}\rangle = \langle W_i|\tilde{N}^{-1/2}\hat{F}\rangle + \sum_{j=AFS}\langle\Psi_i'|H-(E_A+E_B)|\Psi_j'\rangle\langle\chi_j|\hat{F}\rangle \tag{A.25}$$

with i = AFS.

To solve eqs. (A.24) and (A.25), one first makes a unitary transformation to a new set of orthonormal vectors $\tilde{\Psi}_i'$ which span the almost forbidden space. That is, one writes

$$\tilde{\Psi}_i' = \sum_j C_{ij} \Psi_j' \tag{A.26}$$

and determines C_{ij} such that

$$\langle\tilde{\Psi}_i'|\tilde{\Psi}_j'\rangle = \delta_{ij} \ , \tag{A.27}$$

$$\langle\tilde{\Psi}_i'|H|\tilde{\Psi}_j'\rangle = \hat{E}_i \, \delta_{ij} \ . \tag{A.28}$$

By using these new basis vectors, eqs. (A.24) and (A.25) become

$$E\tilde{\Lambda}\hat{F} = \tilde{N}^{-1/2} \hat{H} \tilde{N}^{-1/2} \hat{F} + \sum_{i=AFS} \tilde{N}^{-1/2} |\tilde{W}_i\rangle\langle\tilde{\chi}_i|\hat{F}\rangle \tag{A.29}$$

and

$$E\langle\tilde{\chi}_i|\hat{F}\rangle = \langle\tilde{W}_i|\tilde{N}^{-1/2}\hat{F}\rangle + \left[\hat{E}_i - (E_A+E_B)\right]\langle\tilde{\chi}_i|\hat{F}\rangle \ , \tag{A.30}$$

where

$$\tilde{\chi}_i = \sum_j C_{ij} \chi_j \ , \tag{A.31}$$

$$\tilde{W}_i = \sum_j C_{ij} W_j \ . \tag{A.32}$$

From eq. (A.30), one obtains

$$\langle\tilde{\chi}_i|\hat{F}\rangle = \frac{\langle\tilde{W}_i|\tilde{N}^{-1/2}\hat{F}\rangle}{E_T - \hat{E}_i} \qquad (i = AFS) \ . \tag{A.33}$$

Substituting eq. (A.33) into eq. (A.29) then yields

$$E \tilde{\Lambda} \hat{F} = \tilde{\mathcal{N}}^{-1/2} \hat{\mathcal{H}} \tilde{\mathcal{N}}^{-1/2} \hat{F} + \sum_{i=AFS} \frac{\tilde{\mathcal{N}}^{-1/2} |\tilde{W}_i \rangle \langle \tilde{W}_i | \tilde{\mathcal{N}}^{-1/2}}{E_T - \hat{E}_i} \hat{F} . \qquad (A.34)$$

The second term on the right side of eq. (A.34) represents the coupling between the almost forbidden space and the physical space. This can be seen as follows. By noting that the quantity

$$\tilde{\mathcal{N}}^{-1/2} \hat{F} = \tilde{\mathcal{N}}^{-1/2} \mathcal{N}^{1/2} F = \tilde{\Lambda} F = F_p \qquad (A.35)$$

is a vector in the physical space and by using eq. (A.21), one finds that

$$\langle \tilde{W}_i | \tilde{\mathcal{N}}^{-1/2} \hat{F} \rangle = \int \tilde{W}_i^*(\vec{R}') F_p(\vec{R}') d\vec{R}'$$

$$= \frac{1}{\sqrt{N!}} \langle \tilde{\Psi}_i' | H | \mathcal{A} [\phi(A) \phi(B) F_p(\vec{R}) z(\vec{R}_{cm})] \rangle , \qquad (A.36)$$

which shows clearly the coupling of these two spaces through the action of the Hamiltonian operator H.

From eq. (A.34), one proceeds in exactly the same manner as described in Chapter 9 of ref. [WI 77]. As was discussed there, the presence of the coupling terms in this equation gives rise to resonance structures in the scattering phase shifts. For example, in the ${}^3\text{H} + \alpha$ case shown in fig. 5, such structures are indeed seen at relative energies higher than 20 MeV (see, also, refs. [BO 76, ME 75]). It should be mentioned, however, that these resonances exhibit themselves in such a clear manner only because the elastic channel alone is included in the calculation. In actuality, the coupling to inelastic and other reaction channels is generally so strong that resonance phenomena of this type cannot be observed in elastic-scattering experiments. For this latter reason, these resonances have frequently been labelled as spurious resonances (rather unfortunately, a misnomer) in the literature.

Even though the spurious resonance occurs generally in the high-excitation region [SA 73], it has a rather small level width. Qualitatively, this can be explained by noting that

$$y_i(\vec{R}') = \sqrt{N!} \langle \phi(A) \phi(B) \delta(\vec{R} - \vec{R}') z | \tilde{\Psi}_i' \rangle$$

$$= \sum_{j=AFS} C_{ij} \sqrt{\mu_j} \chi_j(\vec{R}') . \qquad (A.37)$$

Thus, the reduced width amplitude [HO 73, LA 60] for the almost forbidden state $\tilde{\Psi}_i'$

with respect to A + B clustering, given by the radial part of $\mathcal{Y}_c(\vec{R}')$ multiplied by R', becomes vanishingly small as α_A approaches α_B . This indicates that the AFS is coupled weakly to the elastic-scattering channel and, therefore, has a long lifetime when only such channel is included in the calculation.

Now, let us go back to eq. (A.34). We define first an intercluster interaction $\hat{V}$ by the equation

$$\hat{\mathcal{H}} = \mathcal{N}^{1/2}\,(T + \hat{V})\,\mathcal{N}^{1/2} \;, \tag{A.38}$$

where T is the kinetic-energy operator for the relative motion. Then, by using the relation

$$\tilde{\mathcal{N}}^{-1/2}\,\mathcal{N}^{1/2} = \mathcal{N}^{1/2}\,\tilde{\mathcal{N}}^{-1/2} = \tilde{\Lambda} \;, \tag{A.39}$$

one can further write eq. (A.34) as

$$\tilde{\Lambda}\,(E - T - \hat{V})\,\tilde{\Lambda}\,\hat{F} = \sum_{i=AFS} \frac{\tilde{\mathcal{N}}^{-1/2}\,|\,\tilde{W}_i\,\rangle\langle\,\tilde{W}_i\,|\,\tilde{\mathcal{N}}^{-1/2}}{E_T - \hat{E}_i}\,\hat{F} \;. \tag{A.40}$$

In the OCM, the approximations are (i) to omit the coupling terms on the right side of eq. (A.40), and (ii) to replace the complicated nonlocal interaction $\hat{V}$ by an effective local potential V_{eff} which is generally chosen to be energy-independent but may depend on the parity of the relative motion. Thus, the OCM equation of motion is[+]

$$\tilde{\Lambda}\,(E - T - V_{eff})\,\tilde{\Lambda}\,\hat{F} = 0 \;, \tag{A.41}$$

which is an integrodifferential equation for the renormalized function $\tilde{\Lambda}\hat{F} = \tilde{\mathcal{N}}^{1/2}F = \hat{F}_p$.

By the approximation of omitting coupling terms, spurious resonances do not appear in a single-channel OCM calculation. This is of course only a minor drawback, because these resonances are narrow and occur in the high-excitation region. In any case, it is known [CH 73a] that, even in a RGM calculation, such resonance behavior will no longer occur when absorption effects are taken into consideration by, e.g.,

---

[+]Since the operation on χ_{AFS} by $\hat{\mathcal{H}}$ does not yield a null state, one cannot define, in the general case of unequal width parameters, an intercluster interaction $\hat{V}$ by an equation similar to eq. (A.38) but with $\mathcal{N}^{1/2}$ replaced by $\tilde{\mathcal{N}}^{1/2}$ [see eq. (A.17)]. However, if one assumes

$$\hat{\mathcal{H}} \approx \tilde{\mathcal{N}}^{1/2}\,(T + \hat{V})\,\tilde{\mathcal{N}}^{1/2}$$

as an approximate relation, then one obtains the OCM equation (A.41) immediately from eq. (A.13) [note that $\tilde{\Lambda}^2 = \tilde{\Lambda}$]. But, this is a somewhat undesirable procedure, because our understanding about spurious resonances will be lost.

the addition of imaginary potentials into the formulation.

The approximation of replacing $\hat{V}$ by an effective local potential V_{eff} is potentially much more serious. As is well known, $\hat{\mathcal{N}}$ has a highly complicated nonlocal nature because of exchange contributions and the correct treatment of the total c.m. motion. Thus, it is indeed worrisome that, in all cases of interest, the major effects of antisymmetrization could be expected to be well approximated by the relatively simple procedure of using a local potential together with the adoption of the projection operator $\tilde{\Lambda}$ and the renormalized relative-motion wave function $\hat{F}_p$.

The OCM equation (A.41) may be expressed in another way. By operating with $\mathcal{N}^{1/2}$, one obtains

$$\tilde{\mathcal{N}}^{1/2} \left(E - T - V_{eff} \right) \tilde{\mathcal{N}}^{1/2} F = 0 . \tag{A.42}$$

This equation is of course completely equivalent to eq. (A.41), and we mention it here only for the sake of completeness [MA 73].

We should mention that, with a proper choice of V_{eff}, it may be a reasonable approximation to solve, instead of the OCM integrodifferential equation (A.41), but the simpler differential equation

$$\left(E - T - V_{eff} \right) \hat{F}_p = 0 . \tag{A.43}$$

This particular approximation has been named local-OCM by Baye [BA 78] and is expected to yield similar results as the OCM when the local, energy-independent V_{eff} is chosen to support a number of bound states, of which the lowest $\hat{n}$ states, with eigenvalues $\hat{\mathcal{E}}_i$ and eigenfunctions $\hat{\chi}_i$, are sufficiently similar to the $\hat{n}$ almost forbidden states ($\hat{n}$ is the total number of AFS). To show this, we write eq. (A.41) in the form

$$\left(E - T - V_{eff} \right) \hat{F}_p = -\sum_{i=AFS} |\chi_i\rangle\langle\chi_i| T + V_{eff} | \hat{F}_p \rangle . \tag{A.44}$$

Because of the choice of V_{eff} mentioned above, the right side of the above equation can be approximated as

$$-\sum_{i=AFS} |\chi_i\rangle\langle\chi_i| T + V_{eff} | \hat{F}_p \rangle \approx -\sum_{i=1}^{\hat{n}} |\hat{\chi}_i\rangle\langle\hat{\chi}_i| \hat{F}_p \rangle \hat{\mathcal{E}}_i . \tag{A.45}$$

Now, since V_{eff} is energy-independent, all the higher bound and scattering states are orthogonal to the $\hat{n}$ lowest bound states. Consequently, for these states, the right side of eq. (A.44) is nearly equal to zero and the local-OCM becomes a simple and useful model. It is clear of course that, because these $\hat{n}$ lowest states are similar to the AFS, they have no physical significance and, hence, must be disregarded from any consideration concerning the behavior of the compound system.

The local OCM has been successfully applied to the cases of $\alpha + {}^{3}\mathrm{He}$ and $\alpha + \alpha$ scattering [BU 77]. In addition, it has been used to study light-ion clustering states in nuclei such as ${}^{15}\mathrm{N}$, ${}^{16}\mathrm{O}$, ${}^{18}\mathrm{O}$, ${}^{19}\mathrm{F}$, ${}^{20}\mathrm{Ne}$ and ${}^{24}\mathrm{Mg}$ [BU 75, BU 77a, PI 78]. In all these cases, a good agreement with experiment has been obtained, indicating that the local-OCM is rather useful to achieve an understanding of the level structures in light systems.

The effective potential V_{eff} used in the local-OCM is a deep folding-type potential. Thus, the success in the above-mentioned calculations provides some justification for the use of potentials of this kind in optical-model analyses of light-ion (p,d, ${}^{3}\mathrm{He}$, and α) scattering by light nuclei. We must emphasize, however, that the validity of the local-OCM depends on the basic OCM approximation of replacing $\hat{V}$ by an effective local potential. As will be discussed in the next subsection, there is some indication that in heavy-ion scattering, the adoption of folding-type potentials may be rather inappropriate.

For the sake of completeness, we briefly discuss the special case where the width parameters α_A and α_B are exactly equal.[+] In this case, one defines an operator

$$\mathcal{N}_e^{-1/2} = \Lambda - \sum_{i \neq FS} \left(1 - \frac{1}{\sqrt{\mu_i}} \right) |x_i\rangle\langle x_i| \ , \tag{A.46}$$

where

$$\Lambda = 1 - \sum_{i=FS} |x_i\rangle\langle x_i| \ . \tag{A.47}$$

For this operator, one can easily prove the following relations:

$$\mathcal{N}^{1/2} \mathcal{N}_e^{-1/2} = \mathcal{N}_e^{-1/2} \mathcal{N}^{1/2} = \Lambda \ , \tag{A.48}$$

$$\mathcal{N}_e^{-1/2} \mathcal{N} = \mathcal{N}\mathcal{N}_e^{-1/2} = \mathcal{N}^{1/2} \ , \tag{A.49}$$

where $\mathcal{N}$ and $\mathcal{N}^{1/2}$ are given by eqs. (A.2) and (A.3), respectively. Now, by operating on eq. (A.9) with $\mathcal{N}_e^{-1/2}$ and using the fact that

$$\Lambda \hat{\mathcal{H}} = \hat{\mathcal{H}} \Lambda = \hat{\mathcal{H}} \ , \tag{A.50}$$

one obtains

$$E\hat{F} = \mathcal{N}_e^{-1/2} \hat{\mathcal{H}} \mathcal{N}_e^{-1/2} \hat{F} \ , \tag{A.51}$$

[+] It should be noted that certain aspects in the $\alpha_A = \alpha_B$ case cannot be explained by simply considering the unequal-width-parameter case and letting α_A approach α_B .

with $\hat{F}$ defined by eq. (A.12). Substituting eq. (A.38) into eq. (A.51) then yields

$$\Lambda(E-T-\hat{V})\Lambda\hat{F} = 0.$$
(A.52)

Therefore, in this very special case of equal width parameters, the OCM approximation consists only in replacing $\hat{V}$ by an effective local potential V_{eff}. In other words, the OCM equation is

$$\Lambda(E-T-V_{eff})\Lambda\hat{F} = 0.$$
(A.53)

Similar to the unequal-width-parameter case, one can also express the above equation in an alternative but equivalent way. This is achieved by operating on eq. (A.53) with $\mathcal{N}^{1/2}$ and using the relation

$$\Lambda\mathcal{N}^{1/2} = \mathcal{N}^{1/2}\Lambda = \mathcal{N}^{1/2}.$$
(A.54)

The result is

$$\mathcal{N}^{1/2}(E-T-V_{eff})\mathcal{N}^{1/2}F = 0.$$
(A.55)

In addition, it can be easily shown that, in this special case,

$$\Lambda\mathcal{N}\Lambda = \mathcal{N}, \qquad \Lambda\hat{\mathcal{H}}\Lambda = \hat{\mathcal{H}}.$$
(A.56)

Thus, the RGM equation (A.9) can be equivalently written as

$$\Lambda(\hat{\mathcal{H}} - E\mathcal{N})\Lambda F = 0.$$
(A.57)

This particular form will be useful in our consideration to be given in the next subsection.

Generalizations of the OCM to coupled-channel and multi-cluster problems have been discussed and applied to many light systems [FU 79, FU 79a, HO 78, KA 79, NI 79, SA 79]. The effective local interactions used in these calculations are generally chosen to be the direct potentials or a slight modification of such potentials. In view of the complex relation satisfied by the effective interaction as expressed by eq. (A.38), this assumption of adopting direct potentials may be potentially troublesome. Even so, however, it is our opinion that, considering the complicated nature of these many-nucleon problems, the use of the OCM may still be a useful initial step to explore the cluster structures of light nuclear systems.

A.2. Effective potentials in the OCM

Because the interaction $\hat{V}$ is defined by eq. (A.38) which involves the complicated RGM kernels $\hat{\mathcal{H}}$ and $\mathcal{N}$, it is expected that to find an effective OCM potential V_{eff}, which could reasonably represent $\hat{V}$ and thus the RGM result, may not be a trivial procedure. In this subsection, we shall, therefore, consider a simple example of the dineutron-plus-dineutron system where $\hat{V}$ can be exactly determined. This will then enable us to examine the basic features of $\hat{V}$ and, consequently, acquire some general feeling about the required properties of V_{eff}. Also, we shall describe here a procedure, proposed by Friedrich and Canto [FR 77, FR 78], in which V_{eff} is derived by making use of the energy surface obtained in a GCM calculation. Although this procedure is presently rather crude, it is simple enough and can be further refined to possibly become an important part within the OCM framework.

A.2a. A model example of two dineutrons.

We consider here the model example of two dineutrons, originally studied by Kukulin et al. [KU 75] (see also, ref. [SA 77]). As has been mentioned, the main purpose is to achieve some understanding of the features of the interaction $\hat{V}$. Thus, we choose the internal spatial function for the dineutron as simple as possible, namely, a translationally-invariant shell-model function of $(1s)^2$ configuration in a harmonic oscillator well of width parameter α .

The overlap kernel $\mathcal{N}$ can be easily derived [CH 72, KU 75, TH 70]. It is given by

$$\mathcal{N} = 1 - |u_0\rangle\langle u_0| \,, \tag{A.58}$$

where u_0 is the lowest state in a harmonic oscillator well with width parameter also equal to α and is the forbidden state in this system. Thus, in this very simple case where μ_i is equal to either 0 or 1, the relation[+]

$$\mathcal{N} = \Lambda = \mathcal{N}^{1/2} \tag{A.59}$$

holds, and the RGM equation may be written as [see eq. (A.57)]

$$\mathcal{N}^{1/2}(\hat{\mathcal{H}} - E\mathcal{N})\mathcal{N}^{1/2} F = 0. \tag{A.60}$$

Using the expressions derived for the RGM kernels $\hat{\mathcal{H}}$ and $\mathcal{N}$, one may write eq. (A.60) more explicitly in the form

---

[+]The two-dineutron case is a very special one. In general, eq. (A.59) does not hold.

$$\mathcal{N}^{1/2}\left(E - T - V_D - V_{ex}\right)\mathcal{N}^{1/2} F = 0 , \tag{A.61}$$

where V_D is the direct part given by

$$V_D(\vec{R}',\vec{R}'') = -V_0\left(4w - 2m + 2b - 4h\right)\left(\frac{\alpha}{\alpha+K}\right)^{3/2}$$

$$\times \exp\left(-\frac{\alpha K}{\alpha+K}R'^2\right)\delta(\vec{R}'-\vec{R}'') \tag{A.62}$$

and V_{ex} is the exchange part given by

$$V_{ex}(\vec{R}',\vec{R}'') = -V_0\left(-2w + 4m - 4b + 2h\right)\left(\frac{\alpha}{\pi}\right)^{3/2}$$

$$\times \exp\left[-\left(\frac{\alpha}{2}+K\right)(\vec{R}'^2+\vec{R}''^2) + 2K\vec{R}'\cdot\vec{R}''\right]. \tag{A.63}$$

In arriving at the above equations, we have adopted the nucleon-nucleon potential of eq. (4.38). Also, we have used the fact that, because of the presence of the operators $\mathcal{N}^{1/2}$ in eq. (A.61), the potential

$$\hat{V} = V_D + V_{ex} \tag{A.64}$$

is not unique and several terms in $\hat{\mathcal{H}}$ and $\mathcal{N}$ have been eliminated to result in a rather simple expression for V_{ex}.

From eqs. (A.63) and (A.64), it is seen that $\hat{V}$ is, in general, a nonlocal interaction. On the other hand, one notes that in the long-range limit where $K/\alpha \ll 1$, the effect of the term $\mathcal{N}^{1/2} V_{ex} \mathcal{N}^{1/2}$ is vanishingly small and $\hat{V}$ is close to V_D. For the other limit where $K/\alpha \gg 1$, V_{ex} becomes almost local, i.e.,

$$V_{ex}(\vec{R}',\vec{R}'') \approx -V_0\left(-2w + 4m - 4b + 2h\right)\left(\frac{\alpha}{K}\right)^{3/2}$$

$$\times \exp(-\alpha R'^2)\delta(\vec{R}'-\vec{R}'') , \tag{A.65}$$

and

$$\hat{V}(\vec{R}',\vec{R}'') \approx -V_0\left(2w + 2m - 2b - 2h\right)\left(\frac{\alpha}{K}\right)^{3/2}$$

$$\times \exp(-\alpha R'^2)\delta(\vec{R}'-\vec{R}'') . \tag{A.66}$$

Thus, in this short-range limit, $\hat{V}$ is again almost a local potential and has the same form as V_D. However, $\hat{V}$ and V_D can have quite different strength; e.g., assuming a reasonable choice of a Serber exchange mixture in the nucleon-nucleon potential with $w = m = 0.5$ and $b = h = 0$, one obtains

$$\hat{V} = 2\,V_D \tag{A.67}$$

in this limit. Therefore, this model example serves to show that the effective potential V_{eff} in the OCM equation (A.53) or (A.55) should generally be considered as a phenomenological entity and a simple choice for it as being just the direct potential V_D will, in many cases, be inappropriate from a quantitative viewpoint.

The fact that, in the two-dineutron case, $\hat{V}$ has the same form as V_D at both short-range and long-range limits is an interesting finding. This suggests that one might appropriately choose V_{eff} to be simply V_D multiplied by an adjustable constant. We must emphasize, however, that this particular parametrization for V_{eff} may be reasonable only in very light systems. In the $^{16}O + ^{16}O$ case to be discussed below, it will be noted that V_D and V_{eff} may indeed have quite different shapes.

A.2b. <u>Derivation of the effective potential from GCM matrix elements.</u>

By using eq. (5.28), one can further write eq. (A.38) in the equal-width-parameter case as

$$\tilde{\mathcal{H}} - (E_A + E_B)\,\tilde{\mathcal{N}} = \tilde{\Gamma}_e\,\mathcal{N}^{1/2}\,(T + \hat{V})\,\mathcal{N}^{1/2}\,\tilde{\Gamma}_e \tag{A.68}$$

which, in the generator-coordinate represention, has the form

$$\tilde{\mathcal{H}}(\vec{X}', \vec{X}'') - (E_A + E_B)\,\tilde{\mathcal{N}}(\vec{X}', \vec{X}'')$$

$$= \int \tilde{\Gamma}_e(\vec{R}', \vec{X}')\,\langle \vec{R}' \,|\, \mathcal{N}^{1/2}\,(T + \hat{V})\,\mathcal{N}^{1/2} \,|\, \vec{R}''\rangle\,\tilde{\Gamma}_e(\vec{R}'', \vec{X}'')\,d\vec{R}'d\vec{R}'' \,. \tag{A.69}$$

The advantage of this equation is that, especially for equal width parameters, the GCM kernels $\tilde{\mathcal{H}}(\vec{X}', \vec{X}'')$ and $\tilde{\mathcal{N}}(\vec{X}', \vec{X}'')$, given by eq. (5.23), can be rather straight-forwardly evaluated.

In the OCM, $\hat{V}$ is replaced by V_{eff}. If one now makes a simple ansatz for V_{eff} containing a number of adjustable parameters, then eq. (A.69) may be used to determine these parameters by considering appropriate GCM matrix elements.

The procedure suggested by Friedrich and Canto [FR 77] is to assume a local potential V_{eff} of the form

$$V_{eff}(\vec{R}', \vec{R}'') = \left[\sum_{i=1}^{m} V_i\,exp(-\beta_i R'^2) \right] \delta(\vec{R}' - \vec{R}'') \,. \tag{A.70}$$

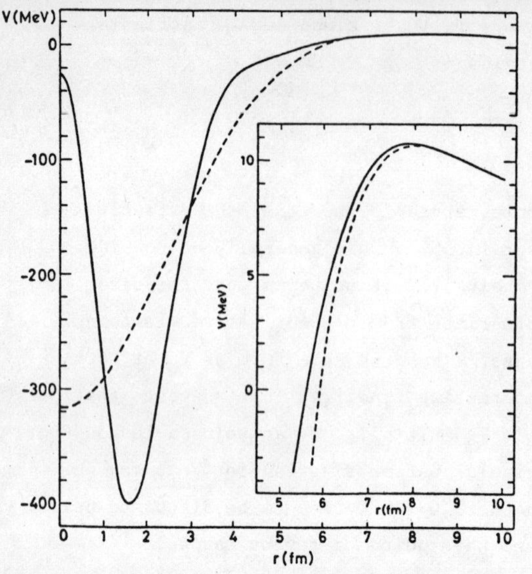

Fig. A1. Potentials V_{eff}(solid curve) and V_D(dashed curve) in the $^{16}O + ^{16}O$ case. (From ref.[FR 77])

The parameters V_i and β_i are then adjusted such that the quantity

$$I(\vec{X}') = \frac{1}{\tilde{\mathcal{N}}(\vec{X}',\vec{X}')} \int \tilde{\Gamma}_e(\vec{R}',\vec{X}') \langle \vec{R}' | \mathcal{N}^{1/2}(T + V_{eff}) \mathcal{N}^{1/2} | \vec{R}'' \rangle$$

$$\times \tilde{\Gamma}_e(\vec{R}'',\vec{X}') d\vec{R}' d\vec{R}'' \tag{A.71}$$

fits as well as possible the energy surface

$$\hat{E}(\vec{X}') = \frac{\tilde{\mathcal{H}}(\vec{X}',\vec{X}')}{\tilde{\mathcal{N}}(\vec{X}',\vec{X}')} - (E_A + E_B) \tag{A.72}$$

obtained with the diagonal elements of the GCM kernels.

In the $\alpha + \alpha$ case, the phase shifts obtained by solving the OCM equation with V_{eff} determined by the procedure described above agree well with those calculated by the GCM or RGM [FR 77]. Comparing with the direct potential, V_{eff} was found to be somewhat stronger, especially at small distances. The basic features of these two potentials are, however, rather similar, thus strengthening our belief that, in very light systems, a simple assumption of V_{eff} being proportional to V_D may likely lead to quantitatively useful results.

The situation in the $^{16}O + ^{16}O$ case is vastly different. In fig. A1, we show a comparison between V_{eff} and V_D [FR 77]. Here one sees that, although these two potentials are rather similar in the long-range part, they have quite different behavior in the region of strong interaction; in particular, V_{eff} exhibits a repulsive core at small distances. This indicates therefore that, in heavy-ion scattering, the choice of V_{eff} is not at all simple and the OCM should be considered only as a refined optical model which may be adopted to fit experimental data through the adjustment of a sufficient number of phenomenological parameters.

An extension of the above-described procedure has been made by Friedrich [FR 78] to determine parity-dependent effective potentials from parity-projected energy surfaces. In the $n + \alpha$ and $^3H + \alpha$ cases, the use of the resultant potentials in the OCM was found to be able to reproduce reasonably well the phase-shift results of the corresponding RGM calculations.

A.3. Comparison with RGM and experimental results

In fig. A2, we show a comparison between RGM and OCM results for some low-lying levels in ^{12}C treated as a 3α system [FU 78]. The effective potential used in the OCM is a sum of $\alpha + \alpha$ direct potentials. Considering the fact that the OCM calculation is substantially simpler than the RGM calculation, one sees from this figure that the agreement can be viewed as reasonably satisfactory. For the positive-parity levels, the OCM result can be somewhat improved if the effective potential is chosen to be slightly more attractive. This is consistent with the $\alpha + \alpha$ scattering calculation described in subsect. A.2b, where a similar finding concerning the

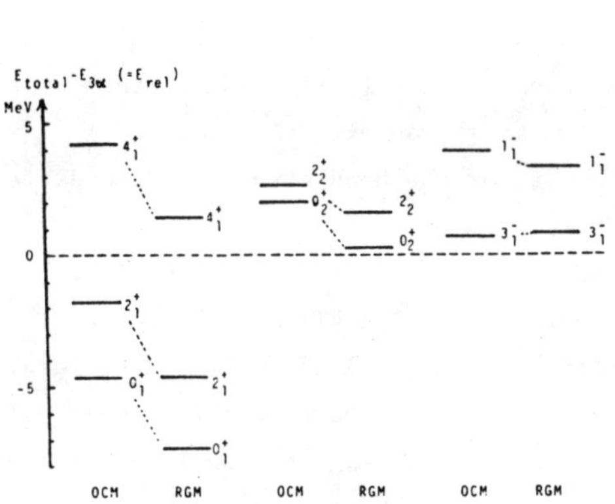

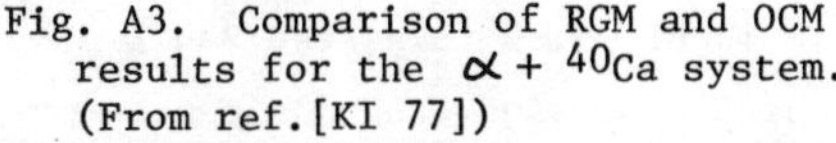

Fig. A2. Comparison of RGM and OCM results for the 3α system. (From ref. [FU 78])

Fig. A3. Comparison of RGM and OCM results for the $\alpha + {}^{40}Ca$ system. (From ref. [KI 77])

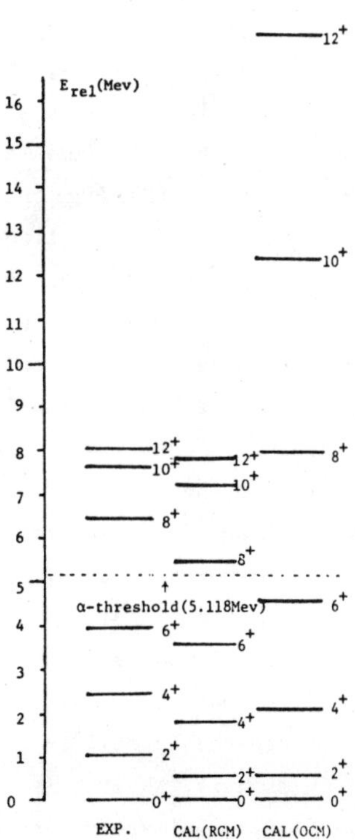

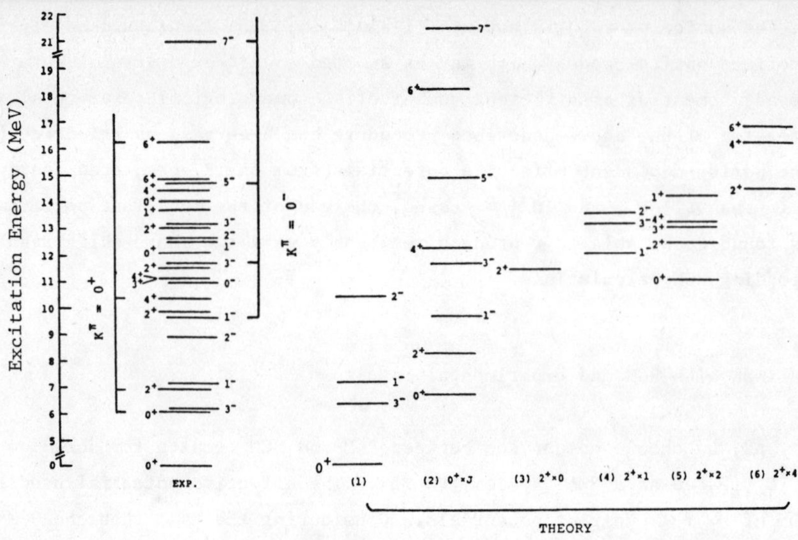

Fig. A4. α + ^{12}C coupled-channel OCM calculation of the energy levels in ^{16}O. (From ref.[HO 78a])

effective potential has also been obtained.

The satisfactory comparison obtained in the 3α system is possibly related to the fact that very light clusters are involved. When one considers a system which contains heavier clusters, the discussion of subsect. A.2b gives some indication that OCM calculations with simple choices for V_{eff} may no longer be sufficiently accurate. That this seems indeed to be the case is shown in fig. A3 for the α + ^{40}Ca system [KI 77]. Here it is found that, with an effective potential assumed to be proportional to V_D, the OCM result for the band structure does differ quite appreciably from the corresponding RGM result.

One of the notable successes for the OCM approach was the coupled-channel α + ^{12}C cluster-model calculation for ^{16}O performed by Suzuki [SU 76], where internal excitations of the ^{12}C cluster to its 2^+ (4.44 MeV) and 4^+ (14 MeV) levels are allowed to couple with the α + ^{12}C relative motion. The result for the level structure in the low-excitation region of ^{16}O is given in fig. A4, where experimental data are also shown. As is seen, the agreement between theory and experiment is rather remarkable. Suzuki has further used the resultant wave functions to compute electromagnetic transition rates (see Table A1) and α-decay reduced widths, and found generally satisfactory agreement with experiment. In addition, it is worth mentioning that, with this simple model, the low-energy α + ^{12}C scattering data can also be reasonably well explained [SU 77].

In summary, the OCM is a semi-microscopic model which takes the Pauli principle approximately into account. As has been discussed, the basic assumption of

Table A1

Electric transition probabilities in ^{16}O

(From ref.[HO78a])

Transition	Exp.	Cal.
Monopole matrix element (fm²)		
$0_1^+(0) \rightarrow 0_2^+(6.05)$	3.66 ± 0.55	3.88
$0_1^+(0) \rightarrow 0_3^+(12.05)$	4.40 ± 0.44	3.50
B(E2) (e² fm⁴)		
$2_1^+(6.92) \rightarrow 0_1^+(0)$	8.6 ± 0.4	2.48
$2_1^+(6.92) \rightarrow 0_2^+(6.05)$	40 ± 15	60.1
$1_1^-(7.12) \rightarrow 3_1^-(6.13)$	69 ± 27	27.6
$2_1^-(8.87) \rightarrow 3_1^-(6.13)$	14 ± 4	9.74
$2_1^-(8.87) \rightarrow 1_1^-(7.12)$	36 ± 5	17.5
$2_2^+(9.85) \rightarrow 0_1^+(0)$	0.082 ± 0.007	0.489
$2_2^+(9.85) \rightarrow 0_2^+(6.05)$	3.0 ± 0.7	4.64
$4_1^+(10.35) \rightarrow 2_1^+(6.92)$	112 ± 12	96.2
$3_1^+(11.08) \rightarrow 2_1^+(6.92)$	40	2.43
$2_3^+(11.52) \rightarrow 0_1^+(0)$	4.4	1.43
$2_3^+(11.52) \rightarrow 0_2^+(6.05)$	7.6 ± 2	1.38
B(E3) (e² fm⁶)		
$3_1^-(6.13) \rightarrow 0_1^+(0)$	230 ± 20	29.6

this model lies in the replacement of the complicated nonlocal interaction $\widehat{V}$, defined by eq. (A.38), by an effective _local_ potential V_{eff}. This replacement seems to be reasonably valid when the system under consideration consists of very light clusters. On the other hand, there is some indications that, when relatively heavy clusters are involved, this assumption may not be well established.[+] In this latter situation, there is in fact the additional problem that, even if it is a reasonable approximation to replace $\widehat{V}$ by a local V_{eff}, one must still have some rather detailed understanding concerning the properties of this effective potential. Until now, such an understanding has not been achieved. Therefore, the OCM should be considered only as an interesting phenomenological model at the present moment. For the establishment of this model as a simplified substitute for the RGM, systematic investigations must still be carried out to examine the validity of the above-mentioned assumption and to determine the main characteristics of the effective potential.

[+] In the equal-width-parameter case, it is interesting to note that, for very light systems, the norm eigenvalues μ_i for non-forbidden states are all rather close to 1. For relatively heavy systems, on the other hand, there are a number of non-forbidden norm eigenvalues which are close to zero (see, e.g., fig. 3 of ref. [BA 78]).

References

[AB 72]   Abulaffio, C. and J. M. Irvine, Phys. Lett. $\underline{38B}$ (1972) 492 (1).
[AB 72a] Abramowitz, M. and I. A. Stegun (editors), Handbook of Mathematical Functions (National Bureau of Standards, 1972) (5.4).
[AJ 80]   Ajzenberg-Selove, F. and C. L. Busch, Nucl. Phys. $\underline{A336}$ (1980) 1 (6.1).
[AN 75]   Ando, T., K. Ikeda, and Y. Suzuki, in ref. [KA 75], p. 192 (6.2d).
[AN 79]   Ando, T., K. Ikeda, and A. Tohsaki-Suzuki, Prog. Theor. Phys. $\underline{61}$ (1979) 101 (6.2e).
[AR 72]   Arima, A., H. Horiuchi, K. Kubodera, and N. Takigawa, in Advances in Nuclear Physics, edited by M. Baranger and E. Vogt (Plenum, New York, 1972), vol. 5, p. 345 (1).
[AU 65]   Austern, N., Phys. Rev. $\underline{137}$ (1965) B752 (4.2a).
[BA 58]   Bayman, B. F. and A. Bohr, Nucl. Phys. $\underline{9}$ (1958/59) 596 (1).
[BA 71]   Baker, S. D., T. A. Cahill, P. Catillon, J. Durnad, and D. Garreta, Nucl. Phys. $\underline{A160}$ (1971) 428 (6.2a).
[BA 71a] Barnes, C. A., in Advances in Nucl. Phys. $\underline{4}$ (1971) 133 (8.1b).
[BA 72]   Bachelier, D., M. Bernas, J. L. Boyard, H. L. Harney, J. C. Jourdain, P. Radvanyi, M. Roy-Stephan, and R. DeVries, Nucl. Phys. $\underline{A195}$ (1972) 361 (6.2c, 7.1c).
[BA 74]   Baye, D. and P. H. Heenen, Nucl. Phys. $\underline{A233}$ (1974) 304 (5.4).
[BA 76]   Baye, D., Nucl. Phys. $\underline{A272}$ (1976) 445 (7.1c).
[BA 77]   Baye, D. and P. H. Heenen, Nucl. Phys. $\underline{A283}$ (1977) 176, $\underline{A276}$ (1977) 354 (5.1).
[BA 77a] Baye, D. and P. H. Heenen, in ref. [MI 77], p. 1 (5.4).
[BA 77b] Baye, D., P. H. Heenen, and M. Libert-Heinemann, Nucl. Phys. $\underline{A291}$ (1977) 230 (5.4).
[BA 77c] Baye, D., J. Deenan, and Y. Salmon, Nucl. Phys. $\underline{A289}$ (1977) 511 (7.1c).
[BA 78]   Baye, D., in Proc. of 2nd Louvain-Krakow Seminar on the Alpha-Nucleus Interaction, edited by G. Gregoire and K. Grotowski, Louvain-la-Neuve, Belgium (1978), p. 330 (A.1, A.3).
[BA 78a] Baye, D., P. H. Heenen, and M. Libert-Heinemann, Nucl. Phys. $\underline{A308}$ (1978) 229 (6.3).
[BA 79]   Baye, D. and Y. Salmon, Nucl. Phys. $\underline{A323}$ (1979) 521 (5.1).
[BA 79a] Baye, D. and Y. Salmon, Nucl. Phys. $\underline{A331}$ (1979) 254 (5.1).
[BA 80]   Bauhoff, W., H. Schultheis, and R. Schultheis, Phys. Rev. $\underline{C22}$ (1980) 861 (1).
[BA 80a] Baye, D., in Proc. Intern. Conf. on the Resonant Behaviour of Heavy-Ion Systems, Aegean Sea, Greece (1980) (1, 5.1, 6.4).
[BA 80b] Bahcall, J. N., S. H. Lubow, W. F. Huebner, N. H. Magee, Jr., A. L. Merts, M. F. Argo, P. D. Parker, B. Rozsnyai, and R. K. Ulrich, Phys. Rev. Lett. $\underline{45}$ (1980) 945 (8.1b).
[BA 81]   Baldock, R. A., B. A. Robson, and R. F. Barrett, Nucl. Phys. $\underline{A351}$ (1981) 157 (7.1).
[BE 69]   Beck, E.R.A. (editor), Proc. Intern. Conf. on Clustering Phenomena in Nuclei, Bochum, Germany, 1969 (International Atomic Energy Agency, 1969) (1).
[BE 75]   Beck, R., J. Borysowicz, D. M. Brink, and M. V. Mihailović, Nucl. Phys. $\underline{A244}$ (1975) 45, 58 (5.4).
[BO 36]   Bohr, N., Nature $\underline{137}$ (1936) 344 (1).
[BO 62]   Bouten, M., Nuovo Cimento $\underline{26}$ (1962) 63 (1).
[BO 76]   Böinghoff, A., H. Husken, and A. Weiguny, Phys. Lett. $\underline{61B}$ (1976) 9 (A.1).
[BR 57]   Brink, D. M., Nucl. Phys. $\underline{4}$ (1957) 215 (1).
[BR 66]   Brink, D. M., in Many-Body Description of Nuclear Structure and Reactions, Proc. of the Intern. School of Physics, "Enrico Fermi", Course 36, Varenna, edited by C. Bloch (Academic, New York, 1966), p. 247 (1, 2.2, 5.1, 5.2).
[BR 68]   Brink, D. M. and A. Weiguny, Nucl. Phys. $\underline{A120}$ (1968) 59 (1).
[BR 68a] Brown, R. E. and Y. C. Tang, Phys. Rev. $\underline{176}$ (1968) 1235 (6.2b).
[BR 70]   Brink, D. M., H. Friedrich, A. Weiguny, and C. W. Wong, Phys. Lett. $\underline{33B}$ (1970) 143 (1).
[BR 71]   Brown, R. E. and Y. C. Tang, Nucl. Phys. $\underline{A170}$ (1971) 225 (4.1, 6.4).

[BR 77] Brink, D. M. and N. Takigawa, Nucl. Phys. A279 (1977) 159 (6.2d).
[BR 78] Brink, D. M., J. Grabowski, and E. Vogt, Nucl. Phys. A309 (1978) 359 (6.2d).
[BU 72] Bumiller, F. A., F. R. Buskirk, J. N. Dyer, and W. A. Monson, Phys. Rev. C5 (1972) 391 (8.1a).
[BU 75] Buck, B., C. B. Dover, and J. P. Vary, Phys. Rev. C11 (1975) 1803 (A.1).
[BU 77] Buck, B., H. Friedrich, and C. Wheatley, Nucl. Phys. A275 (1977) 246 (A.1).
[BU 77a] Buck, B., in ref. [MI 77], p. 215 (A.1).
[CA 77] Canto, L. F. and D. M. Brink, Nucl. Phys. A279 (1977) 85 (5.4).
[CH 67] Chevallier, P., F. Scheibling, G. Goldring, I. Plesser, and M. W. Sachs, Phys. Rev. 160 (1967) 827 (2.1).
[CH 72] Chwieroth, F. S., Y. C. Tang, and D. R. Thompson, Nucl. Phys. A189 (1972) 1 (4.4, A.2a).
[CH 73] Chwieroth, F. S., Ph.D. thesis, University of Minnesota (1973) (1).
[CH 73a] Chwieroth, F. S., Y. C. Tang, and D. R. Thompson, Phys. Lett. 46B (1973) 301 (4.4).
[CH 73b] Chwieroth, F. S., R. E. Brown, Y. C. Tang, and D. R. Thompson, Phys. Rev. C8 (1973) 938 (6.3).
[CH 74] Chwieroth, F. S., Y. C. Tang, and D. R. Thompson, Phys. Rev. C9 (1974) 56 (4.1, 6.3).
[CH 74a] Chien, W. S. and R. E. Brown, Phys. Rev. C10 (1974) 1767 (6.2b).
[CH 77] Chwieroth, F. S., Y. C. Tang, and D. R. Thompson, Fizika (Suppl. 2) 9 (1977) 13 (7.2).
[CL 64] Clegg, T. B., A. C. L. Barnard, J. B. Swint, and J. L. Weil, Nucl. Phys. 50 (1964) 621 (6.2a).
[CL 75] Clement, D. and E. Schmid, in ref. [GO 75], p. 148 (4.4).
[DE 54] Dennison, D. M., Phys. Rev. 96 (1954) 378 (1).
[DE 63] DeShalit, A. and I. Talmi, Nuclear Shell Theory (Academic, New York, 1963) (2.1).
[DE 72] De Takacsy, Phys. Rev. C5 (1972) 1883 (1, 5.2, 5.4).
[DE 78] Delbar, Th., Gh. Gregoire, G. Paic, R. Ceuleneer, F. Michel, R. Vanderpoorten, A. Budzanowski, H. Dabrowski, L. Freindl, K. Grotowski, S. Micek, R. Planeta, A. Strzalkowski, and K. A. Eberhard, Phys. Rev. C18 (1978) 1237 (6.2d).
[EB 79] Eberhard, K. A., Ch. Appel, R. Bangert, L. Cleemann, J. Eberth, and V. Zobel, Phys. Rev. Lett. 43 (1979) 107 (6.2d).
[EC 75] Eck, J. S., W. J. Thompson, K. A. Eberhard, J. Schiele, and W. Trombik, Nucl. Phys. A255 (1975) 157 (6.2d).
[ED 60] Edmonds, A. R., Angular Momentum in Quantum Mechanics (Princeton University Press, Princeton, New Jersey, 1960) (5.1).
[EG 80] Egelhof, P., R. Böttger, W. Dreves, K. H. Möbius, F. Roesel, E. Steffens, G. Tungate, P. Zupranski, and D. Fick, in Proc. Intern. Conf. on Nuclear Physics (Abstracts), Berkeley, California (1980), p. 105 (8.1a).
[EL 55] Elliott, J. P. and T.H.R. Skyrme, Proc. Roy. Soc. (London) A232 (1955) 561 (1, 2.1).
[EL 57] Elliott, J. P. and B. H. Flowers, Proc. Roy. Soc. (London) A242 (1957) 57 (2.3).
[FI 67] Fiedelday, H., Nucl. Phys. A96 (1967) 463 (4.2a).
[FL 75] Fliessbach, T., Z. Phys. A272 (1975) 39 (1).
[FL 76] Fliessbach, T., Z. Phys. A277 (1976) 151 (1).
[FL 77] Fliessbach, T. and P. Manakos, J. Phys. G3 (1977) 643 (4.4).
[FR 57] Frahn, W. E. and R. H. Lemmer, Nuovo Cimento 5 (1957) 523, 1564 (4.2a).
[FR 71] Friedrich, H. and A. Weiguny, Phys. Lett. 35B (1971) 105 (1).
[FR 72] Friedrich, H., H. Hüsken, and A. Weiguny, Phys. Lett. 38B (1972) 199 (1).
[FR 74] Friedrich, H., H. Hüsken, and A. Weiguny, Nucl. Phys. A220 (1974) 125 (5.4).
[FR 75] Friedrich, H. and K. Langanke, Nucl. Phys. A252 (1975) 47 (5.1).
[FR 75a] Friedrich, H., K. Langanke, A. Weiguny, and R. Santo, Phys. Lett. 55B (1975) 345 (5.1).
[FR 77] Friedrich, H. and L. F. Canto, Nucl. Phys. A291 (1977) 249 (A.2, A.2b).
[FR 78] Friedrich, H., Nucl. Phys. A294 (1978) 81 (A.2, A.2b).

[FU 75] Fujii, K., J. Schimizu, K. Takimoto, Y. Kondō, S. Ohkubo, and J. Muto, in ref. [KA 75], p. 711 (7.1c).

[FU 76] Furber, R. D., Ph.D. thesis, University of Minnesota (1976) (6.2c, 8.1a).

[FU 76a] Fukushima, Y., M. Kamimura, and T. Matsuse, Prog. Theor. Phys. $\underline{55}$ (1976) 1310 (8.2).

[FU 78] Fukushima, Y. and M. Kamimura, J. Phys. Soc. Japan (Suppl.) $\underline{44}$ (1978) 225 (2.3, 4.2a, 6.1, 8.2, A.3).

[FU 79] Fujiwara, Y., H. Horiuchi, and R. Tamagaki, Prog. Theor. Phys. $\underline{61}$ (1979) 1629 (A.1).

[FU 79a] Fujiwara, Y., Prog. Theor. Phys. $\underline{62}$ (1979) 122, 138 (A.1).

[FU 79b] Furutani, H., H. Horiuchi, and R. Tamagaki, Prog. Theor. Phys. $\underline{62}$ (1979) 981 (6.2a).

[GA 69] Gambarini, G., I. Iori, S. Micheletti, N. Molho, M. Pignanelli, and G. Tagliaferri, Nucl. Phys. $\underline{A126}$ (1969) 562 (7.2).

[GI 71] Giraud, B. and D. Zaikine, Phys. Lett. $\underline{37B}$ (1971) 25 (5.2).

[GI 73] Giraud, B., J. C. Hocquenghem, and A. Lumbroso, Phys. Rev. $\underline{C7}$ (1973) 2274 (1, 5.2).

[GI 73a] Giraud, B. and J. LeTourneux, Phys. Rev. Lett. $\underline{31}$ (1973) 399 (5.1).

[GI 75] Giraud, B., J. LeTourneux, and E. Osnes, Ann. Phys. (N.Y.) $\underline{89}$ (1975) 359 (5.2).

[GI 75a] Giraud, B. and J. LeTourneux, Nucl. Phys. $\underline{A240}$ (1975) 365 (5.3).

[GO 75] Goldberg, D. A., J. B. Marion, and S. J. Wallace (editors), Proc. Intern. Conf. on Clustering Phenomena in Nuclei: II, College Park, Maryland, 1975 (National Technical Information Service, U. S. Department of Commerce, Springfield, Virginia, 1975) (1).

[GO 79] Golin, M. B., Phys. Lett. $\underline{81B}$ (1979) 5 (1).

[GR 57] Griffin, J. J. and J. A. Wheeler, Phys. Rev. $\underline{108}$ (1957) 311 (1, 4.3b, 5.2).

[GR 60] Griffith, T. C. and E. A. Power (editors), Proc. Intern. Conf. on Nuclear Forces and the Few-Nucleon Problem (Pergamon, London, 1960), vol. II (1).

[GR 66] Gruhn, C. R. and N. S. Wall, Nucl. Phys. $\underline{81}$ (1966) 161 (6.2d).

[GR 68] Greenlees, G. W., G. J. Pyle and Y. C. Tang, Phys. Rev. $\underline{171}$ (1968) 1115 (4.2a).

[GR 74] Greenfield, M. B., M. F. Werby, and R. J. Philpott, Phys. Rev. $\underline{C10}$ (1974) 564 (7.2).

[GU 81] Gubler, H. P., V. Kiebele, H. O. Meyer, G. R. Plattner, and I. Sick, Nucl. Phys. $\underline{A351}$ (1981) 29 (6.2d).

[HA 67] Hackenbroich, H. H., Forschungsbericht K67-93 (Frankfurt, ZAED, 1967) (1).

[HA 69] Hackenbroich, H. H., in ref. [BE 69], p. 129 (5.4).

[HA 69a] Hauser, G., R. Löhken, H. Rebel, G. Schatz, G. W. Schweimer, and J. Specht, Nucl. Phys. $\underline{A128}$ (1969) 81 (6.2c).

[HA 71] Hauge, P. S., S. A. Williams, and G. H. Duffey, Phys. Rev. $\underline{C4}$ (1971) 1044 (1).

[HE 56] Heydenberg, N. P. and G. M. Temmer, Phys. Rev. $\underline{104}$ (1956) 123 (6.2b).

[HE 57] Herzenberg, A., Nucl. Phys. $\underline{3}$ (1957) 1 (1).

[HE 70] Heiss, P. and H. H. Hackenbroich, Z. Phys. $\underline{235}$ (1970) 422 (5.4).

[HE 76] Heenen, P. H., Nucl. Phys. $\underline{A272}$ (1976) 399 (5.4).

[HI 53] Hill, D. L. and J. A. Wheeler, Phys. Rev. $\underline{89}$ (1953) 1102 (1, 4.3b, 5.2).

[HO 54] Hocheberg, S., H. S. W. Massey, and L. H. Underhill, Proc. Phys. Soc. (London) $\underline{A67}$ (1954) 957 (4.4).

[HO 68] Horiuchi, H. and K. Ikeda, Prog. Theor. Phys. $\underline{40}$ (1968) 277 (2.3).

[HO 70] Horiuchi, H., Progr. Theor. Phys. $\underline{43}$ (1970) 375 (5.4).

[HO 72] Horiuchi, H., Prog. Theor. Phys. $\underline{47}$ (1972) 1058 (4.3).

[HO 73] Horiuchi, H. and Y. Suzuki, Prog. Theor. Phys. $\underline{49}$ (1973) 1974 (A.1).

[HO 75] Horiuchi, H., in ref. [KA 75], p. 41 (2.3).

[HO 77] Horiuchi, H., in ref. [MI 77], p. 251 (1).

[HO 77a] Horiuchi, H., Prog. Theor. Phys. $\underline{58}$ (1977) 204 (4.4).

[HO 77b] Horiuchi, H., Prog. Theor. Phys. (Suppl.) $\underline{62}$ (1977) 90 (4.4, 5.2, 5.3a, 5.3b, A.1).

[HO 78] Horiuchi, H., in ref. [VA 78], p. 144 (1, 2.1, 6.1, A.1).

[HO 78a] Horiuchi, H., J. Phys. Soc. Japan (Suppl.) $\underline{44}$ (1978) 85 (A.3).
[HO 80]   Horiuchi, H., Prog. Theor. Phys. $\underline{64}$ (1980) 184 (4.2a, 7.1d).
[HU 71]   Hutson, R. L., N. Jarmie, J. L.Detch, Jr., and J. H. Jett, Phys. Rev.
          $\underline{C4}$ (1971) 17 (6.2a).
[HU 72]   Hub, R., D. Clement, and K. Wildermuth, Z. Phys. $\underline{252}$ (1972) 324 (7.2).
[IK 68]   Ikeda, K., N. Takigawa, and H. Horiuchi, Prog. Theor. Phys. (Suppl.)
          extra number (1968) 464 (2.1).
[IK 75]   Ikeda, K. in ref. [KA 75], p. 23 (2.1, 2.3).
[IV 68]   Ivanovich, M., P. G. Young, and G. G. Ohlsen, Nucl. Phys. $\underline{A110}$ (1968)
          441 (6.3).
[JA 64]   Jancovici, B. and D. Schiff, Nucl. Phys. $\underline{58}$ (1964) 678 (1).
[JA 66]   Janssens, T., R. Hofstadter, E. B. Hughes, and M. R. Yearian,
          Phys. Rev. $\underline{142}$ (1966) 922 (8.1a).
[JA 69]   Jacobs, H., K. Wildermuth, and E. Wurster, Phys. Lett. $\underline{29B}$ (1969)
          455 (4.2b).
[KA 56]   Kameny, S. L., Phys. Rev. $\underline{103}$ (1956) 358 (1).
[KA 59]   Kanellopoulos, Th. and K. Wildermuth, Nucl. Phys. $\underline{14}$ (1959) 349 (2.1).
[KA 74]   Kamimura, M. and T. Matsuse, Prog. Theor. Phys. $\underline{51}$ (1974) 438
          (5.2, 5.3, 5.3b).
[KA 75]   Kamitsubo, H., I. Kohno, and T. Marumori (editors), Proc. of the
          INS-IPCR Symposium on Cluster Structure of Nuclei and Transfer Reactions
          induced by Heavy Ions (IPCR Cyclotron Progress Report, Supplement 4,
          1975) (1).
[KA 77]   Kamimura, M., Prog. Theor. Phys. (Suppl.) $\underline{62}$ (1977) 236 (5.4).
[KA 77a] Kamimura, M., in ref. [MI 77], p. 159 (5.4).
[KA 79]   Katō, K. and A. Bandō, Prog. Theor. Phys. $\underline{62}$ (1979) 644 (A.1).
[KA 79a] Kanada, H., T. Kaneko, S. Nagata, and M. Nomoto, Prog. Theor. Phys. $\underline{61}$
          (1979) 1327 (6.2a).
[KA 80]   Kanada, H., T. Kaneko, H. Nishioka, and S. Saito, Prog. Theor.
          Phys. $\underline{63}$ (1980) 842 (4.2b).
[KA 80a] Kanada, H., Q. K. K. Liu, and Y. C. Tang, Phys. Rev. $\underline{C22}$ (1980) 813
          (8.1a).
[KE 77]   Kerman, A. K. and N. Onishi, Nucl. Phys. $\underline{A281}$ (1977) 373 (1).
[KH 61]   Khanna, F. C., Y. C. Tang, and K. Wildermuth, Phys. Rev. $\underline{124}$ (1961) 515
          (4.2a).
[KH 71]   Khadkikar, S. B., Phys. Lett. $\underline{36B}$ (1971) 451 (1).
[KI 77]   Kihara, H., M. Kamimura, and A. Tohsaki-Suzuki, in Proc. Intern.
          Conf. on Nuclear Structure, Tokyo, 1977 (Contributed papers),
          p. 234 (A.3).
[KI 81]   Kim, B. T., T. Izumoto, and K. Nagatani, Phys. Rev. $\underline{C23}$ (1981) 33 (8.1b).
[KO 74]   Koepke, J. A., R. E. Brown, Y. C. Tang, and D. R. Thompson, Phys. Rev.
          $\underline{C9}$ (1974) 823 (4.4, 6.2c, 7.1c).
[KR 73]   Kramer, P., K. Wildermuth, and J. Beam, Part. and Nucl. $\underline{5}$ (1973) 145 (5.1).
[KR 79]   Kräwinkel, H., K. U. Kettner, W. E. Kieser, C. Rolfs, R. Santo,
          P. Schmalbrock, H. P. Trautvetter, R. E. Azuma, and J. W. Hammer,
          Bereich Physiks Jahresbericht, Münster Universität (1978, 1979) (8.1b).
[KU 75]   Kukulin, V. I., V. G. Neudatchin, and Yu. F. Smirnov, Nucl. Phys.
          $\underline{A245}$ (1975) 429 (A.2a).
[LA 60]   Lane, A. M., Rev. Mod. Phys. $\underline{32}$ (1960) 519 (A.1).
[LA 62]   Laskar, W., Ann. Phys. (N.Y.) $\underline{17}$ (1962) 436 (1).
[LA 69]   Lane, A. M. and D. Robson, Phys. Rev. $\underline{178}$ (1969) 1715 (5.4).
[LA 78]   Langanke, K. and D. Frekers, Nucl. Phys. $\underline{A302}$ (1978) 134 (6.4).
[LA 80]   Langanke, K., Ph.D. thesis, Universität Münster (1980) (6.2d, 6.2e, 6.3).
[LE 75]   LeMere, M., R. E. Brown, Y. C. Tang, and D. R. Thompson, Phys. Rev.
          $\underline{C12}$ (1975) 1140 (6.2b).
[LE 76]   LeMere, M., Y. C. Tang, and D. R. Thompson, Phys. Rev. $\underline{C14}$ (1976) 23;
          Phys. Lett. $\underline{63B}$ (1976) 1; Phys. Rev. $\underline{C14}$ (1976) 1715 (4.3, 8.2).
[LE 77]   LeMere, M., Ph.D. thesis, University of Minnesota (1977) (1, 4.3, 4.3b,
          7.1a).
[LE 77a] LeMere, M., R. E. Brown, Y. C. Tang, and D. R. Thompson, Phys. Rev. $\underline{C15}$
          (1977) 1191 (7.1c).

[LE 79]   LeMere, M. and Y. C. Tang, Phys. Rev. C20 (1979) 2003 (4.3).
[LE 79a]   LeMere, M. and Y. C. Tang, Phys. Rev. C19 (1979) 391 (6.3c, 7.1).
[LE 79b]   LeMere, M., D. J. Stubeda, H. Horiuchi, and Y. C. Tang, Nucl. Phys. A320
           (1979) 449 (7.1, 7.1a, 7.1c, 7.1d).
[LE 79c]   LeMere, M., E. J. Kanellopoulos, W. Sünkel, and Y. C. Tang, Phys.
           Lett. 87B (1979) 311 (7.2).
[LE 80]   LeMere, M. and Y. C. Tang, Phys. Rev. C21 (1980) 1170; Nucl. Phys.
           A339 (1980) 43 (4.3).
[LE 80a]   LeMere, M., Y. C. Tang, E. J. Kanellopoulos, and W. Sünkel,
           Nucl. Phys. A348 (1980) 321 (7.2).
[LI 80]   Libert-Heinemann, M., D. Baye, and P. H. Heenen, Nucl. Phys.
           A339 (1980) 429 (6.3).
[LI 81]   Liu, Q. K. K., H. Kanada, and Y. C. Tang, Phys. Rev. C23 (1981) 645 (8.1b).
[MA 41]   Margenau, H., Phys. Rev. 59 (1941) 37 (1).
[MA 73]   Matsuse, T. and M. Kamimura, Prog. Theor. Phys. 49 (1973) 1765 (1, A.1).
[MA 75]   Matsuse, T., M. Kamimura, and Y. Fukushima, Prog. Theor. Phys. 53 (1975)
           706 (5.4, 8.2).
[MC 64]   McDonald, D. G., W. Haeberli, and L. W. Morrow, Phys. Rev. 133 (1964)
           B1178 (6.2a).
[ME 75]   Meier, W. and W. Glöckle, Nucl. Phys. A255 (1975) 21 (A.1).
[MI 73]   Mihailović, M. V. and M. Rosina (editors), Generator-Coordinate
           Method for Nuclear Bound States and Reactions, Fizika (Suppl.) 5 (1973) (1).
[MI 76]   Mihailović, M., L. Goldfarb, and M. Nagarajan, Nucl. Phys. A273 (1976)
           207 (5.4).
[MI 76a]   Mito, Y. and M. Kamimura, Prog. Theor. Phys. 56 (1976) 583 (5.4).
[MI 76b]   Michel, F. and R. Vanderpoorten, Phys. Rev. C16 (1977) 142 (6.2d).
[MI 77]   Mihailović, M. V. and M. Poljšak (editors), Proc. Intern. Symposium on
           Nuclear Collisions and their Microscopic Description, Fizika (Suppl.3)
           (1977) (1).
[MI 77a]   Mihailović, M. V. and M. Poljšak, Phys. Lett. 66B (1977) 209 (5.1).
[MO 56]   Morinaga, H., Phys. Rev. 101 (1956) 254 (2.1).
[MO 66]   Morinaga, H., Phys. Lett. 21 (1966) 78 (2.1, 6.1).
[MO 69]   Morrow, L. W. and W. Haeberli, Nucl. Phys. A126 (1969) 225 (6.2a).
[MU 76]   Mulligan, B., L. G. Arnold, B. Bagchi, and T. O. Krause, Phys. Rev. C13
           (1976) 2131 (4.4).
[NA 69]   Nagatani, K., M. R. Dwarakanath, and D. Ashery, Nucl. Phys. A128 (1969)
           325 (8.1b).
[NA 77]   Nagata, S. and Y. Yamamoto, Prog. Theor. Phys. 57 (1977) 1088 (5.4).
[NE 69]   Neudatchin, V. G. and Yu F. Smirnov, in Prog. Nucl. Phys., edited by
           D. M. Brink and J. H. Mulvey (Pergamon, London, 1969), vol. 10,
           p. 275 (2.1).
[NI 58]   Nilsson, R., W. K. Jenschke, G. R. Briggs, R. O. Kerman, and
           J. N. Snyder, Phys. Rev. 109 (1958) 850 (6.2b).
[NI 79]   Nishioka, H., S. Saito, and M. Yasuno, Prog. Theor. Phys. 62 (1979)
           424 (A.1).
[NO 66]   Nordstrom, D. L., J. A. Tunheim, and G. H. Duffey, Phys. Rev. 145
           (1966) 727 (1).
[OR 75]   Orth, H., H. Ackermann, and E. W. Otten, Z. Phys. A273 (1975) 221 (8.1a).
[PA 63]   Parker, P. D. and R. W. Kavanagh, Phys. Rev. 131 (1963) 2578 (8.1b).
[PE 56]   Perring, J. K. and T. H. R. Skyrme, Proc. Phys. Soc. (London) 69 (1956)
           600 (1).
[PE 57]   Peierls, R. E. and J. Yoccoz, Proc. Phys. Soc. (London) A70 (1957) 381 (1).
[PE 60]   Pearlstein, L. D., Y. C. Tang, and K. Wildermuth, Nucl. Phys. 18 (1960)
           23 (6.1).
[PE 62]   Perey, F. and B. Buck, Nucl. Phys. 32 (1962) 353 (4.2a).
[PI 78]   Pilt, A. A., in ref. [VA 78], p. 173 (A.1).
[RE 70]   Reichstein, I. and Y. C. Tang, Nucl. Phys. A158 (1970) 529 (4.1).
[RE 72]   Reidemeister, G., Nucl. Phys. A197 (1972) 631 (5.1).
[RE 73]   Resonating Group, Phys. Lett. 43B (1973) 165 (5.4).
[RO 56]   Robertson, H. H., Proc. Cambridge Phil. Soc. 52 (1956) 538 (5.4).

[RO 80]   Roos, P. G., A. Nadasen, P. E. Frisbee, N. S. Chant, T. A. Carey,
          M. T. Collins, B. Th. Leeman, P. J. Griffin, and R. D. Koshel,
          Phys. Rev. C21 (1980) 799 (6.2c).
[RO 81]   Rolfs, C., private communcation (1981) (8.1b).
[SA 68]   Saito, S., Prog. Theor. Phys. 40 (1968) 893 (1).
[SA 69]   Saito, S., Prog. Theor. Phys. 41 (1969) 705 (1, A.1).
[SA 73]   Saito, S., S. Okai, R. Tamagaki, and M. Yasuno, Prog. Theor. Phys.
          50 (1973) 1561 (4.4, A.1).
[SA 77]   Saito, S., Prog. Theor. Phys. (Suppl.) 62 (1977) 11 (1, A.1, A.2a).
[SA 77a] Saloner, D. A. and C. Toepffer, Nucl. Phys. A283 (1977) 108 (6.4).
[SA 77b] Saloner, D. A., C. Toepffer, and B. Fink, Nucl. Phys. A283 (1977) 131 (6.4).
[SA 79]   Sakuda, T. and F. Nemoto, Prog. Theor. Phys. 62 (1979) 1274, 1606 (A.1).
[SC 73]   Schmid, E. W., Nuovo Cim. 18A (1973) 771 (4.4).
[SC 73a] Schwager, J., Nuovo Cim. 18A (1973) 787 (4.4).
[SC 80]   Schmid, E. W., Phys. Rev. C21 (1980) 691 (4.2a).
[SO 76]   Sourkes, A. M., A. Houdayer, W. T. H. van Oers, R. F. Carlson,
          and R.E. Brown, Phys. Rev. C13 (1976) 451 (6.2a).
[ST 78]   Stubeda, D. J., M. LeMere, and Y. C. Tang, Phys. Rev. C17 (1978) 447 (4.3,
          6.2c).
[SU 67]   Suelzle, L. R., M. R. Yearian, and H. Crannell, Phys. Rev. 162 (1967)
          992 (8.1a).
[SU 72]   Sünkel, W. and K. Wildermuth, Phys. Lett. 41B (1972) 439 (4.3).
[SU 75]   Sünkel, W. and K. Wildermuth, in ref. [VA 78], p. 156 (4.3).
[SU 76]   Suzuki, Y., Prog. Theor. Phys. 55 (1976) 1751, 56 (1976) 111
          (2.1, 2.3, A.3).
[SU 76a] Sünkel, W., Phys. Lett. 65B (1976) 419 (4.3, 5.4).
[SU 77]   Suzuki, Y., T. Ando, and B. Imanishi, in Proc. Intern. Conf.
          on Nuclear Structure , Tokyo, 1977 (contributed papers), p. 515 (A.3).
[SU 79]   Sünkel, W. and Y. C. Tang, Nucl. Phys. A329 (1979) 10 (4.3, 6.2c).
[SZ 78]   Szaloky, G., F. Seiler, W. Grüebler, and V. König, Nucl. Phys.
          A303 (1978) 51 (6.2a).
[TA 62]   Tang, Y. C., K. Wildermuth, and L. D. Pearlstein, Nucl. Phys. 32 (1962)
          504 (1).
[TA 62a] Tang, Y. C., F. C. Khanna, R. C. Herndon, and K. Wildermuth, Nucl.
          Phys. 35 (1962) 421 (4.2a).
[TA 63]   Tang, Y. C., E. Schmid, and K. Wildermuth, Phys. Rev. 131 (1963) 2631 (4.4).
[TA 71]   Takigawa, N. and A. Arima, Nucl. Phys. A168 (1971) 593 (6.1).
[TA 72]   Tabakin, F., Nucl. Phys. A182 (1972) 497 (1).
[TA 73]   Tanabe, F., A. Tohsaki, and R. Tamagaki, Prog. Theor. Phys. 50 (1973)
          1774 (5.4).
[TA 74]   Tanabe, F. and F. Nemoto, Prog. Theor. Phys. 51 (1974) 2009 (5.4).
[TA 75]   Tamagaki, R., in ref. [KA 75], p. 321 (1).
[TA 77]   Tang, Y. C., Fizika (Suppl. 3) 9 (1977) 91 (1, 4.3).
[TA 78]   Tang, Y. C., M. LeMere, and D. R. Thompson, Phys. Reports 47 (1978) 167
          (1, 4.2a, 4.3, 4.3b, 5.4, 6.3, 7.1c).
[TE 77]   Teplov, I. B., N. S. Zelenskaya, V. M. Levedev, and A. V. Spasskiĭ,
          Sov. J. Part. Nucl. 8 (1977) 310 (7.2).
[TH 70]   Thompson, D. R., Nucl. Phys. A143 (1970) 304 (A.2a).
[TH 71]   Thompson, D. R. and Y. C. Tang, Phys. Rev. C4 (1971) 306 (4.2a).
[TH 73]   Thompson, D. R. and Y. C. Tang, Phys. Rev. C8 (1973) 1649 (4.2b, 6.2b).
[TH 75]   Thompson, D. R. and Y. C. Tang, Phys. Rev. C12 (1975) 1432, C13 (1976)
          2597 (1, 4.3, 4.3b).
[TH 76]   Thompson, D. R., M. LeMere, and Y. C. Tang, Nucl. Phys. A270 (1976)
          211 (4.3).
[TH 77]   Thompson, D. R., M. LeMere, and Y. C. Tang, Phys. Lett. 69B (1977) 1
          (1, 4.3, 4.3b).
[TH 77a] Thompson, D. R., M. LeMere, and Y. C. Tang, Nucl. Phys. A286 (1977)
          53 (6.2b).
[TH 78]   Thompson, D. R., in ref. [VA 78], p. 69 (4.2).
[TO 63]   Tombrello, T. A. and L. S. Senhouse, Phys. Rev. 129 (1963) 2252 (6.2b).

[TO 63a] Tombrello, T. A. and P. D. Parker, Phys. Rev. $\underline{131}$ (1963) 2582 (8.1b).
[TO 75]   Tohsaki, A., F. Tanabe, and R. Tamagaki, Prog. Theor. Phys. $\underline{53}$ (1975) 1022 (5.2, 5.4).
[TO 75a] Tomoda, T. and A. Arima, in ref. [KA 75], p. 90 (8.2).
[TO 77]   Tohsaki-Suzuki, A., Prog. Theor. Phys. (Suppl.) $\underline{62}$ (1977) 191 (5.2).
[TO 78]   Tohsaki-Suzuki, A., Prog. Theor. Phys. $\underline{59}$ (1978) 1261 (5.3, 5.3a).
[UE 75]   Uegaki, E., S. Okabe, Y. Abe, and H. Tanaka, in ref. [KA 75], p. 682 (2.1).
[VA 59]   Van der Spuy, E., Nucl. Phys. $\underline{11}$ (1959) 615 (1).
[VA 78]   Van Oers, W. T. H., J. P. Svenne, J. S. C. McKee, and W. R. Falk (editors), in Proc. Intern. Conf. on Clustering Aspects of Nuclear Structure and Nuclear Reactions, Winnipeg, Canada, 1978 (AIP Conf. Proc. Number 47) (1).
[VE 63]   Verhaar, H. J., Nucl. Phys. $\underline{45}$ (1963) 129 (1).
[VO 70]   Von Oertzen, W., Nucl. Phys. $\underline{A148}$ (1970) 529 (7.1c).
[VO 74]   Votta, L. G., P. G. Roos, N. S. Chant, and R. Woody, III, Phys. Rev. $\underline{C10}$ (1974) 520 (6.3).
[WE 65]   Werner, J. and J. Zimmerer, in Proc. Intern. Conf. on Nuclear Physics, Paris, 1964 (Editions du Centre National de la Recherche Scientifique, Paris, 1965), p. 241 (6.2b).
[WE 77]   Weiguny, A., Fizika (Suppl. 3) $\underline{9}$ (1977) 115 (4.4, 6.4).
[WH 37]   Wheeler, J. A., Phys. Rev. $\underline{52}$ (1937) 1083, 1107 (1).
[WI 58]   Wildermuth K. and Th. Kanellopoulos, Nucl. Phys. $\underline{7}$ (1958) 150 (1).
[WI 62]   Wildermuth, K., Nucl. Phys. $\underline{31}$ (1962) 478 (2.3).
[WI 66]   Wildermuth, K. and W. McClure, Cluster Representation of Nuclei (Springer-Verlag, Berlin, 1966) (1).
[WI 77]   Wildermuth, K. and Y. C. Tang, A Unified Theory of the Nucleus (Vieweg, Braunschweig, Germany, 1977) (1, 2.2, 2.3, 2.4, 3, 4.1, 4.2b, 4.2c, 5.1, A.1).
[WI 79]   Wildermuth, K. and E. J. Kanellopoulos, Reports on Progress in Physics $\underline{42}$ (1979) 1719 (1).
[WO 70]   Wong, C. W., Nucl. Phys. $\underline{A147}$ (1970) 545 (1).
[WO 75]   Wong, C. W., Phys. Reports $\underline{15C}$ (1975) 285 (1).
[YU 72]   Yukawa, T., Phys. Lett. $\underline{38B}$ (1972) 1 (1, 5.2).
[ZA 71]   Zaikin, Z., Nucl. Phys. $\underline{A170}$ (1971) 584 (1).

Chapter VI

Heavy-ion direct reactions

Q. K. K. Liu

School of Physics and Astronomy
University of Minnesota
Minneapolis, Minnesota 55455

and

Bereich Kern- und Strahlenphysik
Hahn-Meitner-Institut für Kernforschung
1000 Berlin 39, Germany*

Abstract: We discuss in this chapter three subjects in heavy-ion direct reactions. They are (i) one-nucleon transfer reaction, (ii) inelastic scattering and (iii) spin-orbit effects. One nucleon transfer reaction is still one of the best understood rearrangement collision phenomena, while inelastic Coulomb excitations dominate most of the heavy-ion peripheral reactions. Spin-orbit effects in heavy-ion reaction is a relatively new topic of interest. We attempt to render these notes self-contained for graduate students by including an appendix on the elements of collision theory.

*Permanent address.

1. One-nucleon transfer reaction

One nucleon transfer reaction is most often analyzed by DWBA discussed in section A.16, in which the exact stationary solution of the Schrödinger equation is replaced by an elastic scattering wave function which describes as accurately as possible the elastic scattering cross section. Therefore, some understanding of the elastic scattering of heavy-ion collision is part and parcel of the DWBA analysis.

1.1 Elastic scattering

We describe in this subsection some general features of elastic scattering of heavy-ions. Recent review articles should be consulted for more extensive discussions [GE 79, Nö 80, SI 74].

Many notions of classical physics seem to be valid in the heavy-ion domain. This is understandable if we consider the ratio of the typical nuclear dimension in a heavy-ion system and the deBroglie wavelength of the translational motion. A particular useful and convenient ratio is in fact the Coulomb parameter.

$$\eta = \frac{Z_1 Z_2 e^2}{\hbar v}$$

$$= \frac{\text{classical distance of closest approach in a head-on collision}}{2 \times \text{deBroglie wavelength of the asymptotic relative motion}} \cdot \quad (1.1)$$

A large Coulomb parameter means that one can construct wave packets which are small in size in comparison with the nuclear interaction region. For example, in an ^{18}O + ^{58}Ni collision at E(LAB) = 60 MeV, $\eta \approx 44$. Even if we consider the local wavelength

$$\lambdabar = k^{-1} = \sqrt{\frac{A_1 + A_2}{A_1 A_2} \cdot \frac{20}{E - E_{coul}}} \quad , \quad (1.2)$$

where A_1, A_2 are the mass numbers of the collision partners and E_{coul} is the local electrostatic Coulomb potential energy, we have at a separation of $R_1 + R_2$

$$\frac{R_1 + R_2}{\lambdabar} \approx 30 \quad , \quad (1.3)$$

where $R_1 + R_2$ is the sum of the nuclear radii. Therefore, a fair amount of classical terminology and concepts is used in the qualitative discussions. This does not mean, however, that quantum mechanics are abandoned.

In classical language, the incident particles approach the target
with a certain impact parameter which is directly connected with a partial wave. Therefore, scattering of high partial waves into the forward
scattering angles means that the scattering partners are far apart and
is described by Coulomb scattering. For lower partial waves, i.e. scattering into the more backward angles, the nuclei come closer together
and nuclear reactions occur which remove nuclear flux from the elastic
channel, i.e. absorption takes place. Schematically, angular distribution of elastic scattering has the form displayed in Fig.1.1. The quantity which is usually plotted is the ratio of elastic scattering to
Rutherford (Coulomb) cross section. This removal of the dominant Coulomb
scattering is done to highlight any nuclear effect. The approximate separation of those partial waves which are mostly absorbed and those which

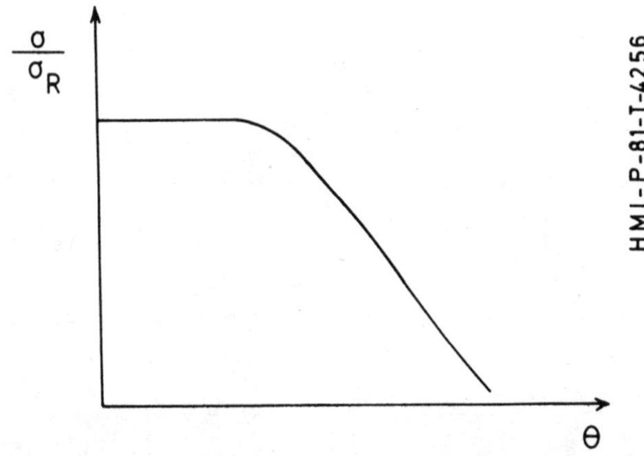

Fig.1.1. Schematic diagram of the angular distribution of elastic
         scattering of heavy ions.

are mostly scattered elastically is given by the grazing partial wave
ℓ_{gr}, whose corresponding scattering angle is named the grazing angle
θ_{gr}.

Usually, the rather flat region of $[\sigma / \sigma_R]$ at small scattering
angles is modified by some oscillations due to diffraction of the matter
waves [FR 72]. With reference to Fig.1.2, a grazing partial wave of impact parameter b is scattered through an angle θ_{gr}. The nuclei on this
trajectory seem to have been emitted from a source at a distance d behind the target, where d could be small compared with the size of the
scatterer, i.e. Fresnel diffraction, instead of from the actual source
which is infinitely far away from the target, i.e. Fraunhofer diffrac-

tion. In the nuclear context, the idealized Fresnel diffraction occurs in the limit of high energies and strong Coulomb interaction, while the Fraunhofer diffraction occurs in the limit of high energies and negligible Coulomb interaction. Since the nuclear charges involved in heavy-

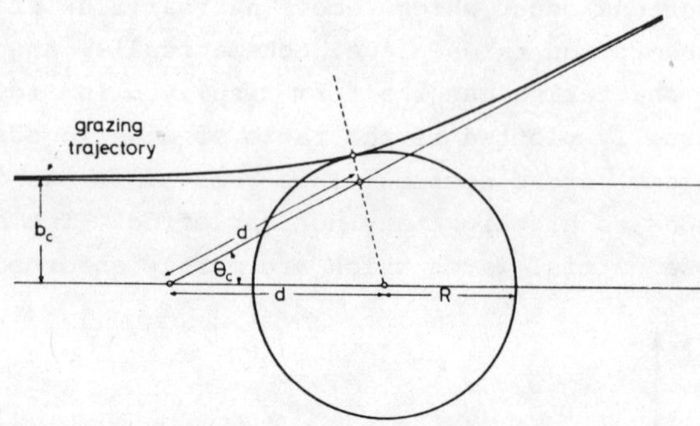

Fig.1.2. The divergent lens action of the Coulomb field and the virtual
source point [FR 72].

ion scatterings are usually large, Fresnel diffraction is the prevalent one [FR 72]. In Fig. 1.3 we show an illustrative example of this phenomenon. The broken curve is the idealized Fresnel diffraction pattern,

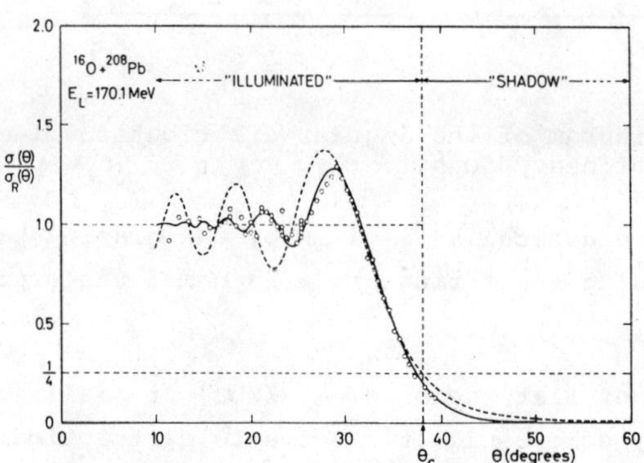

Fig.1.3. The description of $^{16}O + ^{208}Pb$ at E(LAB) = 170.1 MeV by a
Fresnel diffraction formula [FR 72].

while the solid curve contains further nuclear model dependent parameters. The salient point in Fig.1.3 is how much of the idealized Fresnel diffraction pattern can be found in the actual experimental data at finite energy.

One would like to describe the physics of the scattering in terms of an internuclear potential. For phenomenological analysis, a local potential with the shape of a Fermi distribution is often assumed. Also, an imaginary part is included to account for the absorption of flux from the elastic channel, see section (A.15). Hence, one writes the potential as

$$
\begin{aligned}
U(r) &= V(r) + i\,W(r) \\
&= V_0\,f(r) + i\,W_0\,f_w(r) \quad,
\end{aligned}
\tag{1.4}
$$

where $f(r)$ and $f_w(r)$ may or may not be the same. For the most commonly used Woods-Saxon potential, $f(r)$ and $f_w(r)$ have the form of a Fermi distribution,

$$
f(r) = \frac{1}{1 + \exp\left(\frac{r-R}{a}\right)} \quad,
\tag{1.5}
$$

with R and a being the radius and diffuseness parameters. Unfortunately, when there is strong absorption such that the depth of the real potential V_0 is immaterial, in other words, the long range part of the potential is responsible for the elastic scattering, this potential form leads to the so-called Igo ambiguity. Examining the behaviour of (1.5) at large r, we see easily that

$$
V(r) \xrightarrow[r\to\infty]{} V_0\,\exp\left(\frac{R}{a}\right)\exp\left(-\frac{r}{a}\right) \quad.
\tag{1.6}
$$

Therefore, potentials with the same diffuseness a, but different combinations of V_0 and R, such that

$$
V_0\,\exp\left(\frac{R}{a}\right) = \text{constant}
\tag{1.7}
$$

are completely identical in the long range region. Although this continuous ambiguity was first discussed in the analysis of light-ion scattering, it also surfaces in the analysis of heavy-ion elastic scattering, and recently in the discussion of spin-orbit effects in heavy-ion inelastic scattering [LI 80].

There are nuclear systems in which the angular distribution can be measured in the entire range of 0° to 180°. One may see the surprising

features which the schematic sketch of Fig.1.1 does not display, i.e.
the ratio $[\sigma / \sigma_R]$ beyond 90° exhibit oscillations and the envelope of
these oscillations rises in the backward direction. In Fig.1.4,we display such an example in the elastic scattering of ^{28}Si + ^{16}O at E(LAB)
= 41 MeV. To achieve such a good fit over the whole angular range, a
much more complicated potential than Eq.(1.4) had to be employed. It is
displayed in Fig.1.5. It contains a volume absorption, a surface absorption and slightly different potentials for odd and even partial waves,
i.e. parity dependence. [This particular potential has the desirable
property which often eludes other parametrized optical potential, that
it varies smoothly with energy.] To arrive at such a good fit is without
a doubt an achievement.

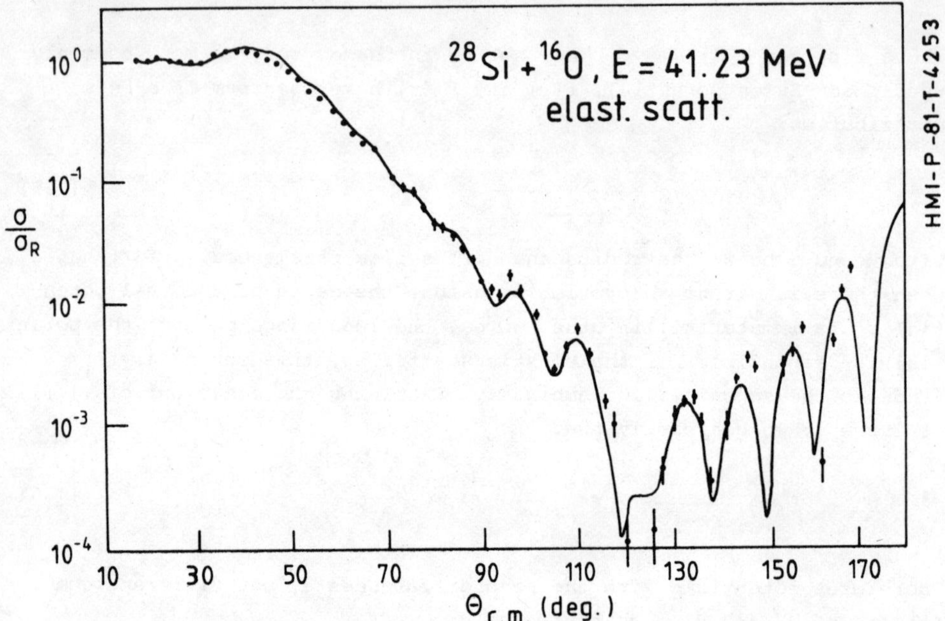

Fig.1.4.   Elastic scattering of ^{28}Si + ^{16}O. The experimental data are
taken from [GE 78] and the theoretical analysis is from [SH 80].

However, one would like to have a physical explanation for the necessity
of such complications. Not only in this case, but in general, we do not
have ready explanations for them.

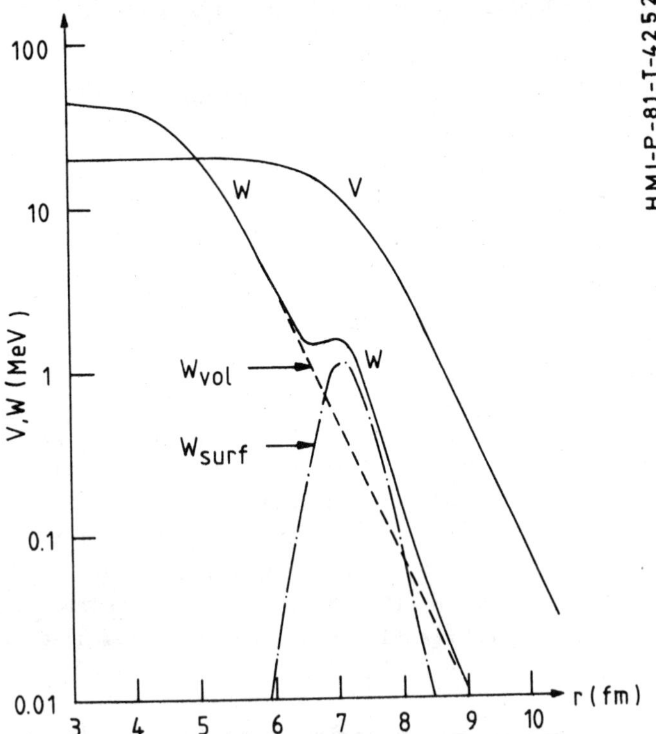

Fig.1.5. The radial shape cf the potential used by [SH 80] to analyse the elastic scattering of ^{28}Si + ^{16}O at E(LAB) = 41.23 MeV.

1.2 Qualitative features of one-nucleon transfer reactions

One-nucleon transfer reactions initiated by light-ion (4 $\lesssim$ A) have
served as a very useful tool in the study of single particle structures
of nuclei. The analyses of heavy-ion induced transfer reactions contain
a great deal more complexities, hence the spectroscopic informations
they deduce are weighted down by greater uncertainties. However, physi-
cists are rewarded by a greater variety of reaction mechanisms. Thus,
we see the emergence of elastic transfer reaction A(a,a)A, where
A = a + 1, in which one-nucleon transfer from A to a to form the ground
state is indistinguishable from elastic scattering [Oe 75]. In other
transfer reactions or inelastic reactions, they can take place via intermediate steps which, in the case of heavy-ions, can have quite a plethora
of possibilities [BO 79]. The large Coulomb barrier associated with
heavy-ion can also be exploited to extract more accurate rms radii of

neutron orbits, in neutron transfer reactions, than in light-
ion [PH 77].

Consider a simple classical model of a uniform cylindrical beam of
particles being accelerated towards the target and each particle ini-
tiates a nuclear reaction. The distances of the particles from the axis
of symmetry are the impact parameters b, and N is the flux of the particles in the beam. Therefore, the number of nuclear reactions per unit
flux, given as a function of b, is

$$d\sigma_R = \frac{2\pi b\,db \cdot N}{N} , \qquad\qquad (1.8)$$

i.e. it is a linear function of b. Using the classical relation of b
being proportional to the angular momentum $\ell\hbar$, one can plot the reaction
cross section as a function of ℓ. We depict an illustrative example of
such a plot in Fig.1.6. The reaction cross section, indicated by the
solid line, becomes zero beyond a certain grazing angular momentum ℓ_{gr} ,i.e.
the incident particles move by the target undisturbed. Quantal corrections of Eq. (1.8) smear out the sharp edge of the solid line in Fig.
(1.6). Qualitatively, the reaction cross section underneath the curve

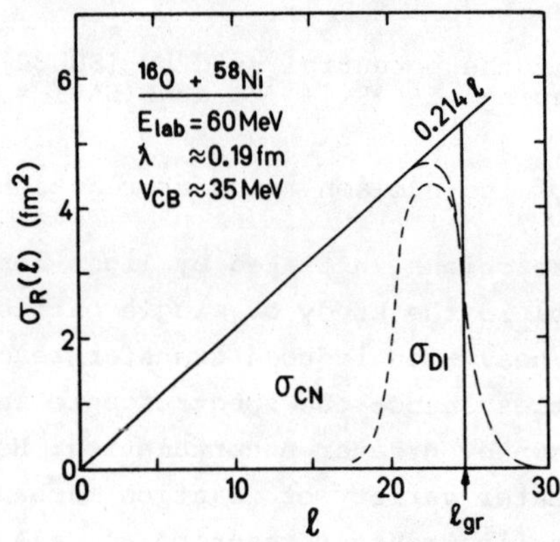

Fig.1.6. Reaction cross section as a function of angular momentum [Nö 80].

can be decomposed into a part which is attributed to direction reaction
(DI) and a part to compound nuclear reaction (compound-fission, compound-
evaporation and fusion reaction). Direction reactions occur in the vicinity of the ℓ_{gr}.

Continuing with our semiclassical picture of the reaction $A(a,b)B$ at a c.m. energy E_i, which is below the Coulomb barrier, in the incoming channel, in which Z_a, Z_A and Z_b, Z_B are the atomic numbers of the nuclear fragments in the incoming and outgoing channels, respectively, we describe the relative motion by trajectories which are determined by the wave number k, the Coulomb parameter η and the partial wave number ℓ (i.e. the angular momentum). The incoming channel parameters (k_i, η_i, ℓ_i) are usually not the same as the outgoing channel parameters (k_f, η_f, ℓ_f). To obtain large reaction cross section, the semi-classical picture is expected to remain intact, i.e. the incoming trajectories must be joined as smoothly as possible onto the outgoing trajectories. If there are kinks in the trajectories caused by a change the parameters (k, η, ℓ), they must be small in dimension compared to the wavelength $\lambdabar = 1/k$. In quantitative terms, this could mean that the distances of closest approach given by the incoming and outgoing trajectories parameters are equal, i.e.

$$R^i_{min} = R^f_{min} \qquad (1.9)$$

or

$$\frac{\eta_i}{k_i}\left\{1+\left[1+\frac{\ell_i(\ell_i+1)}{\eta_i^2}\right]^{1/2}\right\} = \frac{\eta_f}{k_f}\left\{1+\left[1+\frac{\ell_f(\ell_f+1)}{\eta_f^2}\right]^{1/2}\right\} . \qquad (1.10)$$

From Eq. (1.10), one can obtain easily the optimal Q-value at which the reaction cross section is expected to attain a maximum,

$$Q_{opt} = E_i\left[\frac{Z_b Z_B}{Z_a Z_A}\cdot\rho - 1\right] , \qquad (1.11)$$

where

$$\rho = \left\{1+\left[1+\frac{\ell_f(\ell_f+1)}{\eta_f^2}\right]^{1/2}\right\} \Bigg/ \left\{1+\left[1+\frac{\ell_i(\ell_i+1)}{\eta_i^2}\right]^{1/2}\right\} . \qquad (1.12)$$

For an approximate estimate of Q_{opt}, one can set $\rho = 1$, which corresponds to reaction taking place in a head-on collision, an estimate first given by Buttle and Goldfarb [BU 71],

$$Q_{opt} = E_i \left(\frac{Z_b Z_B}{Z_a Z_A} - 1 \right) . \tag{1.13}$$

One derives from Eq.(1.13) immediately the qualitative difference between a neutron and a proton transfer reaction. The value for Q_{opt} for neutron transfer reactions is always approximately zero, while for proton, it is negative when the proton is transferred from the lighter nucleus to the heavier nucleus, and positive when the proton is transferred from the heavier nucleus to the lighter one. Furthermore, for proton transfer, Q_{opt} is energy dependent.

The validity of the estimate of Q_{opt} for neutron transfer by Buttle and Goldfarb is demonstrated in Fig.1.7. The two reactions $^9\text{Be}(^{16}\text{O},^{17}\text{O})^8\text{Be}$ and $^{13}\text{C}(^{16}\text{O},^{17}\text{O})^{12}\text{C}$ lead to the ground and first excited state of ^{17}O. For the former reaction, the Q-values for the ground state and first excited state of ^{17}O are 2.47 MeV (solid circle) and 1.6 MeV (open circle), respectively. The corresponding Q-values in the latter reaction are -0.88 MeV (solid circle) and -1.67 MeV (open circle). These values are situated on opposite slopes of a curve plotted

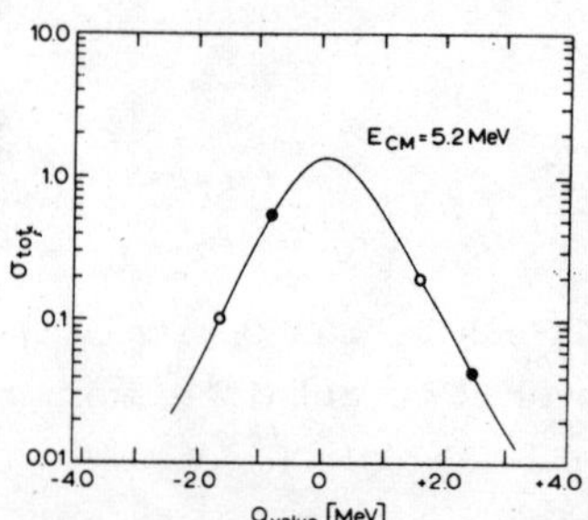

Fig.1.7. Q-values for the reactions $^9\text{Be}(^{16}\text{O},^{17}\text{O})^8\text{Be}$ and $^{13}\text{C}(^{16}\text{O},^{17}\text{O})^{12}\text{C}$. See text for a discussion [Oe 74].

from a hypothetical calculation of the reaction cross section by the quantal DWBA (see section 1.3) as a function of Q-value for the reaction $^{13}\text{C}(^{12}\text{C},^{13}\text{C})^{12}\text{C}$. The optimal Q-value occurs at $Q \approx 0$. This figure has been taken from [Oe 74].

The two types of proton transfer, i.e. the transferred proton originates from (i) the lighter nucleus and (ii) the heavier nucleus, were studied by Barnett et al. [BA 71], in the reaction $^{208}\text{Pb}(^{16}\text{O},^{15}\text{N})^{209}\text{Bi}$ and $^{208}\text{Pb}(^{16}\text{O},^{17}\text{F})^{208}\text{Tl}$ with an incident ^{16}O projectile beam of 69.1 MeV.

The experimental Q-values for the ground state transitions for both of these reactions are approximately -8 MeV. The values of Q_{opt}, according to Eq.(1.13), are -7.34 MeV and 8.28 MeV, respectively. This means that semi-classical arguments predict that the former reaction cross section is much larger than the latter. This was substantiated by the fact that Barnett et al. [BA 71] could observe the first reaction but not the second. A theoretical calculation by DWBA (see subsection 1.3) of the first reaction at 180° as a function of the Q-value is plotted in Fig. (1.8) [PH 77], with incident ^{16}O energy being 69.1 MeV. The final state considered was a $7/2^-$ level with a variable binding energy for the transferred proton. The quantal prediction of Q_{opt} is a value around -8 MeV which agrees remarkably well with the semi-classical prediction of -7.34 MeV.

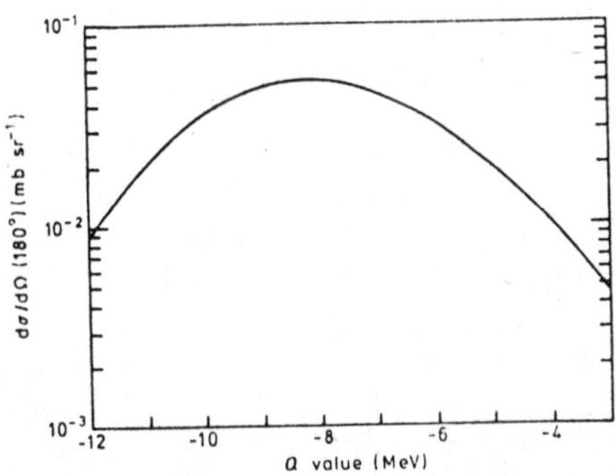

Fig.1.8. Variation of the 180° differential cross section with Q-values for the reaction ^{208}Pb(^{16}O,^{15}N)^{209}Bi [PH 77].

The variation of Q_{opt} as a function of incident energy has also been verified by Schiffer et al. [SC 73], who took into account of ρ not equal to 1.

As the energy is increased above the Coulomb barrier, the description of the relative motion by Coulomb trajectories is no longer valid. However, in the regime of incident energy very high above the Coulomb barrier, such that in the transfer region, the relative velocity is not affected by the change in the potential barrier, Brink [BR 72a] has derived another set of kinematical conditions which must be satisfied if the transfer probability is to be large.

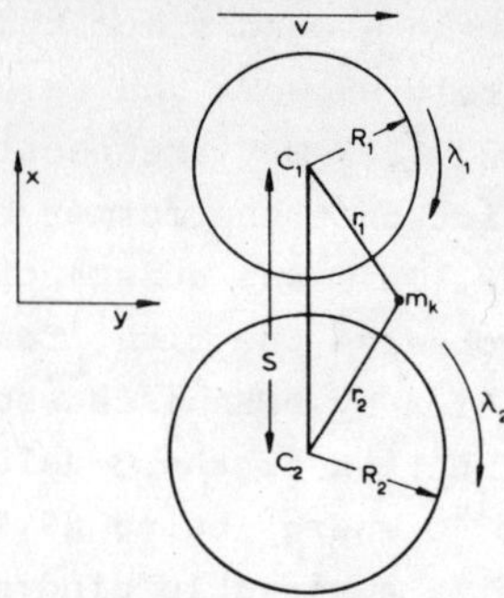

Fig.1.9. The system of coordinates used in the discussion of the semi-
classical transfer reaction theory [AN 74].

In the process

$$(c_1 + k) + C_2 \longrightarrow C_1 + (k + C_2) \quad ,$$
$$\qquad a_1 \qquad\qquad\qquad\qquad a_2$$

the k nucleons transferred are treated as a cluster which occupies a
state

$$\psi_1(\vec{r}_1) = u_1(r_1)\, Y_{L_1}^{\lambda_1}(\theta_1, \varphi_1) \tag{1.14}$$

in the initial nucleus a_1 and a state

$$\psi_2(\vec{r}_2) = u_2(r_2)\, Y_{L_2}^{\lambda_2}(\theta_2, \varphi_2) \tag{1.15}$$

in the final nucleus a_2. The nuclei are assumed to move along well defined classical trajectories. The z-axis is chosen to be perpendicular
to the reaction plane, see Fig.(1.9). A semi-classical evaluation of
the transfer probability [BR 72b, BR 72a, AN 74] gives, for the transition from the initial state $(L_1\ \lambda_1)$ to the final state $(L_2 \lambda_2)$,

$$P(L_1\lambda_1, L_2\lambda_2) = P_0(R) \, | Y_{L_1}^{\lambda_1}(\pi/2, 0) \, Y_{L_2}^{\lambda_2}(\pi/2, 0) |^2$$

$$\times \exp\left[-\left(\frac{R\Delta k}{\sigma_1}\right)^2 - \left(\frac{\Delta L}{\sigma_2}\right)^2\right] \quad , \tag{1.16}$$

with Δk and ΔL defined by

$$\Delta k = k_0 - \lambda_1/R_1 - \lambda_2/R_2 \quad , \tag{1.17}$$

$$\Delta L = \lambda_2 - \lambda_1 + \frac{1}{2} k_0 (R_1 - R_2) + Q_{eff}\, R/(\hbar v) \quad , \tag{1.18}$$

where v is the local relative velocity in the region of transfer, R_1 and R_2 are respectively the approximate radii of nuclei a_1 and a_2, R is the distance of closest approach. The quantity Q_{eff} is the effective Q-value,

$$Q_{eff} = Q - (z_1^f z_2^f - z_1^i z_2^i)e^2/R \quad , \tag{1.19}$$

in which Q is the reaction Q-value, z_1^i, z_2^i and z_1^f, z_2^f are the atomic numbers of the nuclei in initial and final channels, respectively. In the transfer of uncharged particles, $Q_{eff} = Q$. The factor $P_o(R)$, which is a function of the distance of closest approach R, depends on the radial functions $u_1(r_1)$ and $u_2(r_2)$. The quantities σ_1 and σ_2 are not known precisely, but can be inferred to by uncertainty relations which suggest that $\sigma_1 \approx \pi$ and $\sigma_2 \approx \sqrt{\gamma R}$, where $\gamma^2 = 2m_k \epsilon/\hbar^2$, with ϵ being some average of the binding energies of the cluster in the initial and final nuclei.

One deduces immediately from Eq.(1.16) that the transfer probability is small unless

(i) $R\Delta k \ll \sigma_1$, $\tag{1.20}$

(ii) $\Delta L \ll \sigma_2$, $\tag{1.21}$

(iii) $L_1 + \lambda_1 = $ even , $\tag{1.22}$

$$L_2 + \lambda_2 = \text{even} \quad .$$

A classical consideration of the kinematics leads to $\Delta k \approx 0$ and $\Delta L \approx 0$. Quantum mechanics is responsible for smearing them out to conditions (1.20) and (1.21). We derive presently the conditions

$$\Delta k = k_o - \lambda_1/R_1 - \lambda_2/R_2 \approx 0 , \tag{1.20a}$$

$$\Delta L = \lambda_2 - \lambda_1 + \frac{1}{2}k_o(R_1 - R_2) + Q_{eff} R/(\hbar v) \approx 0 , \tag{1.21a}$$

by classical mechanics.

(i) With reference to Fig. (1.9), approximate conservation of the y-component of linear momentum of the transferred cluster before and after the transfer yields condition (2.20a)

$$m_k v - \lambda_1 \hbar/R_1 \approx \lambda_2 \hbar/R_2 .$$

The expression is only approximate because the exact location where the transfer takes place is unknown.

(ii) Conservation of the z-component of the total angular momentum yields

$$(\lambda_2 - \lambda_1)\hbar + \delta(\mu v R) = 0 , \tag{1.23}$$

in which the first term gives the change in the z-component of the angular momentum of m_k which results from the transfer, and $\delta(\mu v R)$ represents the change in the z-component of the angular momentum of the relative motion of the two nuclei. For this latter expression, we assume that the transfer takes place at the distance of closest approach R. Here, μ is the reduced mass. One can further write

$$\delta(\mu v R) = \frac{1}{2} v R \, \delta(\mu) + \mu v \, \delta(R) + \frac{R}{v} \delta(\frac{1}{2}\mu v^2) . \tag{1.24}$$

The change in the reduced mass is

$$\delta(\mu) = \frac{(M_2 + m_k)M_1}{M_2 + M_1 + m_k} - \frac{M_2(m_k + M_1)}{M_2 + M_1 + m_k}$$

$$= \frac{m_k(M_1 - M_2)}{M_2 + M_1 + m_k} . \tag{1.25}$$

If one assumes the nucleon cluster is transferred near the line joining the two nuclear centres, at a distance R_1 from the centre of C_1 and R_2 from the centre of C_2, the change δR in the separation of the centre-of-mass of the two nuclei is

$$\delta R = \left(\frac{M_2}{M_2 + m_k} R_2 + R_1 \right) - \left(R_2 + \frac{M_1}{M_1 + m_k} R_1 \right)$$

$$= m_k \left(\frac{R_1}{M_1 + m_k} - \frac{R_2}{M_2 + m_k} \right)$$

$$- \frac{m_k}{2} \left(\frac{1}{M_1 + m_k} + \frac{1}{M_2 + m_k} \right) (R_1 - R_2)$$

$$+ \frac{m_k}{2} \left(\frac{1}{M_1 + m_k} - \frac{1}{M_2 + m_k} \right) (R_1 + R_2)$$

$$= \frac{m_k}{2} \cdot \frac{(R_1 - R_2)}{\mu} - \frac{\delta(\mu)}{2} \cdot \frac{(R_1 + R_2)}{\mu} \, . \tag{1.26}$$

The last line of Eq.(1.26) is derived to the first order of (m_k/M_1) and (m_k/M_2). The change in the kinetic energy of the relative motion $\delta(\frac{1}{2}\mu v^2)$ is just Q_{eff} of the reaction. Therefore, on substitution of Eqs.(1.24), (1.25) and (1.26) into (1.23), we obtain condition (1.21a).

(iii) The transfer probability is expected to be largest when the transfer cluster is near the reaction plane, i.e. when $\Theta_1 = \Theta_2 \approx \pi/2$ in $Y_{L_1}^{\lambda_1}(\Theta_1, \varphi_1)$ and $Y_{L_2}^{\lambda_2}(\Theta_2, \varphi_2)$. Since $Y_L^{\lambda}(\pi/2, \varphi) = 0$ unless $L + \lambda =$ even, therefore we deduced the necessary conditions (1.22).

The conditions (1.20) to (1.22) can be used to identify the spins of certain states which are populated, or used to explain why certain states are populated and others are not. Assuming that one knows L_1, Q_{eff}, R_1, R_2 and R, one can choose λ_1, λ_2 which satisfy conditions (1.20) to (1.22). The knowledge of λ_2 sets a lower limit for L_2. Since transfer probability is largest when the transfer cluster is in the reaction plane, i.e. $|\lambda_2| = L_2$, this sometimes helps to identify the spin of the final nucleus.

In the situation where L_1 is small, i.e. $|\lambda_1|$ is also small, the kinematic condition (1.20a) requires λ_2 to be large when the linear momentum transfer $m_k v = \hbar k_0$ is large. This is realized at high incident energy when the relative velocity v is large and particularly when the number of nucleons in the cluster is large. Also, when $|\lambda_1|$ is small, condition (1.21a) requires λ_2 to be large when the value for Q_{eff} is large and negative. In short, the kinematic conditions (1.20) and (1.21) indicate that with high incident energy, transfer reactions favour population of high-spin states at high excitation energy.

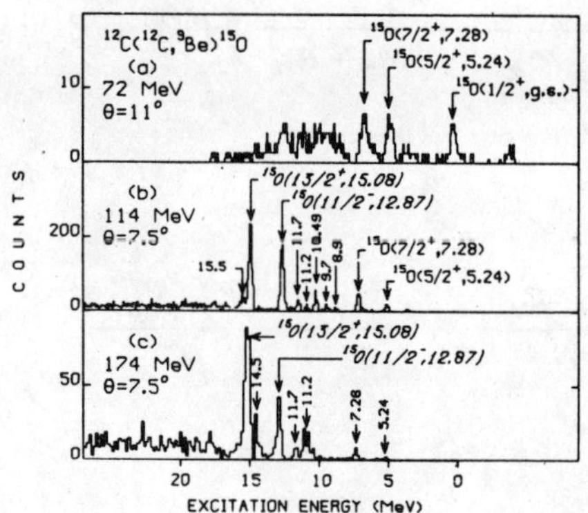

Fig.1.10. States excited in ^{15}O by the reaction $^{12}C(^{12}C,^9Be)^{15}O$ at different energies [AN 74].

The validity of conditions (1.20) and (1.21) is nicely demonstrated in a series of experiments by Anyas-Weiss et al. [AN 74] on cluster transfer. In Fig.(1.10), we reproduce one of their spectra of the reaction $^{12}C(^{12}C,^9Be)^{15}O$. One notes that the low spin states at low excitation energies gradually disappear from the spectrum as the energy is increased, as predicted by conditions (1.20) - (1.21).

For the cases where transfers occur between similar nuclei such that we can approximate $R_1 \approx R_2 \approx R/2$, and $z_1^i z_2^i \approx z_1^f z_2^f$, conditions (1.20a) and (1.21a) reduce to

$$\lambda_2 + \lambda_1 = 2m_k v R / \hbar \; , \tag{1.27}$$

$$\lambda_2 - \lambda_1 = -2QR / (\hbar v) \; . \tag{1.28}$$

From these, we obtain

$$\lambda_2 = \frac{2m_k v R}{\hbar} - 2QR/(\hbar v) \; , \tag{1.29}$$

which, at high energy, becomes

$$\lambda_2 \hbar \approx 2m_k v R \; . \tag{1.30}$$

From the argument given previously, high transfer probability occurs when $\lambda_2 \hbar = L_2 \hbar$, i.e.

$$L_2 \hbar \approx 2m_k v R \; . \tag{1.31}$$

The transfer cluster carries into the final nucleus an amount of orbital angular momentum which corresponds to the cluster's share of the angular momentum of relative motion of the scattering nuclei.

We have used the kinematic conditions (1.20a) - (1.22a) to explain the selective population of certain states. We have not discussed, but must bear in mind, nuclear structure arguments which also play a part in the selectivity.

For the rest of this section, we examine the qualitative features one expects of the angular distribution of transfer reaction.

When the incident energy is below the Coulomb barrier, semi-classical theories [BR 56, BR 72b] give the differential cross section as

$$\frac{d\sigma}{d\Omega} = \left(\frac{d\sigma}{d\Omega}\right)_{Rutherford} \cdot P \quad , \tag{1.32}$$

where P is the transfer probability along the trajectory appropriate for the scattering angle θ. The transfer probability depends on the distance of closest approach which is related directly to the scattering angle. Therefore, one expects a rise in transfer cross section at the backward angle. In Fig. (1.11), we show the differential cross section for the reaction $^{208}Pb(^{16}O.^{15}N)^{209}Bi$ in which the experimental points show quite nicely this rise towards 180°.

As the incident energy is increased above the Coulomb barrier, a quantal approach is more appropriate. From Eq.(A.91), the differential cross section is

$$\frac{d\sigma}{d\Omega} = \frac{v_f}{v_i}\left| f_{i\to f}(\theta) \right|^2 \quad . \tag{1.33}$$

For simplicity, we assume $(v_f/v_i) \approx 1$. We assume further that all the nuclei involved are spinless. In this instance, one may write the scattering amplitude, in the form of Eq. (A.35), as

$$f_{i\to f}(\theta) = \frac{1}{k}\sum_{\ell}(2\ell+1)\, g_\ell\, P_\ell(\cos\theta) \quad . \tag{1.34}$$

According to Strutinskii [ST 64], the following parametrization is introduced:

$$\sqrt{2\ell+1}\; g_\ell = exp\left[-(\ell-\ell_0)^2/\Gamma^2\right] exp\left[i\,2\delta_\ell\right] \quad . \tag{1.35}$$

The quantity $\sqrt{2\ell+1}\; g_\ell$ peaks at a certain partial ℓ_0, which may be the grazing partial wave ℓ_{gr} in elastic scattering but not necessarily so. At low ℓ, $\sqrt{2\ell+1}\; g_\ell$ vanishes because of absorption into other more complicated reaction channels; at high ℓ, $\sqrt{2\ell+1}\; g_\ell$ vanishes because the two nuclei are too far apart for transfer to take place. The peak of $\left|\sqrt{2\ell+1}\; g_\ell\right|$ has a width of 2Γ which may be adjusted to fit the experiment. The "phase shift" δ_ℓ may be expanded in power series of ℓ about ℓ_0,

$$\delta_\ell = \delta_{\ell_0} + \frac{\omega}{2}(\ell-\ell_0) + \cdots \quad , \tag{1.36}$$

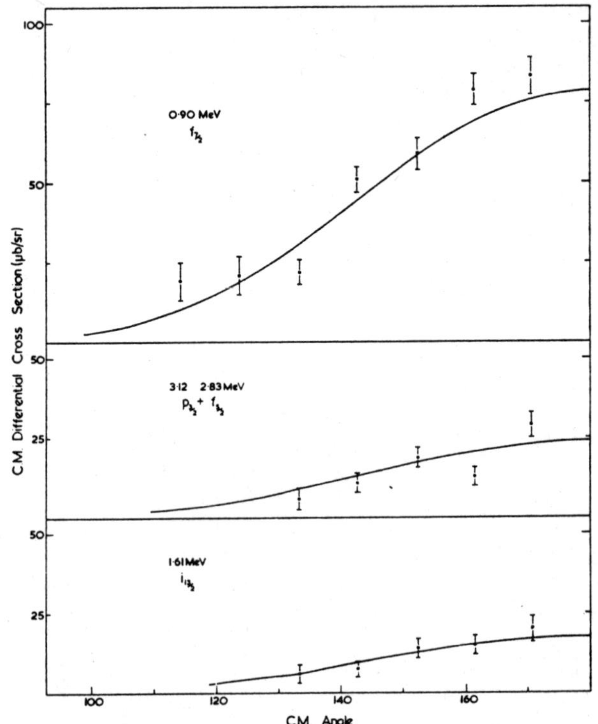

Fig.1.11. The differential cross section for the reaction
^{208}Pb(^{16}O,^{15}N)^{209}Bi taken at a bombarding energy
69.1 MeV. The data shown are transitions to the
2f 7/2 level, the 1i 13/2 level and the unresolved
3p 3/2 and 2f 5/2 levels. The full curves drawn
are the results of a DWBA calculation [BA 71].

where δ_{ℓ_0} is the "phase shift" of the ℓ_0th-partial wave. In Eq.(1.36),
the parameter $\textcircled{H}$ is

$$\textcircled{H} = 2\frac{\partial \delta_\ell}{\partial \ell} \quad . \tag{1.37}$$

In heavy-ion transfer reaction, ℓ_0 is usually large ($\ell_0 \gtrsim 20$), therefore one may use the asymptotic formula

$$P_\ell(\cos\theta) \xrightarrow[\theta > \frac{1}{\ell}]{} \frac{\sin\left[(\ell+\frac{1}{2})\theta + \frac{\pi}{4}\right]}{\sqrt{\frac{\pi}{2}(\ell+\frac{1}{2})\sin\theta}} \quad . \tag{1.38}$$

Substituting Eqs. (1.35), (1.36) and (1.38) into (1.34), the scattering amplitude can be evaluated in closed form after the sum over ℓ is converted into an integral over ℓ $(0 \leqslant \ell \leqslant \infty)$,

$$f_{i\to f}(\theta) = \frac{e^{i2\delta_{\ell_0}}}{k} \int_0^\infty \sqrt{2\ell+1}\, \exp\left[-(\ell-\ell_0)^2/\Gamma^2\right]$$

$$\times \exp\left[i\Theta(\ell-\ell_0)\right] \cdot \frac{\sin\left[(\ell+\frac{1}{2})\theta + \frac{\pi}{4}\right]}{\sqrt{\frac{\pi}{2}(\ell+\frac{1}{2})\sin\theta}} \cdot d\ell \ . \tag{1.39}$$

After some algebra, the expression (1.39) becomes just the difference of two Gaussian integrals. The final expression of the differential cross section is

$$\frac{d\sigma}{d\Omega} = \frac{\Gamma^2}{4k^2\sin^2\theta}\, \exp\left[-(\Theta-\theta)^2\Gamma^2/2\right]$$

$$\times \left\{1 + \exp(-2\Gamma^2\Theta\theta) - 2\exp(-\Gamma^2\Theta\theta)\cos\left[(2\ell_0+1)\theta + \frac{\pi}{2}\right]\right\} \ . \tag{1.40}$$

The first term is a Gaussian function centered around $\theta = \Theta$, with a width of about $2(2/\Gamma^2)^{1/2}$. The second term is usually much smaller in magnitude than the first, but is responsible for a slight skewing of the centre. The third term, which is more important on the low-θ side of the Gaussian than on the high-θ side, introduces oscillations into the Gaussian envelope. In Fig.(1.12), we sketch the angular distribution given by Eq.(1.40). The period of oscillations due to the third term is approximately $\Delta\theta = \pi/\ell_0$. These oscillations can be interpreted as interference between waves scattered from the opposite sides of the nucleus, see Fig.(1.13), or as the interference pattern from a 2-slits experiment. In the latter instance, the period of the oscillations

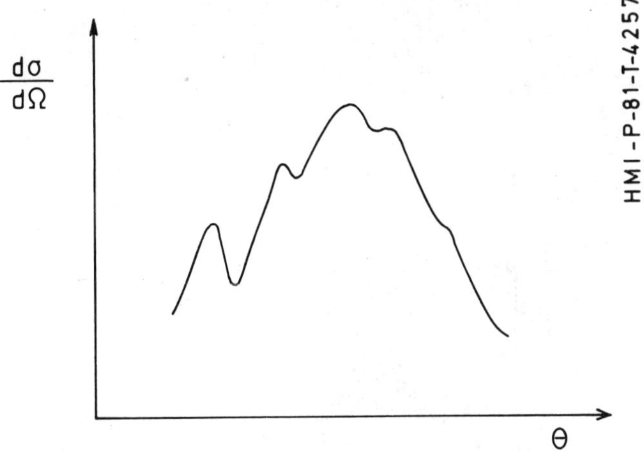

Fig.1.12.   Schematic plot of dσ/dΩ given by Eq.(1.33).

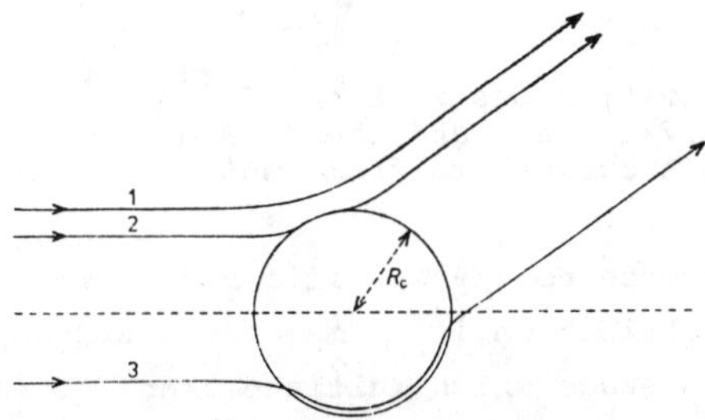

Fig.1.13.   The classical trajectories labelled 1, 2 and 3 are scattered
to the same scattering angle [PH 77].

should be

$$\Delta\theta = \frac{\lambda}{2(R_1+R_2)} = \frac{\pi}{k(R_1+R_2)} \approx \frac{\pi}{\ell_o} \quad , \qquad (1.41)$$

which agrees with that given by Eq. (1.40). In some experiment, the angular distributions do show a remarkable resemblance to the prediction of Eq.(1.40) (compare the experimental results in Fig.(1.14) with the schematic plot of Fig.(1.13)).

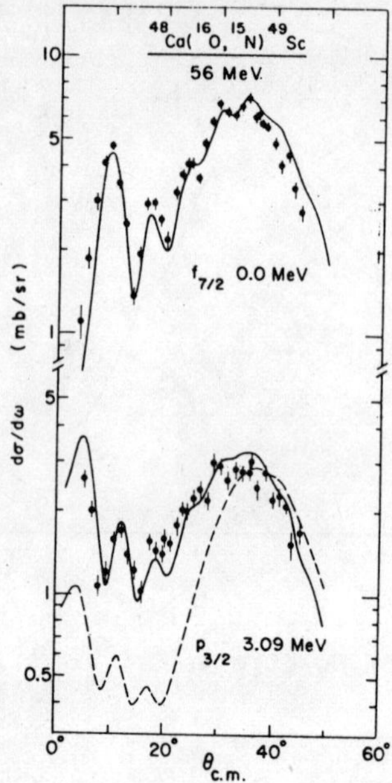

Fig.1.14. Differential cross section of ^{48}Ca(^{16}O, ^{15}N)^{49}Sc to the
ground f7/2 state and the 3.09 MeV p 3/2 state [Ko 74].
The drawn curves are from DWBA calculations.

Without reference to any specific model, we can also view the width
of the angular distribution in terms of the spread of the classical trajectories in the ℓ-space which initiate transfer reaction. Suppose this
bundle of trajectories has a width 2Γ. They are responsible for a spread
in the angular distribution $(\Delta\theta)_{classical}$. Therefore the observed width
in the angular distribution is

$$\Delta\theta \approx \sqrt{(\Delta\theta)^2_{classical} + \left(\frac{1}{2\Gamma}\right)^2} \quad , \tag{1.42}$$

in which $[1/2\Gamma)]$ is due to the uncertainty relation. The dominance of
trajectories dispersion over quantal dispersion results if the inequality

$$(\Delta\theta)_{classical} \gg \frac{1}{2\Gamma} \tag{1.43}$$

is true. To illustrate the usefulness of this criteria, let us assume that the trajectories are Coulomb trajectories, of which the grazing trajectory is of orbital angular momentum ℓ_o. Then we have

$$\ell_o = \eta \cot \frac{\theta_o}{2} \quad ,$$

$$2\Gamma = \Delta\ell_o = \eta \csc^2\left(\frac{\theta_o}{2}\right) \cdot \Delta\theta \quad ,$$

$$(\Delta\theta)_{classical} = \frac{2\Gamma}{\eta \csc^2\left(\frac{\theta_o}{2}\right)} \quad . \tag{1.44}$$

Using Eq. (1.44), the inequality of (1.43) is

$$\frac{2\Gamma}{\eta \csc^2\left(\frac{\theta_o}{2}\right)} > \frac{1}{2\Gamma} \quad ,$$

$$(2\Gamma)^2 > \eta \csc^2\left(\frac{\theta_o}{2}\right) \quad . \tag{1.45}$$

This criteria is satisfied in the reaction $^{208}\text{Pb}(^{11}\text{B},^{10}\text{Be})^{209}\text{Bi}$ with incident energy 72.2 MeV [FO 74], see Fig. (1.15). Based on a quantal calculation of the scattering amplitude, one may set $\ell_o \approx 35$ and $2\Gamma = 12$. With $\eta \approx 27$, hence $\theta_o = 74°$, we obtain $\eta\csc^2(\theta_o/2) \approx 75$, which is smaller than $(2\Gamma)^2 \approx 150$. The width of the angular distribution predicted by Eq.(1.42) is $\Delta\theta \approx 10°$, which is smaller than observed, Fig. (1.15), but is still reasonable in view of the fact that the trajectories are not Coulomb trajectories.

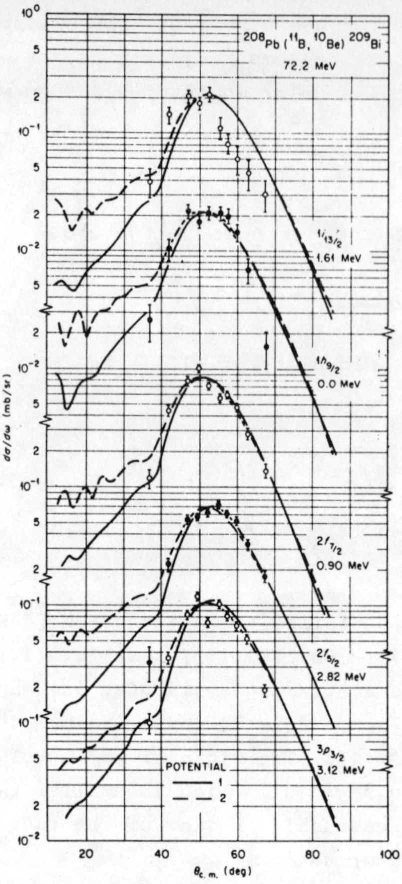

Fig.1.15. The differential cross sections for the reaction
^{208}Pb(^{11}B, ^{10}Be)^{209}Bi with incident energy 72.2 MeV.
Data are shown for the single-particle levels in
^{209}Bi up to the 3p 3/2 level at 3.12 MeV. Drawn
curves are DWBA calculations [Fo 74].

A different situation arises in the reaction ^{26}Mg(^{11}B, ^{10}B)^{27}Mg at
incident energy 114 MeV, see Fig. (1.16) [PA 75]. A quantal estimate of
ℓ_o yields $\ell_o \approx 38$. With $\eta = 3$, this leads to $\theta_o \approx 10°$. In this instance,
we have $\eta\csc^2(\theta_o/2) \approx 400$. An estimate of 2Γ from [PA 75] gives $2\Gamma \approx 12$,
such that $(2\Gamma)^2 < \eta\csc^2(\theta_o/2)$. The structures of the angular distributions in Fig.(1.16) are therefore entirely quantal in origin. In the context of the Strutinskii model, they may be the remnant of the third term,
the interference term, of Eq.(1.40), which does have a period of
$\Delta\theta \approx \pi/\ell_o \approx 4.5°$.

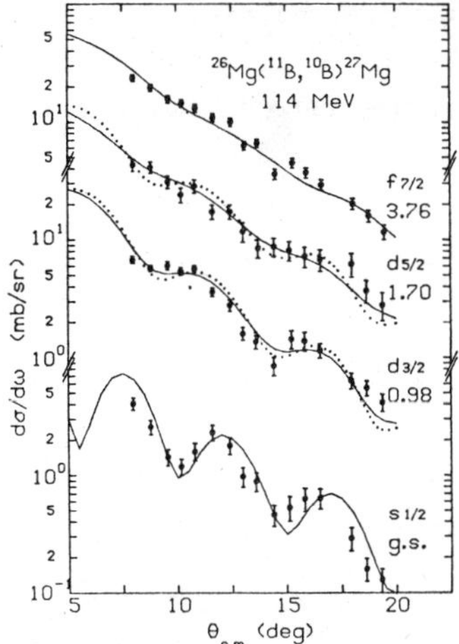

Fig.1.16.   The differential cross sections of the reaction
^{26}Mg(^{11}B,^{10}B)^{27}Mg with incident energy 114 MeV.
Data are shown for the single particle levels of
^{27}Mg up to f 7/2 level at 3.76 MeV. Drawn curves
are DWBA calculations[PA 75].

1.3 Exact DWBA Heavy-Ion Transfer Reaction

In this subsection, we discuss the theory of distorted wave Born
approximation (DWBA) for the analysis of heavy-ion single nucleon transfer reaction. We concentrate on the method proposed by Buttle [BU 78],
whose computer code is available to the general public and is easy to
use. Other unpublished codes due to P. Kunz of the University of Colo-
rado and due to R.M. DeVries of Los Alamos have also been widely used.
In contrast to the approximate method being described in the next subsection, the DWBA T-matrix is evaluated exactly in these codes and it
is usually referred to as the exact finite-range (EFR) calculation.

We consider the reaction at energy E

$$\underset{a_1}{(C_1 + t)} + C_2 \longrightarrow C_1 + \underset{a_2}{(C_2 + t)} , \qquad (1.46)$$

in which a nucleus a_1 collides with nucleus c_2. The nucleon t, which is initially bound to the core c_1 to form the nucleus a_1, is transferred to the core c_2 to form the nucleus a_2. For the initial channel, the following Hamiltonian is introduced

$$H = h_o(a_1) + h_o(c_2) + T_{c_2 a_1} + V_{c_2 a_1}$$

$$= h_o(a_1) + h_o(c_2) + T_{c_2 a_1} + V_{c_2 c_1} + V_{c_2 t}$$

$$= H_o + U^{opt}_{c_2 a_1} + (V_{c_2 c_1} + V_{c_2 t} - U^{opt}_{c_2 a_1}) , \qquad (1.47)$$

where the unperturbed Hamiltonian H_o is

$$H_o = h_o(a_1) + h_o(c_2) + T_{c_2 a_1} , \qquad (1.48)$$

with $h_o(a_1)$ and $h_o(c_2)$ representing the intrinsic Hamiltonians of the nuclei a_1 and c_2; $T_{c_2 a_1}$ is the kinetic energy operator of the relative motion between a_1 and c_2. The interaction between the two nuclei $V_{c_2 a_1}$ is assumed to be made up of two pieces, $V_{c_2 c_1}$ and $V_{c_2 t}$, representing respectively the interaction between the cores c_2 and c_1, and core c_2 with the nucleon t. The optical potential $U^{opt}_{c_2 a_1}$ is usually chosen to be a complex phenomenological potential which describes as accurately as possible the elastic scattering in this channel between the nuclei c_2 and a_1. By the same argument, the Hamiltonian for the final (rearrangement) channel is represented by

$$H = H'_o + U^{opt}_{c_1 a_2} + (V_{c_1 c_2} + V_{c_1 t} - U^{opt}_{c_1 a_2}) , \qquad (1.49)$$

in which the unperturbed Hamiltonian H'_o is

$$H'_o = h_o(c_1) + h_o(a_2) + T_{c_1 a_2} . \qquad (1.50)$$

The optical potential $U^{opt}_{c_1 a_2}$ describes the elastic scattering between the nuclei c_1 and a_2 in the final channel.

The differential cross section of the transfer reaction is given, as in Eq. (A.108), by

$$\frac{d\sigma}{d\Omega} = \frac{\mu_i \mu_f}{(2\pi\hbar^2)^2} \cdot \frac{k_f}{k_i} \cdot \frac{1}{(2a_1+1)(2c_1+1)} \sum_{\substack{\alpha_1 \alpha_2 \\ \gamma_1 \gamma_2}} \left| T_{\gamma_1 \alpha_2 \alpha_1 \gamma_2} \right|^2 , \qquad (1.51)$$

in which the T-matrix in the DWBA approximation is written as

$$T_{\gamma_1 \alpha_2 \alpha_1 \gamma_2} = \left\langle \chi_f^{(-)} ; \gamma_1 \alpha_2 \alpha_1 \gamma_2 \left| \Delta V \right| \chi_i^{(+)} ; \gamma_1 \alpha_2 \alpha_1 \gamma_2 \right\rangle , \qquad (1.52)$$

with the interaction ΔV written in either the post- or prior-form,

$$\Delta V(\text{post}) = V_{c_1 c_2} + V_{c_1 t} - U_{c_1 a_2}^{opt} , \qquad (1.53)$$

$$\Delta V(\text{prior}) = V_{c_2 c_1} + V_{c_2 t} - U_{c_2 a_1}^{opt} . \qquad (1.54)$$

In Eq.(1.51), μ_i and μ_f are the reduced masses, k_i and k_f are the wave numbers in the initial and final channels. Since there is very little chance of confusion, we label the spins of the nuclei by a_1, a_2 and c_1, c_2 whose z-components are α_1, α_2 and γ_1, γ_2. In Eq.(1.51), the spin components of the final nuclei are summed and the spin components of the initial nuclei are averaged.

Assuming that the nuclei a_1 is made up of an inert core c_1 plus a nucleon, the overlap between the nuclei a_1 and c_1 may be written as

$$\langle c_1 \gamma_1 | a_1 \alpha_1 \rangle = \sum_{j_1 \ell_1} \Theta_{j_1 \ell_1} \sum_{m_1 \sigma_1 \mathcal{S}_1} (c_1 \gamma_1 j_1 \mathcal{S}_1 | a_1 \alpha_1)$$

$$\times (\ell_1 m_1 s \sigma_1 | j_1 \mathcal{S}_1) \phi_{\ell_1 m_1}(\vec{r}_1) \chi_{\sigma_1} , \qquad (1.55)$$

where χ_{σ_1} is the spin function of transferred nucleon with z-component of the spin equals to σ_1. The spin function is coupled to the orbital motion of the nucleon, with quantum number (ℓ_1, m_1), to give the total angular momentum of the nucleon as j_1 with z-component $\mathcal{S}_1$. The $(j_1 \mathcal{S}_1)$ state is coupled to the core state $(c_1 \gamma_1)$ to yield the $(a_1 \alpha_1)$

ground state. The reduced width amplitudes $\theta_{j_1\ell_1}$ gives the probability amplitude of finding a particular $(j_1\ell_1)$ nucleon state. One can write down a similar overlap between c_2 and a_2.

In terms of these quantities, the sum of the modulus squared of the T-matrix in Eq. (1.51) becomes

$$\sum_{\substack{\gamma_1\alpha_2 \\ \gamma_2\alpha_1}} \left| T_{\gamma_1\alpha_2,\,\gamma_2\alpha_1} \right|^2 = \sum_{\substack{\gamma_1\alpha_2 \\ \gamma_2\alpha_1}} \left| \sum_{\substack{\ell_1 m_1 \\ \ell_2 m_2}} \sum_{\substack{j_1 \mathcal{I}_1 \\ j_2 \mathcal{I}_2}} \sum_{\sigma_1} \theta_{j_1\ell_1}\,\theta_{j_2\ell_2} \right.$$

$$\times\, (\ell_1 m_1\, S\sigma_1 \,|\, j_1 \mathcal{I}_1)(C_1 \gamma_1\, j_1 \mathcal{I}_1 \,|\, a_1\alpha_1)$$

$$\times\, (\ell_2 m_2\, S\sigma_1 \,|\, j_2 \mathcal{I}_2)(C_2 \gamma_2\, j_2 \mathcal{I}_2 \,|\, a_2\alpha_2)$$

$$\times \left. \int \chi_f^{(-)*}(\vec{r}_f)\, \phi_{\ell_2 m_2}(\vec{r}_2)\, \Delta V\, \phi_{\ell_1 m_1}(\vec{r}_1)\, \chi_i^{(+)}(\vec{r}_i)\, d\vec{r}_f\, d\vec{r}_i \right|^2 . \qquad (1.56)$$

As in light-ion transfer reaction theory, the Clebsch-Gordan coefficients in Eq.(1.56) can be manipulated to yield simplified expression. Thus, using standard formulae [RO 57], we have

$$(\ell_1 m_1\, S\sigma_1 \,|\, j_1 \mathcal{I}_1)(\ell_2 m_2\, S\sigma_1 \,|\, j_2 \mathcal{I}_2)$$

$$= (-)^{S+\sigma_1} \sqrt{(2j_1+1)(2j_2+1)} \sum_{L} (j_1 -\mathcal{I}_1\, j_2 \mathcal{I}_2 \,|\, LM)$$

$$\times (\ell_1 m_1\, \ell_2 -m_2 \,|\, L-M)\, W(\ell_1 j_1\, \ell_2 j_2\, ;\, SL) . \qquad (1.57)$$

Furthermore, the following contractions can be carried out,

$$\sum_{\alpha_2 \gamma_2} (C_2 \gamma_2\, j_2 \mathcal{I}_2 \,|\, a_2\alpha_2)(C_2 \gamma_2\, j_2' \mathcal{I}_2' \,|\, a_2\alpha_2) \sqrt{(2j_2+1)(2j_2'+1)}$$

$$= (2a_2+1)\, \delta_{j_2 j_2'}\, \delta_{\mathcal{I}_2 \mathcal{I}_2'} , \qquad (1.58)$$

and similarly

$$\sum_{\alpha,\gamma_1} (c,\gamma_1 j_1 s_1 \mid a_1 \alpha_1)(c,\gamma_1 j_1' s_1' \mid a_1 \alpha_1)\sqrt{(2j_1+1)(2j_1'+1)}$$

$$= (2a_1+1)\,\delta_{j_1 j_1'}\,\delta_{s_1 s_1'}\quad, \tag{1.59}$$

and

$$\sum_{s_1 s_2} (j_1 -s_1\, j_2 s_2 \mid L M)(j_1 -s_1\, j_2 s_2 \mid L'M') = \delta_{LL'}\,\delta_{MM'}. \tag{1.60}$$

Making use of these contractions, we can simplify Eq.(1.56) to the
following form,

$$\sum_{\substack{\gamma_1 \alpha_2,\,\gamma_2 \alpha_1}} \left| T_{\gamma_1 \alpha_2,\gamma_2 \alpha_1} \right|^2 = (2a_1+1)(2a_2+1)\sum_{j_1 j_2}\sum_{L,M}$$

$$\times \left| \sum_{\substack{\ell_1 m_1 \\ \ell_2 m_2}} (-)^{-M-m_1}\,\Theta_{j_1 \ell_1}\,\Theta_{j_2 \ell_2}\,(\ell_1 m_1\, \ell_2 -m_2 \mid L -M)\,W(\ell_1 j_1\, \ell_2 j_2\,;\,SL)\right.$$

$$\left.\times \int \chi_f^{(-)*}(\vec{r}_f)\,\phi_{\ell_2 m_2}(\vec{r}_2)\,\Delta V\,\phi_{\ell_1 m_1}(\vec{r}_1)\,\chi_i^{(+)}(\vec{r}_i)\,d\vec{r}_i\,d\vec{r}_i \right|^2 . \tag{1.61}$$

Assuming that there is no spin-orbit force for the relative motion of
the nuclei in the initial and final channels, we may write the distorted
waves in partial wave expansions as (cf. Eq. (A.24),

$$\chi_i^{(+)}(\vec{r}_i) = 4\pi \sum_{\ell_i m_i} i^{\ell_i} f_{\ell_i}(r_i)\, Y_{\ell_i}^{m_i*}(\hat{k}_i)\, Y_{\ell_i}^{m_i}(\hat{r}_i) , \tag{1.62}$$

$$\chi_f^{(-)}(\vec{r}_f) = 4\pi \sum_{\ell_f m_f} i^{\ell_f} f_{\ell_f}^*(r_f)\, Y_{\ell_f}^{m_f*}(\hat{k}_f)\, Y_{\ell_f}^{m_f}(\hat{r}_f) . \tag{1.63}$$

On substitution of Eqs. (1.62), (1.63) into Eq. (1.61), the differential cross section of Eq. (1.51) for a specific transition from a $(j_1\ell_1)$ state to a $(j_2\ell_2)$ state is

$$\frac{d\sigma}{d\Omega} = \left(\frac{4\pi}{\hbar}\right)^4 \mu_i \mu_f \frac{k_f}{k_i} \cdot \frac{2a_2+1}{2a_1+1}\,\Theta_{j_1\ell_1}^2\,\Theta_{j_2\ell_2}^2 \sum_{L,M} \left| T_{LM} \right|^2 \,, \qquad (1.64)$$

where

$$T_{LM} = \frac{W(\ell_1 j_1 \ell_2 j_2 ; SL)}{2\pi} \sum_{\substack{\ell_i\, m_i\, m_1 \\ \ell_f\, m_f\, m_2}} i^{\ell_i - \ell_f}\,(-)^{-m_1}\,(\ell_1 m_1\, \ell_2 - m_2 \mid L\text{-}M)$$

$$\times Y_{\ell_i}^{m_i\,*}(\hat{k}_i)\, Y_{\ell_f}^{m_f}(\hat{k}_f)$$

$$\times \int d\vec{r_i}\, d\vec{r_f}\, f_{\ell_f}(r_f)\, Y_{\ell_f}^{m_f\,*}(\hat{r}_f)\, \phi_{\ell_2 m_2}(\vec{r}_2)\, \Delta V\, \phi_{\ell_1 m_1}(\vec{r}_1)\, f_{\ell_i}(r_i)\, Y_{\ell_i}^{m_i}(\hat{r}_i) \,. \qquad (1.65)$$

The coordinates $\vec{r}_1$, $\vec{r}_2$, $\vec{r}_i$ and $\vec{r}_f$ are illustrated in Fig. (1.17). The integration over $\vec{r}_1$ may be changed to an integration over $\vec{r}_f$ by putting $d\vec{r}_1 = J d\vec{r}_f$, where J is the Jacobian,

$$J = \frac{m_{a_1} + m_{a_2}}{m_t (m_{a_1} + m_{c_2})} \,, \qquad (1.66)$$

with the m's denoting the masses of the nuclei.

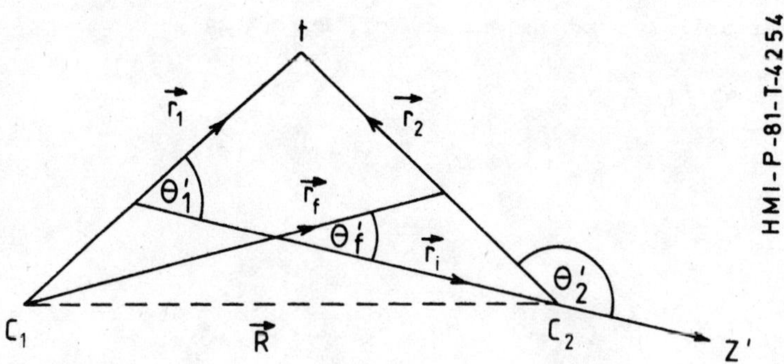

Fig.1.17. The coordinate vectors.

An intuitive interpretation of the integral over $\vec{r}_1$ and $\vec{r}_i$ (or $\vec{r}_i$ and $\vec{r}_f$) is that the contributions to the integral are being summed over all possible triangular configurations $(C_1 C_2 t)$ as shown in Fig. (1.17) and over all possible orientations, with respect to a set of laboratory axes, of these triangular configurations. The coordinates which define the orientations are the "external" coordinates which, as we shall see shortly, can be integrated over analytically.

The polar coordinates $(r_i, \theta_i, \phi_i, r_f, \theta_f, \phi_f)$ are measured relative to a set of laboratory axes. To facilitate the separation of the "external" coordinates, we consider a rotation of the laboratory axes (oXYZ) to a new set of axes (oX'Y'Z') in which the Z'-axis is in the direction of $\vec{r}_i$ and $\vec{r}_f$ (and therefore also $\vec{r}_1$ and $\vec{r}_2$) is in the oX'Y' plane. The rotation $\hat{R}$ to achieve this is through the Euler angles

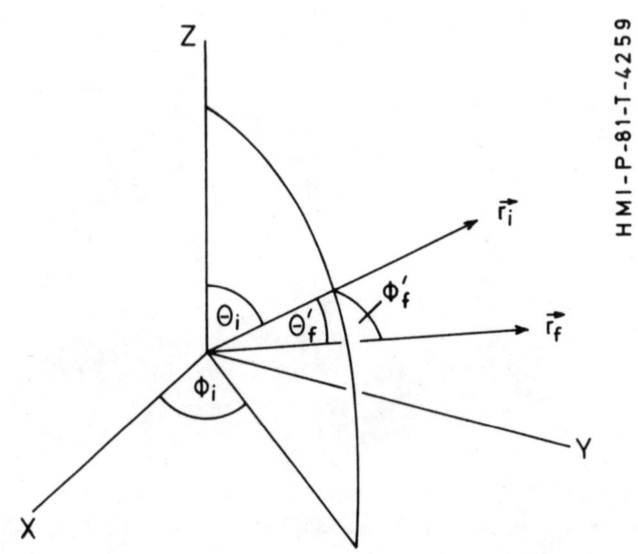

Fig.1.18.   The Euler angles involved in the rotation $\hat{R}$.

$(\phi_i, \theta_i, \phi'_f)$ as indicated in Fig. (1.18). The $\vec{r}_f$ integral may be done in terms of r_f, θ'_f, ϕ'_f, with θ'_f and ϕ'_f defined as in Figs. (1.17) and (1.18), such that

$$d\vec{r_i}\, d\vec{r_f} \equiv r_i^2\, dr_i\, d\cos\theta_i\, d\phi_i\; r_f^2\, dr_f\, d\cos\theta_f'\, d\phi_f'$$

$$= r_i^2\, dr_i\; r_f^2\, dr_f\, d\cos\theta_f' \cdot d\phi_i\, d\cos\theta_i\, d\phi_f' \tag{1.67}$$

The integration of the Euler angles $d\phi_i\, d\cos\theta_i d\phi_f'$ in Eq. (1.67) is the integration over the "external" orientation coordinates we alluded to in the previous paragraph. The functions in Eq. (1.65) which depend on $\vec{r}_i$, $\vec{r}_f$, $\vec{r}_1$ and $\vec{r}_2$ may be expressed relative to the new coordinate axes as follows [BR 62]:

$$\phi_{\ell_1 m_1}(\vec{r_i}) = \sum_{m_i'} \phi_{\ell_1 m_i'}(\vec{r_i'})\, D^{\ell_1}_{m_i' m_1}(\hat{\mathcal{R}}^{-1}) \quad,$$

$$\phi_{\ell_2 m_2}(\vec{r_2}) = \sum_{m_2'} \phi_{\ell_2 m_2'}(\vec{r_2'})\, D^{\ell_2}_{m_2' m_2}(\hat{\mathcal{R}}^{-1}) \quad,$$

$$Y_{\ell_i}^{m_i}(\hat{r_i}) = \sum_{m_i'} Y_{\ell_i}^{m_i'}(\hat{r_i'})\, D^{\ell_i}_{m_i' m_i}(\hat{\mathcal{R}}^{-1}) \quad,$$

$$Y_{\ell_f}^{m_f}(\hat{r_f}) = \sum_{m_f'} Y_{\ell_f}^{m_f'}(\hat{r_f'})\, D^{\ell_f}_{m_f' m_f}(\hat{\mathcal{R}}^{-1}) \quad. \tag{1.68}$$

Since $\vec{r}_1'$, $\vec{r}_2'$, $\vec{r}_i'$ and $\vec{r}_f'$ in Eq. (1.68) depend on r_i, r_f and θ_f' only, the integration of Eq. (1.65) can be separated into two parts: an integral over the "external" orientation coordinates $R = (\phi_i,\ \theta_i,\ \phi_f')$ and an integral over the "intrinsic" coordinates r_i, r_f and θ_f'. To integrate over the orientations $R\ (\phi_i,\ \theta_i,\ \phi_f')$, we make use of the following contractions [RO 57],

$$D^{\ell_2}_{m_2' m_2}(\hat{R}^{-1})\, D^{\ell_1}_{m_1' m_1}(\hat{R}^{-1})$$

$$= \sum_{L_1} (-)^{m_1-m_1'} (\ell_1 -m_1\, \ell_2\, m_2\,|\,L,M_1)(\ell_1 -m_1'\, \ell_2\, m_2'\,|\,L,M_1')\, D^{L_1}_{M,M_1'}(\hat{R})\ , \qquad (1.69)$$

and the following integration property of the D-functions [RO 57]:

$$\int D^{\ell_f *}_{m_f' m_f}(\hat{R}^{-1})\, D^{\ell_i}_{m_i' m_i}(\hat{R}^{-1})\, D^{L_1}_{M,M_1'}(\hat{R})\, d\phi_i\, d\cos\theta_i\, d\phi_f'$$

$$\qquad\qquad (1.70)$$

$$= \frac{8\pi^2}{2\ell_i+1}\ (\ell_f m_f\, L,M_1\,|\,\ell_i m_i)(\ell_f m_f'\, L,M_1'\,|\,\ell_i m_i')\ .$$

With the application of Eqs. (1.69) and (1.70), Eq. (1.65) becomes

$$T_{LM} = 4\pi\cdot JW(\ell_1 j_1\, \ell_2 j_2; SL) \sum_{\substack{\ell_i \\ m_i m_i'}} \sum_{\substack{\ell_f \\ m_f m_f'}} \sum_{\substack{L_1 \\ m_1\, m_2 \\ m_1'\, m_2'}} Y^{m_f}_{\ell_f}(\hat{k}_f)\, Y^{m_i *}_{\ell_i}(0,0)$$

$$\times i^{\ell_i-\ell_f}\, \frac{1}{2\ell_i+1}\ (\ell_1 m_1\, \ell_2 -m_2\,|\,L-M)(\ell_1 -m_1\, \ell_2\, m_2\,|\,L,M_1)$$

$$\times (\ell_1 -m_1'\, \ell_2\, m_2'\,|\,L,M_1')(\ell_f m_f\, L,M_1\,|\,\ell_i m_i)(\ell_f m_f'\, L,M_1'\,|\,\ell_i m_i')$$

$$\times \int d\cos\theta_f'\, r_i^2\, dr_i\, r_f^2\, dr_f\, f_f(r_f)\, Y^{m_f' *}_{\ell_f}(\theta_f',0)\, \phi_{\ell_2}(r_2)\, Y^{m_2'}_{\ell_2}(\theta_2',0)$$

$$\times \Delta V\, \phi_{\ell_1}(r_2)(-)^{m_1'}\, Y^{m_1'}_{\ell_1}(\theta_1',0)\, Y^{m_i}_{\ell_i}(0,0)\, f_i(r_i)\ . \qquad (1.71)$$

After some trivial manipulation of the Clebsch-Gordan coefficients, which includes the contraction

$$\sum_{m_1 m_2} (\ell_1 m_1\, \ell_2 -m_2\,|\,L,-M_1)(\ell_1 m_1\, \ell_2 -m_2\,|\,L-M) = \delta_{L_1 L}\, \delta_{M_1 M}\ , \qquad (1.72)$$

the final expression for $T_{LM}(\theta,0)$ is

$$T_{LM}(\theta,0) = J W(\ell_1 j_1 \ell_2 j_2; SL) \sum_{\ell_i \ell_f} i^{\ell_i - \ell_f} (-)^M Y_{\ell_f}^M (\theta,0)$$

$$\times (\ell_f - M\, LM \mid \ell_i\, 0)\, X_{\ell_i \ell_f L} \quad, \tag{1.73}$$

with

$$X_{\ell_i \ell_f L} = \sum_{m_1' m_2'} (\ell_1 m_1'\, \ell_2 - m_2' \mid LM_i')(\ell_f - M_i'\, LM_i' \mid \ell_i\, 0)\, F_{\ell_i \ell_f}^{m_1' m_2'}, \tag{1.74}$$

where

$$F_{\ell_i \ell_f}^{m_1' m_2'} = F_{\ell_i \ell_f}^{-m_1' -m_2'} = \int_0^\infty r_i^2\, dr_i\, f_{\ell_i}(r_i)\, G_{\ell_f}^{m_1' m_2'}(r_i) \quad, \tag{1.75}$$

$$G_{\ell_f}^{m_1' m_2'} = \int_0^\infty r_f^2\, dr_f\, f_{\ell_f}(r_f)\, H_{\ell_f}^{m_1' m_2'}(r_i, r_f) \quad, \tag{1.76}$$

$$H_{\ell_f}^{m_1' m_2'}(r_i, r_f) = \int_{-1}^{1} d\cos\theta_f'\, Y_{\ell_2}^{m_2'}(\theta_2',0)\, Y_{\ell_f}^{M_i'}(\theta_f',0)\, \phi_{\ell_2}(r_2)$$

$$\times \Delta V\, \phi_{\ell_1}(r_i)(-)^{m_1'}\, Y_{\ell_1}^{m_1'}(\theta_i',0) \quad. \tag{1.77}$$

In [BU 78], a function λ was introduced into Eq. (1.74), such that only those terms with non-negative values of m_2' and M_1' were summed. In view of the time consuming numerical integration for $F_{\ell_i \ell_f}^{m_1' m_2'}$, this device is indeed an advantage. The interaction ΔV has the form

$$\Delta V(post) = V_{c_1 c_2}(R) + V_{c,t}(r_i) - U_{c,a_2}^{opt}(r_f) \quad, \tag{1.53a}$$

$$\Delta V(prior) = V_{c_1 c_2}(R) + V_{c_2 t}(r_2) - U_{c_2 a_1}^{opt}(r_i) \quad. \tag{1.54a}$$

To complete the numerical integration for $F_{\ell_i \ell_f}^{m_1' m_2'}$, the following relations, which express r_1, r_2, θ_1', θ_2', in terms of r_i, r_f and θ_f' are required:

$$r_1^2 = \tilde{\beta}^2 r_i^2 + \tilde{\alpha}^2 r_f^2 - 2\tilde{\alpha}\tilde{\beta}\, r_i r_f \cos\theta_f' \quad, \tag{1.78}$$

$$r_2^2 = \tilde{\alpha}^2 r_i^2 + \tilde{\gamma}^2 r_f^2 - 2\tilde{\alpha}\tilde{\gamma}\, r_i r_f \cos\theta_f' \quad, \tag{1.79}$$

$$R^2 = \left(m_{a_1}^2 r_i^2 + m_{a_2}^2 r_f^2 + 2 m_{a_1} m_{a_2} r_i r_f \cos\theta_f' \right) / (m_{a_1} + m_{c_2})^2 \quad, \tag{1.80}$$

$$\cos\theta_1' = \left(-\tilde{\beta} r_i + \tilde{\alpha} r_f \cos\theta_f' \right) / r_1 \quad, \tag{1.81}$$

$$\cos\theta_2' = \left(-\tilde{\alpha} r_i + \tilde{\gamma} r_f \cos\theta_f' \right) / r_2 \quad, \tag{1.82}$$

$$\tilde{\alpha} = \frac{m_{a_1} m_{a_2}}{m_t (m_{a_1} + m_{c_2})} \quad, \qquad \tilde{\beta} = \frac{m_{a_1} m_{c_2}}{m_t (m_{a_1} + m_{c_2})} \quad,$$

$$\tilde{\gamma} = \frac{m_{c_1} m_{a_2}}{m_t (m_{a_1} + m_{c_2})} \quad. \tag{1.83}$$

The final results for the differential cross section and the total cross section, allowing for antisymmetrization of the nucleons as done in light-ion transfer reaction [AU 70], so that $\theta_{j\ell}^2$ becomes the spectroscopic factor $S_{j\ell}$, are

$$\frac{d\sigma}{d\Omega} = \left(\frac{4\pi}{\hbar}\right)^4 \mu_i \mu_f \frac{k_f}{k_i} S_{j_1\ell_1}^{(1)} S_{j_2\ell_2}^{(2)} \frac{2a_2+1}{2a_1+1} \sum_{L,M} |T_{LM}|^2 , \tag{1.84}$$

and

$$\sigma_T = \left(\frac{4\pi}{\hbar}\right)^4 \mu_i \mu_f \frac{k_f}{k_i} S_{j_1\ell_1}^{(1)} S_{j_2\ell_2}^{(2)} \frac{2a_2+1}{2a_1+1} \sum_{\ell_i \ell_f L} J^2 W(\ell_1 j_1, \ell_2 j_2; SL)^2$$

$$\times |X_{\ell_i \ell_f L}|^2 \quad. \tag{1.85}$$

The quantum number L, which is being summed over in Eqs. (1.84), (1.85), has its origin in the contraction of the Clebsch-Gordan coefficients of Eq. (1.57). Its physical origin can easily be interpreted as the angular momentum transfer in the reaction.The Racah coefficient $W(\ell_1 j, \ell_2 j_2; SL)$

guarantees that

$$|\ell_1 - \ell_2| \leq L \leq \ell_1 + \ell_2 \quad , \tag{1.86}$$

$$|j_1 - j_2| \leq L \leq j_1 + j_2 \quad . \tag{1.87}$$

1.4 Approximate DWBA for heavy-ion transfer reaction

The exact finite range DWBA (EFR-DWBA) is routinely evaluated nowadays. This has not always been the case. In the early days, approximations were introduced to simplify the six-dimensional integral of Eq. (1.65) [BU 66, BU 71, SC 70, BR 74]. They were remarkably accurate and at the same time yielded some interesting insight into the nature of the T-matrix element of Eq.(1.65) and into the physics of the transfer reaction. The approximate calculation of the T-matrix is now generally known as no-recoil DWBA (or NR-DWBA). The reason for this label will be apparent shortly.

The integrand of the six-dimensional integral in Eq.(1.65) is written in terms of the vectors $\vec{r}_i$, $\vec{r}_f$, $\vec{r}_1$, $\vec{r}_2$ and $\vec{R}$ displayed in Fig. (1.17). Only two of these are independent quantities and these two independent vectors are integrated over. Suppose we choose them to be $\vec{R}$ and $\vec{r}_1$ for this discussion. Hence, we have

$$\vec{r}_2 = \vec{r}_1 - \vec{R} \quad , \tag{1.88}$$

$$\vec{r}_i = \vec{R} - \frac{m_t}{m_{a_1}} \vec{r}_1 \quad , \tag{1.89}$$

$$\vec{r}_f = \frac{m_{c_2}}{m_{a_2}} \vec{R} + \frac{m_t}{m_{a_2}} \vec{r}_1 \quad . \tag{1.90}$$

In a no-recoil calculation, we substitute $\hat{r}_i \approx \hat{R}$ and $\hat{r}_f \approx \hat{R}$, which amounts to assuming $m_t \ll m_{a_1}$ and $m_t \ll m_{a_2}$, i.e. a no-recoil situation in the transfer of a practically massless particle m_t. As a result of the substitution, we can carry out the addition of the spherical harmonics,

$$Y_{\ell_f}^{m_f*}(\hat{r}_f)\, Y_{\ell_i}^{m_i}(\hat{r}_i) \approx Y_{\ell_f}^{m_f*}(\hat{R})\, Y_{\ell_i}^{m_i}(\hat{R})$$

$$= \sum_{L,M} Y_L^M(\hat{R})\,(-)^{m_f}\sqrt{\frac{(2\ell_i+1)(2\ell_f+1)}{4\pi(2L+1)}}\;(\ell_f\,{-m_f}\;\ell_i\,m_i\,|\,LM)(\ell_f\,0\;\ell_i\,0\,|\,L\,0)\,. \tag{1.91}$$

The triangular relation $\vec{\ell}_f + \vec{\ell}_i = \vec{L}$ implies that $\vec{L}$ is the angular momentum transfer in the reaction. The second Clebsch-Gordan coefficient of Eq. (1.91) requires that $\ell_i + \ell_f + L =$ even. However, as a result of parity conservation, i.e. $(-)^{\ell_i + \ell_1} = (-)^{\ell_f + \ell_2}$, we obtain a parity selection rule

$$\ell_1 + \ell_2 + L = \text{even} \;. \tag{1.92}$$

In the earlier calculations [BU 66], such an approximation was in fact introduced for nucleon transfer well below the Coulomb barrier. In those discussions, $\Delta V(\text{post}) = V_{c_1 t}(r_1)$ and $\Delta V(\text{prior}) = V_{c_2 t}(r_2)$ are assumed. To be specific, we consider $\Delta V(\text{post})$ in

$$\int d\vec{r}_f\, d\vec{r}_i\, f_{\ell_f}(r_f)\, Y_{\ell_f}^{m_f*}(\hat{r}_f)\, \phi_{\ell_2 m_2}(\vec{r}_2)\, V_{c_2 t}(r_1)\, \phi_{\ell_1 m_1}(\vec{r}_1)\, Y_{\ell_i}^{m_i}(\hat{r}_i)\, f_{\ell_i}(r_i) \tag{1.93}$$

of Eq. (1.65). The strategy of [BU 66] had been to rewrite $\phi_{\ell_2 m_2}(\vec{r}_2)$ as a separable function of $\vec{r}_1$ and $\vec{R}$. Together with the no-recoil approximation,

$$Y_{\ell_f}^{m_f*}(\hat{r}_f) \approx Y_{\ell_f}^{m_f*}(\hat{R}) \quad \text{and} \quad Y_{\ell_i}^{m_i}(\hat{r}_i) \approx Y_{\ell_i}^{m_i}(\hat{R}) \;,$$

the six-dimensional integral of the expression (1.93) becomes a product of two three-dimensional integrals. Since the Coulomb barrier keeps the two nuclei well apart, the transferred nucleon is in the external region of the core c_2, where $\phi_{\ell_2 m_2}(\vec{r}_2)$ is well approximated by the spherical Hankel function in the surface region,

$$\phi_{\ell_2 m_2}(\vec{r}_2) \approx N_2\, h_{\ell_2}^{(1)}(i\kappa r_2)\, Y_{\ell_2}^{m_2}(\hat{r}_2) \;, \tag{1.94}$$

$$= N_2 \sqrt{4\pi} \sum_{\substack{L,M \\ L',M'}} i^{\ell_2 + L - L'} \sqrt{\frac{(2L+1)(2\ell_2+1)}{(2L'+1)}}$$

$$\times (LM\,\ell_2 m_2 \,|\, L'M')(L0\,\ell_2 0\,|\,L'0)$$

$$\times h_L^{(1)}(i\kappa R)\, j_{L'}(i\kappa r_1)\, Y_L^{M*}(\hat{R})\, Y_{L'}^{M'}(\hat{r}_1),\ (R \geq r_1), \quad (1.95)$$

where N_2 is a normalization constant, and κ is given by the separation
energy B_2 of the nucleon t in nucleus a_2, $\kappa = (2m_t B_2/\hbar^2)^{1/2}$. Eq.(1.95)
is the addition theorem of the Hankel function (see Appendix 1 of
[Bu 66]) which accomplishes the separation of $\phi_{\ell_2 m_2}(\vec{r}_2)$ into product
functions of $\vec{R}$ and $\vec{r}_1$. As a result, the T-matrix of Eq.(1.56) becomes

$$T_{r_1\alpha_2,\,r_2\alpha_1} = N_2\sqrt{4\pi} \sum_{\substack{j_1 S_1 \\ j_2 S_2}} \sum_{\ell_1 \ell_2} \sum_{L,M} \Theta_{j_1 \ell_1}\, \Theta_{j_2 \ell_2}\, (LM\,j_1 S_1\,|\,j_2 S_2)$$

$$\times \sqrt{(2j_1+1)(2\ell_1+1)(2L+1)}\,(L0\,\ell_1 0\,|\,\ell_2 0)\, W(L\ell_1 j_2 S ; \ell_2 j_1)$$

$$\times (j_1 S_1 c_1 \gamma_1 \,|\, a_1 \alpha_1)(j_2 S_2 c_2 \gamma_2 \,|\, a_2 \alpha_2)\, A_{\ell_1}\, T_{LM}(\theta), \quad (1.96)$$

where

$$A_{\ell_1} = \int_0^\infty r_1^2 dr_1\, j_{\ell_1}^*(i\kappa r_1)\, V_{c,t}(r_1)\, u_{\ell_1}(r_1), \quad (1.97)$$

$$\phi_{\ell_1 m_1}(\vec{r}_1) = u_{\ell_1}(r_1)\, Y_{\ell_1}^{m_1}(\hat{r}_1), \quad (1.98)$$

$$T_{LM}(\theta) = \int d\vec{R}\, \chi_f^{(-)*}(\vec{R}')\, h_L^{(1)*}(i\kappa R)\, Y_L^{M*}(\hat{R})\, \chi_i^{(+)}(\vec{R}). \quad (1.99)$$

The no-recoil approximation, from Eqs. (1.89) and (1.90),

$$\vec{r}_f \approx \vec{R}' = \frac{m_{c_2}}{m_{a_2}} \vec{R} \quad , \qquad\qquad (1.100)$$

$$\vec{r}_i \approx \vec{R} \quad , \qquad\qquad (1.101)$$

has been introduced into Eq.(1.99). As discussed earlier, the parity selection rule, $\ell_1+\ell_2+L$ = even, is contained in the Clebsch-Gordan coefficient $(L0\ell_10|\ell_20)$.

By inspection, one deduces from Eq.(1.99) that $T_{LM}(\theta)$, hence the reaction cross section, is large if $\chi_f^{(-)*}$ and $\chi_i^{(+)}$ have large overlap. In Fig.(1.19), we depict two contrasting situations of this overlap in the reaction $^{25}\mathrm{Mg}(^{17}\mathrm{O},^{16}\mathrm{O})^{26}\mathrm{Mg}$ at 12 MeV [BU 66]. The distorted waves displayed are partial Coulomb waves with $\ell_i=\ell_f=10$. The distances of closest approach a_i and a_f for head-on collision are also indicated. One sees immediately that when the overlap is large, we have simultaneously

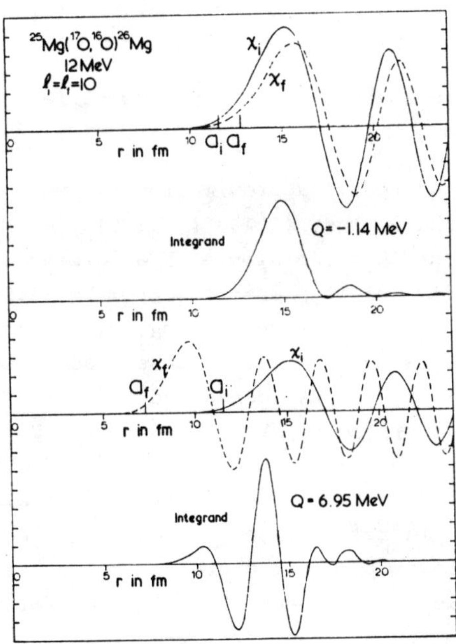

Fig.1.19. The radial part of the Coulomb wave functions and the corresponding integrand of Eq.(1.99) [BU 66].

$a_i \approx a_f$ ($|a_i - a_f| \ll 1/k_i$ or $1/k_f$). Since the Q-value is directly related to the distances of closest approach, hence this discussion on the overlap of χ_i and χ_f is in fact a quantal explanation of the semi-classical formula of the optimal Q-value of Eq.(1.11).

If we had used the prior form of the T-matrix so that $\Delta V(\text{prior}) \approx V_{c_2t}(r_2)$, we could express r_i and r_f in terms of R and r_2,

$$\vec{r_i} = \frac{m_{c_1}}{m_{a_1}} \vec{R} - \frac{m_t}{m_{a_1}} \vec{r_2} \quad , \tag{1.102}$$

$$\vec{r_f} = \vec{R} + \frac{m_t}{m_{a_2}} \vec{r_2} \quad , \tag{1.103}$$

which, in the no-recoil approximation, would be reduced to

$$\vec{r_i} \approx \frac{m_{c_1}}{m_{a_1}} \vec{R} \quad , \tag{1.104}$$

$$\vec{r_f} \approx \vec{R} \tag{1.105}$$

In general, a no-recoil approximation yields different results in the post- and prior-form.

In [BU 71] the no-recoil approximation is improved by assuming that the omitted terms in Eqs. (1.100), (1.101), (1.104) and (1.105) are proportional to $\vec{R}$. Thus the simple three-dimensional character of the integral of Eq.(1. 99) is preserved. From Eq.(1.90), the omitted term $(m_t/m_a)\vec{r_1}$ is to be replaced by $(m_t/m_{a_1})R_1\hat{R}$. This is reasonable because the transfer is most likely to take place when the nucleon lies on the line joining the two cores and when it is at the surface of nucleus a_1, at a distance R_1 from the core C_1. Further, we approximate $\hat{R} = \vec{R}/|\vec{R}| \approx \vec{R}/a$, where a is the average of the distances of closest approach in the initial and final channels. This is also reasonable for our purpose because, in terms of the R integral, the most likely region where the transfer takes place is where the cores are separated by their distance of closest approach $|\vec{R}| \approx a$. Therefore, the improved no-recoil approximation for the post-form T-matrix is

$$\vec{r_i} \approx \left(1 - \frac{m_t}{m_{a_1}} \cdot \frac{R_1}{a}\right)\vec{R} \quad , \quad \vec{r_f} \approx \left(\frac{m_{c_2}}{m_{a_2}} + \frac{m_t}{m_{a_2}} \cdot \frac{R_1}{a}\right)\vec{R} \quad . \quad (1.106)$$

Similarly, the improved no-recoil approximation for the prior-form T-matrix is

$$\vec{r_i} \approx \left(\frac{m_{c_1}}{m_{a_1}} + \frac{m_t}{m_{a_1}} \cdot \frac{R_2}{a}\right)\vec{R} \quad , \quad \vec{r_f} \approx \left(1 - \frac{m_t}{m_{a_2}} \cdot \frac{R_2}{a}\right)\vec{R} \quad . \quad (1.107)$$

Since the distorted waves are Coulomb functions in which the arguments are written in terms of $k_i r_i$ and $k_f r_f$, the improved no-recoil approximation can be incorporated by replacing $k_i r_i$ and $k_f r_f$ by $k_i' r_i$ and $k_f' r_f$, respectively, where in the post-form

$$k_i' = \left(1 - \frac{m_t}{m_{a_1}} \cdot \frac{R_1}{a}\right) k_i \quad , \quad k_f' = \left(\frac{m_{c_2}}{m_{a_2}} + \frac{m_t}{m_{a_2}} \cdot \frac{R_1}{a}\right) k_f \quad , \quad (1.108)$$

and in the prior-form

$$k_i' = \left(\frac{m_{c_1}}{m_{a_1}} + \frac{m_t}{m_{a_1}} \cdot \frac{R_2}{a}\right) k_i , \quad k_f' = \left(1 - \frac{m_t}{m_{a_2}} \cdot \frac{R_2}{a}\right) k_f \quad . \quad (1.109)$$

The usefulness of these improvements is apparent when we compare the differential cross sections calculated from the prior- and post-form T-matrix, see (Fig.1.20). The improved post and prior caluclations show that our initial conjecture, that the no-recoil approximation is reasonable, since the error committed is $A^{-1} \ll 1$, where A is the mass number of one of the nuclei involved, is in fact too optimistic. This is because one ought to compare the error with the wavelength of the local relative motion in the region of transfer (or the wavelength of the asymptotic relative motion in sub-Coulomb transfer). In the reaction $^{208}Pb(^{16}O,^{17}O)^{207}Pb$, calculated for Fig. (1.20), the error $R_1(Pb)/A_2(^{16}O)$ can be as large as 0.5 fm, which is considerable when the asymptotic wavelength at 70 MeV is only 1 fm.

Such error consideration can also serve as a guide to our choice of post or prior calculation in the no-recoil approximation. In the post calculation, the errors are R_1/A_1 and R_1/A_2; in the prior calculation, they are R_2/A_1 and R_2/A_2. Clearly, if the transferred nucleon originates

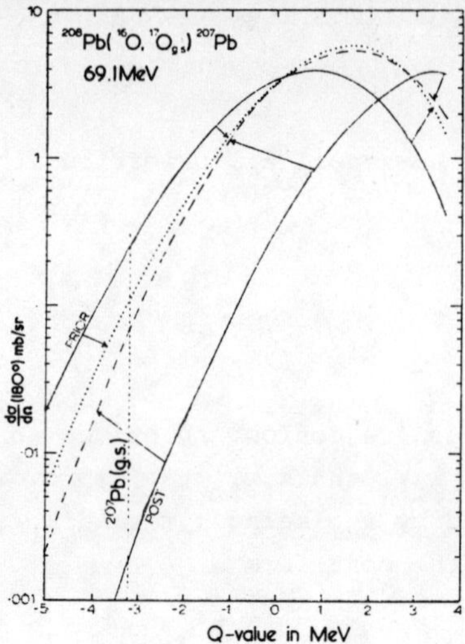

Fig.1.20. Comparison of post and prior calculations for the L=3
reaction ^{208}Pb(^{16}O,^{17}O)^{207}Pb. Recoil corrections are omitted
for the solid curves and are included for the broken curves.
The variation of Q-value is produced by varying the binding
energy in ^{208}Pb and the physical value for the ground state
transition is indicated [BU 71].

from the lighter nucleus, then the post form is preferred because of
the relatively smaller errors involved. For the same reason, if the
nucleon originates from the heavier nucleus, the prior treatment ought
to be adopted. Although recoil correction improves both treatments,
this guide should still remain valid.

As bombarding energy is increased, the inaccuracy of approximating
the tail of the bound state function $\phi_{\ell_2 m_2}(\vec{r}_2)$ by a Hankel function is
exposed. This can be remedied by expressing $\phi_{\ell_2 m_2}(\vec{r}_2)$ as a linear com-
bination of Hankel functions $\left[\sum_i b_i h_{\ell_2}^{(i)}(i\kappa_i r_2) \right] Y_{\ell_2}^{m_2}(\hat{r}_2)$ [LI 75], in which
b_i and κ_i are independent parameters. However, in practice, such elabo-
rate procedure was never needed. More seriously, as energy is increased,
hence local wavelength of the relative motion in the region of transfer
decreases, the errors of the no-recoil approximation with respect to
this wavelength increase. The L-transfers which are exluded by the parity selection rule, usually referred to as the non-normal parity transition, become more important. This shortcoming cannot be remedied by
the no-recoil correction discussed in the previous paragraph. In [DE 73],
a series of EFR-DWBA were performed for the reaction ^{12}C(^{14}N,^{13}C)^{13}N at

incident energies 21.16 MeV, 39.72 MeV, 61.25 MeV, and 99.36 MeV which
are well above the Coulomb barrier. The no-recoil approximation incurs
an error of $(M_t/m_a)R \approx 0.3$ fm, while the local wavelengths are approximately 3 fm, 2 fm, 1.5 fm and 1 fm, respectively. At the two highest
energies, the normal parity and non-normal parity transitions contribute
equally to the cross sections. More importantly, their oscillations in
the differential cross section are out of phase with each other so
that their sum results in a structureless differential cross section.
In Fig.(1.21), the experimental data at 78 MeV are reproduced by an
EFR-DWBA calculation in which normal and non-normal parity transitions
are equally important.

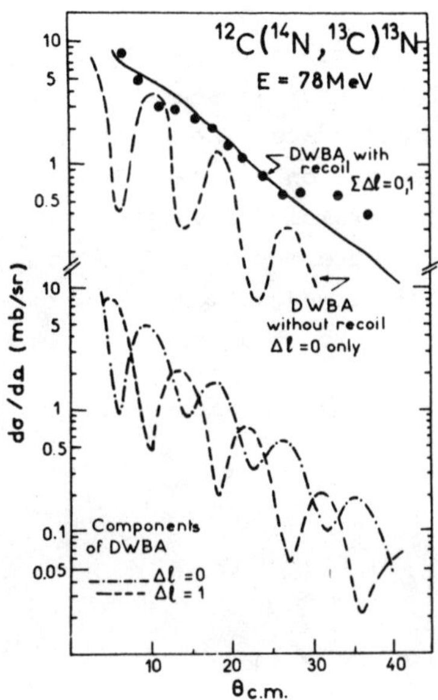

Fig.1.21. EFR-DWBA calculation of the reaction $^{12}C(^{14}N,^{13}C)^{13}N$ at
78 MeV [DE 73].

2. Inelastic scattering

An important theoretical tool in the analysis of inelastic scattering is the coupled channels formalism, which encompasses DWBA as a special case. The basic formalism is essentially that developed for nucleon-nucleus and α-nucleus inelastic scattering [YO 56, MA 57, CH 58, BU 63, TA 65, GL 69]. We give a brief sketch of the formalism and discuss a few modifications relevant to inelastic scattering of heavy-ions.

2.1. The coupled equations

In the course of scattering of two nuclei, one or both of the nuclei may emerge in its excited state. In our present discussion, we restrict ourselves to the excitation of only one nucleus, an analogue of the more familiar nucleon-nucleus inelastic scattering. The intrinsic structure of the low-lying excited states are assumed to be not very different from the ground state. The implication is that in exciting them from the ground state, only a very limited number of degrees of freedom is involved. For example, a valence nucleon in a single particle state outside an inert core may be lifted to a higher single particle state; or a strongly deformed nucleus is induced to have a faster bulk rotation while its nucleonic structure remains the same.

The scattering $A + B \longrightarrow A^* + B$ is assumed to be described by the model Hamiltonian H such that

$$\left(H - E \right) \Psi \left(\vec{r}, \xi_A, \xi_B \right) = 0 \quad ,$$

$$\left[H_A + H_B + T + V(\vec{r}, \xi_A) - E \right] \Psi (\vec{r}, \xi_A, \xi_B) = 0 \quad , \qquad (2.1)$$

where H_A and H_B are the intrinsic Hamiltonians of the nuclei A and B, T is the kinetic energy operator of the relative motion. The eigenstates of H_A are Φ_{α, I_A}

$$\left(H_A - E_{\alpha, I_A} \right) \Phi_{\alpha, I_A} (\xi_A) = 0 \quad , \qquad (2.2)$$

which are labelled by α and the spin I_A. The former represents a collection of other identifying quantum numbers. Similarly, the eigenstates of H_B are $\Phi_{\beta, I_B}(\xi_B)$,

$$\left(H_B - E_{\beta, I_B} \right) \Phi(\xi_B) = 0 \quad . \tag{2.3}$$

By writing the interaction between the nuclei as $V(\vec{r}, \xi_A)$, we assume that only the excited states of A play a part in the inelastic scattering, while B remains in its ground state.

We expand the exact scattering solution $\psi^{(+)}(\vec{r}, \xi_A, \xi_B)$ in terms of Φ_{α, I_A} and Φ_{β, I_B}. To this end, we introduce a spin-angle function for the nucleus B,

$$\mathcal{Y}_{\ell I_B, jm} = \left[Y_\ell(\hat{r}) \otimes \Phi_{\beta, I_B} \right]_{jm} \quad . \tag{2.4}$$

We couple the spin-angle function with the eigenstate Φ_{α, I_A} to form states of total angular momentum J and parity π,

$$\vec{J} = \vec{j} + \vec{I}_A \quad , \qquad \pi = (-)^\ell \pi_\alpha \pi_\beta \quad , \tag{2.5}$$

such that

$$\varphi^M_{c\pi J}(\hat{r}, \xi_A, \xi_B) = \left[\mathcal{Y}_{\ell I_B, j} \otimes \Phi_{\alpha, I_A} \right]^M_J \quad . \tag{2.6}$$

We introduce in Eq.(2.6) the channel label $C \equiv (\alpha\beta I_A I_B \ell j)$ to represent the collection of quantum numbers which define the intrinsic state of the nuclei A and B, and the relative orbital quantum numbers. In terms of these quantities, we write down an expansion for $\psi^{(+)}(\vec{r}, \xi_A, \xi_B)$, for a given total angular momentum and parity J and π,

$$\Psi^{M\,(+)}_{c\pi J}(\vec{r}, \xi_A, \xi_B) = \frac{1}{r} \sum_{C'} u^{c\pi J}_{C'}(r)\, \varphi^M_{c'\pi J}(\hat{r}, \xi_A, \xi_B) \tag{2.7}$$

We insert the label c in the state function Ψ and the radial function $u(r)$ to remind us that the initial state of the system is labelled by c.

Our expansion in Eq.(2.7) is an ansatz which contains two further important approximations. The joint nuclear system of A+B nucleons is assumed to be made up of two fragments A and B at all times. Effects on the inelastic scattering due to reaction such as A+B $\longrightarrow$ (A+1) + (B-1) $\longrightarrow$ A*+B are neglected. Furthermore, the state function Ψ is not antisymmetric with respect to the exchange of nucleons between the two fragments. We shall discuss the first approximation later. The second approximation is discussed in detail in the framework of the microscopic cluster theory by Tang in this winter school.

We insert the expansion (2.7) in (2.1), multiply on the left with $\varphi^{M'\,*}_{c'\pi'J'}$ and integrate over $\hat{r}$, ξ_A and ξ_B. By making use of the orthonormal property of φ's,

$$\langle \varphi^{M'}_{c'\pi'J'} \,|\, \varphi^{M}_{c\pi J} \rangle = \delta_{cc'}\,\delta_{\pi\pi'}\,\delta_{JJ'}\,\delta_{MM'} \quad , \tag{2.8}$$

the result is a set of coupled equations (coupled channels equations). For each channel c', it has the form

$$\left[T_{c'} + V^{J}_{c'c'}(r) - E_{c'} \right] u_{c'}(r) = - \sum_{c''\neq c'} V^{J}_{c'c''}(r)\, u_{c''}(r) \quad , \tag{2.9}$$

where

$$T_{c'} = \frac{\hbar^2}{2\mu}\left(-\frac{d^2}{dr^2} + \frac{\ell(\ell+1)}{r^2} \right) \quad , \quad k_{c'}^2 = \frac{2\mu}{\hbar^2} E_{c'}$$

$$E_{c'} = E - E_{\alpha, I_A} - E_{\beta, I_B} \quad . \tag{2.10}$$

The coupling between different channels is represented by

$$V^{J}_{c'c''}(r) = \left(\varphi^{M}_{c'\pi J}(\hat{r},\xi_A,\xi_B) \,|\, V(\vec{r},\xi_A) \,|\, \varphi^{M}_{c''\pi J}(\hat{r},\xi_A,\xi_B) \right), \tag{2.11}$$

in which the round bracket indicates that $\hat{r}$, ξ_A and ξ_B are being integrated over. As it stands, the system of equations is infinite in number. In practical calculations, we neglect those excited states of A which, according to our opinion, do not affect the elastic scattering and inelastic scattering to the first few excited states.

2.2 Numerical solution of the coupled equations

Similar in spirit to the discussion of partial waves in elastic scattering in subsection (A.5), the radial functions $u_{c'}(r)$ are subjected to two boundary conditions. (i) The radial functions $u_{c'}(r)$ must vanish at the origin. Given any slope at the origin, the coupled equations (2.9) can be integrated outwards to large r where $V_{c'c''}^{J}(r)$'s essentially vanish. In the exterior region where the nuclear force has become negligible, the Schrödinger equation describes pure Coulomb scattering. Therefore, the radial function $u_{c'}(r)$ is a linear combination of the regular and irregular Coulomb functions, $F_\ell(\eta_{c'},k_{c'}r)$ and $G_\ell(\eta_{c'},k_{c'}r)$, respectively, cf. Eq. (A.27) (in Eq.(A.27), F_ℓ and G_ℓ are related to the spherical Bessel function and Neumann function),

$$u_{c'}(r) \xrightarrow[r\to\infty]{} a\,\bar{F}_\ell(\eta_{c'},k_{c'}r) + b\,G_\ell(\eta_{c'},k_{c'}r) \,, \quad (2.12)$$

where $\eta_{c'} = Z_A Z_B e^2/(\hbar^2 k_{c'})$ is the Coulomb parameter. However, a more helpful way to write the asymptotic form of $u_{c'}(r)$ is to express it in terms of combinations of F_ℓ's and G_ℓ's which behave asymptotically as incoming and outgoing waves,

$$I_\ell^* = O_\ell = G_\ell(\eta_{c'},k_{c'}r) + i\,F_\ell(\eta_{c'},k_{c'}r)$$
$$\xrightarrow[r\to\infty]{} exp\left[i(k_c r - \eta_{c'} \ln 2k_{c'}r - \ell\pi/2 - \sigma_\ell)\right] \,, \quad (2.13)$$

where σ_ℓ is the Coulomb phase shift, see subsection (A.11). In the initial channel c, there are incoming and outgoing spherical waves in the asymptotic region, while in the other channels c', there are only outgoing spherical waves. Hence, the second boundary condition (ii) may be expressed as

$$u_{c'}(r) \xrightarrow[r\to\infty]{} \delta_{c'c}\,I_c(k_c r) - \left(\frac{k_c}{k_{c'}}\right)^{\frac{1}{2}} S_{c'c}^{J}\,O_{c'}(k_{c'}r), \quad (2.14)$$

where $S_{c'c}^{J}$ is the scattering matrix element.

Let us suppose that for the excitation of a particular state Φ_{α,I_A}, we have a system of N channels being coupled together, where N=1 denotes the initial channel c. Eq.(2.12) tells us that there are N coefficients of a and N coefficients to b to be determined, i.e. a total of 2N unknowns. The numerical integration of the N coupled equations can be repeated N times, each time with a different set of boundary conditions. Let us label the solutions by two indices: k denotes the solution in the k-th channel, p denotes which of the N different sets of boundary conditions is used to generate the solutions. One of these boundary conditions may be

$$u_{kp}(0) = \begin{cases} 0 \\ 0 \end{cases}, \quad \lim_{r \to 0} u'_{kp}(r) = \begin{cases} 1, & k = p \\ 0, & k \neq p \end{cases}, \quad (2.15)$$

for an arbitrary k. By letting k run from 1 to N, we obtain N different boundary conditions. However, a more sensible set of boundary conditions for numerical analysis is

$$\lim_{r \to 0} u_{kp}(r) = \begin{cases} \dfrac{r^{\ell_k + 1}}{(2\ell + 1)!!}, & k = p, \\[2mm] 0, & k \neq p. \end{cases} \quad (2.16)$$

In any event, we require that linear combinations of u_{kp} generated from the N boundary conditions do result in solutions which satisfy Eq.(2.14). This is expressed by the linear equations,

$$a_1 u_{11} + a_2 u_{12} + \cdots + a_N u_{1N} + S_1 O_{\ell_1} + 0 + \cdots + 0 = I_{\ell_1},$$

$$a_1 u_{21} + a_2 u_{22} + \cdots + a_N u_{2N} + 0 + \left(\frac{k_1}{k_2}\right)^{\frac{1}{2}} S_2 O_{\ell_2} + \cdots + 0 = 0,$$

$$\vdots$$

$$a_1 u_{N1} + a_2 u_{N2} + \cdots + a_N u_{NN} + 0 + \cdots + \left(\frac{k_1}{k_N}\right)^{\frac{1}{2}} S_N O_{\ell_N} = 0, \quad (2.17)$$

where all the functions are evaluated at R. By writing down the derivative of these equations, we obtain in total 2N equations to solve for the 2N unknown coefficients: the coefficients $a_1, \ldots, a_N$ and the scattering matrix element $S_1, \ldots, S_N$. The elastic and inelastic cross section can easily be written in terms of the scattering matrix $S_1, \ldots, S_N$ [NE 66].

One should note that the number of coupled equations far exceeds the number of nuclear states considered, since c labels not only the nuclear state (α, I_A) but also the orbital angular momentum of the scattering particles. As an illustration, let us consider $I_B = 0$ and a given nuclear state I_A and π_A. For a particular total angular momentum and parity (J, π), the orbital angular momenta ℓ which are coupled are

$$|J - I_A| \leqslant \ell \leqslant J + I_A, \quad (2.18)$$

and in general there are $2I_A+1$ of them. Therefore for <u>each</u> (J,π), there are N coupled equations, where

$$N = \sum_{\substack{nuclear \\ states}} (2I_A + 1)$$

(2.19)

in which the sum runs over the nuclear states included. In light-ion inelastic scattering, the upper limit of ℓ may be set as 2kR, where R is the radius beyond which the nuclear interaction is effectively zero. Therefore, the total number of equations to be solved is $2kRN^2$.

2.3. Distorted wave Born approximation.

We show in this subsection that the DWBA for inelastic scattering is a special case in the coupled channels formalism. With reference to Fig.(2.1) we limit the type of transition to those of 1 and 2. In such circumstances, the coupled equations, Eq. (2.9) are reduced to

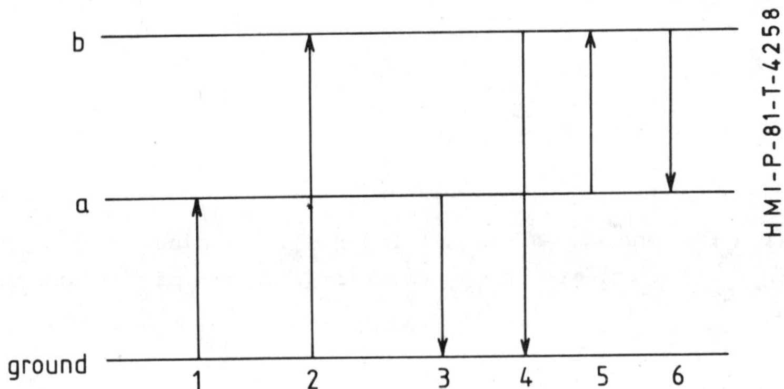

Fig.2.1. The DWBA includes transitions 1 and 2 only. The coupled
channels formalism takes into account the other routes also.

$$\left(T_c + V_{cc}^{J} - E_c \right) u_c^o = 0 \qquad \text{,(target channel)} \qquad (2.20a)$$

$$\left(T_{c'} + V_{c'c'}^{J} - E_{c'} \right) u_{c'}' = - V_{c'c}^{J} u_c^o \qquad \text{,(excited states channel)} \quad (2.20b)$$

which are the first order approximations in V of the coupled equations. The superscript "0" and "1" designate the zero-th and first order (in V) solutions of the coupled equations. The subscript c labels those channels which have the nuclei in the ground states, while channels c' contain excited state of A. The coupling matrix elements $V^I_{c'c}$ connect the ground state and the excited states in one direction only.

We note immediately, by inspection of Eq.(2.20), that, (i) the coupled equations are now uncoupled, (ii) the zero-th order solution u^o_c is just the distorted wave we discussed in subsection 1.1. By virtue of (i), the number of differential equations to be solved numerically is reduced to 2kRN.

To calculate the scattering matrix element S in the first order approximation with Eq.(2.20), the DWBA limit, only the distorted waves u^o_c and $u^o_{c'}$ are required. To illustrate this, we compare Eq.(2.20b) with Eq.(A.43), where $V^J_{c'c}u^o_c/r$ in the former may be regarded as the coordinate representation of F in the latter. The whole of subsection A.10 on the partial wave expansion of the Green's function can be carried over to our present discussion with slight modification due to the insertion of a potential term on the left hand side of Eq.(A.68). Therefore, the distorted wave Green's function, as a counterpart of Eq.(A.74), is

$$G^{(+)}(\vec{r},\vec{r}') = -\frac{2\mu k}{\hbar^2}\sum_{\ell,m}\frac{u^o_\ell(\eta,\delta_\ell,kr_<)}{kr_<}\cdot\frac{H_\ell(\eta,\delta_\ell,kr_>)}{kr_>}$$

$$\times Y^{m\,*}_\ell(\hat{r}')Y^m_\ell(\hat{r})\ , \tag{2.21}$$

where, with the generalized channel label c' replacing ℓ, $u^o_{c'}(\eta_{c'},\delta_{c'},k_{c'}r)$ and $H_{c'}(\eta_{c'},\delta_{c'},k_{c'}r)$ are the particular solutions of the homogeneous equation

$$(T_{c'} + V^J_{c'c'} - E_{c'})y = 0\ , \tag{2.22}$$

which are, respectively, regular and irregular at the origin and have the asymptotic forms

$$u^o_{c'}(\eta_{c'},\delta_{c'},k_{c'}r)\longrightarrow \{I_{c'} - exp[2i\delta_{c'}]O_{c'}\}/(2i)$$

$$= exp(i\delta_{c'})\sin(k_{c'}r - \eta_{c'}\ln 2k_{c'}r - \ell'\pi/2 + \sigma_{\ell'} + \delta_{c'}) \tag{2.23}$$

$$H_{c'}(\eta_{c'}, \delta_{c'}, k_{c'}r) \longrightarrow O_{c'}$$

$$= \exp\left[i\left(k_{c'}r - \eta_{c'}\ln 2k_{c'}r - \ell'\pi/2 + \sigma_{\ell'} + \delta_{c'}\right)\right] \quad . \qquad (2.24)$$

By making use of Eq.(A.76), whose first term on the right hand side has to be omitted because of the absence of incoming wave in the channel c', the solution $u_{c'}^{!}$ of Eq.(2.20b) is

$$u_{c'}^{'}(r) = -\frac{2\mu k_{c'}}{\hbar^2}\int_0^\infty u_{c'}^{\circ}(r_<)\, H_{c'}(r_>)\, V_{c'c}(r')\, u_c^{\circ}(r')\, dr' \, . \quad (2.25)$$

This is the correct solution for $u_{c'}^{!}$, because as r goes to infinity, it is proportional to the outgoing waves,

$$u_{c'}^{'}(r) \longrightarrow -\frac{2\mu k_{c'}}{\hbar^2}\cdot O_{c'}\int_0^\infty u_{c'}^{\circ}(r')\, V_{c'c}(r')\, u_c^{\circ}(r')\, dr'$$

$$= -\left(\frac{k_c}{k_{c'}}\right)^{\frac{1}{2}} S_{c'c}\, O_{c'}(r) \, . \qquad (2.26)$$

Hence, the scattering matrix element is expressed in terms of the distorted waves $u_{c'}^{\circ}$ and u_c°,

$$S_{c'c} = \frac{2\mu k_{c'}}{\hbar^2}\left(\frac{k_{c'}}{k_c}\right)^{\frac{1}{2}}\int_0^\infty u_{c'}^{\circ}(r')\, V_{c'c}(r')\, u_c^{\circ}(r')\, dr' \, , \quad (2.27)$$

which, in fact, is the DWBA limit of the coupled channels result. The distorted waves $u_{c'}^{\circ}(r)$ and $u_c^{\circ}(r)$ are calculated in the optical potentials $V_{c'c'}$ and V_{cc} which are not necessarily identical.

In the discussion of the coupled channels formalism being applied to heavy-ion reaction, a few pertinent remarks must be made concerning the size of the coupled equations system.

Since the Coulomb charges involved in heavy-ion scattering are large, a dominant component of the direct inelastic reaction is Coulomb excitation in which, in terms of a macroscopic model, the deformed charge distribution of the target gives rise to the coupling potential $V_{c'c''}^{J}$. The highest spin of the excited states reached can be very large [CL 79]. For example, in the multiple Coulomb excitation of even-even nuclei,

spins of I=16 is observed in ^{232}Th in the reaction ^{84}Kr + ^{232}Th at 371
and 450 MeV [CO 74]; spin of I $\approx$ 28 is observed in ^{232}Th and
^{234,236,238}U with 5.3 MeV/u ^{208}Pb-ions [FU 80]. Under such circumstances,
the number of coupled equations to be solved, defined as N in Eq.(2.19),
is prohibitively large, and so is the total number of differential equations $2kRN^2$ to be solved numerically. Furthermore, since the coupling
potential is Coulomb in nature, its range, 2kR, is also large. These
factors contribute towards the explosion of the required numerical
effort.

Numerical accuracy is also a problem one has to deal with carefully.
With reference to Eq.(2.16), we see that the boundary conditions of
$u_{kp}(r)$'s, at r=0.1 fm say, can be $(0.1)^1$, $(0.1)^3$, ..., $(0.1)^{20}$, ...
To maintain quantitative numerical accuracy in these numbers, which
range over many orders of magnitude, can be a severe task.

To remedy such shortcomings, authors have proposed methods of performing the numerical integration of the coupled equations more effectively [TO 80, AL 77, IC 77]. Others have extended the semi-classical
method used in light-ion Coulomb excitation to the heavy-ion domain
[AL 75, BO 77].

2.4. The macroscopic interaction

The macroscopic model usually employed is an extension of the phenomenological model used in light-ion scattering,

$$U(r) = V_o f(r) + i\left[W f_w(r) - W_D a_D f'_D(r)\right]$$
$$+ \langle \vec{\ell} \cdot \vec{\sigma} \rangle \left(\frac{\hbar}{m_\pi c}\right)^2 V_{so} \frac{1}{r}\frac{d}{dr} f_{so}(r) + V_{coul} , \qquad (2.28)$$

where

$$f_x(r) = \left[1 + exp\left(\frac{r - R_x}{a_x}\right)\right]^{-1}, \quad f' = \frac{df}{dr} , \qquad (2.29)$$

is the Woods-Saxon form factor. This potential contains a real central
part $V_o f(r)$, a volume absorption part $W f_w(r)$, a surface absorption part
$W_D a_D f'_D(r)$, a spin-orbit term and a Coulomb interaction, which can be
taken as due to a uniformly charged sphere or to a diffused distribution
of charges resembling the Woods-Saxon form factor of the real central

part of the potential f(r). Theoretical prediction is not particularly sensitive, within reasonable limit, to this difference.

In light-ion reaction, one writes for the radius of the target $R_x = r_{xo}A^{1/3}$, where r_{xo} is the radius parameter and A is the mass number of the target. For a permanently deformed rotational nucleus, the radius is replaced by

$$R_x = R_{xo}\left[1 + \sum_\lambda \beta_\lambda Y_\lambda^o(\theta')\right] \quad , \qquad (2.30a)$$

with $R_{xo} = r_{xo}A^{1/3}$. The angular coordinate θ' refers to the body fixed axis of symmetry. Under normal circumstances, only very limited number of deformation parameters β_λ's are needed. For the present discussion we restrict ourselves, for the sake of simplicity, to the quadrupole deformation only, i.e.

$$R_x = R_{xo}\left[1 + \beta_2 Y_2^o(\theta')\right] \quad . \qquad (2.30b)$$

Inserting Eq.(2.30b) into the respective form factors, we can expand them into power series of $\beta_2 Y_2^o(\theta')$. However, such series seem to converge badly when β_2 is of the order of 0.3 [TA 65]. Alternately, one can expand the form factors in terms of $Y_\lambda^o(\theta')$, with the expansion coefficients being functions of r. For example, for the real part of the potential we write

$$V_o \frac{1}{1+exp\left[\dfrac{r-R_o(1+\beta_2 Y_2^o(\theta'))}{a_o}\right]} = \sum_{\lambda'=0} g_{\lambda'}(r) Y_{\lambda'}^o(\theta') \quad , \qquad (2.31)$$

$$g_{\lambda'} = 4\pi V_o \int_0^\infty \frac{1}{1+exp\left[\dfrac{r-R_o(1+\beta_2 Y_2^o(\theta'))}{a_o}\right]} Y_{\lambda'}^o(\theta')d(\cos\theta') \quad . \qquad (2.32)$$

The diagonal potential terms $V_{c'c'}^J(r)$ in the coupled equations correspond to $g_o(r)$, while the higher λ' terms constitute the coupling potential $V_{c''c'}^J(r)$.

For discussions of vibrational nuclei and the explicit evaluation of the coupling potential of Eq.(2.11) by assuming some collective model for the nuclei, one should consult the standard text [BU 63, TA 65, GL 69].

In most cases the rotational and vibrational models give very similar results [BU 63].

The charge distribution is usually assumed to be deformed in a similar manner as the real form factor f(r). On the other hand, the charge deformation parameter β^c can be extracted from B(E2) strength in electromagnetic transition between collective states. In the rotational model [AL 56), for transition within a band, we have

$$B(E2; I_i \rightarrow I_f) = \frac{5}{16\pi^2} e^2 Q_o^2 \langle I_i 2 K0 | I_f K \rangle^2 \; , \qquad (2.33)$$

in which the intrinsic quadrupole moment Q_o is given in terms of the quadrupole deformation parameter β_2 as

$$Q_o = \frac{3}{5\pi} Z R_c^2 \beta_2^c \left(1 + 0.16 \beta_2^c + \cdots \right) , \qquad (2.34)$$

where R_c is the radius of the average charge distribution and Z is the charge number of the target. The inelastic scattering most often encountered is the transition $0^+ \rightarrow 2^+$ of the ground state band, in which case the B(E2) value can be written, to the first order of β_2^c, as

$$B(E2; 0^+ \rightarrow 2^+) = \left(\frac{3}{4\pi} Z e R_c \right)^2 \cdot \left(\beta_2^c R_c \right)^2 \; . \qquad (2.35)$$

The nuclear quadrupole deformation parameter β_2^N is set equal to β_2^c.

In the discussion of heavy-ion inelastic scattering, the radius of the macroscopic optical potential is not equated to the radius of the target, but instead,

$$R_{xo} = r_{xo} \left(A^{\frac{1}{3}} + B^{\frac{1}{3}} \right) \; , \qquad (2.36)$$

in which the finite size of the projectile is included in an empirical fashion. In order to achieve the same deformation of the potential via Eq.(2.30b), the nuclear quadrupole deformation has to be suitably readjusted. We define a quantity βR called the deformation length and equate the nuclear deformation length to the Coulomb deformation length,

$$\beta_2^N R_{xo} = \beta_2^c R_c \quad .$$ (2.37)

The Coulomb deformation length $\beta_2^C R_c$ is still being given by the B(E2) value using Eq.(2.35), while the modified nuclear deformation is β_2^N.

2.5. The nuclear-Coulomb interference

A particularly distinctive feature of heavy-ions inelastic scattering at energy near the Coulomb barrier is the nuclear-Coulomb interference [VI 72]. The theoretical analysis is usually performed by using the coupled channel formalism described in the previous subsection.

In the transition of $0^+ \rightarrow 2^+$, the coupling potential $V_{c''c'}^J(r)$ of Eq.(2.11) contains a nuclear term and a Coulomb term. Below the Coulomb barrier, the Coulomb terms dominates, because of its longer range. However, as the Coulomb barrier is approached, the nuclear coupling becomes increasingly important. The scattering amplitudes due to the Coulomb and the nuclear couplings are added coherently, hence they give rise to the nuclear-Coulomb interference.

Experimental observation of this phenomena can be found in the excitation function [VI 72] as well as the angular distribution [RE 75]. In Fig.(2.2), the excitation functions of elastic scattering and the inelastic process, at 60° lab, of ^{58}Ni(^{16}O,^{16}O')^{58}Ni*(2^+, 1.45 MeV) are shown [VI 72]. In the lower energy region up to about 44 MeV, the elastic scattering is predominantly Rutherford scattering, in which the two nuclei are kept far apart; while in the higher energy region above 50 MeV, the elastic scattering diminishes exponentially because of absorption processes since the two nuclei overlap with each other strongly during the scattering. Hence the energy range of 44 MeV to 50 MeV loosely defines a "surface region" in which Coulomb and nuclear interactions are equally important. The energy range can be expressed directly in terms of distances of closest approach, assuming Coulomb trajectories, via

$$D = \frac{0.72 \, Z_1 Z_2}{E} \cdot \frac{A_1 + A_2}{A_2} \left[1 + \csc\left(\frac{\theta}{2}\right) \right] ,$$ (2.38)

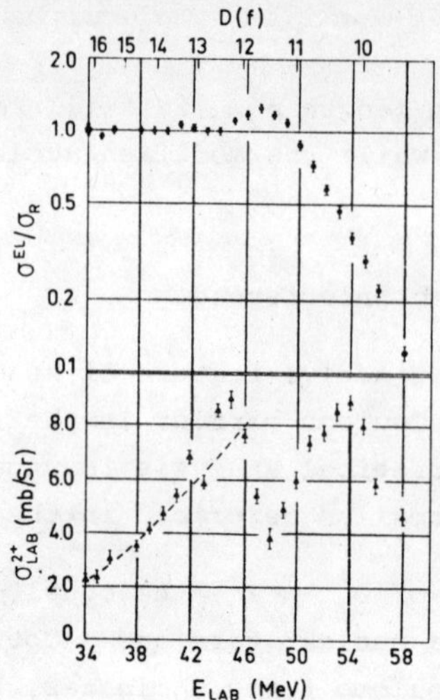

Fig.2.2. Upper figure: Ratio of the elastic cross section to Rutherford cross section, at 60° lab, of ^{16}O as a function of the incident energy. Lower figure: excitation function, at 60° lab, of the inelastic excitation of the 1.45 MeV (2^+) state in ^{58}Ni. The scale at the top shows the distances of closest approach, assuming Coulomb trajectories. Dashed curve shows the calculated cross section for Coulomb excitation for a B(E2) value of $0.066e^2b^2$ [VI 72].

in which E is the incident energy; A_1 and A_2 are the atomic numbers of the nuclei; Z_1 and Z_2 are the nuclear charges; θ is the c.m. scattering angle. The trajectory picture is valid because the Coulomb parameter η is $\approx$ 18; see Eq.(1.1).

In the lower figure of Fig.(2.2), the inelastic excitation in the lower energy region increases as a function of the incident energy,as predicted by Coulomb excitation. However, in the surface region, a rather deep minimum appears, before the excitation function falls off monotonically at higher energies. Since the Coulomb coupling is repulsive while the nuclear coupling is attractive, their scattering amplitudes interfere destructively. This gives rise to the destructive minimum in the "surface region". According to the Coulomb trajectory assumption, this region is confined to a width of 1 fm. The location of the minimum at   11.5 fm is still far from the geometric nuclear surface $R = r_o(A_1^{1/3}+A_2^{1/3}) \approx$ 8.0 fm, where r_o = 1.25 fm.

These characteristic features of the nuclear-Coulomb interference
are also observed in the same reaction at 75° lab and 90° lab. The interpretation with a classical picture of strictly Coulomb trajectories
is later modified [MA 73].

The nuclear-Coulomb effect had been investigated in the earliest
days of light-ion Coulomb excitation [TE 56], but the structures found
in α inelastic scattering were much less pronounced [SA 68, WA 70, PR 70].

Inelastic scattering of ^{58,60,62,64}Ni + ^{18}O at 63 MeV has also been
extensively examined [RE 75]. This reaction has the added feature that
either the Ni nucleus or the ^{18}O nucleus can be excited to their 2^+
states. The simultaneous excitation of both nuclei is in comparison a
second order process. In Fig. (2.3), we show the angular distributions
of the elastic and inelastic scattering of ^{18}O on ^{58,60,64}Ni. The solid
lines are DWBA calculations which reproduce the Ni isotopes' excitations, but not those of ^{18}O. The latter's excitation has subsequently

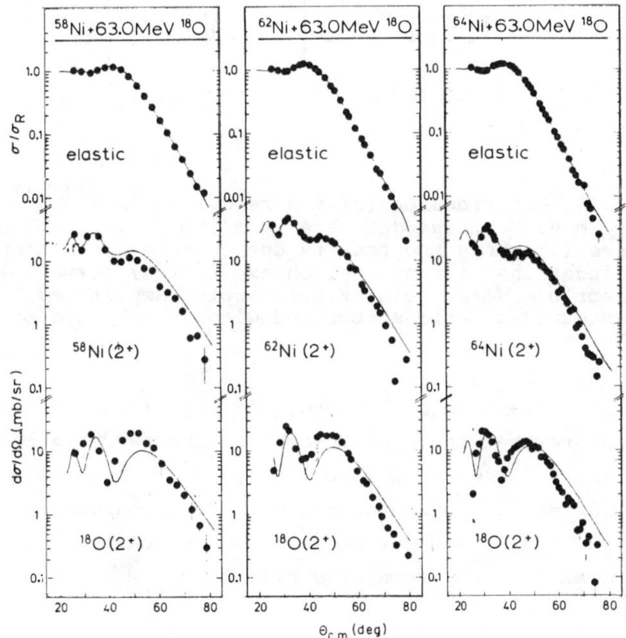

Fig.2.3. The angular distribution of the elastic and inelastic
    scattering of ^{18}O on ^{58,60,64}Ni at incident energy of 63 MeV
    [RE 75].

been the subject of further empirical analyses which are not entirely
satisfactory [GL 76, GL 77, LO 77, BA 78a). A two-step model was also
suggested, in which particle-transfer intermediate channels $^{18}O \rightarrow {}^{17}O$
$\rightarrow {}^{18}O*$ contribute towards the $^{18}O*$ excitation as well [LI 78]. However,
these turned out to be too small.

A model used in light-ion inelastic scattering has recently been
revived to explain the excitation of ^{18}O [LA 79, LO 67]. In this model
the coupling potential contains a core excitation term and a valence
particle excitation term [LA 79]. The long-standing discrepancy between
the experimental data

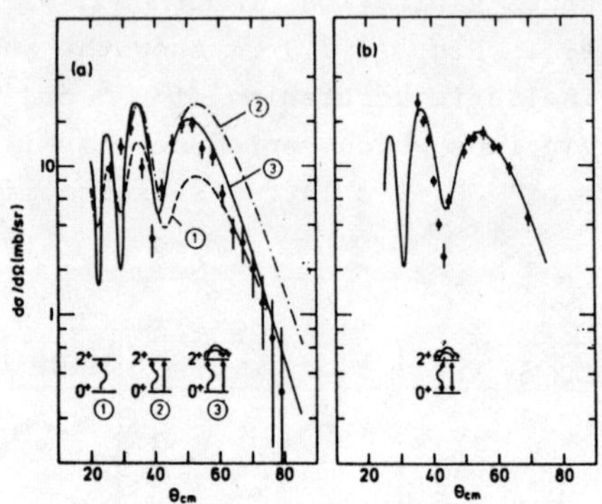

Fig.2.4. Angular distributions of the reaction $^{58}Ni({}^{18}O, {}^{18}O*(2^+){}^{58}Ni)$ at
E_{lab} = 63 MeV. The data are from [RE 75]. (a) DWBA results:
curve (1) shows the results due to core excitation; curve (2)
includes the valence nucleon excitation; curve (3) shows the
effect of adding total reorientation amplitudes. (b) Coupled
channels calculations compared with 60 MeV data of [RE 75]
([LA 79]).

and calculations seems to be removed; see Fig. (2.4). Nevertheless, one
ought to be aware that the spectroscopy of the two valence neutrons
outside the inert ^{16}O core is notoriously complicated; see for example
[EL 77]. The effect of double-counting is not clear, since the core
excitation also contains the same valence neutrons excitation which, in
the model, is explicitely separated off.

3.   Spin-dependent interaction

In some transfer reactions, the discrepancies between theory and the experimental data persisted in spite of the considerable effort put into the theoretical analyses via the ordinary optical potential [KU 77, BA 78b, WU 79]. The theory fails to reproduce the oscillations in the angular distributions. As a result, a spin-orbit potential between heavy ions was suggested, and it did remove the discrepancy in some way. The effect on the angular distribution of a spin-orbit interaction can be understood in terms of an extension of the Strutinski model [BA 78, ST 64], i.e. a parametrization of the scattering amplitude as a function of the orbital angular momentum; cf. Eq.(1.35).

In contrast to these indirect indications, direct evidence for a spin-dependent potential in the inelastic reaction $^{24}Mg(^{13}C,^{13}C')^{24}Mg^*$ (2^+) at 35 MeV has been reported recently [DÜ 79]. From general considerations of symmetry, the observation of γ-rays emitted perpendicular to the reaction plane indicates that spin-flip of ^{13}C has taken place. The $|m|$-substate population probability P_m of $^{24}Mg^*(2^+)$ is deduced, where the quantization axis is perpendicular to the reaction plane. It was found that P_1, which would be zero in the absence of spin-flip, was $(1.7 \pm 0.5)\%$ and less than 0.8% at, respectively, $\theta_{c.m.} = 10°$ and 21°.

In this section, we discuss an analysis of the spin-flip experiment [LI 80].

3.1. Symmetry and spin-flip

We sketch in this subsection a simple rule, given by Bohr [BO 59], which relates the parities and polarizations of particles involved in a collision process.

The reaction plane in a two particles collision process is defined by the incoming relative momentum $\vec{p}_i$ and outgoing relative momentum $\vec{p}_f$. Provided that the interaction is parity conserving, the collision exhibits reflexion symmetry with respect to the reaction plane. Furthermore, the reflexion can be represented by a parity inversion plus a rotation of 180° about an axis $\vec{n}$ perpendicular to the reaction plane. Hence, if R is the reflexion operator, its eigenvalue is given by $\mathscr{p}\,e^{i\pi S_{\vec{n}}}$, where $\mathscr{p}$ is the product of the parities of the particles and $S_{\vec{n}}$ is the sum of the spin components along $\vec{n}$. Thus, we have the conservation law

$$\mathcal{P}_i \, e^{i\pi S_i \vec{n}} = \mathcal{P}_f \, e^{i\pi S_f \vec{n}} \quad . \tag{3.1}$$

In the reaction $^{24}\mathrm{Mg}(^{13}\mathrm{C},^{13}\mathrm{C}')^{24}\mathrm{Mg}^*(2^+)$, in which $\mathcal{P}_i = \mathcal{P}_f$, we obtain from Eq. (3.1),

$$e^{i\pi m_{1/2}} = e^{i\pi(m'_{1/2} + m)} \quad ,$$

$$e^{i\pi(m_{1/2} - m'_{1/2})} = e^{i\pi m} \quad , \tag{3.2}$$

in which $m_{1/2}$ and $m'_{1/2}$ are the initial and final spin projections of $^{13}\mathrm{C}$ along $\vec{n}$, while m is the spin projection of $^{24}\mathrm{Mg}^*(2^+)$ along $\vec{n}$. It follows from Eq.(3.2) that the population of the $|m| = 1$ substate of $^{24}\mathrm{Mg}^*(2^+)$ is entirely associated with the spin-flip of $^{13}\mathrm{C}$.

In the de-excitation of the 2^+ state to the 0^+ ground state, the radiative transition must have $|\Delta m| = |m|$. At $\theta_\gamma = 0°$, i.e. the direction of $\vec{n}$, only the $|m| = 1$ component has a non-vanishing γ-ray intensity. Hence, detecting γ-ray in this direction is a signature of population of $|m| = 1$ substate. However, because of the finite size of the detector, the $|m| = 0,2$ components must also be considered in the analysis.

3.2. Rotation of the quantization axis

The differential cross section expressed in terms of the scattering amplitude $\mathcal{G}(K\mu, K'\mu')$ is

$$\frac{d\sigma}{d\Omega} = \frac{1}{(2I+1)(2S+1)} \sum_{\substack{K,\mu \\ K',\mu'}} \left| \mathcal{G}(K\mu, K'\mu') \right|^2$$

$$= \frac{D}{(2I+1)(2S+1)} \quad , \tag{3.3}$$

where I and S are the spins of the target and projectile, K and μ their respective z-components; K' and μ' are the z-components of the residual nucleus and the ejectile.

In the usual DWBA or coupled channels analysis, the z-axis is taken to be the beams axis, with the x-z axes defining the reaction plane and the y-axis being the reaction normal. We are interested in the probability of obtaining a definite component, K'_y, of I' in the y direction. This may be determined by expressing $\mathcal{I}(K\mu, K'\mu')$ with respect to axes which are rotated by $\pi/2$ about the y-direction and then by $\pi/2$ about the original z-direction. Under these rotations, the scattering amplitude, in terms of the D-matrices $D_{m'm'}$ is

$$R\,\mathcal{I}(K\mu, K'\mu') = \sum_{\substack{\bar{K},\bar{\mu} \\ \bar{K}',\bar{\mu}'}} D_{\bar{K}K}\, D_{\bar{\mu}\mu}\, D_{\bar{K}'K'}\, D_{\bar{\mu}'\mu'}\, \mathcal{I}(\bar{K}\bar{\mu}, \bar{K}'\bar{\mu}') \;, \qquad (3.4)$$

which can be substituted into the probability of the K'_y-substate,

$$\alpha(K'_y) = \frac{\sum_{K,\mu,\mu'} \left| R\,\mathcal{I}(K\mu, K'\mu') \right|^2}{D} \;, \qquad (3.5)$$

where D is defined in Eq.(3.3). By making use of the orthogonality relation

$$\sum_{\mu'} D_{\bar{\mu}'\mu'}\, D_{\bar{\bar{\mu}}'\mu'} = \delta_{\bar{\mu}'\,\bar{\bar{\mu}}'} \;, \qquad (3.6)$$

and similar ones with respect to K and μ, we can simplify Eq.(3.5) to

$$\alpha(K'_y) = \frac{\sum_{K,\mu,\mu'} \left| \sum_{\bar{K}'} D_{\bar{K}'K'}\, \mathcal{I}(K\mu, \bar{K}'\mu') \right|^2}{D}$$

$$= \frac{\sum_{K\mu\mu'} \left| \sum_{K'} (-i)^{K'} d^{I'}_{K'K'_y}(\pi/2)\, \mathcal{I}(K\mu, K'\mu') \right|^2}{D} \;, \qquad (3.7)$$

where $d^{I'}_{K'K'_y}$ is the reduced rotational matrix. We have normalized so that

$$\sum_{K'_y} \alpha(K'_y) = 1 \;. \qquad (3.8)$$

Straightforward evaluation of Eq.(3.7) for I' = 2 yields $P_2 = \alpha(2) + \alpha(-2)$

$$P_2 = \alpha(2) + \alpha(-2)$$

$$= \frac{1}{8} \sum_{K,\mu,\mu'} \left| -\mathcal{G}(K\mu, 2\mu') + \sqrt{6}\, \mathcal{G}(K\mu, 0\mu') - \mathcal{G}(K\mu, -2\mu') \right|^2 / D$$

$$+ \frac{1}{2} \sum_{K,\mu,\mu'} \left| -\mathcal{G}(K\mu, 1\mu') + \mathcal{G}(K\mu, -1\mu') \right|^2 / D \quad , \quad (3.9)$$

$$P_1 = \alpha(1) + \alpha(-1)$$

$$= \frac{1}{2} \sum_{K,\mu,\mu'} \left| \mathcal{G}(K\mu, 2\mu') - \mathcal{G}(K\mu, -2\mu') \right|^2 / D$$

$$+ \frac{1}{2} \sum_{K,\mu,\mu'} \left| \mathcal{G}(K\mu, 1\mu') + \mathcal{G}(K\mu, -1\mu') \right|^2 / D \quad , \quad (3.10)$$

and $P_0 = \alpha(0)$ may be deduced from Eqs. (3.8), (3.9) and (3.10). These results can be checked by using the operator $I'_{y'}$, in turn expressed in terms of the raising and lowering operators, which acts on the spin of the residual nucleus. It is rather obvious that

$$\sum_{K\mu\mu'} \langle \mathcal{G}(K\mu, K'\mu') | (I'_{y'})^2 | \mathcal{G}(K\mu, K'\mu') \rangle / D = \sum_{K'_y} (K'_y)^2 \alpha(K'_y) \quad , \qquad (3.11)$$

and

$$\sum_{K\mu\mu'} \langle \mathcal{G}(K\mu, K'\mu') | (I'_y)^4 | \mathcal{G}(K\mu, K'\mu') \rangle / D = \sum_{K'_y} (K'_y)^4 \alpha(K'_y) \quad . \qquad (3.12)$$

By explicitly evaluating the left-hand side of Eqs. (3.11) and (3.12), the expressions of (3.9 and (3.10) can be verified.

3.3. The systematics of P_1

Although the experimental evidence for spin-flip is unambiguous, the spin-flip probability P_1 is measured only at two scattering angles of $^{13}C'$, $\theta_{c.m.} = 10°$ and $21°$. The experimental errors associated with the latter are rather large. Hence, attempting to pin down a spin-dependent potential by essentially one experimental point, i.e. P_1 at

$\theta_{c.m.} = 10°$, is premature. Instead we investigate the systematic variation of P_1 as the potential parameters are varied. This is done with the hope to encourage further measurements to discriminate the details of any proposed spin-dependent potential.

In [LI 80], an $\vec{\ell} \cdot \vec{s}$ spin-orbit potential is proposed as the mechanism for spin-flip. Out of a great variety of possible spin-dependent potentials, scalar or tensor, between the angular momentum vectors $\vec{I}$, $\vec{S}$ and $\vec{\ell}$, where $\vec{\ell}$ is the orbital momentum of the relative motion, we consider the relative simple model of $\vec{\ell} \cdot \vec{s}$. The model is an extension of the spin-orbit potential used in nucleon-nucleon and nucleon-nucleus scattering. It has also been used in other analyses of heavy-ion transfer reactions [KU 77, BA 78b, WU 79]. However, one ought to be aware that there is no a priori reason to prefer a $\vec{\ell} \cdot \vec{s}$ potential to other possibilities.

Other scalar spin-dependent potentials should, in principle, be considered to be on the same footing as the $\vec{\ell} \cdot \vec{s}$ potential, i.e. $\vec{I} \cdot \vec{S}$ and $\vec{I} \cdot \vec{\ell}$. We discard $\vec{I} \cdot \vec{S}$ as a possibility because in heavy-ion scattering, the high partial waves dominate such that $|\vec{\ell}| >> |\vec{I}|$. The other combination $\vec{I} \cdot \vec{\ell}$ does not lead to the spin-flip of ^{13}C. This is also verified in the calculations.

In the analysis of [LI 80], a potential of the form of Eq.(2.28) was used, with $W_D = 0$. The spin-orbit part has the standard form

$$V_{\ell s}(r) = 2\left(\frac{\hbar}{m_\pi c}\right)^2 \frac{V_{so}}{a_{so} r} \cdot \frac{exp\left[\frac{(r - R_{so})}{a_{so}}\right]}{\left\{1 + exp\left[\frac{(r - R_{so})}{a_{so}}\right]\right\}^2} \cdot \vec{\ell} \cdot \vec{S}$$

$$= \overline{V}_{\ell s} \, \vec{\ell} \cdot \vec{S} \quad , \tag{3.13}$$

where $R_{so} = r_{so}(A_1^{1/3} + A_2^{1/3})$, with $A_1 = 13$ and $A_2 = 24$. The parameters of V_{so}, R_{so} and a_{so} are discussed below.

We show in Fig.(3.1) the angular distributions of the elastic, inelastic scattering and $P_1(\theta_{c.m.})$, using a set of optical potential parameters given in [DÜ 79]. The calculations, which are done in the coupled channels (CC) formalism described in the previous section, reproduce reasonably, though not perfectly, the experimental data. Since we are interested in the systematics, we use this set as a standard and consider variants of it.

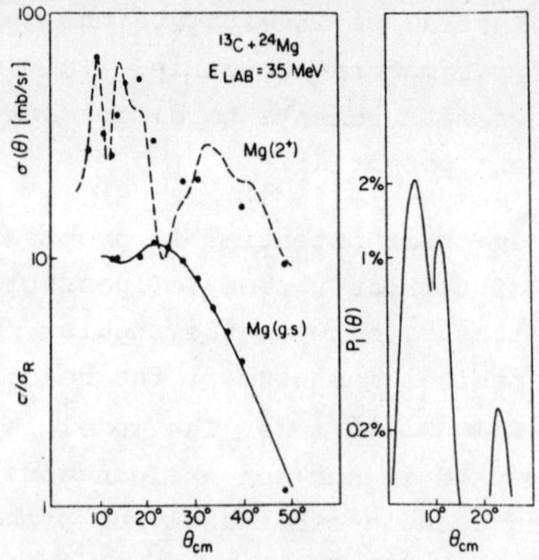

Fig.3.1. Coupled channels calculations of the elastic, inelastic scattering and P_1 ($\theta_{c.m.}$) of $^{24}Mg*(2^+)$ in $^{24}Mg(^{13}C,^{13}C')$ $^{24}Mg*(2^+)$ at E_{lab} = 35 MeV [LI 80].

As far as reproducing the experimental values of P_0, P_1 and P_2 is concerned, we limit ourselves to the $\theta_{c.m.}$ = 10° data, since at $\theta_{c.m.}$ = 21° the data carry very large errors. We tabulate in Fig.(3.2) a comparison of the measured and the calculated values. The marked qualitative difference between the measured P_m's of [DÜ 79] and the calculations is that the observed dominance of P_2 is not reproduced. However, Bybell [BY 80] repeated the experiment and found values of

		P_0	P_1	P_2
	Exp[a]	(4.2 ± 5.5)%	(1.67 ± 0.52)%	(94 ± 5)%
$\theta_{c.m.}$ = 10°	Exp[b]	$\sim$ 32%	$\sim$ 1%	$\sim$ 67%
	DWBA	33%	1%	66%
	CC	36%	0.7%	63%

Fig.3.2. Comparison of P_m from experiments with DWBA and CC calculations. [a]Reference [DÜ 79], [b]Reference [BY 80].

P_m's, which are quite close to the calculated ones. The calculations of [LI 80], which are reproduced in Fig.(3.2), are performed with

V_{so} = 0.7 MeV given in [DÜ 79]. A slight readjustment of this value would certainly yield P_1 = 1% from a CC calculation, with only very small changes in P_0 and P_2.

After a series of calculations, it is found that P_1 is most sensitive to the parameters which enter directly into the spin-orbit potential, i.e. V_{so}, a_{so} and r_{so}. The spin-orbit potential can be chosen to be attractive or repulsive, real or imaginary, but all these lead to similar values of P_m's. Therefore, a real and attractive potential is adopted and two of the parameters are kept constant and equal to those in the standard set [DÜ 79], the third is varied over a reasonable range of values.

It is found that (i) the positions of the three peaks of $P_1(\theta_{c.m.})$, as displayed in Fig.(3.1), do not change in any of the calculations; (ii) when any one of the three parameters is varied, the ratio of the values of the three peaks is unaltered; (iii) for constant a_{so}, the same values of the three peaks of P_1 are produced by various combinations of V_{so} and R_{so} as long as the Igo ambiguity condition $V_{so} \exp(R_{so}/a_{so})$ = constant is satisfied, see Eq.(1.7);(iv) the elastic and inelastic cross sections are only slightly affected by the above variations at the largest angles of interest in Fig. (3.1).

We discuss here only the property (iii), while further discussions of (i), (ii), and (iv) can be found in [LI 80]. We can examine three sample sets of spin-orbit potentials which yield $P_1(6°)$ = 3%. The potential parameters are tabulated in Fig.(3.3) as (I), (II) and (III). The potentials as a function of the radial distance r are plotted in Fig.(3.4). One sees immediately the similarity of these potentials at distances greater than 7.5 fm. This value is smaller than the estimate of the strong absorption radius from 1.5 $(13^{1/3} + 24^{1/3})$ = 7.9 fm, and also the semiclassical estimate of 9.0 fm, by taking ℓ = 18.5 which

	V_{so} (MeV)	a_{so} (fm)	r_{so} (fm)
(I)	0.85	0.658	1.165
(II)	0.7	0.745	1.165
(III)	0.7	0.658	1.195
(IV)	0.215	1.0	1.5
(V)	0.256	1.2	1.5

Fig.3.3. Spin-orbit potential parameters [LI 80].

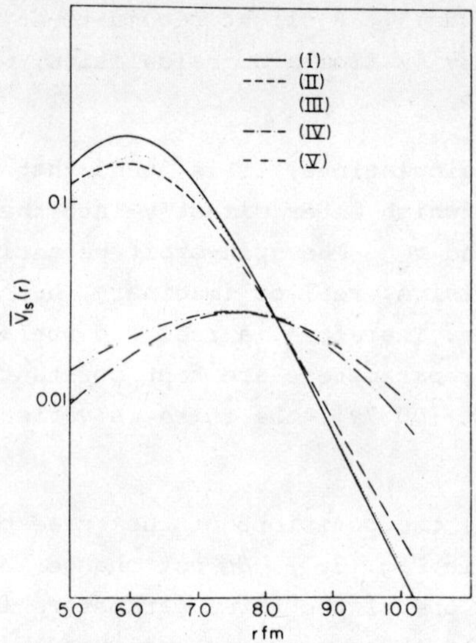

Fig.3.4. Spin-orbit potentials for parameter sets (I)-(V) of Fig.(3.3).
Sets (I)-(III) give $P_1(6°)$ = 3%; set (IV) and (V) yield
$P_1(6°)$ = 5% and 6%, respectively.

corresponds to a transmission coefficient of 1/2. To confirm the specu-
lation that not only the magnitude but also the slope of the spin-orbit
potential are responsible for yielding nearly equal values of P_1, two
other unrealistic spin-orbit potentials, sets (IV) and (V), are also
plotted in Fig.(3.4). Their potential parameters are tabulated in
Fig.(3.3). They both give spin-orbit potential strengths approximately
equal to those of (I), (II) and (III) at 8.2 fm. They lead to $P_1(6°)$
= 5% and 6 %, respectively, about twice the value for sets (I)-(III).
The parameters of (I) and (III) are related to each other via the Igo
ambiguity, $V_{so} \exp(R_{so}/a_{so})$ = constant. This merely means that the long-
range part of the spin-orbit potential is responsible for the spin-flip,
and the potential sets are nearly equal in that region.

In [LI 80], a number of other possible effects are examined and
found to produce very small changes in P_1. They include (i) adding in
the 4.12 MeV 4^+ level and carrying out a three-level CC calculation;
(ii) deforming the spin-orbit potential so that it contributes to the
inelastic excitation; (iii) switching off the reorientation effects
for the 2^+ level; (iv) modifying the spin-independent optical potential

parameters. As a fairly extreme example of (iv), the deformation length βR was increased by a factor of 2. While the inelastic cross section increases by a factor of 5, $P_1(6°)$ remains approximately 2-2.5%.

By this analysis, the CC calculation requires $V_{so} = 1$ MeV (a slight overestimate in view of the latest measurement of [BY 80]), with a 20% uncertainty arising from experimental error. This gives a $\bar{V}_{so}(r)$ in Eq.(3.13) of 4.4 x 10^{-2} MeV at 7.9 fm and 8.0 x 10^{-3} MeV at 9.0 fm. This is comparable to other values used in the analysis of 63 MeV ^{13}C incident of ^{40}Ca [BA 78b], but does not support the folding model [PE 78], which gives, for the ^{13}C + ^{40}Ca system, a value of 1.7 x 10^{-4} MeV at the strong absorption radius (9.5 fm).

4.   Conclusion

From a survey of heavy-ions direct reactions, we may make the observation that, essentially, the theoretical tools, DWBA and coupled-channels formalism, previously used in light-ions reactions,are versatile enough to be applicable in this domain. However, the computational effort is greatly increased, and sometimes the numerical calculation may even be impossible to carry out. Semi-classical methods are often used to circumvent this hurdle. They yield, in many instances, qualitative estimate of the experimental data and afford an intuitively reasonable interpretation of the physics. Heavy-ions direct reactions do yield new information on nuclear structure and nuclear reaction mechanism which are not delivered by light-ions reactions: production of high-spin states and states of unusual spectroscopy, and multi-step reactions. The interpretation of the experimental data by multi-step reaction models is usually an indirect inference. How much of the effects seen are only a consequence of potential parameters variations is often debated. One would welcome an unambiguous recognition of the experimental signatures of multi-step reactions, as, for example, γ-ray emission in the reaction normal is an unambiguous signature of a spin-flip process discussed in section 3.

Acknowledgment

This article was prepared when the author was on an extended
sabbatical leave from the Hahn-Meitner-Institut für Kernforschung,
Berlin-West, Germany, at the School of Physics, University of Minnesota.
I thank my colleagues at the Hahn-Meitner-Institut, whose extraordinary
generosity made this extended sabbatical leave possible. It is my
pleasure to thank B. F. Bayman, P.J. Ellis and Y.C.Tang of the University of Minnesota for numerous helpful discussions and constant en-
couragement, Sandy Smith, Bobbi Olsen and U. Fischer for their valuable
assistance. The work was supported in part by the U.S. Department of
Energy under Contract No. DOE/DE-ACO2-79ER10364.

Appendix

Elements of collision theory

In this section we sketch some elements of collision theory of spinless particles which are relevant to our discussion of direct reactions in heavy-ion collision. For the sake of simplicity, we restrict ourselves, in this section, to the scattering of two structureless particles via a central potential (exception: those subsections on rearrangement collision).

A.1. The Schrödinger equation

We consider collision at relative energy E with the reduced mass of the system defined as μ. The two particles interact through a potential $V(\vec{r})$, where $\vec{r}$ is the relative coordinate between the two masses. The non-relativistic quantal description of the collision is given by the solution of the Schrödinger equation, when solved, of course, with a scattering boundary condition,

$$\left[-\frac{\hbar^2}{2\mu}\nabla^2 + V(\vec{r})\right]\psi(\vec{r}) = E\,\psi(\vec{r}) \;. \tag{A.1}$$

In the absence of the interaction $V(\vec{r})$, the solution of Eq.(A.1) is a plane wave, which, in the laboratory situation, corresponds to a beam of particles moving undisturbed in a particular direction, i.e.,

$$-\frac{\hbar^2}{2\mu}\nabla^2\phi(\vec{r}) = E\phi(\vec{r}) \;, \tag{A.2}$$

where

$$\phi(\vec{r}) = e^{i\vec{k}\cdot\vec{r}} \;, \tag{A.3}$$

and

$$k^2 = \frac{2\mu E}{\hbar^2} \;. \tag{A.4}$$

In the presence of a scatterer, the solution of Eq.(A.1) should have the following form in the asymptotic region,

$$\psi^{(+)}(\vec{r}) \xrightarrow[r \to \infty]{} e^{i\vec{k}\cdot\vec{r}} + f(\theta,\phi)\,\frac{e^{ikr}}{r} \quad . \tag{A.5}$$

This is usually called the outgoing wave boundary condition. In contrast, the incoming wave boundary condition has the form

$$\psi^{(-)}(\vec{r}) \xrightarrow[r \to \infty]{} e^{i\vec{k}\cdot\vec{r}} + f^{(-)}(\theta,\phi)\,\frac{e^{-ikr}}{r} \quad . \tag{A.6}$$

The physical interpretation of Eq.(A.1) can be visualized as the following. Some particles from the monochromatic beam have been scattered by $V(\vec{r})$ as spherical matter wave. In the asymptotic region where the interaction is not directly felt, its effect on the incident plane wave manifests itself as an additional spherical scattered wave. The detail of the effect of the interaction is contained in the scattering amplitude $f(\theta,\phi)$, which is a function of the polar angles (θ,ϕ). This is an analogue of the Huygens' principle in the scattering of electromagnetic waves.

We should keep in mind that the incident beam in the laboratory is not strictly a monochromatic plane wave, and the particles in the beam do not constitute an uninterrupted current as implied by Eq.(A.3). However, a wave-packet treatment of scattering, which takes into account the spread in energy of the incident beam and the bunched nature of the particles in the beam, shows that, in spite of their deficiencies, Eqs. (A.3), (A.5) cause no error [NE 66].

A.2. The cross section

The probability current density connected with the wave function ψ is given by

$$\vec{j}(\vec{r}) = Re\left[\psi^{*}\frac{\hbar}{i\mu}\nabla\psi\right] \quad . \tag{A.7}$$

Therefore, the incident probability current density arising from the incident plane wave (A.3) is,

$$\vec{j}_{inc}(\vec{r}) = \frac{\hbar k}{\mu}\,\hat{z} = j_{inc}\,\hat{z} \quad , \tag{A.8}$$

where the direction of the incident beam is chosen to be the z-axis. The probability current density of the scattered wave,

$$\psi_{sc} = f(\theta,\phi)\, \frac{e^{ikr}}{r} \; , \tag{A.9}$$

is

$$\vec{j}_{sc}(\vec{r}) = \frac{|f(\theta,\phi)|^2}{r^2} \cdot \frac{\hbar k}{\mu}\, \hat{r} + \cdots$$

$$= j_{inc} \cdot \frac{|f(\theta,\phi)|^2}{r^2}\, \hat{r} + \cdots \; , \tag{A.10}$$

where $\hat{r}$ is the unit radial vector. The higher order terms contain higher inverse powers of r and are therefore neglected. Since Eq.(A.5) is valid for large r, the neglect of higher order terms in Eq.(A.10) is justified. Suppoes a detector of detecting surface ΔA is placed at a distance r from the target, where r is much greater than the dimension of the interaction region. The solid angle subtended by ΔA at the target is $\Delta\Omega = \Delta A/r^2$. Therefore, the number of particles scattered into the detecting surface per unit time is

$$\Delta n = \vec{j}_{sc}(\vec{r}) \cdot \vec{\Delta A}$$

$$= j_{inc} \cdot |f(\theta,\phi)|^2 \Delta\Omega \; . \tag{A.11}$$

Therefore, we have

$$\frac{\Delta n}{\Delta\Omega} = j_{inc} \cdot |f(\theta,\phi)|^2 \; . \tag{A.12}$$

We are now in the position to define the differential cross section, which is the number of incident particles scattered into the direction (θ,ϕ), per unit solid angle Ω , per unit incident current, per unit scatterer,

$$\frac{d\sigma}{d\Omega} = \frac{1}{j_{inc}} \lim_{\Delta\Omega\to 0} \frac{\Delta n}{\Delta\Omega} = |f(\theta,\phi)|^2 \; . \tag{A.13}$$

Hence, theoretical calculation of scattering cross section is synonymous with the calculation of the scattering amplitude $f(\theta,\phi)$.

A.3. Partial wave expansion of the plane wave

The plane wave solution (A.3) can be analysed in a different way. It is advantageous to write the Laplacian in Eq.(A.2) in spherical coordinates which is then separable

$$\nabla^2 = \frac{1}{r^2}\frac{\partial}{\partial r}\left(r^2\frac{\partial}{\partial r}\right) - \frac{1}{\hbar^2 r^2}\,L^2(\theta,\phi)\,, \tag{A.14}$$

where

$$L^2(\theta,\phi) = -\hbar^2\left[\frac{1}{\sin\theta}\cdot\frac{\partial}{\partial\theta}\left(\sin\theta\frac{\partial}{\partial\theta}\right) + \frac{1}{\sin^2\theta}\cdot\frac{\partial^2}{\partial\phi^2}\right]\,, \tag{A.15}$$

and

$$L_z = (\vec{r}\times\vec{p})\cdot\hat{z} = -i\hbar\frac{\partial}{\partial\phi}\,. \tag{A.16}$$

The eigenfunctions of the angular momentum operator L^2 are the spherical harmonics $Y_\ell^m(\theta,\phi)$ [NE 66], such that

$$L^2 Y_\ell^m(\theta,\phi) = \ell(\ell+1)\hbar^2\, Y_\ell^m(\theta,\phi)\,, \tag{A.17}$$

and

$$L_z Y_\ell^m(\theta,\phi) = m\hbar\, Y_\ell^m(\theta,\phi)\,. \tag{A.18}$$

The plane wave solution (A.3) is a rotationally invariant function of $\hat{k}$ and $\hat{r}$. Therefore, it can be expanded in general form [NE 66]

$$e^{i\vec{k}\cdot\vec{r}} = \sum_{\ell,m} c_\ell(kr)\, Y_\ell^m(\hat{r})\, Y_\ell^{m\,*}(\hat{k})\,. \tag{A.19}$$

By writing the expansion coefficients as $c_\ell(kr)\propto u_\ell(kr)/(kr)$, $u_\ell(kr)$ satisfies the radial equation

$$\frac{d^2 u_\ell}{dr^2} + \left[k^2 - \frac{\ell(\ell+1)}{r^2}\right]u_\ell = 0\,. \tag{A.20}$$

A convenient form for Eq.(A.19) is

$$e^{i\vec{k}\cdot\vec{r}} = \frac{4\pi}{kr}\sum_{\ell,m} i^{\ell}\, u_{\ell}(kr)\, Y_{\ell}^{m}(\hat{r})\, Y_{\ell}^{m*}(\hat{k}) \quad , \qquad (A.21)$$

where the solution $u_{\ell}(kr)$ of Eq.(A.20) can be written in terms of the spherical Bessel function $j_{\ell}(kr)$

$$u_{\ell}(kr) = kr\, j_{\ell}(kr) \quad . \qquad (A.22)$$

The incident beam of particles has a specific direction and there exists rotational symmetry about this direction. Choosing the $\hat{z}$-direction to be $\hat{k}$, we have

$$e^{i\vec{k}\cdot\vec{r}} = \frac{1}{kr}\sum_{\ell}\sqrt{4\pi(2\ell+1)}\, u_{\ell}(kr)\, Y_{\ell}^{o}(\hat{r}) \quad . \qquad (A.23)$$

The fact that we have m = O is a consequence of the symmetry about the $\hat{z}$- or $\hat{k}$-direction.

A.4. Partial wave expansion of the scattering wave function

If we have a spherical symmetrical potential, i.e. $V(\vec{r}) = V(r)$, it is also useful to write the Laplacian in (A.1) in spherical coordinates because of the separation of the coordinates. Thus, in the same spirit, the scattering solution $\psi^{(+)}(\vec{r})$ of Eq. (A.1) can be written in a similar fashion as (A.23),

$$\psi^{(+)}(\vec{r}) = \frac{1}{kr}\sum_{\ell}\sqrt{4\pi(2\ell+1)}\, \psi_{\ell}(kr)\, Y_{\ell}^{o}(\hat{r}) \quad , \qquad (A.24)$$

where $\psi_{\ell}(kr)$ satisfies the radial equation

$$\frac{d^{2}\psi_{\ell}}{dr^{2}} + \left[k^{2} - U(r) - \frac{\ell(\ell+1)}{r^{2}}\right]\psi_{\ell} = 0 \quad . \qquad (A.25)$$

with $U(r) = 2\mu V(r)/h^{2}$. In Eq.(A.24), the $\hat{z}$-direction has already been chosen to be the direction about which there is rotational symmetry. The expansion is not valid for potentials which vanish as r^{-n} asymptotically and requires special consideration for the important case of the Coulomb potential with n = 1. We shall return to this point later.

A.5. Determination of the scattering amplitude $f(\theta)$

The requirement that $u_\ell(kr)$ and $\psi_\ell(kr)$ be finite everywhere means that $u_\ell(kr)$ and $\psi_\ell(kr)$ vanish at the origin. The asymptotic behaviours of $u_\ell(kr)$ and $\psi_\ell(kr)$ are intimately connected with the scattering amplitude. Asymptotically, $u_\ell(kr)$ is a sinusoidal function,

$$u_\ell(kr) = kr\, j_\ell(kr) \xrightarrow[kr\gg1]{} \sin\left(kr - \tfrac{1}{2}\ell\pi\right)\, , \qquad (A.26)$$

while $\psi_\ell(kr)$ must have the same sinusoidal behaviour with some modification, because the potential function $U(r)$ vanishes in the asymptotic region. Therefore, we can represent the effect of the scattering potential by modifying the amplitude and phase of the sinusoidal function (A.26) so that

$$\psi_\ell(kr) \xrightarrow[kr\gg1]{} A_\ell \sin\left(kr - \tfrac{1}{2}\ell\pi + \delta_\ell\right)$$

$$= A_\ell \cos\delta_\ell \sin\left(kr - \tfrac{1}{2}\ell\pi\right) + A_\ell \sin\delta_\ell \cos\left(kr - \tfrac{1}{2}\ell\pi\right)$$

$$= A_\ell \cos\delta_\ell\, F_\ell(kr) + A_\ell \sin\delta_\ell\, G_\ell(kr)\, , \qquad (A.27)$$

where A_ℓ and δ_ℓ are independent of r in the asymptotic region. Since the oscillatory behaviour of (A.27) differs from the plane wave solution in the asymptotic region by a phase, δ_ℓ is called the phase shift of the ℓ-th partial wave. $F_\ell(kr)$ and $G_\ell(kr)$ are the regular and irregular solutions of Eq.(A.20). These functions have the special properties

$$F_\ell(kr) \xrightarrow[kr\ll\ell]{} \frac{(kr)^{\ell+1}}{(2\ell+1)!!}\, , \qquad (A.28)$$

$$F_\ell(kr) \xrightarrow[kr\gg\ell]{} \sin\left(kr - \tfrac{1}{2}\ell\pi\right)\, , \qquad (A.29)$$

$$G_\ell(kr) \xrightarrow[kr\ll\ell]{} \frac{(2\ell-1)!!}{(kr)^\ell}\, , \qquad (A.30)$$

Using (A.23) and (A.26), we have

$$\psi(\vec{r}) \xrightarrow[kr \gg 1]{} e^{i\vec{k}\cdot\vec{r}}$$

$$+ \frac{e^{ikr}}{r}\left[\frac{1}{k}\sum 4\pi(2\ell+1)\, i^{-\ell} e^{i\delta_\ell} \sin\delta_\ell \, Y_\ell^o(\theta)\right]. \quad \text{(A.34)}$$

Comparing (A.34) with (A.5), with the $\hat{z}$-direction chosen to be $\hat{k}$ in the latter, one obtains immediately

$$f(\theta) = \frac{1}{k}\sum_\ell \sqrt{4\pi(2\ell+1)}\, i^{-\ell} e^{i\delta_\ell} \sin\delta_\ell \, Y_\ell^o(\theta)$$

$$= \frac{1}{k}\sum_\ell (2\ell+1)\, e^{i\delta_\ell} \sin\delta_\ell \, P_\ell(\theta)$$

$$= \sum_\ell f_\ell(\theta) \quad , \qquad\qquad\qquad \text{(A.35)}$$

where $P_\ell(\theta)$ is the Legendre polynomial and $f_\ell(\theta)$ is the ℓ-th partial scattering amplitude.

Normally, one can solve Eq.(A.25) analytically only if the potential function U(r) is simple. In realistic nuclear scattering problems where U(r) is complicated, one solves Eq.(A.25) numerically to deduce the phase shifts. The procedure is the following. That $\psi_\ell(kr)$ vanishes at the origin and its derivative there is given an arbitrary value are sufficient for the numerical integration of (A.25) such that $\psi_\ell(kr)$ is determined up to a multiplicative constant D_ℓ. At a point R in the asymptotic region where U(R) becomes negligible, the value and the logarithmic derivative of the numerical solution at R are equated to their counterparts calculated from the asymptotic form of $\psi_\ell(kr)$, Eq.(A.27), i.e.

$$\lim_{\varepsilon\to o} \psi_\ell(r=R-\varepsilon) = \lim_{\varepsilon\to o} \psi_\ell(r=R+\varepsilon) \quad , \qquad \text{(A.36)}$$

and

$$\lim_{\varepsilon\to o} \frac{\frac{d}{dr}\psi_\ell(r=R-\varepsilon)}{\psi_\ell(r=R-\varepsilon)} = \lim_{\varepsilon\to o} \frac{\frac{d}{dr}\psi_\ell(r=R+\varepsilon)}{\psi_\ell(r=R+\varepsilon)} \qquad \text{(A.37)}$$

where the left-hand side of Eqs.(A.36), (A.37) are derived from the numerical solution, while the right-hand side is calculated from (A.27).

$$G_\ell(kr) \xrightarrow[kr \gg \ell]{} \cos(kr - \tfrac{1}{2}\ell\pi) \tag{A.31}$$

The quantities $j_\ell(kr) = F_\ell(kr)/(kr)$ and $n_\ell(kr) = -G_\ell(kr)/(kr)$ are tabulated as spherical Bessel function and spherical Neumann function, respectively.

We have chosen explicitly a particular normalization for the scattering outgoing wave in Eq.(A.5). Therefore, $\psi_\ell(kr)$ can only have those values of A_ℓ such that the asymptotic form of the expansion (A.24) is exactly that of (A.5). Substituting (A.27) in (A.24) and replacing $Y_\ell^o(\theta)$ for $Y_\ell^o(\hat{r})$, we have

$$\psi(\vec{r}) \xrightarrow[kr \gg 1]{} \frac{1}{kr} \sum_\ell \sqrt{4\pi(2\ell+1)}\, A_\ell \sin(kr - \tfrac{1}{2}\ell\pi + \delta_\ell)\, Y_\ell^o(\theta)$$

$$= \frac{1}{kr} \sum_\ell \sqrt{4\pi(2\ell+)}\, A_\ell \left[\cos\delta_\ell \sin(kr - \tfrac{1}{2}\ell\pi) \right.$$

$$\left. + \sin\delta_\ell \cos(kr - \tfrac{1}{2}\ell\pi) \right] Y_\ell^o(\theta)$$

$$= \frac{1}{kr} \sum_\ell \sqrt{4\pi(2\ell+1)}\, A_\ell \left\{ (\cos\delta_\ell - i\sin\delta_\ell)\sin(kr - \tfrac{1}{2}\ell\pi) \right.$$

$$\left. + \sin\delta_\ell \left[\cos(kr - \tfrac{1}{2}\ell\pi) + i\sin(kr - \tfrac{1}{2}\ell\pi)\right] \right\} Y_\ell^o(\theta)$$

$$= \frac{1}{kr} \sum_\ell \sqrt{4\pi(2\ell+1)}\, A_\ell \left[e^{-i\delta_\ell}\sin(kr - \tfrac{1}{2}\ell\pi) \right.$$

$$\left. + e^{ikr} \cdot i^{-\ell}\sin\delta_\ell \right] Y_\ell^o(\theta) \quad . \tag{A.32}$$

Therefore, by choosing $A_\ell = e^{i\delta_\ell}$, we reduce Eq.(A.32) to

$$\psi(\vec{r}) \xrightarrow[kr \gg 1]{} \frac{1}{kr} \sum_\ell \sqrt{4\pi(2\ell+1)}\, \sin(kr - \tfrac{1}{2}\ell\pi)\, Y_\ell^o(\theta)$$

$$+ \frac{e^{ikr}}{r} \left[\frac{1}{k} \sum_\ell 4\pi(2\ell+1) i^{-\ell} e^{i\delta_\ell} \sin\delta_\ell\, Y_\ell^o(\theta) \right] \quad . \tag{A.33}$$

These two equations determine the two unknown quantities, namely δ_ℓ and D_ℓ. This procedure, with minor variation, is still the prevalent method of solving the radial equation. The reader will observe that the principle one employs is identical to that used in scattering problems in elementary wave mechanics.

A.6 The Lippmann-Schwinger equation (given without proof)

The method to calculate the scattering amplitude in the previous section is essentially a differential equation method, since the differential equation (A.25) is solved and then the appropriate boundary condition is imposed. However, integral equation is a more convenient tool for handling approximations and for the investigation of the formal theory.

The general solution f(x) of an inhomogeneous differential equation

$$Df(x) = F(x), \qquad (A.38)$$

where D is a differential operator, is [PI 29]

$$f = h + \frac{1}{D}\, F, \qquad (A.39)$$

where h is the solution of the homogeneous part of the differential equation, i.e.

$$Dh = O \;, \qquad (A.40)$$

and 1/D is symbolically an integral operator. The Schrödinger equation can be casted in the form of (A.38). By writing the Hamiltonian operator in the following form

$$H = -\frac{\hbar^2}{2\mu}\nabla^2 + V = H_o + V \;, \qquad (A.41)$$

the Schrödinger equation becomes

$$(H_o + V)\psi = E\psi \;, \qquad (A.42)$$

or

$$(E - H_o)\psi = V\psi \;. \qquad (A.43)$$

Therefore, the solution can be written symbolically as

$$\psi^{(\pm)} = \phi + \frac{1}{E \pm i\varepsilon - H_o} V \psi^{(\pm)} \;, \tag{A.44}$$

where ϕ is a solution of the homogeneous part of (A.43), i.e.

$$(E - H_o)\phi = 0 \;. \tag{A.45}$$

One recognizes that ϕ is a plane wave, Eq.(A.3). Equation (A.44) is
known as a Lippmann-Schwinger equation. The energy in (A.44) is given
a small complex part of either sign and is allowed to approach zero.

The inverse operator $G_o = (E-H_0)^{-1}$, sometimes known as the plane
or free wave Green's function, exists everywhere on the complex E-plane
except for a branch cut on the positive real E axis and isolated points
on the negative real E axis. Unfortunately, the physical world of
scattering corresponds to points on the positive real E axis. Therefore,
to obtain a more precise definition of this inverse operator, we must
be careful about how to approach this branch cut. Approaching the branch
cut from above or below will lead to different results, namely the outgoing or the incoming scattering wave of (A.5) and (A.6), respectively.
It is inappropriate to go into detail in these pages the mathematical
structure of the Green's function operator. The reader is referred to
standard text on scattering theory, e.g. [NE 66] and a collection of
original papers edited by Ross [RO 63].

A.7. Coordinate representation of the Green's function

To make practical use of the Lippmann-Schwinger equation, one may
write (A.44) in coordinate representation. To this end, one needs the
Green's function in coordinates representation. We make use of the complete orthonormal set of eigenfunctions of H_o in the momentum representation, such that for complex z, we have

$$\langle \vec{r}\,|(z-H_o)^{-1}|\vec{r}\,'\rangle = \iint \langle \vec{r}\,|\vec{p}\,\rangle \, d^3\vec{p} \, \langle \vec{p}\,|(z-H_o)^{-1}|\vec{p}\,'\rangle \, d^3\vec{p}\,' \, \langle \vec{p}\,'|\vec{r}\,'\rangle \;. \tag{A.46}$$

It is assumed for definiteness that the eigenvalues $\vec{p}$ form a continuous set. The orthonormal and closure relations are given by

$$\langle \vec{p} \mid \vec{p}' \rangle = \delta(\vec{p} - \vec{p}') \; , \tag{A.47}$$

$$\int \mid \vec{p} \rangle \, d^3\vec{p} \, \langle \vec{p} \mid \, = 1 \quad . \tag{A.48}$$

The unitary transform from the $\{\vec{p}\}$ representation to the $\{\vec{r}\}$ representation is

$$\langle \vec{r} \mid \vec{p} \rangle = \frac{1}{(2\pi\hbar)^3} \, e^{i\vec{p}\cdot\vec{r}/\hbar} \quad . \tag{A.49}$$

Therefore, since $(Z = H_o)^{-1}$ is diagonal in the $\{\vec{p}\}$ representation, (A.46) can be written as

$$\langle \vec{r} \mid (Z - H_o)^{-1} \mid \vec{r}' \rangle = \frac{1}{(2\pi)^3} \int e^{i\vec{k}'\cdot(\vec{r}-\vec{r}')} \left(Z - \frac{\hbar^2 k'^2}{2\mu} \right)^{-1} d^3\vec{k}' \; . \tag{A.50}$$

For convenience we may write

$$Z = \frac{\hbar^2 \sigma^2}{2\mu} \quad , \quad Re\,\sigma > 0 \; , \tag{A.51}$$

and

$$\vec{R} = \vec{r} - \vec{r}' \; , \tag{A.52}$$

such that

$$\langle \vec{r} \mid (Z - H_o)^{-1} \mid \vec{r}' \rangle = \frac{\mu}{4\pi^3\hbar^2} \int \frac{e^{i\vec{k}'\cdot\vec{R}}}{\sigma^2 - k'^2} \, d^3\vec{k}' \; . \tag{A.53}$$

Choosing the polar axis to be in the direction of $\hat{R}$, we perform the angular integration and obtain

$$\langle \vec{r}\,|(Z-H_0)^{-1}|\vec{r}'\rangle = \frac{\mu}{4\pi^3\hbar^2}\iiint \frac{e^{ik'R\cos\theta}}{\sigma^2-k'^2}\,k'^2\,dk'\sin\theta\,d\theta\,d\phi$$

$$= \frac{\mu}{\pi^2\hbar^2 R}\int_0^\infty \frac{k'\sin k'R}{\sigma^2-k'^2}\,dk'$$

$$= \frac{\mu}{\pi\hbar^2 R}\cdot\frac{1}{2\pi i}\int_{-\infty}^\infty \frac{k'e^{ik'R}}{\sigma^2-k'^2}\,dk' \quad . \tag{A.54}$$

The integral can be evaluated by contour integration. The contour taken
is a semi-circle either in the upper or lower half k-plane.

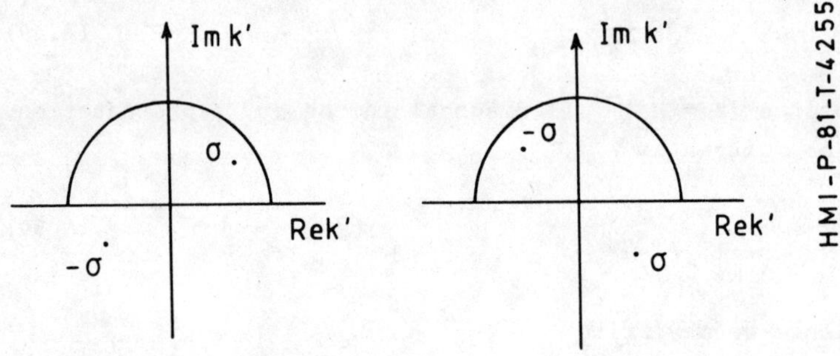

(a) $\mathrm{Im}\,Z > 0\,(\mathrm{Im}\,\sigma>0)$ (b) $\mathrm{Im}\,Z < 0\,(\mathrm{Im}\,\sigma<0)$

Fig. A.1.

(a) $\mathrm{Im}\,Z>0\,(\mathrm{Im}\,\sigma>0)$, Fig.A.1a.

The poles of the integrand occurs at $\pm\sigma$, with $\mathrm{Im}\,\sigma>0$. When the contour is closed in the upper half k-plane, only the residue from the pole $k = \sigma$ in the upper half-plane contributes, and it is $-e^{i\sigma R}/2$.

(b) $\mathrm{Im}\,Z<0\,(\mathrm{Im}\,\sigma<0)$, Fig.A1b.

The poles of the integrand occurs at $\pm\sigma$, with $\mathrm{Im}\,\sigma<0$. When the contour is closed in the upper half-plane, again only the residue from the pole in the upper half plane $k = -\sigma$ contributes, and it is $-e^{-i\sigma R}/2$. Consequently, Eq.(A.54) becomes

$$\langle \vec{r} | (Z - H_0)^{-1} | \vec{r}' \rangle = - \frac{\mu}{2\pi \hbar^2} \times \begin{cases} e^{i\sigma R}/R \,, & \text{if } \operatorname{Im} Z > 0 \\[2mm] e^{-i\sigma R}/R \,, & \text{if } \operatorname{Im} Z < 0 . \end{cases} \quad \text{(A.55)}$$

Therefore, one may now write for the coordinate representation of the Green's function as

$$G_0^{(\pm)}(\vec{r}, \vec{r}') = \operatorname*{Lim}_{\varepsilon \to 0} \langle \vec{r} | (E \pm i\varepsilon - H_0)^{-1} | \vec{r}' \rangle$$

$$= - \frac{\mu}{2\pi \hbar^2} \cdot \frac{e^{\pm ik|\vec{r} - \vec{r}'|}}{|\vec{r} - \vec{r}'|} \qquad \text{(A.56)}$$

A.8. The Lippmann-Schwinger equation in coordinate representation

We are in the position to write the Lippmann-Schwinger equation in coordinate representation using (A.56), i.e.

$$\psi^{(\pm)}(\vec{r}) = \phi(\vec{r}) - \frac{\mu}{2\pi \hbar^2} \int \frac{e^{\pm ik|\vec{r} - \vec{r}'|}}{|\vec{r} - \vec{r}'|} V(\vec{r}') \psi^{(\pm)}(\vec{r}') \, d^3\vec{r}' . \qquad \text{(A.57)}$$

This is sometimes known as the scattering integral equation, and it encompasses both the outgoing wave type and the incoming wave type.

The usefulness of the Lippmann-Schwinger equation is more apparent when we examine (A.57) in the asymptotic region $r \longrightarrow \infty$. For this, we consider first the asymptotic expression

$$|\vec{r} - \vec{r}'| \equiv r \left[1 - 2 \frac{\vec{r} \cdot \vec{r}'}{r^2} + \frac{r'^2}{r^2} \right]^{\frac{1}{2}}$$

$$\xrightarrow[r \to \infty]{} r - \frac{\vec{r}}{r} \cdot \vec{r}' + O\left(\frac{r'}{r}\right) , \qquad \text{(A.58)}$$

then we have

$$\frac{e^{\pm ik|\vec{r} - \vec{r}'|}}{|\vec{r} - \vec{r}'|} \xrightarrow[r \to \infty]{} \frac{e^{\pm ikr \mp ik\hat{r} \cdot \vec{r}'}}{r} . \qquad \text{(A.59)}$$

This leads to (A.57) acquiring the form

$$\psi^{(\pm)}(\vec{r}) \xrightarrow[r\to\infty]{} \phi(\vec{r}) - \frac{\mu}{2\pi\hbar^2}\cdot\frac{e^{ikr}}{r}\int e^{\mp i\vec{k}'\cdot\vec{r}'}V(\vec{r}')\,\psi^{(\pm)}(\vec{r}')\,d^3\vec{r}'\,, \qquad \text{(A.60)}$$

where $\vec{k}' = k\hat{r}$. The vector $\vec{k}'$ is a vector, of magnitude k, which has the direction of the polar angles (θ,ϕ) of the point at which the scattering amplitude is being measured. A comparison of (A.60) with (A.5) gives us immediately the scattering amplitude

$$f(\theta,\phi) = -\frac{\mu}{2\pi\hbar^2}\int e^{-i\vec{k}'\cdot\vec{r}'}V(\vec{r}')\,\psi^{(+)}(\vec{r}')\,d^3\vec{r}' \qquad \text{(A.61)}$$

Equation (A.61) is valid for a short range potential. This is because (A.60) was derived under the assumption of r >> r'. The integration of r' in (A.61) includes region where r' is very large, but because $V(\vec{r}')$ vanishes or becomes negligible beyond a certain range, the violation of the assumption r >> r' does not cause any problem. In contrast to the differential equation method of section (A.5), the scattering amplitudes are obtained in this instance from an integral equation, in which the boundary condition is included.

A.9. The Born expansion

At first sight, Eq.(A.57) or Eq.(A.61) does not appear to be helpful to us because the scattering amplitude is expressed in terms of the scattering function $\psi^{(+)}$ which we intend to solve for. However, it does point a way towards successive iteration. Substituting a plane wave for $\psi^{(+)}(\vec{r}')$ on the right hand side of (A.57), we obtain as the first iteration

$$\psi_1^{(+)}(\vec{r}) = \phi(\vec{r}) + \int G_o^{(+)}(\vec{r},\vec{r}')V(\vec{r}')\,e^{i\vec{k}\cdot\vec{r}'}\,d^3\vec{r}'\,. \qquad \text{(A.62)}$$

Subsequently, inserting (A.62) on the right hand side of (A.57) yields the second iteration,

$$\psi_2^{(+)}(\vec{r}) = \phi(\vec{r}) + \int G_o^{(+)}(\vec{r},\vec{r}')V(\vec{r}')\,\psi_1^{(+)}(\vec{r}')\,d^3\vec{r}'\,, \qquad \text{(A.63)}$$

and so on. Under suitable conditions [NE 66], the series given by this procedure converges to the exact solution $\psi^{(+)}(\vec{r})$ which is represented by an expansion in powers of V, namely

$$\psi^{(+)}(r) = e^{i\vec{k}\cdot\vec{r}} + \sum_{n=1}^{\infty} \int K_n(\vec{r},\vec{r}')\, e^{i\vec{k}\cdot\vec{r}'}\, d^3\vec{r}' \ , \tag{A.64}$$

where

$$K_n(\vec{r},\vec{r}') = \int K_1(\vec{r},\vec{r}'')\, K_{n-1}(\vec{r}'',\vec{r}')\, d^3\vec{r}'' \ ,$$

$$K_1(\vec{r},\vec{r}') = G_0(\vec{r},\vec{r}')\, V(\vec{r}') \ . \tag{A.65}$$

This is the Born expansion of the scattering wave function.

A.10. Partial wave expansion of the Green's function

We first derive the differential equation which the Green's function satisfies. Substituting the Lippmann-Schwinger equation (A.44) into the left hand side of (A.43), we obtain

$$(E-H_0)(\phi + G_0^{(+)} V \psi^{(+)}) = V\psi^{(+)} \ , \tag{A.66}$$

$$(E-H_0)\, G_0^{(+)} V \psi^{(+)} = V\psi^{(+)} \ . \tag{A.67}$$

Therefore, in coordinate representation the Green's function satisfies the differential equation

$$\frac{\hbar^2}{2\mu}(\nabla^2 + k^2)\, G_0^{(+)}(\vec{r},\vec{r}') = \delta(\vec{r}-\vec{r}') \ . \tag{A.68}$$

We may write Eq.(A.68) in spherical coordinates as was done in section A.3 and also expand the Green's function into spherical harmonics,

$$G_0^{(+)}(\vec{r},\vec{r}') = \frac{2\mu}{\hbar^2} \sum_{\ell,m} G_\ell(r,r')\, Y_\ell^m(\hat{r})\, Y_\ell^{m*}(\hat{r}') \ . \tag{A.69}$$

The function $G_\ell(r,r')$ satisfies the radial part of Eq.(A.68), such that

$$\frac{1}{r^2}\frac{d}{dr}\left[r^2\frac{dG_\ell(r,r')}{dr}\right] + \left[k^2 - \frac{\ell(\ell+1)}{r^2}\right]G_\ell(r,r') = \frac{\delta(r-r')}{r'^2} \quad . \qquad (A.70)$$

We note that $G_\ell(r,r')$ is the free wave solution for all r except at $r = r'$, i.e. a solution in terms of $j_\ell(kr)$ and $n_\ell(kr)$. A possible form of $G_\ell(r,r')$ is

$$G_\ell(r,r') = C_\ell\, j_\ell(kr_<)\left[n_\ell(kr_>) + i\,j_\ell(kr_>)\right] \quad , \qquad (A.71)$$

where $r_>$ (and $r_<$) indicates the larger (and smaller) of r and r'. It has the property of being regular at the origin and has the asymptotic property

$$G_\ell(r,r')\xrightarrow[r\to\infty]{} j_\ell(kr')\,\frac{e^{i(kr-\ell\pi/2)}}{kr} \quad , \qquad (A.72)$$

which satisfies the boundary condition imposed earlier for an outgoing scattering wave, Eqs. (A.56), (A.59). In addition, the solution is continuous, but the slope of the solution is not continuous, at $r = r'$. The derivative of the discontinuous slope at $r = r'$ that arises from the left hand side of Eq.(A.70) will give a multiple of a delta-function. We determine now the coefficient C_ℓ which makes (A.71) a solution of (A.70) by integrating (A.70) over an infinitesimal interval centred at $r = r'$. The first term on the left hand side of (A.70) survives in the limit of $\varepsilon \longrightarrow o$, i.e.

$$\lim_{\varepsilon\to o}\int_{r'-\varepsilon}^{r'+\varepsilon}\frac{1}{r^2}\frac{d}{dr}\left[r^2\frac{dG_\ell(r,r')}{dr}\right]dr$$

$$= \frac{1}{r'^2}\lim_{\varepsilon\to o}\left[r^2\frac{dG_\ell(r,r')}{dr}\right]\Bigg|_{r'-\varepsilon}^{r'+\varepsilon}$$

$$= k\,C_\ell\left[j_\ell(kr')\,n'_\ell(kr') - j'_\ell(kr')\,n_\ell(kr')\right]$$

$$= - k\, C_\ell\, \frac{1}{(kr')^2}$$

$$= - \frac{C_\ell}{k} \cdot \frac{1}{r'^2} \quad . \tag{A.73}$$

After integration, the right hand side of (A.70) yields $1/r'^2$. Therefore $G_\ell(r,r')$ of (A.71) is the desired solution, if $C_\ell = -k$, and the Green's function of (A.56), (A.69) may be written as

$$G_o^{(+)}(\vec{r},\vec{r}') = - \frac{\mu}{2\pi \hbar^2} \cdot \frac{e^{ik|\vec{r}-\vec{r}'|}}{|\vec{r}-\vec{r}'|}$$

$$= - \frac{2\mu k}{\hbar^2} \sum_{\ell,m} j_\ell(kr_<)\, h_\ell^{(+)}(kr_>)\, Y_\ell^m(\hat{r})\, Y_\ell^{m\,*}(\hat{r}') \ , \tag{A.74}$$

where $h_\ell^{(+)}(kr)$ is the Hankel function of the first kind

$$h_\ell^{(+)}(kr) = n_\ell(kr) + i\, j_\ell(kr) \ . \tag{A.75}$$

Further, the ℓ-th partial wave of the outgoing wave in the coordinate representation of the Lippmann-Schwinger equation (A.44) is

$$\psi_\ell(kr)/(kr) = j_\ell(kr)$$

$$- \frac{2\mu k}{\hbar^2} \int_0^\infty j_\ell(kr_<)\, h_\ell^{(+)}(kr_>)\, V(r')\, \frac{\psi_\ell(kr')}{kr'}\, r'^2 dr' \ . \tag{A.76}$$

A.11. Coulomb scattering

If the potential in question in (A.1) is a Coulomb potential, the Schrödinger equation can still be solved in terms of hypergeometric functions. The form of the scattering wave function in the asymptotic region is

$$\psi^{(+)}(\vec{r}) \xrightarrow[k(r-z)\to\infty]{} e^{i[kz+\eta\ln k(r-z)]}$$

$$+ f_c(\theta)\,\frac{e^{i(kr-\ln 2kr)}}{r} \quad , \tag{A.77}$$

where the Coulomb scattering amplitude is

$$f_c(\theta) = -\frac{\eta}{2k\sin^2\theta/2}\,e^{-i\left[\eta\ln(\sin^2\theta/2)-2\sigma_o\right]} \quad , \tag{A.78}$$

with the Coulomb parameter $\eta = \dfrac{Z_1 Z_2 e^2}{\hbar v}$ and $\sigma_o = \arg\Gamma(1+i\eta)$. Here Z_1, Z_2 represent the nuclear charges of the two nuclei, v the asymptotic relative velocity.

Coulomb scattering is also amenable to partial wave analysis. The radial equation (A.25) has a regular solution $F_\ell(\eta,kr)$ and an irregular solution $G_\ell(\eta,kr)$ whose asymptotic behaviours are

$$F_\ell(\eta,kr) \xrightarrow[kr\to\infty]{} \sin\left(kr-\tfrac{1}{2}\ell\pi-\eta\ln 2kr+\sigma_\ell\right) \quad , \tag{A.79}$$

$$G_\ell(\eta,kr) \xrightarrow[kr\to\infty]{} \cos\left(kr-\tfrac{1}{2}\ell\pi-\eta\ln 2kr+\sigma_\ell\right) \quad , \tag{A.80}$$

where the Coulomb phase shift σ_ℓ is given by

$$\sigma_\ell = \sigma_o + \sum_{s=1}^{\ell} \tan^{-1}\frac{\eta}{s} \quad . \tag{A.81}$$

However, the situation one encounters most often is scattering in a Coulomb field modified by a short range nuclear potential. Then the scattering wave function in partial wave expansion to replace that of Eq. (A.32), after putting $A_\ell = e^{i(\delta_\ell+\sigma_\ell)}$, is

$$\psi(\vec{r}) \xrightarrow[kr\to\infty]{} \frac{1}{kr}\sum_\ell \sqrt{4\pi(2\ell+1)}\,e^{i(\delta_\ell+\sigma_\ell)}$$

$$\times \sin\left(kr-\tfrac{1}{2}\ell\pi-\eta\ln 2kr+\sigma_\ell+\delta_\ell\right) Y_\ell^0(\theta) \,. \tag{A.82}$$

The additional phase shift δ_ℓ due to the short range potential can be
determined by the matching procedure described in section A.5 and
hence, each partial wave is unambiguously determined for all values
of r. This type of wave function play a crucial role in our later discussion on the distorted wave Born approximation (DWBA).

A.12. Rearrangement collision

We discuss in this subsection collisions in which the outgoing
systems are not identical with the incoming systems. The unperturbed
Hamiltonian for the final state, when the fragments are well separated,
differs from that of the initial state. One often uses the term channel
to denote a particular mode of fragmentation. One considers the rearrangement collisions

$$a + A \rightarrow b + B \quad , \tag{A.83}$$

which are described by the Hamiltonian

$$H = H_i + V_i = H_f + V_f$$
$$= h_i + t_i + V_i = h_f + t_f + V_f \quad , \tag{A.84}$$

where i and f denotes the quantities in the initial and final channels,
respectively. The internal quantum states of the two fragments in the
initial and final channels are $\varphi_i(a,A)$ and $\varphi_f(b,B)$, which are eigenfunctions of h_i and h_f, respectively.

$$h_i \varphi_i(a,A) = e_i \varphi_i(a,A) \quad ,$$
$$h_f \varphi_f(b,B) = e_f \varphi_f(b,B) \quad , \tag{A.85}$$

and t_i and t_f are the kinetic energy operators. In the asymptotic
regions where the two framents are well separated, the total energy
of the system may be written as

$$E = e_i + \frac{\hbar^2 k_i^2}{2\mu} = e_f + \frac{\hbar^2 k_f^2}{2\mu} \quad . \tag{A.86}$$

For particles to escape to infinity in the final channel, hence labelled as an open channel, $E-e_f$ must be positive.

We define a plane wave in the initial channel as

$$\Phi_i = \varphi_i(a,A)\, e^{i\vec{k}_i\cdot\vec{r}_i} \quad , \tag{A.87}$$

with

$$H_i\Phi_i = E\Phi_i \quad , \tag{A.88}$$

and $\Psi_i^{(\pm)}$ as the outgoing (+) and incoming (−) scattering wave such that

$$H\Psi_i^{(\pm)} = E\,\Psi_i^{(\pm)} \quad . \tag{A.89}$$

The asymptotic behaviour of the wave functions in the initial and final channels are

$$\Psi_i \xrightarrow[r_i\to\infty]{} \varphi_i(a,A)\left[e^{i\vec{k}_i\cdot\vec{r}_i} + f_i(\theta_i,\phi_i)\frac{e^{ik_i r_i}}{r_i}\right] \quad ,$$

$$\Psi_f \xrightarrow[r_f\to\infty]{} \varphi_f(b,B)\left[f_f(\theta_f,\phi_f)\frac{e^{ik_f r_f}}{r_f}\right] \quad . \tag{A.90}$$

In the second line of Eq. (A.90), there is no incident plane wave in the final channels. This corresponds to the physical reality that the incident wave exists only in the initial channel. The cross section for emission of particles in the direction of (θ_f,ϕ_f) is

$$\left(\frac{d\sigma}{df}\right)_{i\to f} = \frac{v_f}{v_i}\left| f_f(\theta_f,\phi_f)\right|^2 \tag{A.91}$$

where v_i and v_f are respectively the initial and final asymptotic relative velocities. The derivation of Eq. (A.91) is merely a generalization of the procedure given in section A.2.

We make use of the integral operator method to derive the scattering amplitude $f_f(\theta_f,\phi_f)$ in Eqs. (A.90) and (A.91). For the initial channel, we divide the Hamiltonian into an unperturbed Hamiltonian H_o and a potential V, whose counterparts in the final channel are H'_o and V',

$$H = H_o + V = H_o' + V'$$
(A.92)

The way in which H is separated into H_O and V or H_O' and V' is entirely
arbitrary, not necessarily coinciding with the one of Eq. (A.94), and
depends on the problem at hand. The outgoing scattering solution
in the initial channel is given by the Lippmann-Schwinger equation (A.44),

$$\Psi_i^{(+)} = \Phi_i + \frac{1}{E+i\varepsilon-H_o} V \Psi_i^{(+)} \, ,$$
(A.93)

where Φ_i is the incident wave, an eigenfunction of H_O. The initial
channel unperturbed Green's function $1/(E+i\varepsilon-H_O)$ is related to the final
channel unperturbed Green's function $1/(E+i\varepsilon-H_O')$ in the following way,

$$\frac{1}{E+i\varepsilon-H_o} - \frac{1}{E+i\varepsilon-H_o'} = \frac{1}{E+i\varepsilon-H_o'} (V'-V) \frac{1}{E+i\varepsilon-H_o} .$$
(A.94)

This operator equation is equivalent to the algebraic relation

$$A^{-1} - B^{-1} = A^{-1}(B-A)B^{-1} \, .$$
(A.95)

Inserting the expression for the initial channel Green's function from
Eq.(A.94) into Eq.(A.93), we obtain

$$\Psi_i^{(+)} = \Phi_i + \frac{1}{E+i\varepsilon-H_o'} V \Psi_i^{(+)} + \frac{1}{E+i\varepsilon-H_o'} (V'-V) \frac{1}{E+i\varepsilon-H_o} V \Psi_i^{(+)}$$

$$= \Phi_i + \frac{1}{E+i\varepsilon-H_o'} V \Psi_i^{(+)} + \frac{1}{E+i\varepsilon-H_o'} (V'-V)(\Psi_i^{(+)} - \Phi_i)$$

$$= \Phi_i - \frac{1}{E+i\varepsilon-H_o'} (V'-V) \Phi_i + \frac{1}{E+i\varepsilon-H_o'} V' \Psi^{(+)} \, .$$
(A.96)

If V' = V such that $H_o = H_o'$, Eq. (A.96) reduces to the original Lippmann-Schwinger equation. If there is a true rearrangement, we can write the first two terms as

$$\lambda = \Phi_i - \frac{1}{E+i\varepsilon - H_o'}(V'-V)\Phi_i$$

$$= \left[1 - \frac{1}{E+i\varepsilon - H_o'}(H_o - H_o')\right]\Phi_i$$

$$= \frac{1}{E+i\varepsilon - H_o}\left[E - H_o' - H_o + H_o'\right]\Phi_i \quad . \tag{A.97}$$

Therefore λ vanishes, since Φ_i is an eigenfunction of H_o. The scattering state vector is given solely by the third term of Eq. (A.96),

$$\Psi_i^{(+)} = \frac{1}{E+i\varepsilon - H_o'}V'\Psi_i^{(+)} \quad . \tag{A.98}$$

To obtain the scattering wave function and the scattering amplitude in the final (rearrangement) channel, one can isnert Eq. (A.98) into a time-dependent wave packet formalism and examine the scattering solution at $t \longrightarrow \infty$ [RO 67]. In coordinate representation, the asymptotic behaviour of the stationary solution in the final channel is

$$\Psi_f \xrightarrow[r_f \to \infty]{} \varphi_f(b,B)\left(-\frac{\mu_f}{2\pi\hbar^2}\right)\frac{e^{ik_f r_f}}{r_f}$$

$$\times \int e^{-i\vec{k}_f \cdot \vec{r}'}V'(\vec{r}')\Psi_i^{(+)}(\vec{r}')d^3\vec{r}' \quad , \tag{A.99}$$

where μ_f and k_f are the reduced mass and wave number in the final channel. $\vec{k}_f$ is in the direction in which the scattered particles are being measured. By comparison with Eq. (A90), the scattering amplitude $f_f(\theta_f,\phi_f)$ may be written as

$$f_f(\theta_f,\phi_f) = -\frac{\mu_f}{2\pi\hbar^2}\int e^{-i\vec{k}_f \cdot \vec{r}'}V'(\vec{r}')\Psi_i^{(+)}(\vec{r}')d^3\vec{r}' \quad . \tag{A.100}$$

Rearrangement scattering is also amenable to partial wave analysis. As in Eq. (A.76), the ℓ-th partial wave $\Psi_\ell(k_f r_f)$ of the solution in the final channel may be written as

$$\Psi_\ell(k_f r_f) / (k_f r_f) = \varphi_f(b, B)\left[-\frac{2\mu_f k_f}{\hbar^2}\right.$$

$$\left. \times \int_0^\infty j_\ell(k_f r_<) \, h_\ell^{(+)}(k_f r_>) \, V(r') \, \Psi_\ell(k_f r') / (k_f r') \, r'^2 dr'\right] . \tag{A.101}$$

A.13. The transition (T) matrix

Very often, the scattering amplitude of Eq.(A.61) is discussed in terms of state vectors, such that the scattering amplitude of Eq.(A.61) may be written as

$$f(\theta, \phi) = -\frac{\mu}{2\pi\hbar^2} \int e^{-i\vec{k}'\cdot\vec{r}'} \, V(\vec{r}') \, \psi^{(+)}(\vec{r}') \, d^3\vec{r}'$$

$$= -\frac{\mu}{2\pi\hbar^2} \langle \phi_{\vec{k}'} | V | \psi_{\vec{k}}^{(+)} \rangle , \tag{A.102}$$

and hence the differential cross section becomes

$$\frac{d\sigma}{d\Omega} = \frac{\mu^2}{4\pi^2\hbar^4} |\langle \phi_{\vec{k}'} | V | \psi_{\vec{k}}^{(+)} \rangle|^2 . \tag{A.103}$$

In spite of the fact that $|\phi_{\vec{k}'}\rangle$ and $|\psi_{\vec{k}}^{(+)}\rangle$ cannot be basis vectors in the same representation, a transition matrix element T is often taken to be

$$T = \langle \phi_{\vec{k}'} | V | \psi_{\vec{k}}^{(+)} \rangle , \tag{A.104}$$

where $\vec{k}$ refers to the direction of the incident beam and $\vec{k}'$ has the same magnitude as $\vec{k}$ but is in the direction of the polar angles (θ, ϕ) of the point at which the scattering amplitude is being measured (see Eq.(A.60) and the discussion which follows it).

Furthermore, we introduce the energy density of states at Energy E,

$$\rho(E) = \frac{1}{(2\pi\hbar)^3} \, p^2 \frac{dp}{dE} = \frac{\mu\hbar k}{(2\pi\hbar)^3} , \tag{A.105}$$

so that the differential cross section, written in terms of T and $\rho(E)$ is

$$\frac{d\sigma}{d\Omega} = \frac{2\pi}{\hbar v} |T|^2 \rho(E) \ . \tag{A.106}$$

Other formulation of collision theory leads quite naturally to this form for the differential cross section. It is highly suggestive that the Fermi Golden Rule No.2 for the number of transition between quantum states per unit time [OR 50] is given by

$$\omega = \frac{2\pi}{\hbar} |\mathcal{H}|^2 \tilde{\rho}(E) \ , \tag{A.107}$$

where $\mathcal{H}$ is the matrix element of the perturbation causing the transition and $\tilde{\rho}(E)$ is the energy density of the final states. It is consistent that the differential cross section can be defined as the number of transition per unit incident current. In a similar fashion, the differential cross section for rearrangement collision, Eq. (A.91) may be written as

$$\left(\frac{d\sigma}{d\Omega}\right)_{i \to f} = \frac{v_f}{v_i} \left| f_f(\theta_f, \phi_f) \right|^2$$

$$= \frac{2\pi}{\hbar v_i} |T|^2 \rho(E - e_f) \ , \tag{A.108}$$

where, using Eq. (A.100),

$$T = \langle \Phi_{\vec{k}_f} | V' | \Psi_{\vec{k}_i}^{(+)} \rangle$$

$$= \int e^{-i \vec{k}_f \cdot \vec{r}'} V'(\vec{r}') \Psi_i^{(+)}(\vec{r}') \, d^3\vec{r}' \ . \tag{A.109}$$

A.14. The post- and prior- form of the T-matrix

Consider an operator equation which involves quantities of the incident channel only,

$$\frac{1}{E\pm i\varepsilon-H} - \frac{1}{E\pm i\varepsilon-H_o} = \frac{1}{E\pm i\varepsilon-H}\, V\, \frac{1}{E\pm i\varepsilon-H_o}$$

$$= \frac{1}{E\pm i\varepsilon-H_o}\, V\, \frac{1}{E\pm i\varepsilon-H} \quad , \tag{A.110}$$

the following identities can be established,

$$\left(1+\frac{1}{E\pm i\varepsilon-H}\,V\right)\left(1-\frac{1}{E\pm i\varepsilon-H_o}\,V\right) = 1 \quad , \tag{A.111}$$

$$\left(1-\frac{1}{E\pm i\varepsilon-H_o}\,V\right)\left(1+\frac{1}{E+i\varepsilon-H}\,V\right) = 1 \quad . \tag{A.112}$$

The Lippmann-Schwinger equation (A.93) can be rearranged so that, after the application of the identity (A.111), we obtain

$$\Psi_i^{(+)} = \left(1+\frac{1}{E+i\varepsilon-H}\,V\right)\Phi_i \quad . \tag{A.113}$$

Inserting this into the transition matrix element for rearrangement scattering (A.109)

$$\langle\Phi_{\vec{k}_f}|V'|\Psi_{\vec{k}_i}\rangle = \langle\Phi_{\vec{k}_f}|V'|\Phi_{\vec{k}_i}\rangle$$

$$+ \langle\Phi_{\vec{k}_f}|V'\frac{1}{E+i\varepsilon-H}V|\Phi_{\vec{k}_i}\rangle \quad . \tag{A.114}$$

Replacing V' and V by the expressions

$$V' = \frac{1}{2}(V'+V) + \frac{1}{2}(E-H_o') - \frac{1}{2}(E-H_o) \quad , \tag{A.115}$$

$$V = \frac{1}{2}(V'+V) - \frac{1}{2}(E-H_o') + \frac{1}{2}(E-H_o) \quad , \tag{A.116}$$

we can establish the equality

$$\langle \Phi_{\vec{k}_f} | V' | \Phi_{\vec{k}_i} \rangle = \langle \Phi_{\vec{k}_f} | V | \Phi_{\vec{k}_i} \rangle$$

$$= \frac{1}{2} \langle \Phi_{\vec{k}_f} | V + V' | \Phi_{\vec{k}_i} \rangle \; . \qquad \text{(A.117)}$$

Substituting the first equality of (A.117) into the first term of (A.114), we obtain

$$\langle \Phi_{\vec{k}_f} | V' | \Psi_{\vec{k}_i}^{(+)} \rangle = \langle \Phi_{\vec{k}_f} | (1 + V' \frac{1}{E + i\varepsilon - H}) V | \Phi_{\vec{k}_i} \rangle \; . \qquad \text{(A.118)}$$

An analogous equality of (A.113) is

$$\Psi_{\vec{k}_f}^{(-)} = (1 + \frac{1}{E - i\varepsilon - H} V') \Phi_{\vec{k}_f} \; . \qquad \text{(A.119)}$$

Inserting (A.119) into (A.118) leads to

$$T = \langle \Phi_{\vec{k}_f} | V' | \Psi_{\vec{k}_i}^{(+)} \rangle = \langle \Psi_{\vec{k}_f}^{(-)} | V | \Phi_{\vec{k}_i} \rangle \; , \qquad \text{(A.120)}$$

in which $\langle \Phi_{\vec{k}_f} | V' | \Psi_{\vec{k}_i}^{(+)} \rangle$ is the post-form of the transition matrix element and $\langle \Psi_{\vec{k}_f}^{(-)} | V | \Phi_{\vec{k}_i} \rangle$ is the prior-form. The descriptive terms 'post' and 'prior' derive from the fact that, in the former only the potential of the final channel is involved, while in the latter only the potential of the initial channel appears. The special case for the transition matrix when $H_o = H_o'$ and $V = V'$, i.e. no rearrangement, yields

$$T = \langle \Phi_{\vec{k}_f} | V | \Psi_{\vec{k}_i}^{(+)} \rangle$$

$$= \langle \Psi_{\vec{k}_f}^{(-)} | V | \Phi_{\vec{k}_i} \rangle \qquad \text{(A.121)}$$

A.15. Scattering from a complex potential

It is known since very early days [FE 54] that to describe the gross or averaged properties of nucleon-nucleus scattering, it is necessary to employ a complex potential (optical potential). Up to the present time, although sophisticated theoretical frameworks have been set up to investigate the microscopic origin of these phenomenological potentials [GE 79], most analyses of nuclear reactions are performed

with the help of phenomenological complex potentials. We mention briefly in this section the physical consequence of scattering from a complex potential.

Let us consider a complex potential of the form V(r) + iW(r).
From the Schrödinger equation

$$(\nabla^2 + k^2)\,\psi = \frac{2\mu}{\hbar^2}(V + iW)\psi \quad , \tag{A.122}$$

and its complex conjugate, we can write down the divergence of the
probability current density

$$\mathrm{div}\,\mathbf{j} = \frac{\hbar}{2\mu i}\,\nabla \cdot (\psi^*\nabla\psi - \psi\nabla\psi^*)$$

$$= \frac{2}{\hbar}\,W\psi^*\psi \quad . \tag{A.123}$$

From the classical equation of continuity, a positive W means the presence of a source from which particles are emitted, while a negative W
means the presence of a sink into which particles are absorbed. Therefore, if we postulate that the nuclear elastic scattering is governed
by a phenomenological potential V(r) + iW(r), where W(r) is negative,
then W(r) simulates the disappearance of particles from the elastic
channel. In the actual scattering process, this does occur when inelastic and rearrangement scattering are also initiated such that particles are dissipated from the elastic channel into these other channels.

A.16. The two-potentials formula

In a single channel scattering problem, the interaction between
the target and the projectile, V, may conveniently be divided into two
parts: $V = V_0 + V_1$. This separation may be very useful if the scattering
under the influence of one, V_0, say, can be solved for exactly, and the
additional effect of the other potential term V_1 can be approximated.
For example, V_0 can be taken to be the long range Coulomb potential
whose scattering solution is known analytically, and represent the
short range, usually complex, nuclear potential by V_1.

In a general discussion, we write the T-matrix as in Eq.(A.104)

$$T = \langle \phi | V | \psi^{(+)} \rangle \equiv \langle \phi | V_0 | \psi^{(+)} \rangle + \langle \phi | V_1 | \psi^{(+)} \rangle . \tag{A.124}$$

The outgoing and incoming wave solution due to the potential V_0 is given by

$$\chi^{(\pm)} = \phi + G_0^{(\pm)} V_0 \chi^{(\pm)} , \tag{A.125}$$

where $G_0^{(\pm)}$ is the free wave Green's function discussed in subsection A.6. By making use of the identities (A.111), (A.112), we can rewrite (A.125) as

$$\chi^{(\pm)} = \phi + G_1^{(\pm)} V_0 \phi , \tag{A.126}$$

where $G_1^{(\pm)}$ is the distorted wave Green's function,

$$G_1^{(\pm)} = \frac{1}{E \pm i\varepsilon - H_0 - V_0} . \tag{A.127}$$

The coordinate representation of the state vector $\chi^{(\pm)}$ is known as the distorted wave, while V_0 is the distorting potential. Substituting Eq. (A.126) into the second term of Eq.(A.124), we obtain

$$T = \langle \phi | V_0 | \psi^{(+)} \rangle + \langle \chi^{(-)} | V_1 | \psi^{(+)} \rangle$$

$$- \langle \phi | V_0 G_1^{(+)} V_1 | \psi^{(+)} \rangle$$

$$= \langle \phi | (V_0 - V_0 G_1^{(+)} V_1) | \psi^{(+)} \rangle + \langle \chi^{(-)} | V_1 | \psi^{(+)} \rangle . \tag{A.128}$$

An equality relation equivalent to Eq.(A.110) is

$$G_0^{(+)} = G_1^{(+)} - G_1^{(+)} V_0 G_0^{(+)} . \tag{A.129}$$

Applying this to the Lippmann-Schwinger equation

$$\psi^{(+)} = \phi + G_0^{(+)} V \psi^{(+)}$$

yields

$$\psi^{(+)} = \phi + \left[G_1^{(+)} - G_1^{(+)} V_0 G_0^{(+)} \right] V \psi^{(+)}$$

$$= \phi + G_1^{(+)} V \psi^{(+)} - G_1^{(+)} V_0 (\psi^{(+)} - \phi)$$

$$= \phi + G_1^{(+)} V_0 \phi + G_1^{(+)} (V - V_0) \psi^{(+)}$$

$$= \chi^{(+)} + G_1^{(+)} V_1 \psi^{(+)} . \tag{A.130}$$

Thus, $\chi^{(+)}$ can be interpreted as an incident wave, while the potential V_1 is responsible for the scattering. From Eq.(A.130), we derive easily the relation

$$\langle \phi | V_0 | \chi^{(+)} \rangle = \langle \phi | (V_0 - V_0 G_1^{(+)} V_1) | \psi^{(+)} \rangle , \tag{A.131}$$

which, when substituted into Eq.(A.128), leads to the two-potential formula for the T-matrix,

$$T = \langle \phi | V_0 | \chi^{(+)} \rangle + \langle \chi^{(-)} | V_1 | \psi^{(+)} \rangle . \tag{A.132}$$

As Eq.(A.125) is equivalent to Eq.(A.126), Eq.(A.130) can be rewritten as

$$\psi^{(+)} = \chi^{(+)} + G^{(+)} V_1 \chi^{(+)}, \tag{A.133}$$

where $G^{(+)} = 1/(E+i\varepsilon - H_0 - V_0 - V_1)$. Replacing $\psi^{(+)}$ in (A.132) by (A.133), the T-matrix becomes

$$T = \langle \phi | V_0 | \chi^{(+)} \rangle + \langle \chi^{(-)} | V_1 + V_1 G^{(+)} V_1 | \chi^{(+)} \rangle . \tag{A.134}$$

If the separation of V_0 and V_1 is so propitious that V_1 is weak, we may approximate the T-matrix by

$$T \approx \langle \phi | V_0 | \chi^{(+)} \rangle + \langle \chi^{(-)} | V_1 | \chi^{(+)} \rangle . \tag{A.135}$$

The second term of Eq.(A.135) is often called the distorted wave Born approximation (DWBA), which is a generalization of the plane wave Born approximation, see subsection A.9.

In rearrangement reaction, it is most useful to decompose the Hamiltonian differently in the initial and final channels,

$$H = H_o + V = H_o + V_o + V_1 \quad , \tag{A.136}$$

$$H = H_o' + V' = H_o' + V_o' + V_1' \quad , \tag{A.137}$$

where V_0 and V_1 are the distorting and residual potentials in the initial channel, while V_0' and V_1' are their counterparts in the final channel. The T-matrix for the rearrangement reaction in the post-form, from Eq.(A.120), is

$$T = \langle \Phi_f | V_o' + V_1' | \Psi_i^{(+)} \rangle$$
$$= \langle \Phi_f | V_o' | \Psi_i^{(+)} \rangle + \langle \Phi_f | V_1' | \Psi_i^{(+)} \rangle \quad . \tag{A.138}$$

For convenience we omit the explicit labels of k_i and k_f of Eq.(A.120) By defining the plane wave and distorted wave Green's function of the final channel as

$$G_o'^{(\pm)} = \frac{1}{E \pm i\varepsilon - H_o'} \quad , \tag{A.139}$$

$$G_1'^{(\pm)} = \frac{1}{E \pm i\varepsilon - H_o' - V_o'} \quad , \tag{A.140}$$

we may write, for the final channel, the Lippmann-Schwinger equation of the distorted wave $X_f^{(-)}$

$$X_f^{(-)} = \Phi_f + G_o'^{(-)} V_o' X_f^{(-)} \quad , \tag{A.141}$$

or its equivalent form

$$X_f^{(-)} = \Phi_f + G_1'^{(-)} V_o' \Phi_f \quad , \tag{A.142}$$

(cf Eqs. (A.125), (A.126)). Inserting Eq.(A.142) into the second term of Eq.(A.138), we obtain

$$T = \langle \Phi_f | V_o' | \Psi_i^{(+)} \rangle + \langle X_f^{(-)} | V_i' | \Psi_i^{(+)} \rangle - \langle \Phi_f | V_o' G_i'^{(+)} V_i' | \Psi_i^{(+)} \rangle . \tag{A.143}$$

We proceed now to derive $G_i'^{(+)} V_i' | \Psi_i^{(+)} \rangle$. An operator equation equivalent to Eq.(A.129) is

$$G_o^{(+)} = G_i'^{(+)} - G_i'^{(+)} (V - V_i') G_o^{(+)} . \tag{A.144}$$

We apply this to the Lippmann-Schwinger equation

$$\Psi_i^{(+)} = \Phi_i + G_o^{(+)} V \Psi_i^{(+)}$$

$$= \Phi_i + \left[G_i'^{(+)} - G_i'^{(+)} (V - V_i') G_o^{(+)} \right] V \Psi_i^{(+)}$$

$$= \Phi_i + G_i'^{(+)} V \Psi_i^{(+)} - G_i'^{(+)} (V - V_i') (\Psi_i^{(+)} - \Phi_i)$$

$$= \Phi_i + G_i'^{(+)} (V - V_i') \Phi_i + G_i'^{(+)} V_i' \Psi_i^{(+)} . \tag{A.145}$$

The first two terms of Eq.(A.145) cancel out each other, because

$$\lambda = \Phi_i + G_i'^{(+)} (V - V_i') \Phi_i$$

$$= \frac{1}{E + i\varepsilon - H_o' - V_o'} \left[E - H_o' - V_o' + V - V_i' \right] \Phi_i$$

$$= \frac{1}{E + i\varepsilon - H_o' - V_o'} \left[E - H_o \right] \Phi_i = 0 . \tag{A.146}$$

Therefore, Eq.(A.145) is reduced to

$$\Psi_i^{(+)} = G_i'^{(+)} V_i' \Psi_i^{(+)} , \tag{A.147}$$

which is a two-potential formula generalization of Eq.(A.98). Substituting Eq.(A.147) into the third term of Eq.(A.143), we can cancel the first term of Eq.(A.143) with the third term. Therefore, the two-potential formula for the rearrangement T-matrix in the post-form is

$$T = \langle X_f^{(-)} | V_i' | \Psi_i^{(+)} \rangle . \tag{A.148}$$

It is clear that if we had started with the rearrangement T-matrix in the prior-form, we would have obtained

$$T = \langle \Psi_f^{(-)} | V_i | X_i^{(+)} \rangle . \tag{A.149}$$

If V_1 and V_1' are weak, one may approximate $\Psi_i^{(+)}$ by $X_i^{(+)}$ and $\Psi_f^{(-)}$ by $X_f^{(-)}$, such that the DWBA T-matrix for rearrangement reaction is

$$T \approx \langle X_f^{(-)} | V_i' | X_i^{(+)} \rangle \tag{A.150}$$

$$= \langle X_f^{(-)} | V_i | X_i^{(+)} \rangle . \tag{A.151}$$

Although the post- and prior-form of the DWBA T-matrix, Eqs.(A.150), (A.151), are formally different, they are in fact identical. This can be easily seen if we make the substitution $V_1' = H - H_o' - V_o'$ and $V_1 = H - H_o - V_o$.

References

[AL 56]   K.Alder, A.Bohr, T.Huus, B.Mottelson and A.Winther, Rev.Mod.
          Phys.$\underline{28}$ (1956) 432.

[AL 75]   K.Alder and A.Winther, Electromagnetic Interaction (North-
          Holland, Amsterdam, 1975).

[AL 77]   K.Alder, F.Roesel and R.Morf, Nucl.Phys. $\underline{A284}$ (1977) 145.

[AN 74]   N.Anyas-Weiss, J.C.Cornell, P.S.Fisher, P.N.Hudson,
          A.Menchaca-Rocha, D.J.Millener, A.D.Panagiotou, D.K.Scott,
          D.Strottman, D.M.Brink, B.Buck, P.J.Ellis and T.Engeland,
          Phys.Rep. $\underline{12C}$ (1974) 201.

[BA 71]   A.R.Barnett, W.R.Phillips, P.J.A.Buttle and L.J.B.Goldfarb,
          Nucl.Phys. $\underline{A176}$ (1971) 321.

[BA 78a] A.M.Baltz and S.Kahana, Phys.Rev. $\underline{C17}$ (1978) 555.

[BA 78b] B.F.Bayman, A.Dudek-Ellis, P.J.Ellis, Nucl.Phys. $\underline{A301}$
          (1978) 141.

[BA 59]   A.Bohr, Nucl.Phys. $\underline{10}$ (1959) 486.

[BA 77]   J.deBoer, G.Dannhauser, H.Massmann, F.Roesel, A.Winther,
          J.Phys. G $\underline{3}$ (1977) 889.

[BO 79]   W.Bohne, K.Grabisch, J.Hergesell, Q.K.K.Liu, H.Morgenstern,
          W.von Oertzen, W.Galster, W.Treu, and H.H.Wolter, Nucl.Phys.
          $\underline{A332}$ (1979) 501.

[BR 56]   G.Breit and M.E.Ebel, Phys.Rev.$\underline{103}$ (1956) 679, Phys.Rev.$\underline{104}$
          (1956) 1030.

[BR 62]   D.M.Brink, and G.R.Satchler, Angular Momentum (Oxford, 1962).

[BR 72a] D.M.Brink, Phys.Lett. $\underline{408}$ (1972) 37.

[BR 72b] R.A.Broglia and A.Winther, Phys.Rep. $\underline{4C}$ (1972) 155.

[BR 74]   P.Braun-Munzinger and H.L.Harvey, Nucl.Phys. $\underline{A223}$ (1974) 381.

[BU 63]   B.Buck, Phys.Rev. $\underline{130}$ (1963) 712.

[BU 66]   P.J.A.Buttle and L.J.B.Goldfarb, Nucl.Phys. $\underline{78}$ (1966) 409.

[BU 71]   P.J.A.Buttle and L.J.B.Goldfarb, Nucl.Phys. $\underline{A176}$ (1971) 299.

[BU 78]   P.J.A.Buttle, Computer Physics Communication $\underline{14}$ (1978) 133.

[BU 80]   D.P.Bybell, D.P.Balamuth and W.K.Wells, private communication,
          and Bull. Amer. Phys. Soc., Vol.25, No.7 (1980) 739.

[CH 58]   D.M.Chase, L.Wilets and A.R.Edmunds, Phys.Rev. $\underline{110}$ (1958) 1080.

[CL 78]   D.Cline, in Nuclear Reaction, ed. B.A.Robson (Lecture Notes in Physics 92, Springer Verlag, Berlin, 1979).

[CO 75]   P. Columbani, J.C.Jacmart, N.Poffi, M.Riou, C.Stephan, P.P. Singh, A.Weidinger, Phys.Lett. $\underline{48B}$ (1974) 315.

[DE 73]   R.M.DeVries and K.I.Kubo, Phys.Rev.Lett. $\underline{30}$ (1973) 325.

[DÜ 79]   W.Dünnweber, P.D.Bond, C.Chasman, S.Kubono, Phys.Rev.Lett. $\underline{43}$ (1979) 1642.

[EL 77]   P.J.Ellis and E.Osnes, Rev.Mod.Phys. $\underline{49}$ (1977) 777.

[FE 54]   H.Feshbach, C.E.Porter and V.F.Weiskopf, Phys.Rev. $\underline{96}$ (1954) 448.

[FO 74]   J.C.L.Ford, K.S.Toth, G.R.Satchler, D.C.Hensley, L.W.Owen, R.M.DeVries, R.M.Gaedke, P.J.Riley, and S.T.Thornton, Phys. Rev. $\underline{C10}$ (1974) 1429.

[FR 72]   W.E.Frahn, Ann.Phys. (N.Y.) $\underline{72}$ (1972) 524.

[FU 80]   P.Fuchs, H.Ower, Th.W.Elze, H.Euling, E.Grosse, J.Idzko, N.Kaffrell, D.Schwalen, K.Stelzer, N.Trautmann, H.J.Wollersheim, in Proceedings of the International Conference on Nuclear Physics, Berkeley, 1980, p. 334.

[GE 78]   C.K.Gelbke, T.Awes, U.E.P.Berg, J.Barrette, M.J.LeVine and P.Braun-Munzinger, Phys.Rev.Lett. $\underline{41}$ (1978) 1778.

[GE 79]   Microscopic Optical Potentials, ed. H.V.v.Geramb (Lecture Notes in Physics 89, Springer Verlag, Berlin, 1979).

[GL 69]   N.K.Glendenning, Proceedings of the International School of Physics "Enrico Fermi", Course XL, 1967, ed. M.Jean (Academic Press, New [ork 1969), p.332.

[GL 76]   N.K.Glendenning and G.Wolschin, Phys.Rev.Lett. $\underline{36}$ (1976) 1532.

[GL 77]   N.K.Glendenning and G.Wolschin, Nucl.Phys. $\underline{A281}$ (1977) 486.

[IC 77]   M.Ichimura, M.Igaroshi, S.Landowne, C.H.Dasso, B.S.Nilsson, R.A.Broglia and A.Winther, Phys.Lett. $\underline{67B}$ (1977) 129.

[KO 74]   D.G.Kovar, Proceedings of the International Conference on Reaction between Complex Nuclei, Nashville (1974), p.235.

[KU 77]   S.Kubono, D.Dehnhard, D.A.Lewis, T.K.Li, J.L.Artz, D.J.Weber, P.J.Ellis and A.Dudek-Ellis, Phys.Rev.Lett. $\underline{38}$ (1977) 817.

[LA 79]   S.Landowne and H.H.Wolter, Phys.Rev.Lett. $\underline{43}$ (1979) 1233.

[LI 75]   Q.K.K.Liu and P.J.A.Buttle, Z. Phys. $\underline{A275}$ (1975) 283.

[LI 78]   Q.K.K.Liu, W.von Oertzen and H.H.Wolter, in Proceedings of
          the Conference on Aspects of Nuclear Structure and Nuclear
          Reactions, Winnipeg, Canada, 1978.

[LI 80]   Q.K.K.Liu and P.J.Ellis, Phys.Rev.$\underline{C22}$ (1980) 540.

[LO 67]   W.G.Love and G.R.Satchler, Nucl.Phys.$\underline{A101}$ (1967) 424.

[LO 77]   K.S.Low, in Proceedings of the International Conference on
          Nuclear Structure, Tokyo, Japan, 1977 (International Academic
          Printing Co. Ltd. Tokyo).

[MA 57]   B.Margolis and E.S.Troubetzkoy, Phys.Rev. $\underline{106}$ (1957) 105.

[MA 73]   R.A.Malfliet, S.Landowne and V.Rostokin, Phys.Lett. $\underline{44B}$
          (1973) 238.

[NE 66]   R.G.Newton, Scattering Theory of Waves and Particles (McGraw-
          Hill, New York, 1966).

[NÖ 80]   W.Nörenberg and H.A.Weidenmüller, Introduction to the Theory
          of Heavy-Ion Collision (Lecture Notes in Physics 51, Springer-
          Verlag, Berlin, 1980).

[OE 74]   W. von Oertzen, in Nuclear Spectroscopy and Reactions, ed.
          Joseph Cerny (Academic Press, New York, 1974).

[OE 75]   W. von Oertzen and H.G.Bohlen, Phys.Rep. $\underline{19C}$ (1975) 1.

[OR 50]   J.Orear, A.H.Rosenfeld, and R.A.Schluter, Nuclear Physics
          (notes compiled of a course given by E. Fermi) (The Univer-
          sity of Chicago Press, 1950).

[PA 75]   I.Paschopoulos, P.S.Fisher, N.A.Jelly, S.Kahana, A.A.Pilt,
          W.D.M.Rae and D.Sinclair, Nucl.Phys. $\underline{A252}$ (1975) 173.

[PH 77]   W.R.Phillips, Rep. Prog. Phys. $\underline{40}$ (1977) 345.

[PI 29]   H.T.H.Piaggio, Elementary Treatise on Differential Equations
          and Their Applications (Bell, London, 1929).

[PR 70]   R.J.Pryor, F.Rösel, J.X.Saladin and K.Alder, Phys.Lett. $\underline{32B}$
          (1970) 26.

[RE 75]   K.E.Rehm, H.J.Körner, M.Richter, H.P.Rother, J.P.Schiffer,
          H.Spieler, Phys.Rev. $\underline{C12}$ (1975) 1945.

[RO 57]   M.E.Rose, Elementary Theory of Angular Momentum, (John Wiley
          and Sons, Inc., New York, 1957).

[RO 63]   Quantum Scattering Theory, ed. M.H.Ross (Indiana University
          Press, Bloomington, Ind., 1963).

[RO 67]   L.S.Rodberg and R.M.Thaler, Introduction to the Quantum
          Theory of Scattering (Academic Press, New York, 1967).

[SA 68]   M.Samuel and U.Smilansky, Phys.Lett. $\underline{28B}$ (1968) 318.

[SC 70]   F.Schmittroth, W.Tobocman and A.A.Golestaneh, Phys.Rev.$\underline{C1}$
          (1970) 377.

[SC 73]   J.P.Schiffer, H.J.Körner, R.H.Siemssen, K.W.Jones, and
          A.Schwarzschild, Phys.Lett. $\underline{44B}$ (1973) 47.

[SH 80]   V.Shkolnik, M.A.Franey and D.Dehnhard, Proceedings of the
          International Conference on Nuclear Physics, Berkeley, 1980,
          p.588.

[SI 74]   R.H.Siemssen, in Nuclear Spectroscopy and Reaction, Part B,
          ed. by Joseph Cerny (Academic Press, New York, 1974).

[SI 64]   V.M.Strutinskii, Soviet Physics - JETP $\underline{19}$ (1964) 1401.

[TA 65]   T.Tamura, Rev. Mod. Phys. $\underline{37}$ (1965) 679.

[TE 56]   G.M.Temmer and N.P.Heydenburg, Phys.Rev. $\underline{104}$ (1956) 989.

[TO 80]   L.D.Tolsma, in Proceedings of the International Conference
          on Nuclear Physics, Berkeley, 1980, p.637.

[VI 72]   F.Videbaek, I.Chernov, P.R.Christensen, and E.E.Gross, Phys.
          Rev.Lett. $\underline{28}$ (1972) 1072.

[WA 70]   B.Wakefield, I.M.Nagib, R.P.Harper, I.Hall, and A.Christy,
          Phys.Lett. $\underline{31B}$ (1970) 56.

[WU 79]   P.Wust, W.von Oertzen, H.Ossenbrink, H.Lettau, H.G.Bohlen,
          W.Saathoff, C.A.Wiedner, Z.Phys. $\underline{A291}$ (1979) 151.

[YO 56]   S.Yoshida, Proc. Phys. Soc. (London) $\underline{A69}$ (1956) 668.

Chapter VII

HEAVY ION COLLISIONS[+]

Che Ming Ko[*]
Cyclotron Institute and Physics Department
Texas A&M University
College Station, Texas 77843, U.S.A.

Abstract

The study of heavy ion collisions is motivated by the desire to probe properties of nuclear matter under unusual conditions. Depending on the energy of the
beam and the impact parameter of the collision, different phenomena are observed. A
survey of typical experimental data and their interpretations are presented. For
heavy ion collisions at low energies, the reaction is dominated by irreversible,
non-equilibrium processes. Both the time-dependent mean-field theory and the
statistical tranport theory are employed for understanding these reactions. At high
energies, the main part of heavy ion reactions is due to the multiple scatterings
between the projectile nucleons and the target nucleons. We discuss both the linear
cascade model and the statistical model.

[+]Work supported in part by the Department of Energy
[*]The author is grateful to Mrs. Nancy Lehne for her patience in typing the
 manuscript.

I. INTRODUCTION

In the past ten years or so, the study of collisions between two heavy nuclei has become one of the main research areas in nuclear physics. In view of the complexity of such reactions involving many nucleons, comparing to nuclear reactions induced by light projectiles such as nucleon, deuteron, or triton, one might wonder what motivates such studies. Indeed, heavy ion collisions do not seem to be a useful way of studying nuclear properties at low excitations and low spins. But reactions using heavy ions offer unique means for exploring the properties of nuclei at extreme conditions, such as high excitations, large angular momenta, and high densities. Tremendous progress has been achieved in the past decade in both the experimental and the theoretical aspects of heavy ion collisions. For summaries of these developments, one is referred to the many review articles and conferences proceedings.[1-9].

In these lectures, we shall discuss in Chapter I the motivations for studying heavy ion collisions, the classifications of heavy ion collisions, and the experimental observations and their theoretical interpretations. Theories for describing heavy ion collisions at both the low and high energy are discussed in the other chapters. In Chapter II, we treat the time-dependent Hartree-Fock theory in which the motions of individual nucleons are followed under the influence of the average field of the two ions. In Chapter III, the statistical transport theory is developed. Unlike in the TDHF theory, the internal motions of the two ions are treated statistically in the transport theory. Both the TDHF theory and the transport theory have been used to study heavy ion collisions at low energies. Finally in Chapter IV, models for describing high energy heavy ion collisions are discussed.

I.1 <u>Motivations for Studying Heavy-Ion Collisions</u>

A. Production of Superheavy Elements

Based on the liquid drop model and the Strutinsky shell correction, many theoretical calculations have predicted the existence of superheavy elements with the proton and the neutron number around 114 and 184, respectively [10]. The search for superheavy elements in heavy-ion reactions has been one of the main reasons for carrying out such experiments. From the collisions of a uranium target with a uranium beam of ≈ 8 MeV/nucleon, it was found that the cross section limit in producing a superheavy element with a half-life of 1 ms is $\approx 10^{-32}$ cm^2 [11]. Efforts are continued in the search of superheavy elements.

B.   Probing the Limit of Nuclear Stability

As summarized by Scott [12], the number of stable nuclei existing naturally is about 300. Including the radio-isotope produced artificially in the laboratory, this number increases to $\approx$1600. Theoretically, it was estimated from various nuclear models that about another 6000 nulei could exist, although with relatively short half-lives. These nuclei are away from the beta-stability and are therefore either neutron rich or proton rich. An example of producing neutron rich isotopes in heavy ion reactions is given in Ref. [13], in which a beam of ^{40}Ar at E = 205 MeV/nucleon collides with a carbon target. From the fragmentation of ^{40}Ar, three new isotopes ^{28}Ne, ^{33}Mg, and ^{35}Al were discovered. We expect to find more such isotopes when heavier projectiles are used.

C.   Probing the limit of nuclear rotation and deformation

The study of nuclear deformation in the fission process and that of backbending in high spin states have provided us with a wealth of knowledge concerning the nuclear properties under large deformation and high angular momentum. Collisions between heavy ions can be considered as an inverse process with respect to the nuclear fission. Because of the large angular momentum involved in heavy ion collisions, one expects that these reactions will be useful in finding the limit of stability of a nucleus under fast rotation and strong deformation. As an example, Diamond and Stephens [14], in bombarding ^{40}Ar with ^{124}Sn at 192 MeV, have found an excited state of the compound nucleus ^{164}Er, which has a spin I = 64 and a moment of inertia $2I/h^2 \approx 180$ MeV^{-1}. Assuming a rigid ellipsoidal deformation, the ratio of the major to the minor axis is ≈ 2.

D.   Study of the nuclear equation of state

The nuclear equation of state $W(\rho,T)$ is defined as the energy per nucleon of the nuclear matter at a given density ρ and temperature T. Besides its intrinsic interest, the nuclear equation of state is essential for the understanding of the properties of neutron star. Our knowledge about the nuclear equation of state is, however, very limited. We know that it has a value $\approx$-16 MeV at normal nuclear matter density ρ_0 = 0.17 fm^{-3} and zero temperature. We also know about the curvature of the nuclear equation of state in this case from the recent experimental information about the isoscalar monopole excitation energy in nuclei [15]. This is because the monopole excitation energy is related to the nuclear compression modulus, which is given by the curvature of the nuclear equation of state, i.e.

$$K = 9\rho_0^2 \frac{d^2(E/A)}{d\rho_0^2}.$$

On the theoretical side, there are predictions that when the density of the nuclear matter increases, the nuclear equation of state will be softened through the appearance of other nuclear phases. At about twice the normal nuclear matter density, the nuclear matter becomes unstable against transition to the pion condensate ⌐16], a state of spin-isospin lattice of quantum numbers similar to that of a pion. At even higher density, the nuclear matter would make a transition to the abnormal state, i.e. the density isomer [17]. Finally, if the nuclear matter could be compressed to about nine times the normal density, then nucleons would loose their identities and dissociate into quarks [18]. In particular, it has been shown that a multiquark state with great strangeness may be metastable [19].

In heavy ion collisions, the density of the interaction region in the initial stage of the reaction may exceed substantially that of the normal nuclear matter. Model calculations have shown that up to five times the normal nuclear density can be reached in high energy heavy-ion collisions [20]. Therefore, high energy nuclear projectiles offer a unique tool for probing exotic nuclear phases in the laboratory.

I.2. <u>Classifications of Heavy-Ion Collisions</u>

As we are mainly interested in collisions between heavy nuclei with energies well above the Coulomb barrier, the wave length of the relative motion is small compared to the characteristic lengths of the interaction potential. Therefore, heavy-ion collisions can frequently be treated semi-classically. In terms of the impact parameter, which is defined as the distance between the two ions in the tranverse plane, heavy-ion reactions can be classified as either peripheral collisions for large impact parameters or central collisions for small impact parameters. In the peripheral collisions, only a few nucleons from the projectile and the target participate in the reaction and the process is therefore relatively gentle and has the features of direct reactions with relatively small momentum and energy transfers. In the central collision, however, more nucleons are involved and the reaction is more violent, characterized by an almost complete destruction of both the projectile and target nuclei.

One can also classify the heavy-ion reactions according to the energy of the projectile. For incident energies below the Fermi energy $E_F \approx 35$ MeV one expects to observe phenomena dominated by the nuclear mean field. Above this energy, two-body nucleon-nucleon collisions become important. When the incident energy per nucleon is above the pion mass ≈ 140 MeV, pion production will become increasingly appreciable. If the two-body collisions are frequent that the local equilibrium can be achieved during the reaction, then we can describe the process in terms of the hydrodynamics and expect to observe phenomena of such features. It has been argued that because of the instability of the nuclear matter against the pionic mode when its density is higher than the normal density, the two-body nucleon-nucleon cross section in heavy-ion collisions would be enhanced compared to that in the free space [21]. Therefore, we are close to the hydrodynamical regime in the intermediate energy heavy-ion collisions. For even higher incident energies above ≈ 1 GeV, the total collision times are so short that local equilibrium can not be reached and we are thus dealing with systems which can be described using the Boltzmann equation. In this energy region, we also reach the threshold $E_{th} \approx 1.6$ GeV for K^+ meson production. Eventually, we encounter the ultra-relativistic regime in which many other particles can be produced. Because of the increasing numbers of degrees of freedom, the thermodynamical concept such as the entropy may be useful.

Another way of classifying the heavy-ion collisions is according to the reaction mechanisms. Depending on the impact parameter and the incident energy per nucleon, heavy ion collisions can be classified roughly into five categories. For low incident energies and small impact parameters, the two ions fuse into a compound system. This is the fusion processes [22]. For intermediate impact parameters, the dominant process is the deep inelastic collisions [1] in which the projectile after the reaction loses a substantial fraction of its kinetic energy and angular momentum to the internal motion in the two ions. Also a large number of nucleons are exchanged between the projectile and the target. It has been quite established that the deep inelastic collisions are a multi-step process. For impact parameters close to the sum of the radii of the ions, the reaction is direct and involves the transfer of only a few nucleons from one ion to the other. The kinetic energy loss of the projectile is also relatively small. This process is usually called grazing collisions or quasi-elastic reactions.

For high incident energies, total explosion may occur if the impact parameter is small. But more likely the nucleons are divided into two groups; the participants that interact violently and the spectators that do not. This is especially so for relatively large impact parameters.

Of course, all the classifications discussed above are not sharp and well defined. There exist boundaries between them and the transition from one type of reaction to the other is itself an interesting subject.

I.3 Experimental Observations and Theoretical Interpretations

Since the grazing collisions are described in detail in the lectures by Dr. Q. Liu, we shall not include any discussions on these collisions. Rather, we shall discuss heavy-ion collision processes which are more macroscopic and involve higher energies.

A. Deep Inelastic Collision

The pioneering experiment by Artukh et al. [23] of the reaction ^{40}Ar + ^{233}Th at

an incident energy of 388 MeV/nucleon opened a new field in nuclear research that is
the deep inelastic heavy-ion collisions. As pointed out in the previous section,
the deep inelastic collision is essentially a binary process, characterized by the
large loss of the relative kinetic energy and angular momentum to the internal
motion of the ions, and by the transfer of many nucleons between the projectile and
the target. A typical experimental cross section of observing a ^{39}K ion as a function of its kinetic energy and scattering angle is shown in Fig. I.1, known as the
Wilczynski plot [24]. According to Wilczynski, the existence of low energy ions is
due to the deflection of the projectile to the negative scattering angles. This
involves a longer interaction time between the two ions than in quasielastic collisions. If frictional forces exist between the ions, an appreciable amount of the
relative kinetic energy is then expected to be dissipated into the internal degrees
of freedom.

Gross and Kalinowski [25] were first to apply the concept of frictional force
to study the deep inelastic collisions. In their model, two collective degrees are
considered, namely the distance r between the center of mass of the two ions and its
polar angle θ with respect to the incident direction. The shapes of both nuclei are

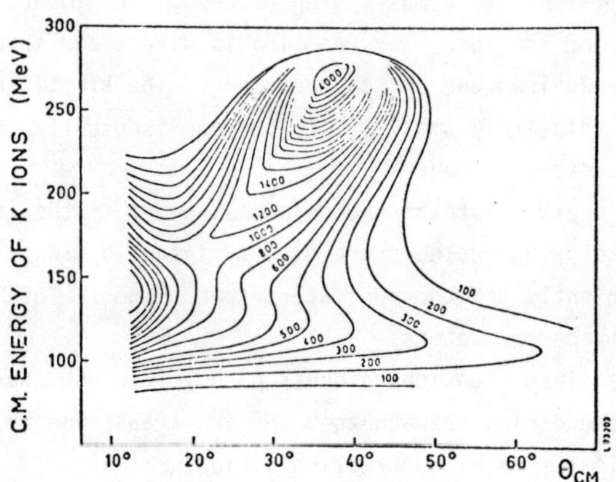

Fig. 1.1. The double differential cross section $d^2\sigma/dEd\theta$ in μb/(MeV rad) versus
c.m. scattering angle θ_{cm} and c.m. energy of the K ions from the
reaction ^{40}Ar + ^{232}Th at 388 MeV lab energy.

assumed to be spherical in the reaction. The equations of motion are then postu-
lated to follow the Newtonian equations with frictional forces:

$$\mu\ddot{r} = - \frac{dV(r)}{dr} + \mu r\dot{\theta}^2 - C_r f(r)\dot{r} \qquad\qquad (I.1.a)$$

$$\frac{d}{dt}(\mu r^2\theta) = -C_\theta\, r^2 f(r)\dot{\theta} \qquad\qquad (I.1.b)$$

In the above, μ is the reduced mass of the two ions. The interaction potential $V(r)$ is the sum of a coulomb potential $V_c(r)$ and a nuclear potential $V_N(r)$. The coulomb potential is normally taken to be the one between two point charges. As to the nuclear potential, there are a variety of models such as the proximity potential [26], energy density potential [27], and the folding potential [28]. All these potentials are based on the sudden approximation, i.e. the densitites of the two nuclei are assumed to remain unchanged during the collision. The sudden approximation is reasonable if the two ions do not overlap strongly which is the case for deep inelastic collisions.

The form factor $f(r)$ in Eqs. (I.1) expresses the spatial dependence of the frictional force. In the case that $V_N(r)$ is a folding potential, Gross and Kalinowski used the form

$$f(r) = (\vec{\nabla} V_N(r))^2 \qquad\qquad (I.2)$$

The coefficients C_r and C_θ denote, respectively, the radial and tangential frictional coefficients.

One can solve the two equations of motion for different values of the relative angular momentum ℓ or the impact parameter b and obtain the corresponding classical trajectories followed by the system. For a given ℓ value, when the trajectory enters the interaction region where the frictional force acts, some relative kinetic energy and orbital angular momentum are lost in a continuous manner. The trajectory is thus modified compared to the case without the presence of the frictional force. Some of these trajectories can escape from the interaction region, leading to deep inelastic collisions. The kinetic energy in this case is then lower than the original energy in the entrance channel.

This classical model can also allow one to calculate the differential cross section from the deflection function $\theta(b)$,

$$\frac{d\sigma}{d\theta} = 2\pi b \, \frac{db}{d\theta} = \frac{2\pi}{k^2} \ell \, \frac{d\ell}{d\theta} \qquad\qquad (I.3)$$

Here $\hbar k$ is the relative momentum. The result of such a calculation is given in Fig. I.2. where the experimental energy integrated cross section, obtained from Fig. I.1, is also shown.

However, the classical fricitional model with only the relative degrees of freedom cannot explain the low energy ridge in the Wilzinski plot. This corresponds to energy below that of two touching spheres. To understand the low energy ridge, it was suggested that deformations of the ions must be included [29]. Calculations including such deformation degrees of freedom have been recently carried out [30].

Since the deep inelastic collisions appear as dissipative processes, it is therefore expected that statistical fluctuations are importnat as well following the well-known example of the Brownian motion. Nörenberg [31] was the first one who considers the deep inelastic collisions as a diffusion process, and used the Fokker-

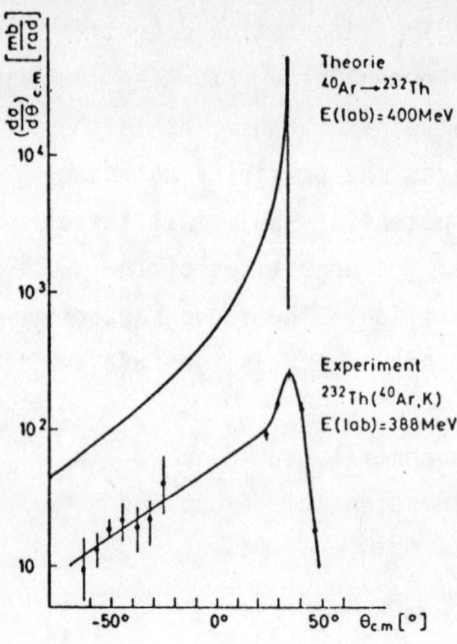

Fig. I.2. Energy integrated angular distribution for the reaction of ^{40}Ar + ^{232}Th. The
theoretical calculation is from Ref. [25], and
the experimental data is from Ref. [23].

Planck equation to understand the results of such reactions. Assuming that x represents some macroscopic measurable quantities such as the final kinetic energy E or
the charge Z of the projectile-like fragments, one can introduce the probability
function P(x,t) that the variable x takes a value in between x and x + dx at time
t. The Fokker-Planck equation for this probability function is

$$\frac{\partial P(x,t)}{\partial t} = -v\,\frac{\partial P(x,t)}{\partial x} + D\,\frac{\partial^2 P(x,t)}{\partial x^2} \tag{I.4}$$

Here v and D are, respectively, the drift and the diffusion coefficients and are
assumed to be constant. If at time t = 0, the variable x is at x_0, i.e. P(x,0) =
$\delta(x-x_0)$, then the solution to the Fokker-Planck equation is given by the following
Gaussian function

$$P(x,t) = \frac{1}{(2\pi Dt)^{1/2}}e^{-(x-vt)^2/2Dt} \tag{I.5}$$

This shows that the maximum of the Gaussian distribution increases linearly in time
with velocity equal to v, i.e. <x> = vt. The square of the finite width at half
maximum (FWHF) of the distribution also increases linearly in time, i.e. Γ^2 = 16(log
2)Dt.

This diffusion model has been used by Nörenberg to study the reaction of Artukh
et al.[23] We show in Fig. I.3 the normalized energy distributions for the element
Cl for different deflection angles. These curves are obtained from the experimental
data of Artukh et al. by interpreting the lower energy peak as a contribution from
the negative scattering angles. The data is consistent with a constant energy drift
coefficient of $V_E \approx 1 \times 10^{23}$ MeV/sec if the interaction time for scattering to -35°

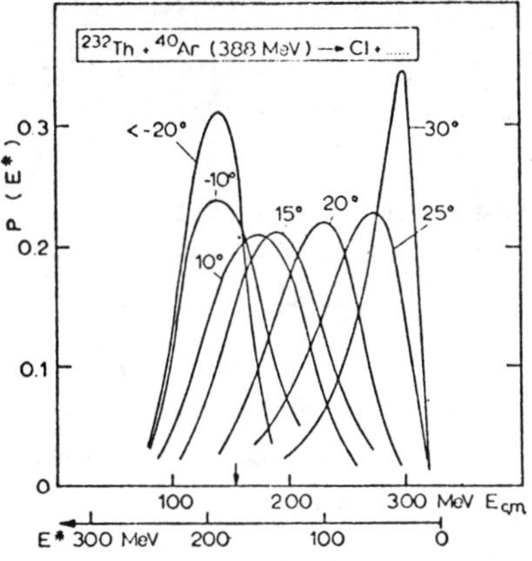

Fig. I.3. Normalized energy distributions of the element Cl for different lab deflection angles in the collision ^{40}Ar + ^{232}Ar at 388 MeV lab energy. (From Ref. [31].)

is taken to be $2 \cdot 10^{-21}$ sec. We also show in Fig. I.4 the FWHM Γ^2 of the element distribution as a function of the deflection angle for two incident energies. The straight lines are obtained with the same element diffusion coefficient $D_Z \approx 3 \cdot 10^{22}$

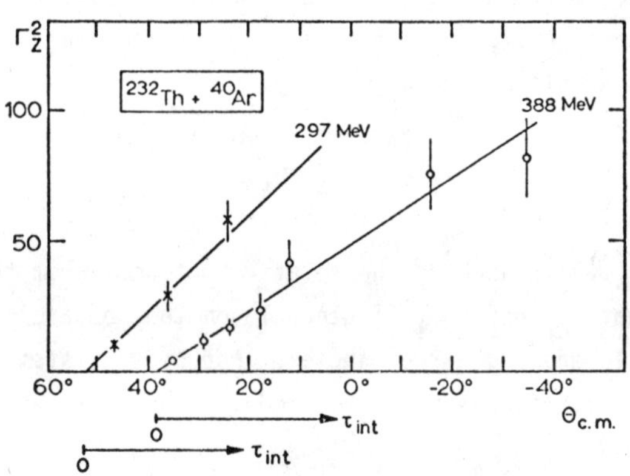

Fig. I.4. Dependence of the squared width (FWHF) of the element distribution on the deflection angle or the interaction time. (From Ref. [31])

charge^2/sec and by assuming that the rotational velocity of the interacting nuclei is proportional to the grazing angular momentum.

Another collective variable which has been received extensive study is the isospin motion. The usual way to study this collective mode is to measure the atomic number distribution for fixed mass asymmetry [32]. It has been observed that the isospin mode reaches equilibrium in a time much shorter than the collision time. Also, the width of the atomic number distribution for fixed mass assymmetry saturates quickly when the kinetic energy loss becomes large. Berlanger et al. [32] interpreted the experimental results in terms of the zero point motion of the giant isospin mode. When the phonon energy is much greater than the temperature of the intrinsic system, the width of the atomic number distribution is governed by the quantum fluctuation and is therefore independent of the temperature.

Many theories have been developed to understand the deep inelastic collisions. In Chapter III, we shall discuss one of the theories in which it is shown that both the classical friction model and the diffusion model can be derived in a unified fashion from the general statistical theory of nuclear reactions.

B. Fusion process

In this process, all the kinetic energy in the entrance channel is converted into the excitation energy of the compound nucleus. It is usually assumed that trajectories due to the small angular momentum ℓ lead to fusion. The largest ℓ value leading to fusion is called the critical angular momentum ℓ_{cr}. In this model, the total fusion cross section is then given by

$$\sigma_f = \frac{\pi}{k^2} \sum_{\ell=0}^{\ell_{cr}} (2\ell+1) = \frac{\pi}{k^2} (\ell_{cr} + 1)^2 \qquad (I.5)$$

where k is the initial wave number in the relative motion. From the experimental data, one discovers that ℓ_{cr} depends, in general, on the bombarding energy and the nature of the projectile and the target nuclei. For heavy systems, the fusion process does not exist. This happens when the product of the two atomic numbers $Z_1 Z_2$ is larger than $\approx 2500\text{-}3000$.

As a function of the incident energy, the total fusion cross section exhibits three types of behavior.

(a) In region 1, just above the threshold, the fusion cross section is a linear function of $1/E_{cm}$. This is illustrated in Fig. I.5 for the reaction of Ni + ^{35}Cl systems. It is seen, however, that close to the fusion threshold, i.e. for large values of $1/E_{cm}$ the experimental fusion cross section deviates from a straight line.

(b) In region 2 of intermediate energies σ_f is also a linear function of $1/E_{cm}$ but with a different slope. This is illustrated in Fig. I.6 for the ^{16}O + ^{27}Al system. Both regions 1 and 2 are seen here and the second region corresponds to a positve slope.

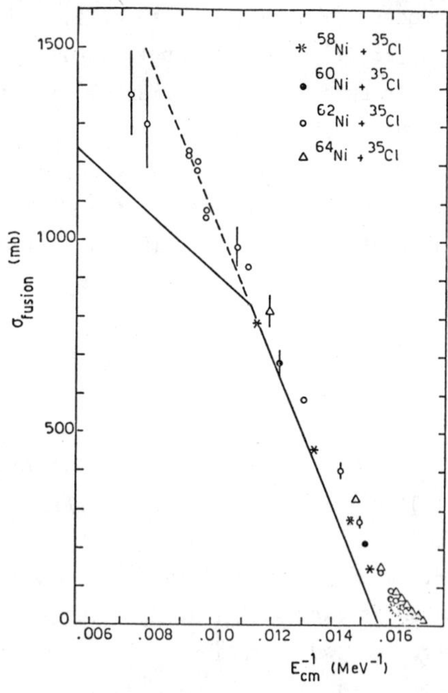

Fig. I.5. Fusion cross section as a function of the inverse of c.m. energy for the system Ni + ^{35}Cl. The solid line is from the static calculation of Ref. [40]. The data are from Ref. [33,34].

Fig. I.6. Same as Fig. I.5 for the system ^{27}Al + ^{16}O. The data are from Ref. [35-38].

c)   In region 3 when the bombarding energies are relatively high, the critical angular momentum saturates at a constant value and the fusion cross section is therefore again a linear function of $1/E_{cm}$ with positive slope. In Fig. I.7, the reaction of ^{24}Mg + ^{63}Cu is shown for two experimental data. It

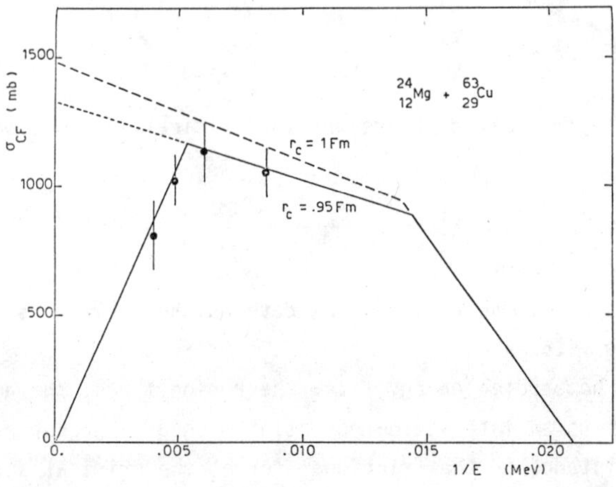

Fig. I.7. Same as Fig. I.5 for the system ^{24}Mg + ^{63}Cu. The data are from Ref. [39].

is not clear at what incident energy the fusion process finally ceases to exist.

Both the static [40] and the dynamical models [25,41] have been applied to study theoretically the heavy ion fusion process. In both approaches, fusion occurs when the projectile is trapped into the pocket of the interaction potential. In the static model, the critical angular momentum for fusion is the smallest ℓ value for which the ions can overcome the fusion barrier, i.e. the outer barrier of the ion-ion potential. For a given ℓ value, the total interaction potential $V_\ell(R)$ including the centrifugal potential is given by

$$V_\ell(R) = V(R) + \frac{\ell(\ell+1)h^2}{2\mu R^2} \quad . \tag{I.6}$$

The critical angular momentum is thus

$$\ell_{cr}(\ell_{cr}+1) = \frac{2\mu R_I^2(\ell)}{h^2} \, (E - V(R(\ell))), \tag{I.7}$$

where $R_I(\ell)$ is the position of the fusion barrier, and E is the center-of-mass energy. If we assume that the critical angular momentum is large and that $R_I(\ell)$ does not depend on ℓ strongly, then we can deduce the fusion cross section

$$\sigma_f = \pi R_C^2(1 - \frac{V_0}{E}) \quad . \tag{I.8}$$

It shows that the fusion cross section is a linear function of 1/E with a negative slope as it is observed experimentally. This explains the region one mentioned previously. Near the fusion threshold, one does not expect the above equation based on the classical picture to be valid. There it is important to take into account the effect of the quantum mechanical penetration.

When the energy is increased we are in the region two and the above experession Eq. (I.8) is still capable of parametrizing the experimental fusion cross section but with different parameters, i.e.

$$\sigma_f = \pi R_C^2(1 - \frac{V_C}{E}) \quad . \tag{I.9}$$

Here R_C and V_C are the critical distance and the critical interaction potential. It is found that

$$R_C \approx A_1^{1/3} + A_2^{1/3} \quad , \tag{I.10}$$

gives a good description of the experimental data as shown by the solid line with a positive slope in Fig. I.6.

For even higher bombarding energy, i.e. the region three, the angular momentum ℓ_I, which corresponds to an interaction potential without a pocket remains the same. Due to the existance of the frictional force, the critical angular momentum

ℓ_{cr} in the entrance channel is much larger than ℓ_I. In the sticking limit [41], the ratio $f = \ell_I/\ell_{cr}$ is given by

$$f = \frac{\mu R^2}{\mu R^2 + I_1 + I_2} \qquad , \qquad (I.11)$$

where I_1 and I_2 are the moment of inertia of the two nuclei, and R is the distance between the two centers. Using the rigid body value for I_1 and I_2, the model prediction for the fusion cross section in the region three is shown in Fig. I.7.

In the dynamical approach [25,41], classical equations of motion such as Eq. (I.1) are solved. A trajectory is considered to lead to fusion if it is trapped in the potential pocket. When applied to many heavy ion systems, Gross and Kalinowski [25] were able to reproduce sufficiently well the experimental critical angular momentum. More recently, Birkeleund et al. [41] have carried out quite complete dynamical calculations of the fusion cross section using the proximity potential [26] and the one-body dissipation [43]. One of their results is shown in Fig. I.8 together with the experimental data from Ref. [35-38].

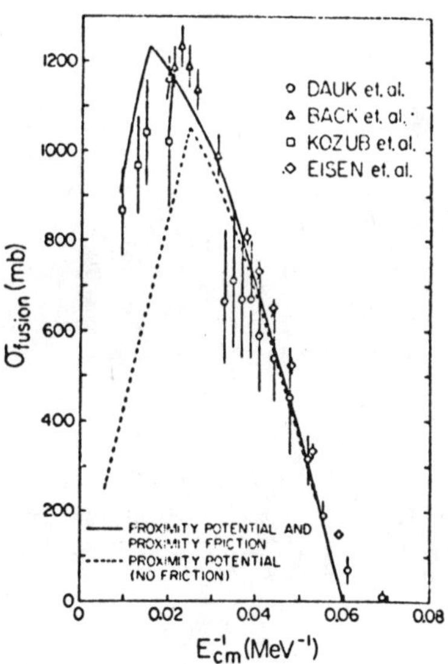

Fig. I.8. Fusion cross section as a function of the inverse of the c.m. energy. Lines are from the dynamical calculations of Ref. [41]. Data are from Ref. [35-38].

Recently, the prompt emission of energetic light particles is seen to play an important role in limiting the fusion cross sections of high energy heavy ion collisions [44]. One therefore must be cautious in assessing the success of both the staic and the dynamical approaches discussed in the above.

A microscopic theory which accounts quite well for the low energy heavy-ion fusion cross sections is the time-dependent Hartree-Fock theory. In Chapter II, we shall discuss how the theory is constructed and show the interesting results obtained from calculations with the TDHF.

C.   Projectile fragmentation

This process has been studied for various bombarding energies, ranging from 20 MeV per nucleon to 2 GeV per nucleon. In Fig. I.9, we show the energy spectra of various fragments at 15° from the reaction $^{16}O + ^{208}Pb$ at 315 MeV [45]. All these spectra are dominated by a Gaussian form. It peaks at an energy corresponding to the fragment travelling with a velocity close to that of the incident beam for fragments with mass close to the projectile. For fragments with mass much less than

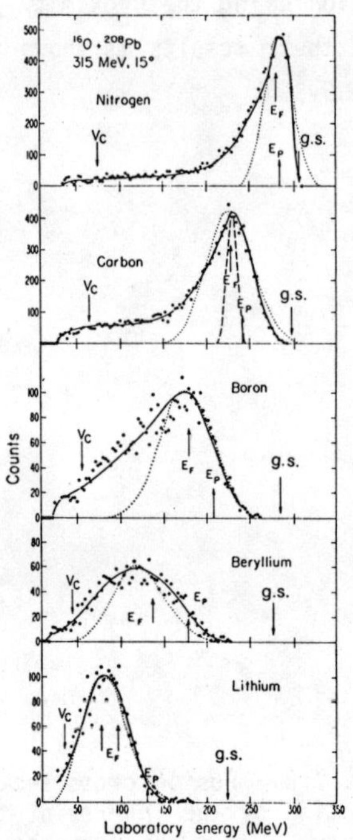

Fig. I.9. Energy distribution of fragments at 15° from the reaction $^{16}O + ^{208}Pb$ at 315 MeV [45].

that of the projectile, the velocity is correspondingly smaller. This is because more energy is required for the projectile to dissociate into a smaller fragment. The low energy tails in the spectra are probably due to contributions from the deep inelastic process.

Similar spectra are also observed at high energies. In Fig. I.10, we show the longitudinal momentum distribution of the fragment ^{10}Be in the rest frame of the projectile from the reaction ^{12}C on Be at an energy 2.1 GeV/nucleon [46]. Again, the spectrum is well fitted by a Gaussian; namely

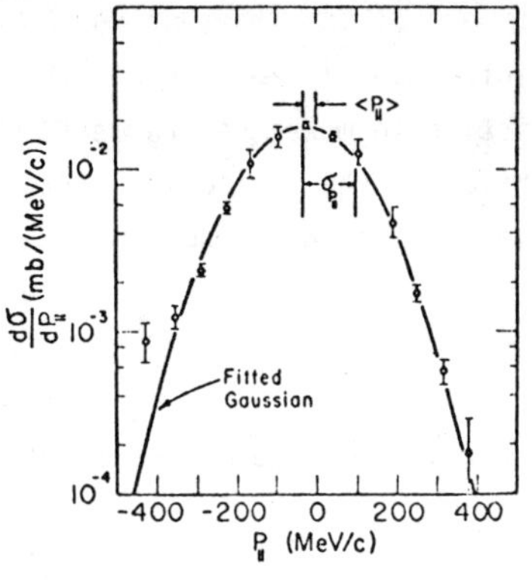

Fig. I.10. Longitudinal momentum distribution of the fragment ^{10}Be in the rest frame of the projectile from the reaction ^{12}C on Be at 2.1 GeV/nucleon [46].

$$\frac{d\sigma}{dP_{\shortparallel}} \approx e^{-\frac{p_{\shortparallel}^2}{2\sigma_{P_{\shortparallel}}^2}} \tag{I.12}$$

The width of the momentum distribution is shown as a function of the fragment mass in Fig. I.11. The experimental data is consistent with

$$\sigma_{P_{\shortparallel}}^2 \approx F(A - F) \tag{I.13}$$

where A and F denote the projectile and fragment mass, respectively.

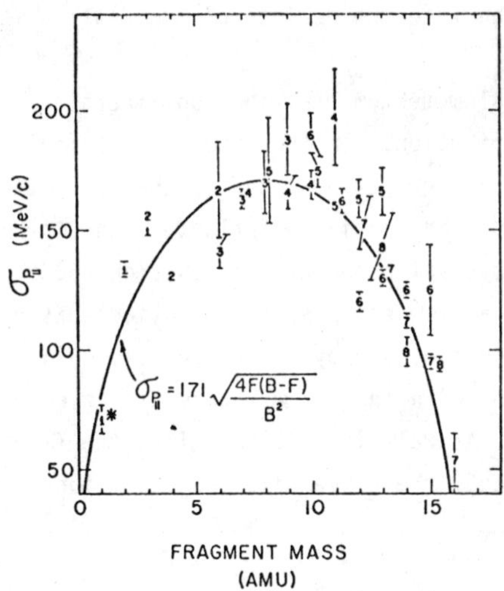

Fig. 1.11. Width of the fragment longitudinal momentum distribution as a function of the fragment mass [46].

To interpret such a mass dependence of the width of the fragment momentum distribution, a simple statistical model has been introduced by Goldhaber [47]. In this model, F nucleons are picked from the projectile to make the fragment. The dispersion of the longitudinal momentum distribution in the projectile frame is given by the expectation value of the squared momentum of these nucleons; i.e.

$$\sigma_{P_\parallel}^2 = \langle (\sum_i^F P_Z(i))^2 \rangle = F\langle P_Z^2(1)\rangle + F(F-1)\langle P_Z(1)P_Z(2)\rangle \ . \qquad (I.14)$$

The first term in the above equation can be estimated in the Fermi gas model as

$$\langle P_Z^2 \rangle = \frac{P_F^2}{5} \ , \qquad (I.15)$$

where P_F is the Fermi momentum in the projectile.

To calculate the second term in Eq. (I.14), Goldhaber uses the fact that the total momentum of the projectile is zero,

$$\langle (\sum_i^A P_Z(i))^2 \rangle = A\langle P_Z^2(1)\rangle + A(A-1)\langle P_Z(1)P_Z(2)\rangle = 0 \ . \qquad (I.16)$$

Combining Eqs. (I.14) and (I.16), we find for the momentum dispersion

$$\sigma_{P_\parallel}^2 = \frac{F(A-F)}{A-1}\frac{P_F^2}{5} \qquad (I.17)$$

Using the Fermi momentum determined from the inelastic electron scattering, it is found that the experimental dispersion is 30% - 50% lower than predicted by Eq. (I.17). Bertsch [48] attributes such discrepancy to the neglect of the Pauli effect in the Goldhaber model.

Besides the dispersion of the longitudinal momentum distribution, there is also a dispersion in the transversal momentum distribution. In the case of intermediate incident energies, the transversal dispersion is further enhanced by the deflection of the orbital motion [49]. It is pointed out that statistical fluctuations may also contribute to the dispersion of the transversal momentum distribution [50].

To describe the projectile fragmentation at low energies, formulations based on the distorted wave Born approximation have been proposed by Udagawa et al. [51]. Including the fusion of the projectile fragment with the target, they are able to describe the energy spectra of the fragments. At relativistic energies, the Glauber theory was used by Hüfner et al. [52,53] to study the projectile fragmentations.

D.   Central Collision

The study of the central collisions of high energy heavy ions was first carried out at Berkeley using high energy ^{4}He and ^{20}Ne beams[54]. Inclusive spectra of light particles such as p, d, t, and ^{4}He were measured. Typical data are shown in Fig. I.12. These are the proton inclusive spectra from a uranium target at

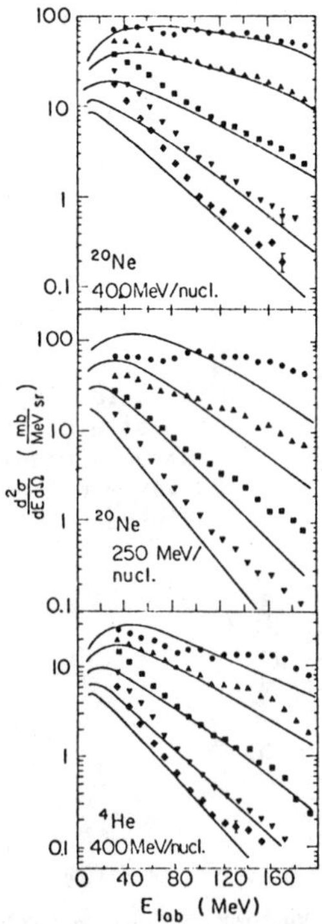

Fig. I.12. Proton energy spectrum from a uranium target at 30°, 60°, 90°, 120°, and 150°, in the laboratory. The solid lines are calculated with the fireball model [54].

various angles in the laboratory. We also show in Fig. I.13 the double differential cross section for hydrogen and helium isotopes from the same reaction. All spectra are characterized by an exponenential form, its slope increases with increasing scattering angle. To understand the experimental data, the fireball model was introduced [54]. According to this model, the observed light particles are emitted from the participants which include nucleons from both the projectile and the target, and are assumed to be in thermal equilibrium.

For a given impact parameter b, the number of participating nucleons can be obtained from simple geometrical consideration if we assume that the projectile nucleons move in a straight trajectory. Let the number of participating nucleons from the projectile and the target be $N_P(b)$ and $N_T(b)$, respectively. If all the participating nucleons form a compound system called the fireball, then the mean momentum $\vec{P}_{cm}(b)$ and the total excitation energy $E^*(b)$ of the fireball is given by

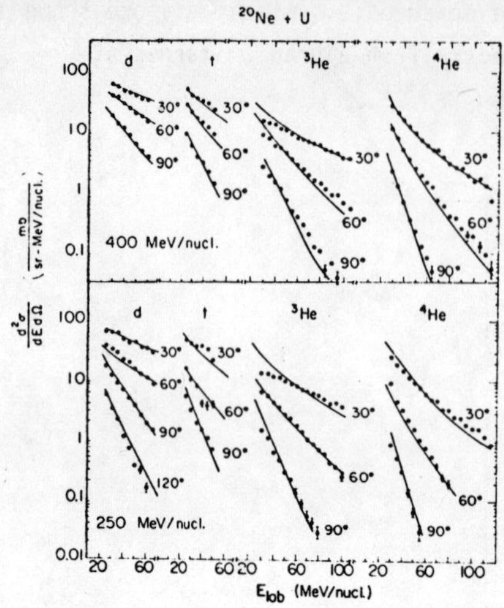

Fig. I.13 Inclusive energy spectra of composite particles from the reaction ^{20}Ne on a uranium target. The solid curves are from the calculations of coalescence model [54].

$$\vec{P}_{cm}(b) = \frac{N_p(b)}{N(b)}\,\vec{P}_0 \qquad\qquad (I.18.a)$$

$$E^*(b) = \frac{N_T(b)\,N_p(b)}{N(b)}\,E_0 \quad, \qquad\qquad (I.18.b)$$

where $\vec{P}_0$ and E_0 are, respectively, the initial momentum and energy of the incident nucleon in the projectile; $N(b)$ is the number of participating nucleons, $N(b) = N_p(b) + N_T(b)$.

In the fireball model, it is assumed that all the nucleons are in thermal equilibrium and their momentum distribution is given by the Maxwellian form

$$f(\vec{P},b) = \frac{1}{[2\pi mT(b)]^{3/2}}\,\exp\left[-\frac{(\vec{P}-\vec{P}_{cm})^2}{2mT(b)}\right] \quad, \qquad\qquad (I.19)$$

with the temperature

$$\frac{3}{2}\,T(b) = E^*(b)/N(b) \quad . \qquad\qquad (I.20)$$

The nucleon inclusive cross section is then

$$\frac{d^2\sigma}{d\vec{p}^3} = \int_0^{R_P+R_T} 2\pi b\;db\;N(b)\;f(\vec{P},b) \qquad\qquad (I.21)$$

Results from the fireball calculations are shown by the solid lines in Fig. I.12, and are in fair agreement with the data.

The fireball model has been extended to study the production of pions [55] and kaons [56] in high energy heavy ion collisions. Siemens and Rasmussen [57] have argued that a fireball would be expected to explode, leading to a blast wave of nucleons and pions. The energy for the blast wave comes from both the compressional and thermal energy of the fireball. Experimental data for inclusive pion and proton cross sections from the reaction Ne on NaF at 0.8 GeV/nucleon suggest that about 40% of the available energy appears as translational kinetic energy of the blast.

As to the composite particles observed experimentally two simple models have been put forward for their explanation. The first attempt was the coalescence model [54] in which the light composite particles are formed from the coalescence of emitted nucleons via final state interactions. In this model, a nucleus is formed when a group of nucleons corresponding to this nucleus are emitted with momenta differing by less than a coalescence radius P_0. The cross section for the emission of light nuclei are therefore related to the cross section for the emission of nucleons at the same momentum per nucleon; namely

$$\frac{d^2\sigma_A}{d\vec{P}^3} = \frac{1}{A!} \left(\frac{4\pi P_0^3 \gamma}{3\sigma_0}\right)^{A-1} \left(\frac{d^2\sigma_1}{d\vec{P}^3}\right)^A \tag{I.22}$$

Here σ_0 is the total reaction cross section and γ is the Lorentz factor. Calculated results are shown as solid lines in Fig. I.13, treating the coalescence radius P_0 as an adjustable parameter.

Another model for the production of the light composite particles is to assume that thermadynamical equilibrium is established among the various species in the fireball [58].

From the observed ratio of deuterons to protons at large transverse momentum, Siemens and Kapusta [59] have estimated the entropy of the fireball and found that more entropy is generated during the collision than one would naively expect. This excess of entropy in fireballs from heavy ion collisions can be explained by a modified pionic spectrum [60].

In Chapter IV, we shall discuss in detail the linear cascade model of Hüfner and Knoll [61] which describes microscopically the evolution of the momentum distribution of the nucleons during the course of the collision. In this way, one is able to assess the validity of the fireball model.

II.   TIME-DEPENDENT HARTREE-FOCK THEORY

II.1 Derivation of the TDHF Equation

For incident energy a few MeV per nucleon, the mean free path of a nucleon in nucleus is large or at least comparable to the size of the nucleus due to the effect of Pauli blocking. We therefore expect that the mean field will be the dominant factor in governing the evolution of the collision process. There are many ways of deriving the time-dependent Hartree-Fock equations from the nuclear many-body Shrödinger equation. In the following, we shall follow Negele [6] and perform a variational derivation. We shall work in a one-dimensional space for simplicity. The generalization to the three-dimension is obvious.

Let the nuclear Hamiltonian be

$$H = -\sum_i \frac{\hbar^2}{2m} \frac{d^2}{dx^2} + \sum_{i<j} v(x_i - x_j) \tag{II.1}$$

The interaction potential between nucleons $v(x_i - x_j)$ should, in principle, be a realistic one such as the Raid potential. But for practical reasons, an effective interaction is used, e.g. the Skyrme interaction [62].

For a determinantal wave function, the action may be written as

$$I = \int dt \ \langle \psi(t) \ | i\hbar \frac{\partial}{\partial t} - H | \ \psi(t) \rangle \tag{II.2}$$

$$= \int dt \ \{\sum_i \int dx \ \phi_i^*(x) \ (ih \frac{\partial}{\partial t} + \frac{\hbar^2}{2m} \frac{d^2}{dx^2}) \phi_i(x) - \frac{1}{2} \sum_{i,j} \int dx_1 dx_2 \ \phi_i^*(x_1) \ \phi_j^*(x_2) \ v(x_1-x_2)$$

$$* \ [\phi_i(x_1) \ \phi_j(x_2) - \phi_i(x_2) \ \phi_j(x_1)] \}$$

Stationarity of the action with respect to $\delta\phi_i(x)$ yields

$$i\hbar \ \dot{\phi}_i(x) = - \frac{\hbar^2}{2m} \frac{d^2}{dx^2} \ \phi_i(x) + \sum_j \int dx' \ v(x-x') \ |\phi_j(x')|^2 \ \phi_i(x) \tag{II.3}$$

$$- \sum_j \int dx' \ v(x-x') \ \phi_j^*(x') \ \phi_j(x) \ \phi_i(x')$$

In order to write the above equation more compactly, we introduce

$$h(x,x') = - \frac{\hbar^2}{2m} \delta(x-x') \frac{d^2}{dx'^2} + \delta(x-x') \int dx'' \ v(x-x'') \ \rho(x'') \tag{II.4}$$

$$- \rho(x,x') \ v(x-x') \equiv T(x,x') + W(x,x')$$

with
$$\rho(x,x',t) \equiv \sum_i \phi_i^*(x',t) \ \phi_i(x,t) \tag{II.5}$$

and
$$\rho(x) \equiv \rho(x,x),$$ (II.6)

then Eq. (II.3) becomes

$$i\dot{\phi_i}(x) = \int dx' \; h(x,x') \; \phi_i(x')$$ (II.7)

We see that the 2nd term in Eq. (II.4) is just the local Hartree potential, while the 3rd term is the non-local exchange potential. The density matrix of Eq. (II.5) has a simple form in nuclear matter

$$\rho(x,x') = \rho_0 \frac{3j_1(k_F|x-x'|)}{k_F|x-x'|}$$ (II.8)

with ρ_0 the nuclear matter density and k_F the Fermi-momentum.

From Eq. (II.7), we can obtain the equation of motion for the density matrix

$$i\dot{\rho}(x,x') = \sum_i \phi_i^*(x') \int dx'' \; h(x,x'') \; \phi_i(x'') - \sum_i \int dx'' \; \phi_i^*(x'') \; h^*(x',x'') \; \phi_i(x)$$

$$= \int dx'' \; \{h(x,x'') \; \rho(x'',x') - \rho(x,x'') \; h(x'',x')\}$$ (II.9)

Here we have made use of the property that $h(x,x')$ is hermitian, i.e. $h^*(x',x'') = h(x'',x')$. Eq. (II.9) can be more compactly written as

$$i\dot{\rho} = [h,\rho]$$ (II.10)

There are a few important features associated with the TDHF. From Eq. (II.5), one can easily see that for determinantal wave function, the density matrix is idempotent, i.e.

$$\rho^2 = \rho$$ (II.11)

Also the scalar product of two single-particle wave functions remains constant in time, since

$$\frac{d}{dt} \int dx \; \phi_i^*(x) \; \phi_j(x) = -i \iint dx \; dx' \; [\phi_i^*(x) \; h(x,x') \; \phi_j(x') - \phi_i^*(x) \; h^*(x,x') \; \phi_j(x)]$$

$$= -i \iint dx \; dx' \; \phi_i^*(x) \; [h(x,x') - h^*(x',x)]\phi_j(x') = 0$$ (II.12)

This implies that if one starts with an orthonormal basis, the wave functions will remain orthonormal in later times.

The total energy of the system is conserved in TDHF. This is because

$$\frac{d}{dt} \langle H \rangle = \frac{d}{dt} \, \mathrm{tr} \; (T + \frac{1}{2} W(\rho))\rho$$

$$= \mathrm{tr} \; (T + W)i\dot{\rho}$$
$$= \mathrm{tr} \; h[h,\rho] = 0$$ (II.13)

II.2 Solution of TDHF Equation and Results

In the following, we shall apply the TDHF equation to study the collision of two semi-infinite slabs of spin-isospin symmetric nuclear matter. Since the system is translationally invarient in the two transverse directions, the most general structure of the single-particle wave functions is

$$\Psi_{n,\vec{k}_\perp}(\vec{r},t) = \frac{1}{\sqrt{S}}\, e^{i\vec{k}_\perp \cdot \vec{r}_\perp}\, e^{-i\frac{k_\perp^2 t}{2m}}\, \phi_n(z,t) \tag{II.14}$$

where $\vec{r}_\perp = (x,y)$, $\vec{k}_\perp = (k_x,k_y)$ and S is the transverse normalization area. The effective interaction is taken to be

$$v(\vec{r},\vec{r}') = t_0\, \delta(\vec{r}-\vec{r}') + \frac{t_3}{6}\, \delta(\vec{r}-\vec{r}')\, \rho(\vec{r}) + V_0\, \frac{e^{-|\vec{r}-\vec{r}'|/a}}{|\vec{r}-\vec{r}'|/a}\left(\frac{16}{15} + \frac{4}{15}\, P_x\right) \tag{II.15}$$

where P_x denotes the space exchange operator. This interaction is closely related to the Skymre interaction [62] except that the momentum dependent part is replaced by a finite-range Yukawa interaction. The parameters t_0, t_3, V_0, and a are adjusted to yield a nuclear matter binding energy of 15.77 MeV per particle at a saturation of $\rho = 0.145$ fm^{-3} corresponding to $k_F = 1.29$ fm^{-1}.

Substituting the interaction Eq. (II.15) into Eq. (II.2) for the action integral and performing the variation, we obtain the TDHF equation for $\phi_n(z,t)$

$$i\dot{\phi}_n(z,t) = \left[-\frac{\hbar^2}{2m}\frac{d^2}{dz^2} + W(z,t)\right]\phi_n(z,t) \tag{II.16}$$

where

$$W(z,t) = \frac{3}{4}t_0\,\rho(z) + \frac{3}{16}\rho^2(z) + 2\pi a^3 V_0 \int_{-\infty}^{\infty} dz'\,\rho(z')\, e^{-|z-z'|/a} \tag{II.17}$$

To solve Eq. (II.16), the initial states have to be specified. For the case of two slabs approaching each other with relative velocity 2v, the initial wave functions are

$$\phi_n(z,0) = \phi_n^{HF}(z)\, e^{imvz} . \tag{II.18}$$

The wave function $\phi_n^{HF}(z)$ satisfies the static HF equation

$$\left[-\frac{\hbar^2}{2m}\frac{d^2}{dz^2} + W(z)\right]\phi_n^{HF}(z) = e_n \phi_n^{HF}(z). \tag{II.19}$$

The TDHF equation Eq. (II.16) with the initial conditions Eq. (II.18) has been solved numerically by Bonche et al. [63] In Fig. II.1, the ratio of the final to initial translational kinetic energy is plotted as a function of c.m. incident energy per particle. Below 1 MeV, the two slabs fuse into one compound system. Between 1 MeV and 2 MeV, there are some resonance structures as a result of the multiple reflections of the single-particle wave functions. Above 2 MeV, more than 90% of the initial kinetic energy is converted into the internal excitation energy. This is the typical feature of the deeply inelastic reactions.

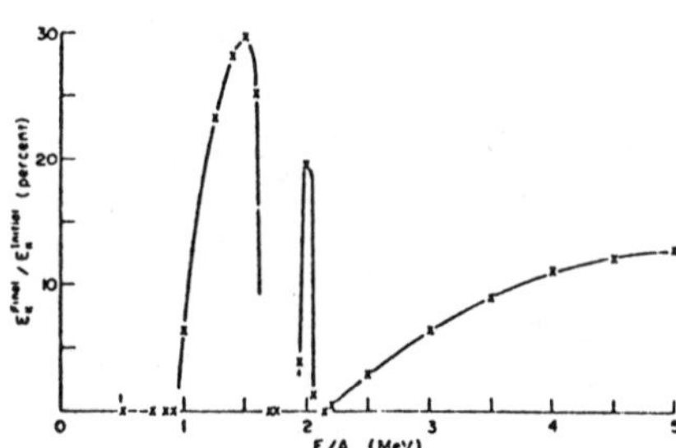

Fig. II.1. Ratio of the final to initial translational kinetic energy as a function of c.m. incident energy per particle [63].

Since the TDHF calculations involve large amounts of numerical computations, it was initially not attempted to perform a complete 3-dimensional calculation. In Ref. [64], the full time-dependent Hartree-Fock problem is reduced to two dimensions by treating the relative orbital motion of the ions in the rotating frame approximation. For mass symmetric systems, if one works in the intrinsic, body fixed frame, a classical energy of rotation is added to the nuclear Hamiltonian,

$$H' = H + \frac{L^2}{2I[\rho]} \tag{II.20}$$

Here, $L = (\ell + \frac{1}{2})\hbar$ is the conserved relative angular momentum along the rotation axis perpendicular to the scattering plane and the moment of inertia is a prescribed functional of the density. The variation of the energy functional with respect to the single-particle wave functions yields the TDHF equations with the HF potential

$$W'(\vec{r}) = W(\vec{r}) - \frac{1}{2}\omega^2 \frac{\partial I}{\partial \rho(\vec{r})} \tag{II.21}$$

$$\omega = \frac{d\theta}{dt} = \frac{L}{I[\rho]} \qquad\qquad (II.22)$$

In Ref. [64], computations have been performed for $^{16}O + ^{16}O$ and $^{40}Ca + ^{40}Ca$. Both the deflection function and the energy loss as a function of the impact parameter are obtained. While the deflection functions are consistent with experiment, the energy dissipation is too small. Also, the charge distribution appears to be narrower than experimental results.

The same approximation has been applied to collisions between heavier nuclei of unequal masses. In Ref. [65], the Kr-induced deep inelastic collisions were studied. It was found that the calculated fragment energies, mean masses, and scattering angles for deep inelastic collisions are in good agreement with experiment. In Fig. II.2, the calculated points, labeled by the orbital angular momentum are compared with the experimental Wilczynski plot from the reaction $^{84}Kr + ^{208}Pb$ at $E_{lab} =$ 494 MeV [66]. Similar to the mass symmetric case, the mass distribution widths are an order of magnitude too small.

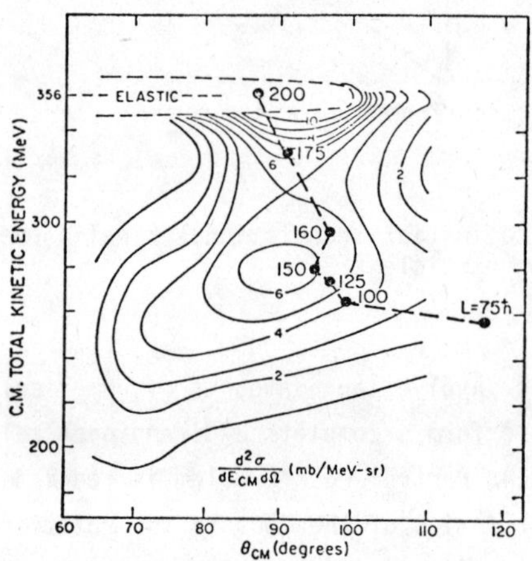

Fig. II.2. Comparison of TDHF calculations [65], labelled by the orbital angular momentum, with the experimental Wilczynski plot from the reaction $^{84}Kr + ^{208}Pb$ at E_{lab} = 494 MeV [66].

An alternative approximation to three-dimensional TDHF calculations for heavy-ion reactions were introduced in Refs. [67,68]. This is the separable approximation, in which the motion normal to the scattering plane is neglected. Let the coordinate normal to the reaction plane be z, then the time-dependent single-particle wave-functions are written in a factorized form

$$\Psi_i(\vec{r},t) = \phi_i(x,y,t) \, \chi_i(z) \qquad\qquad (II.23)$$

The equations of motion for the function ϕ follow from the TDHF equations. For an effective interaction which results in a local HF potential W, they are of the form

$$i\hbar \frac{\partial \phi_i}{\partial t} = \left[- \frac{\hbar^2}{2m} \left(\frac{\partial^2}{\partial x^2} + \frac{\partial^2}{\partial y^2}\right) + \langle T_z \rangle_j + W_j(x,y)\right]\phi_j \tag{II.24}$$

Here, W_j is the projected HF potential

$$W_j(x,y) = \int_{-\infty}^{\infty} dz \, |\chi_j(z)|^2 \, W(x,y,z) \tag{II.25}$$

and $\langle T_z \rangle_j$ is the kinetic energy of χ_j

$$\langle T_z \rangle_j = \frac{\hbar^2}{2m} \int_{-\infty}^{\infty} dz \, |d\chi_j/dz|^2 \tag{II.26}$$

This approximation leads to an order-of-magnitude reduction in computation time.

More recently, full three-dimensional TDHF calculations have been carried out by many groups [69-73]. It was found that both the separable approximation and the rotating frame approximation are good except for very high energy collisions [73].

Both the rotational frame approximation and the separable approximation have been used to calculate the heavy-ion fusion excitation functions. Fig. II.3 shows the fusion cross sections as a function of the bombarding energy for the reaction $^{16}O + {}^{40}Ca$ [73]. They are calculated with the sharp cutoff formula

$$\sigma_{fus}(E_{lab}) \approx \frac{\pi\hbar^2}{\mu E_{lab}} \left[(\ell_> + 1)^2 - (\ell_< + 1)^2\right] \tag{II.27}$$

Here, $\ell_>$ and $\ell_<$ are the lower and upper angular momentum limits of the angular momentum window which leads to fusion of the two nuclei. The calculated results

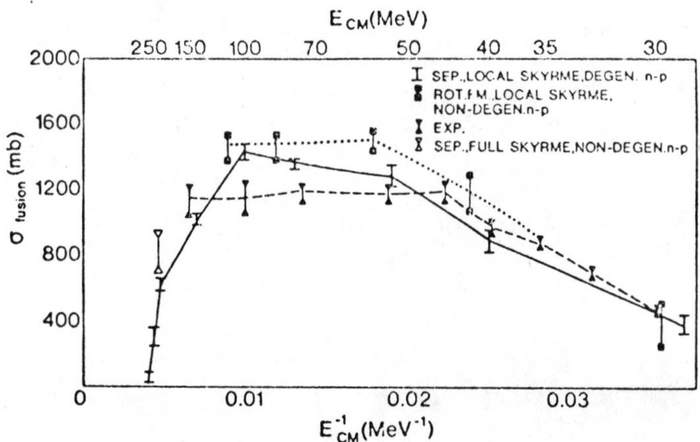

Fig. II.3. Comparison of the calculated fusion cross section from the TDHF [73] with the experimental results [74] for the reaction $^{16}O + {}^{40}Ca$.

reproduce the overall magnitude of experimental data. The theoretical results are not sensitive to the non-local part of the effective interaction used in the TDHF equations. Also, the effect of including the proton and neutron asymmetry is not appreciable.

III. STATISTICAL TRANSPORT THEORY

III.1. Introduction

In the previous chapter, we have derived the time-dependent Hartree-Fock
approximation to describe low energy heavy-ion collisions. There, the motions of
the nucleons are followed according to the TDHF equations. It is seen from the calculated results that a large fraction of the initial kinetic energy of the relative
motion is converted into the internal energy of the nuclei. To what extent the
internal degrees of freedom can be treated statistically is very important for the
understanding of deep inelastic collisions. In the TDHF approximation, the internal
degrees of freedom are followed explicitly. In the following, we shall present the
other extreme case of treating the internal degrees of freedom entirely statistically. It will be seen later that the equation of motion in this approach is the
transport equation and that the deeply inelastic collisions between heavy ions can
be described as transport phenomena.

There are a variety of ways of deriving a transport equation for heavy ion
collisions, based on different assumptions and approximations. The approach of
Swiatecki and Randrup et al. [75,76] emphasizes the one-body aspect of nuclear
dissipation. In their model, the single-particle potentials of the two fragments
are viewed as containers for the thermalized Fermi gas of nucleons. As the two
potentials overlap, a window opens between the two containers and nucleons in one
fragment may move freely into the other. Since the two potentials are in motion
relative to each other, the exchange of nucleons results in transforming the relative kinetic energy into the intrinsic excitations. This transformation of energy
is irreversible if one assumes that the transferred nucleon becomes thermally
equilibrated with the nucleus in the receptor nucleons. They assume that the
equilibration of the nucleon is due to the collision with the wall of the potential. In the linear response theory of Hofmann and Siemens [77,78], the couplings
between the collective degrees of freedom and the non-collective degrees of freedom
are treated using first-order perturbation theory. It is assumed that the equilibration time for the non-collective degrees of freedom is much shorter than the
charcteristic time over which the collective degrees of freedom change appreciably.
This assumption leads to the postulation that the non-collective degrees of freedom
can be considered as a heat bath with a temperature determined by the excitation
energy. The collective energy is therefore transformed irreversibly into the intrinsic excitation. The theory of Norenberg and Ayik et al., [79-82], on the other
hand, treats the intrinsic degrees of freedom statistically. Assuming that the collective degrees of freedom couples strongly with the non-collective degrees of freedom, they derive the transport coefficients which describe the relaxation of various
collective degrees of freedom. Another approach has been considered by Agassi, Ko,

and Weidenmüller [83,84]. Using a random matrix model for the matrix elements of the coupling between the collective and the non-collective degrees of freedom, they derive a transport equation from the coupled-channel reaction theory. A critical review of these approaches can be found in a recent article by Weidenmuller.[2] We shall in the following describe the approach taken by Aggasi, Ko, and Weidenmüller [84].

III.2 The Random Matrix Model

For simplicity, we consider only one collective degree of freedom, namely the relative coordinate $\vec{r}$ of the two heavy ions. All other degrees of freedom ζ are treated as intrinsic coordinates. Let the Hamiltonian of the system be

$$H(\vec{r},\zeta) = H(\vec{r}) + H_0(\zeta) + V(\vec{r},\zeta) \tag{III.1}$$

$$= H_1(\vec{r},\zeta) + V(\vec{r},\zeta)$$

Here, $H(\vec{r})$ is the Hamiltonian of the relative motion, $H_0(\zeta)$ the intrinsic Hamiltonian, and $V(\vec{r},\zeta)$ the coupling between the relative and the intrinsic coordinates. The Hamiltonians $H_0(\zeta)$ and $H(\vec{r})$ have the eigenstates

$$H_0|s\rangle = \varepsilon_s|s\rangle \ , \ \langle s|m\rangle = \delta_{sm} \tag{III.2}$$

and

$$H|\chi(\varepsilon)\rangle = \varepsilon|\chi(\varepsilon)\rangle, \ \langle\chi(\varepsilon)|\chi(\varepsilon')\rangle = \delta(\varepsilon-\varepsilon') \tag{III.3}$$

Let us define the coupling matrix element by

$$V_{sm}(\vec{r}) = \langle s|V(\vec{r},\zeta)|m\rangle \tag{III.4}$$

Since we are interested in the average of physical quantities over a sufficiently large number of levels, we therefore assume that $V_{sm}(\vec{r})$ have a Gaussian distribution with zero mean value and non-vanishing second moment. Denoting the energy average by a bar, then

$$\overline{V_{sm}(\vec{r})} = 0 \tag{III.5}$$

and

$$\overline{V_{sm}(\vec{r}) V_{s'm'}(\vec{r}')} = \left(\delta_{ss'}\delta_{mm'} + \delta_{sm'}\delta_{ms'}\right) \tag{III.6}$$

$$* D_s^{1/2}D_m^{1/2} W_0 \, f(\frac{r+r'}{2}) \exp(-(\varepsilon_s-\varepsilon_m)/2\Delta^2) * \exp(-(\vec{r}-\vec{r}')^2/2\sigma^2)$$

This form of the second moment has been justified microscopically by Barrett <u>et al</u>. [85]. The parameter D_s is the mean spacing of levels around the state s, the factor $(D_s^{1/2} D_m^{1/2})$ takes account of the increasing complexity of nuclear levels with increasing excitation energy, which reduces the overlap between levels s and m. The strength of the interaction W_o is typically 10 MeV. The form factor f describes the overlap of the two density distributions. The exponential in energies retricts the energy transfer to values $< \Delta$. The magnitude of Δ can be estimated using the picture that the interaction potential $V(\vec{r}, \zeta)$ is essentially the average potential felt by the nucleon in one nucleus due to all nucleons in the other nucleus. The effect of this interaction is then to produce particle-hole excitatons. Significant contributions result only if $\Delta \leq 7$ MeV. The exponential in space implies that the the mean value tends to zero as $|\vec{r} - \vec{r}'|$ increases. This is so because then the interaction takes place at different parts of the nucleus and becomes out of phase. This correlation length σ was found to have a typical value of 3.5 fm on the nuclear surface.

It is worthwhile to describe briefly the various time scales involved in the collision. The collision time τ_{coll} is roughly given by the size of the nucleus divided by the relative velocity, i.e.

$$\tau_{coll} = \text{a few } 10^{-21} \text{ sec} \tag{III.7}$$

From Eqs. (III.6), we obtain the time for transferring an energy Δ,

$$\tau_\Delta = h/\Delta \approx 10^{-22} \text{ sec} \tag{III.8}$$

A correlation time is given by

$$\tau_\sigma = \sigma/v \simeq 3. * 10^{-22} \text{ sec} \tag{III.9}$$

Here v is the relative velocity. Another time, which is important, is the time between two successive collisions τ_λ. This is given by the mean free path divided by the relative velocity v. Since the typical energy loss in deeply inelastic heavy ion collision is ≈ 200 MeV and the energy loss per collision $\approx \Delta = 7$ MeV, there is about 30 collisions. Using the total distance $\approx 2R \approx 15$ fm, we have the mean free path $\lambda \approx 0.5$ fm. Therefore

$$\tau_\lambda = \lambda/v \simeq 4. * 10^{-23} \text{ sec} \tag{III.10}$$

Finally, we need to know the time needed for the system to equilibrate after a single action of the interaction. This is given by $\tau_{equ} = h/\Gamma_\downarrow$ with the spreading width $\Gamma_\downarrow = 2\pi V^2 \rho(E)$. Using $V \simeq 200$ keV and $\rho^{-1} \approx 5$ to 10 keV, we have

$$\tau_{equ} \lesssim 3 \ldots .7 \times 10^{-23} \text{ sec} \tag{III.11}$$

The fact that $\tau_\lambda \approx \tau_{equ}$ implies that after each action of the interaction, the system barely has sufficient time to equilibrate. From the above equations, we see that

$$\tau_\lambda < \tau_\Delta, \tau_\sigma \ll \tau_{coll} \qquad (III.12)$$

This means that the duration of each action of the interaction is shorter than the time needed for transferring energy of the amount Δ or momentum of $\hbar/\sigma$. The implication of this inequality is that the collision process is non-Markoffian and we are dealing with a strong coupling theory. This is in contrast with the situation encountered in a macroscopic system in contact with a heat bath, there the process is Markoffian.

II.3 <u>Derivation of the Transport Equation</u>

Let us consider the scattering matrix element S_{ab} for the transition from the channel a to the channel b. For $a \neq b$, we have in the Born series

$$S_{ba} = \langle \chi(E-\varepsilon_b) \, | \{ V_{ba} + \sum_m V_{bm} G_m V_{ma}$$

$$+ \sum_{m,n} V_{bm} G_m V_{mn} G_n V_{na} + \ldots \} \; \chi(E-\varepsilon_a) \rangle, \qquad (III.13)$$

where E is the total energy of the system and

$$G_m = [E + i\varepsilon - \varepsilon_m - H(r)]^{-1} \qquad (III.14)$$

the Green's function. For simplicity, we rewrite Eq. (III.13) as

$$S_{ba} = \langle \chi_b \, | V \sum_{n=0}^{\infty} (GV)^n | \, \chi_a \rangle \qquad (III.15)$$

The cross section is proportional to $|S_{ba}|^2$. But we need to evaluate $\overline{|S_{ba}|^2}$ as we are only interested in energy averaged quantities. From Eq. (III.15), it is seen that we need to evaluate the energy average of the products of V's. In order to carry out such calculations, we make use of the following two properties of Gaussian distributed random variables with zero mean value. Let $Z_1, Z_2, \ldots Z_j$ be these j variables, then

$$\overline{Z_1 Z_2 \ldots Z_j} = 0 \text{ if j is odd} \qquad (III.16)$$

and $\overline{Z_1 Z_2 \ldots Z_j} = \sum \overline{Z_{\alpha_1} Z_{\alpha_2}} \; \overline{Z_{\alpha_3} Z_{\alpha_4}} \ldots \overline{Z_{\alpha_{j-1}} Z_{\alpha_j}}$

if j is even. The sum extends over all possible ways of arranging the Z's in pairs. As an example, we consider the ensemble average of four V's.

$$\overline{VVVV} = \underbrace{V\ V}\ \underbrace{V\ V} + \underbrace{V\ \underbrace{V\ V}\ V} + \underbrace{V\ \overbrace{V\ V}\ V} \qquad (III.17)$$

This can be written explicitly as

$$\sum_{m\ell k} \overline{V_{jm}\ G_m\ V_{m\ell}\ G_\ell\ V_{\ell k}\ G_k\ V_{ki}} = \delta_{ij}(\sum_m \overline{V^2_{im}}\ G_m)\ G_i\ (\sum_k \overline{V^2_{ik}}\ G_k) \qquad (III.18)$$

$$+ \delta_{ij} \sum_m (\overline{V^2}_{im}\ G^2_m(\sum_\ell \overline{V^2}_{m\ell}\ G_\ell)) + \delta_{ij}(\sum_m \overline{(V^2}_{im}G_m)^2)G_i$$

For sufficiently high energy, we replace the summation over the intrinsic states by

$$\sum_m \rightarrow \int d\varepsilon_m\ \rho(\varepsilon_m) \qquad (III.19)$$

with $\rho(\varepsilon_m)$ the level density at energy ε_m. Each integral gives a contribution of $\rho(\varepsilon_m)\Delta E$, with $\hbar/\Delta E$ being one of the time τ_Δ, τ_σ, or τ_λ introduced previously. Another time scale of interest is the Poincare recurrence time given by $\tau_{poincare} \approx h\rho(\varepsilon)$. For heavy nuclei, the level density at neutron thershold is $\approx 10^5$/MeV and it increases exponentially. It is therefore easy to see that $\tau_{poincare} >> max(\tau_\lambda, \tau_\Delta, \tau_\sigma)$. This immediately implies that the last term in Eq. (III.18) is negligibly small in comparison with the first two terms, i.e. we can safely neglect in Eq. (III.17) the term where two contraction lines interesct.

Let us consider how the free Green's function is modified when energy averaging is carried out. Collecting the series

$$G + G\overbrace{V\ G\ V}G + \overline{GVGVGVGVG} + \ldots \qquad (III.20)$$

and using the rule mentioned above, we find that the series in Eq. (III.20), denoting as the optical-model Green's function G^{opt}, satisfies the equation

$$G^{opt} = G + G\ \overbrace{V\ G^{opt}\ V}\ G^{opt} \qquad (III.21)$$

Similarly, we collect the series of scattering wave functions and obtain optical-model wave function

$$\chi_a^{opt} = \chi_a + G\ \overbrace{V\ G^{opt}\ V}\ \chi_a^{opt} \qquad (III.22)$$

Eqs. (III.21) and (III.22) show that the optical-model potential has the form

$$V^{opt} = \overbrace{V\ G^{opt}\ V} \qquad (III.23)$$

In terms of χ^{opt} and G^{opt}, the energy average $\overline{|S_{ba}|^2}$ can be expressed as

$$|S_{ba}|^2 = \langle\chi_b^{opt}| V \{1 + G^{opt} V + G^{opt} V G^{opt} V + \ldots\}|\chi_a^{opt}\rangle \qquad (III.24)$$

$$* \langle\chi_a^{opt}|\{\ldots + V G^{opt*} V G^{opt*} + V G^{opt*} + 1\} V|\chi_b^{opt}\rangle$$

Defining an average density matrix $\bar{\rho}$ by the equation

$$\bar{\rho}_b = \delta_{ba}|\chi_a^{opt}\rangle\langle\chi_a^{opt}| + G_b^{opt} V \bar{\rho} V G_b^{opt*} \qquad (III.25)$$

then the average cross section is proportional to

$$|S_{ba}|^2 = \langle\chi_a^{opt}| V \bar{\rho} V | \chi_b^{opt}\rangle \qquad (III.26)$$

The evaluation of the average cross-section has thus been reduced to two integral equations, Eqs. (III.21) and (III.25). It can also be obtained by taking the asymptotic value of the average density matrix $\bar{\rho}$.

To derive a transport equation from (III.25), we write it explicitly in the coordinate space, i.e.

$$\bar{\rho}_b(\vec{r},\vec{r}') = \delta_{ba}|\chi_a^{opt}(\vec{r})\rangle\langle\chi_a^{opt}(\vec{r}')| \qquad (III.27)$$

$$+ \sum_s \int d^3\vec{r}_1 \int d^3\vec{r}_1' \, G_b(\vec{r},\vec{r}_1) V_{bs}(\vec{r}_1)\bar{\rho}_s(\vec{r}_1,\vec{r}_1')V_{sb}(\vec{r}_1')G_b^*(\vec{r}_1',\vec{r}')$$

We apply the operator $(E - \varepsilon_b - H_o)$ to the right and to the left of Eq. (III.27), and take the difference of the resulting equations. Making use of the defining equations for the optical-model Green's function and the optical-model wave, function, we find

$$\{-\frac{\hbar^2}{2\mu}(\nabla^2_{\vec{r}_1} - \nabla^2_{\vec{r}_2}) + V(\vec{r}_1) - V(\vec{r}_2)\} \, \bar{\rho}_s(\vec{r}_1,\vec{r}_2)$$

$$= \sum_t \int d^3\vec{r}_1' \, G_s(\vec{r}_1,\vec{r}_1') V_{st}(\vec{r}_1') \bar{\rho}_t(\vec{r}_1',\vec{r}_2) V_{ts}(\vec{r}_2)$$

$$- \sum_t \int d^3\vec{r}_2' \, V_{st}(\vec{r}_1) \bar{\rho}_t(\vec{r}_1,\vec{r}_2') V_{ts}(\vec{r}_2') G_s^*(\vec{r}_2',\vec{r}_2)$$

$$- \sum_t \int d^3\vec{r}_1' \, V_{st}(\vec{r}_1) G_t(\vec{r}_1,\vec{r}_1') V_{ts}(\vec{r}_1') \bar{\rho}_s(\vec{r}_1',\vec{r}_1)$$

$$+ \sum_t \int d^3\vec{r}_2' \, \bar{\rho}_s(\vec{r}_1,\vec{r}_1') V_{st}(\vec{r}_2') G_t^*(\vec{r}_2',\vec{r}_2) V_{ts}(\vec{r}_2) \qquad (III.28)$$

We now make the following transformation of variables

$$\vec{R} = \frac{1}{2}\,(\vec{r}_1 + \vec{r}_2)\;,\quad \vec{r} = \vec{r}_1 - \vec{r}_2 \tag{III.29}$$

and introduce the Wigner transformation of the density matirx

$$F_s(\vec{R},\vec{k}) = (2\pi)^{-3} \int d^3\vec{r}\; e^{-i\vec{k}\cdot\vec{r}}\; \rho_s(\vec{R} + \frac{1}{2}\vec{r}, \vec{R} - \frac{1}{2}\vec{r}) \tag{III.30}$$

Replacing in Eq. (III.28) $\vec{r}_1$ and $\vec{r}_2$ by $\vec{R}$ and $\vec{r}$, and performing the Fourier transform of the resulting equation with respect to $\vec{r}$, we obtain the transport equation for F

$$\{\frac{\hbar^2}{\mu}\,\vec{k}\cdot\vec{\nabla}_R - (\vec{\nabla}_R V(\vec{R}))\cdot\vec{\nabla}_k\}\, F_s(\vec{R},\vec{k}) = \sum_t \int d^3\vec{R'} \int d^3\vec{k'}\; \{G_{st}(\vec{R},,\vec{k};\vec{R'},\vec{k'})\, F_t(\vec{R'},\vec{k'})$$

$$- L_{st}(\vec{R},\vec{k'}\vec{R'},\vec{k'})\, F_s(\vec{R'},\vec{k'})\} \tag{III.31}$$

Here,

$$G_{st}(\vec{R},\vec{k};\vec{R'},\vec{k'}) = 2\pi^{-3}\,\mathrm{Im}\,\{\int d^3\vec{r'}\; e^{i\vec{r'}\cdot(\vec{k}-\vec{k'})}\; e^{2i\,\vec{k}\cdot(\vec{R}-\vec{R'})}$$

$$*\; V_{st}(\vec{R'} + \frac{1}{2}\vec{r'})\, V_{ts}(\vec{R'} - \frac{1}{2}\vec{r'})\, G_s{}^*(\vec{R'} + \frac{1}{2}\vec{r'}, 2\vec{R}-\vec{R'} + \frac{1}{2}\vec{r'})\} \tag{III.32}$$

and

$$L_{st}(\vec{R},\vec{k};\vec{R'},\vec{k'}) = 2\pi^{-3}\,\mathrm{Im}\{\int d^3\vec{r'}\; e^{i\vec{r'}\cdot(\vec{k}-\vec{k'})}\; e^{2i\vec{k}\cdot(\vec{R}-\vec{R'})}$$

$$*\; V_{st}(2\vec{R}-\vec{R'} + \frac{1}{2}\vec{r'})\, V_{ts}(\vec{R'} + \frac{1}{2}r')\, G_t{}^*(2\vec{R}-\vec{R'} + \frac{1}{2}\vec{r'}, \vec{R'} + \frac{1}{2}\vec{r'})\} \tag{III.33}$$

and are denoted as the "gain" and the "loss" terms, respectively. In obtaining Eq. (III.31), we have neglected all derivatives of $V(\vec{R})$ higher than the first. This is justified if the wavelength of the relative motion is shorter in comparison with the typical distance over which the potential changes. This condition is well satisfied in heavy-ion reactions.

In the semiclassical limit, the Wigner distribution function F is equal to the joint distribution probability for finding the two heavy ions with relative distance $\vec{R}$ and momentum $\hbar\vec{k}$, and with the intrinsic excitation at state s. Eq. (III.31) is a stationary transport equation for F. The boundary condition to be supplemented is that a beam of particles impinging with the initial momentum $\hbar\vec{k}_a$, both fragments being in a state of excitation a. At the beginning of the process, all Coulomb trajectories, which are asymptotically in parallel with $\hbar\vec{k}_a$ and have energy $E-\varepsilon_a$, are populated with equal weight. In the interaction region, these trajectories are depopulated while other trajectories corresponding to different intrinsic excitation are populated. Due to the density-of-state factor, trajectories corresponding to higher intrinsic excitation energy are favored, leading to the loss of the relative kinetic energy. At the same time, the original well-defined impact parameter and momentum become more spread. At the end of the interaction region, we have therefore a distribution spread in intrinsic energy, over many R-values, and over many values of $h\vec{k}$.

It is easy to verify that

$$\sum_s \int d^3\vec{k}\; \mu^{-1}\,\vec{k}\cdot\nabla_R\,F_s(\vec{R},\vec{k}) = 0 \tag{III.34}$$

This states that the divergence of the current of the probability density is zero. This is the same as overall probability conservation since we deal with a stationary problem.

III.4 Solution of the Transport Equation in One-Dimension

In order to exhibit explicitly the structure of the transport equation Eq. (III.31), we shall in the following consider an one-dimensional model and assume that the conservative potential is absent [86,87]. If we make use of the approximation that the density overlap function $f(z)$ changes slowly with z, then the transport equation takes the form

$$k\frac{\partial}{\partial z}F_s(z,k) = \frac{4\mu W_0}{h^2}\int d\varepsilon_t\left(\frac{D_s}{D_t}\right)^{1/2} e^{-(\varepsilon_s-\varepsilon_t)/2\Delta^2} \tag{III.35}$$

$$*\int dk'\left(\frac{\sigma}{(2\pi)^{1/2}}\right) e^{-(k-k')^2\sigma^2/2}\int_{-\infty}^z dz'\, f(z')$$

$$*\left[F_t(z',k')\mathrm{Im}\left\{e^{2ik(z-z')}\,G_s^{opt^*}(2(z-z'))\right\}\right.$$

$$\left.-F_s(z',k')\,\mathrm{Im}\left\{e^{2ik'(z-z')}G_t^{opt^*}(2(z-z'))\right\}\right]$$

The optical-model Green's function is denoted by $G_s^{opt}(z,z')$ and satisfies the equation

$$\left(E-\varepsilon_s+\frac{\hbar^2}{2\mu}\frac{\partial^2}{\partial z^2}\right)G_s^{opt}(z,z') = \delta(z-z')$$

$$+ W_0\int d\varepsilon_t(D_s/D_t)^{1/2}\,e^{-(\varepsilon_s-\varepsilon_t)^2/2\Delta^2}\int_{-\infty}^z dz''\, f\left(\frac{z+z''}{2}\right)$$

$$*\,e^{-(z-z'')^2/2\sigma^2}\,G_t^{opt}(z,z'')\,G_s^{opt}(z'',z') \tag{III.36}$$

Let us assume that the optical-model Green's function is a product of two terms

$$G_s^{opt}(z,z') = G_s^0(z,z') \, g(z,z') \tag{III.37}$$

where $G_s^0(z,z')$ is the free Green's function

$$G_s^0(z,z') = - \frac{i\mu}{\hbar^2 k_s} e^{ik_s|z-z'|} \tag{III.38}$$

with $E = \varepsilon_s + \hbar^2 k_s^2/2\mu$. We first expand $f(z'')$ in a Taylor series about z and keep only the leading term. We insert the ansatz Eq. (III.37) into Eq. (III.36) and multiply it by $(E - \varepsilon_s + (\hbar^2/2\mu)(\partial^2/\partial z^2))^{-1}$. Then we divide the resulting equation by $G_s^0(z,z')$. Differentiating the result with respect to z, we obtain, for $z > z'$, the following equation for $g(z,z')$

$$\frac{\partial}{\partial z} g(z,z') = - \frac{\mu^2 W_0}{\hbar^4} f(z) \int d\varepsilon_t \, (D_s/D_t)^{1/2} \, e^{-(\varepsilon_s-\varepsilon_t)^2/2\Delta^2} \tag{III.39}$$

$$* \, (k_s k_t)^{-1} \int_{z'}^z dz'' \, e^{-(z-z'')^2/2\sigma^2} \, g(z,z'')g(z'',z') * e^{i(k_t-k_s)(z-z'')}$$

In deriving the above equation, we have neglected terms like $e^{i(k_s+k_t)(z-z'')}$, which oscillate rapidly in comparison with terms like $e^{i(k_s-k_t)(z-z'')}$. The ratio D_s/D_t can be approximated by $e^{\beta(\varepsilon_s-\varepsilon_t)}$ where $\beta^{-1} = kT$ with T the nuclear temperature. Putting $k_s k_t \simeq k_s^2$ and $k_t - k_s \simeq \gamma(\varepsilon_s-\varepsilon_t)$ with $\gamma = \mu/(h^2 k_s)$, then

$$\frac{d}{dz} g(z,z') = - (2\pi)^{1/2} \mu^2 W_0 \Delta f(z)/(\hbar^4 k_s^2) \tag{III.40}$$

$$* \int_{z'}^z e^{-(z-z'')^2/2\sigma^2} e^{\frac{\Delta^2}{2}(i\frac{1}{2}\beta+\gamma(z-z''))^2} g(z,z'')g(z'',z')dz''$$

The solution to Eq. (III.40) can be found for two limits. In the limit of weak coupling, i.e. $W_0 \to 0$, $g(z,z')$ decays exponentionally over distances larger than the width of the kernel. Writing

$$g(z,z') = e^{-\eta|z-z'|} \tag{III.41}$$

then we find

$$\mathrm{Re}\ \eta = \frac{\pi W_0 \tilde{\Delta} f(z)}{\hbar^4 k_s^2} e^{\frac{1}{8}(\beta\tilde{\Delta})^2} \tag{III.42}$$

$$\tilde{\Delta} = \Delta^2(1 + \sigma^2\Delta^2\gamma^2)^{-1}$$

The imaginary part of η modified only slightly the value of k_s and is neglected. In the strong-coupling limit, $g(z,z')$ is narrower than the width of the kernel, and we replace the two Gaussians by unity. This yields

$$g(z,z') = e^{-(z-z')^2/2\Lambda^2} \tag{III.43}$$

with

$$\Lambda^{-1} = \mu\left[\left(\frac{\pi}{2}\right)^{1/4} (\pi\Delta W_0 f(z))\right]^{1/2} e^{\beta^2\Delta^2/16} /\hbar^2 k_s \tag{III.44}$$

The conditions for weak and for strong coupling are, $\eta^{-1} \gg \min\left[\sigma, \Delta^{-1}\gamma^{-1}\right]$ and $\Lambda \ll \min\left[\sigma, \Delta^{-1}\gamma^{-1}\right]$, respectively. In heavy ion collisions, using $f(z) = 1$. $\beta \simeq 1$ MeV^{-1}, $\Delta = 7$ MeV, $\sigma = 3.5$ fm, and $W_0 = 10$ MeV, we have $\lambda \approx 0.4$ fm while $(\Delta\gamma)^{-1} \approx 4$ fm. Even for $f(z) \approx 0.1$, we still have the strong coupling limit because of the exponential factor in Eq. (III.42).

To find the solution of the transport equation Eq. (III.35) we introduce the function

$$\Pi_s(z,k) = k\,\rho(\varepsilon_s)\,F_s(z,k) \tag{III.45}$$

where $\rho(\varepsilon_s) = D_s^{-1}$ is the level density. In the strong-coupling limit, we have

$$\frac{\partial}{\partial z} \Pi(z,k) = \frac{\mu^2 W_0 \Lambda\sigma}{\hbar^4} f(z) \int d\varepsilon\ e^{-(\varepsilon_s-\varepsilon_t)^2/2\Delta^2}$$

$$* \int dk'\ e^{-(k-k')^2\sigma^2/2} (k_s k_t)^{-1} *$$

$$* \left\{\frac{k_t}{k}\Pi_t(z,k')\left(\frac{\rho(\varepsilon_s)}{\rho(\varepsilon_t)}\right)^{1/2} e^{-(k-k_s)^2\Lambda^2/2}\right.$$

$$\left. - \frac{k_s}{k}\Pi_s(z,k)\left(\frac{\rho(\varepsilon_t)}{\rho(\varepsilon_s)}\right)^{1/2} e^{-(k-k_t)^2\Lambda^2/2}\right\} \tag{III.46}$$

Assuming $\Pi_s(z,k)$ has Gaussian distribution, then its first and second moment can be solved. Multiplying Eq. (III.46) by k and integrating over ε_s and k, we find,

$$\frac{d}{dz}\bar{k}(z) = f(z)\,\bar{v}_k(z) \tag{III.47}$$

Here,

$$\bar{k}(z) = \int d\varepsilon_s\, dk\, \Pi_s(z,k)k \tag{III.48}$$

and

$$\vec{v}_k(z) = \left(\mu^2 W_0 \Lambda\, \sigma/\hbar^4\right) \int d\varepsilon_t \int dk' e^{-(\varepsilon_s-\varepsilon_t)^2/2\Delta^2}$$

$$* e^{-(k-k')^2\sigma^2/2} k_t^{-1} e^{-(k'-k_t)^2\Lambda^2/2}$$

$$* \ \left(\frac{\rho(\varepsilon_t)}{\rho(\varepsilon_s)}\right)^{1/2} \ \frac{1}{k} \ (k'-k) \tag{III.49}$$

We can similarly obtain equations for $\overline{\varepsilon_s}(z)$, $\overline{k^2}(z)$, $\overline{k\varepsilon_s}(z)$, $\overline{\varepsilon_s^2}(z)$ and calculate the corresponding transport coefficients, the cross-section from the initial ground state to the final state s is then given by

$$\sigma_{0\to s} = k_0^{-1} \lim_{z\to\infty} \int dk \ kF_s(z,k) \tag{III.50}$$

The behavior of the solutions of the transport equation is determined by three factors.

1. The level-density factor favors the population of high intrinsic states, but limited by Δ because of the factor $e^{-(\varepsilon_s-\varepsilon_t)^2/2\Delta^2}$.

2. The distribution in relative momentum k is determined by the exponential $e^{-(k-k')^2\sigma^2/2}$. This only produces a widening of the distribution but no shift is expected because of the lack of a phase factor as in the case of ε_s.

3. The distributions of k and ε_s are coupled by the exponential $e^{-(k-k_s)^2\Lambda^2/2}$. In the weak coupling limit, it becomes a delta function and the two distributions coincide. However, for strong coupling, the two distributions are coupled. The increase of the centroid of the k_s distribution, due to the increasing level density with the excitation energy, then pulls with it the centroid of of the k distribution. But the increase of the latter is restricted by σ and thus exerts a drag on the k_s distribution. The larger σ is, the stronger is the restriction on the increase of the relative momentum and the larger is the effect of this drag on the k_s distribution.

It was shown numerically that the mean free path Λ depends most strongly on the parameter Δ. On the other hand, the transport coefficients and the cross sections depend more strongly on the parameter σ.

Due to the strong coupling in the process, the Einstein relation between the diffusion and drift coefficients is badly violated. This relation reads

$$\overline{kD_{kk}} = -\frac{2\mu}{\hbar^2} \ T \ \overline{v_k} \tag{III.51}$$

where the temperature T is given by $T = \bar{\varepsilon}/a$ with a the level density parameter. Numerical calculation indicates that the reaction $-\overline{kD_{kk}}/\overline{v_k}$ can be a factor of 5 to 10 larger than $2\mu/\hbar^2\ T$.

III.5   Calculation of Cross Sections

The transport Eq. (III.31) can be generalized to include the charge exchange between projectile and target [84,88]. This is done by introducing a variable

$x = (Z_1-Z_2)/(Z_1+Z_2)$ in terms of the charge Z_1 and Z_2 of the projectile and target, respectively. The statistical assumptions on the form factors given by Eq. (III.6) can be extended to depend on x. A further parameter δ is therefore introduced to determine the width of the second moment of the form factors with respect to x. The resulting transport equation for the distribution function $F_s(\vec{R},\vec{k},x)$ is very similar to Eq. (III.31) except that the right hand side also involves a summation over x', with G_{st} and L_{st} depending upon x and x'.

In order to compare the predictions of the statistical transport theory with experimental observations, we need to calculate the cross section from the transport equation (III.31). If we normalize the distribution function $F_s(\vec{R},\vec{k},x)$ to unit incident flux, we find for the cross section in an energy interval between ε_b and $\varepsilon_b + d\varepsilon_b$ and with a charge asymmetry betwen x and x + dx

$$\frac{d^3\sigma}{d\varepsilon_b d\Omega dx} = \left(\frac{\hbar}{\mu_b v_a}\right) \rho_0(\varepsilon_b) \lim_{R\to\infty} R^2 \int d^3\vec{k}\ (\vec{e}_R \cdot \vec{k})\ F_b(\vec{R},\vec{k},x) \qquad (III.52)$$

Here, μ_b is the reduced mass in channel b. The unit vector $\vec{e}_R = \vec{R}/R$ points in the direction of Ω_b, and v_a is the initial relative velocity.

The distribution function $F_b(\vec{R},\vec{k},x)$ satisfies the transport equation Eq. (III.31). As discussed in the previous section, deep inelastic heavy-ion collisions are described by a strong coupling theory. In this case, we should use the optical-model Green's function given by Eqs. (III.37) and (III.43). This has not been carried out for the three dimensional case. In Ref. [89], the optical-model Green's function in the weak coupling limit was used with an appropriate renormalization of the coupling strength W_o defined in Eq. (III.6). It was demonstrated that in one-dimensional case this procedure could reproduce reasonably well the results one would obtain with a strong coupling optical-model Green's function. Three approximations have been introduced in in order to solve the transport equation. (i) It is assumed that on the right hand side of Eq. (III.31), $F_s(\vec{R},\vec{k}')$ and and $F_s(\vec{R}',\vec{k}')$ may be replaced by $F_t(\vec{R},\vec{k}')$ and $F_t(\vec{R},\vec{k}')$, respectively. (ii) It is assumed that $F_s(R,k)$ is on shell throughout so that it is proportional to a δ-function of $E - \frac{\hbar^2 k^2}{2\mu} - \varepsilon_s - V(R)$, where E is the incident center-of-mass energy. This assumption is equivalent to a semiclassical approximation and thus holds only if the quantum-mechanical uncertainty is sufficiently small. (iii) It is assumed that $F_s(\vec{R},\vec{k})$ has a Gaussian spread about the mean values, so that only the mean values and the second moments of momentum and position need be calculated. All three approximations were tested and found very satisfactory in one-dimension.

With these approximations, the transport equation can be transformed into the Fokker-Planck equation

$$\left(\frac{\hbar^2}{\mu} \vec{k}\cdot\vec{\nabla}_R - \vec{\nabla}_R V\cdot\vec{\nabla}_k\right) P(\vec{R},\vec{k},x) =$$

$$- \sum_{i=1}^{3} f(R) \frac{\partial}{\partial k_i} (v_i P) + \frac{1}{2} \sum_{i,j=1}^{3} f(R) \frac{\partial^2}{\partial k_i \partial k_j} (D_{ij} P)$$

$$+ \frac{1}{2} f(R) \frac{\partial^2}{\partial x^2} (D_{xx} P) \qquad\qquad (III.53)$$

The distribution function $P(\vec{R},\vec{k},x)$ is related to $F_s(\vec{R},\vec{k},x)$ by the relation

$$F_s(\vec{R},\vec{k},x) = \frac{\mu}{\hbar^2 k_s} \delta(k-k_s) \rho_0^{-1}(\varepsilon_s) P(\vec{R},\vec{k},x) \qquad\qquad (III.54)$$

with k_s defined by $E = \varepsilon_s + \dfrac{\hbar^2 k_s^2}{2\mu} + V(R)$. The drift coefficient $\vec{v}$ and the diffusion coefficients D_{ij} are given by

$$\vec{v} = \vec{k}\, v(k) \qquad\qquad (III.55.a)$$

$$D_{ij} = \delta_{ij} D_0(k) + (k_i k_j - \tfrac{1}{3} k^2) k^{-2} D_2(k) \qquad\qquad (III.55.b)$$

with

$$V(k) = -(2\pi)^{1/2} W_0 \sigma(\sigma k)^{-2} \int dk' A(k,k') \{1 + \sigma^2 k(k-k')\} \qquad\qquad (III.56.a)$$

$$D_0(k) = (2\pi)^{1/2} W_0 \sigma^{-1} \tfrac{1}{3} h(k) \qquad\qquad (III.56.b)$$

$$D_2(k) = (2\pi)^{1/2} W_0 \sigma^{-1} \{ \tfrac{2}{3} (\sigma k)^{-2} g(k) - \tfrac{1}{2} h(k)\} \qquad\qquad (III.56.c)$$

In the above, we have defined the functions

$$A(k,k') = \frac{k'}{k} \left[\frac{\rho_0(\varepsilon(k'))}{\rho_0(\varepsilon(k))} \right]^{1/2} * \exp\{-\tfrac{1}{2} \sigma^2 (k-k')^2\} \exp\{-(\varepsilon(k)-\varepsilon(k'))^2/2\Delta^2\} \quad (III.57.a)$$

$$h(k) = \int dk' A(k,k') \{2 + \sigma^2 (k-k')^2\} \qquad\qquad (III.57.b)$$

$$g(k) = \int dk' A(k,k') \{\sigma^4 k^2 k'^2 - 2\sigma^2 kk' + 2 + \sigma^2 k^2 (\sigma^2 k^2 - 2(\sigma^2 kk' - 1))\} \qquad (III.57.c)$$

In addition, the charge drift coefficient D_{ij} is given by

$$D_{xx} = (2\pi)^{1/2} W_0 \delta^2 \int dk' A(k,k') \qquad\qquad (III.58)$$

Using the spherical coordinates (r,θ,ϕ), we define

$$P_r = (\hbar \vec{k} \cdot \vec{e}_r), \quad P_\theta = r(\hbar \vec{k} \cdot \vec{e}_\theta), \quad P_\phi = r \sin\theta (\hbar \vec{k} \cdot \vec{e}_\phi) \qquad\qquad (III.59)$$

where $\vec{e}_r$, $\vec{e}_\theta$, and $\vec{e}_\phi$ are the three unit vectors. Since the problem possesses cylindrical symmetry, we introduce a new distribution function

$$\Pi(r,\theta;P_r,P_\theta,P_\phi;x) = P_r \int_0^{2\pi} d\phi \ P(r,\theta,\phi;P_r,P_\theta,P_\phi;x) \tag{III.60}$$

This definition is motivated by the fact that for large R, the momentum $\vec{P}$ attains a purely radial direction. Therefore, the conservation law Eq. (III.34) implies that for $r \rightarrow \infty$,

$$\frac{d}{dr} \int dV \ \Pi(r,\theta;P_r,P_\theta,P_\phi;x) = 0 \tag{III.61}$$

where the volume element dV is given by $d\theta \ dP_r \ dP_\theta \ dP_\phi \ dx$.

We define the mean values $\bar{u}$ and the half variances Δ_{uv} by

$$\bar{u} = \int dV \ u \ \Pi \tag{III.62.a}$$

$$\Delta_{uv} = \frac{1}{2} \int dV \ (u-\bar{u})(v-\bar{v})\Pi \tag{III.62.b}$$

It is straightforward to find the equations of motion for the mean values

$$\frac{dr}{dt} = \frac{\bar{P}_r(r)}{\mu} \tag{III.63.a}$$

$$\frac{d\bar{\theta}}{dt} = \frac{\bar{P}_\theta}{\mu r^2} \tag{III.63.b}$$

$$\frac{d\bar{P}_r}{dt} = - \frac{dV}{dr} + \frac{\bar{P}_\theta^{\ 2}}{\mu r^2} + \frac{\bar{v}}{\hbar} f(r)\bar{P}_r \tag{III.63.c}$$

$$\frac{d\bar{P}_\theta}{dt} = \frac{\bar{v}}{\hbar} f(r) \ \bar{P}_\theta \tag{III.63.d}$$

$$\frac{d\bar{P}_\phi}{dt} = 0 \tag{III.63.e}$$

$$\frac{dx}{dt} = 0 \tag{III.63.f}$$

In deriving the above equations, we have approximated $\int dV \ \Pi \ v \ P_r$ by $\bar{v} \ \bar{P}_r = v(\bar{R})\bar{P}_r$ and similarly for P_θ. We have used the fact that $\bar{P}_\phi = 0$ at $t = 0$ because of the initial conditions and it therefore stays zero. We have also neglected terms involving $\bar{P}_\phi^{\ 2}$ since $\bar{P}_\phi = 0$ and the variance introduces only a small correction.

Equation (III.63.a) defines the time parameter for the process. The next four equations have the form of classical Newtonian equations of motion with both radial and tangential frictional forces. The last equation shows that the charge drift vanishes. This is approximately true for the reactions we shall discuss later.

Using simlar approximations introduced above, we find the equations of motion for the half variances

$$\frac{d\Delta_{\theta\theta}}{dt} = \frac{2\Delta_{\theta P_\theta}}{\mu r^2} \tag{III.64.a}$$

$$\frac{d\Delta_{\theta P_r}}{dt} = \frac{\Delta_{P_r P_\theta}}{\mu r^2} + \frac{2\bar{P}_\theta \Delta_{\theta P_\theta}}{\mu r^3} + \frac{\bar{v}}{\hbar}\, f(r)\, \Delta_{\theta P_r} \tag{III.64.b}$$

$$\frac{d\Delta_{\theta P_\theta}}{dt} = \frac{\Delta_{P_\theta P_\theta}}{\mu r^2} + \frac{\bar{v}}{\hbar}\, f(r)\, \Delta_{\theta P_\theta} \tag{III.64.c}$$

$$\frac{d\Delta_{P_r P_r}}{dt} = \frac{4\bar{P}_\theta \Delta_{P_r P_\theta}}{\mu r^3} + 2\frac{\bar{v}}{\hbar}\, f(r)\, \Delta_{P_r P_r} + \frac{1}{2} f(r)\left(\bar{D}_0 + P_r^{\,2}\bar{D}_2\right) \tag{III.64.d}$$

$$\frac{d\Delta_{P_r P_\theta}}{dt} = \frac{2\bar{P}_\theta \Delta_{P_\theta P_\theta}}{\mu r^3} + 2\,\frac{\bar{v}}{\hbar}\, f(r)\, \Delta_{P_r P_\theta} + \frac{1}{2} D_2 f(r)\bar{P}_r\bar{P}_\theta \tag{III.64.e}$$

$$\frac{d\Delta_{P_\theta P_\theta}}{dt} = 2\,\frac{\bar{v}}{\hbar}\, f(r)\, \Delta_{P_\theta P_\theta} + \frac{1}{2} f(r)\left(\bar{D}_0 r^2 + \bar{D}_2 \bar{P}_\theta^{\,2}\right) \tag{III.64.f}$$

$$\frac{d\Delta_{P_\phi P_\phi}}{dt} = 2\,\frac{\bar{v}}{\hbar}\, f(r)\, \Delta_{P_\phi P_\phi} + \frac{1}{2} f(r)\left(\bar{D}_0 r^2 \sin^2\bar{\theta} + \bar{D}_2 \bar{P}_\phi^{\,2}\right) \tag{III.64.g}$$

$$\frac{d\Delta_{\theta P_\phi}}{dt} = \frac{d\Delta_{P_r P_\phi}}{dt} = \frac{d\Delta_{P_\theta P_\phi}}{dt} = 0 \tag{III.64.h}$$

$$\frac{d\Delta_{xx}}{dt} = \frac{1}{2} f(r)\, \frac{\bar{D}_{xx}}{\hbar} \tag{III.64.i}$$

Equation (III.64.h) is a consequence of the initial conditions. We also have $\Delta_{x\alpha} = 0$ for $\alpha = (\theta, P_r, P_\theta, P_\phi)$.

In terms of the solutions of Eqs. (III.63) and (III.64) at time $t = +\infty$, the differential cross section in the c.m. system given by Eq. (III.52) can be transformed into the form

$$\frac{d^3\sigma}{dE_f d\Omega dx} = 2\pi\, v_f^{-1} \int_0^\infty dbb\, (4\pi\Delta_{xx})^{-1/2}\, \exp\left\{-\frac{(x-\bar{x})^2}{4\Delta_{xx}}\right\} \tag{III.65}$$

$$* \; [4\pi\Delta_{\theta\theta}\Delta_{P_r P_r} - \Delta_{\theta P_r}^2)^{1/2}]^{-1} \exp\{- \; \tfrac{1}{4}(\Delta_{\theta\theta}\Delta_{P_r P_r} - \Delta_{\theta P_r}^2)^{-1}$$

$$* \; [\Delta_{\theta\theta}((2\mu E_f)^{1/2} - \bar{P}_r)^2 + \Delta_{P_r P_r}(\theta - \bar{\theta})^2 - 2\Delta_{P_r\theta}((2\mu E_f)^{1/2} - \bar{P}_r)(\theta - \bar{\theta})]\}$$

Here, b is the impact parameter and v_f is the final velocity of the relative motion corresponding to a kinetic of E_f.

In Refs. [89-91], Eq. (III.63) and (III.64) were solved using the proximity nuclear potential [26] and a form factor f(r) which is given by the overlap of the two nuclear density distributions. In Fig. III.1, we show the calculated double differential cross section for K ion from the reaction ^{40}Ar $\cdot + ^{232}$Th at 388 MeV. Comparing with the experimental Wilzinski plot shown in Fig. I.1, we see that the overall agreement is good. The double ridge is reproduced, and so are the values of the cross sections in the upper ridge. The forward peak occurs at too high an energy and has too low a cross section. This is because we are close to the Coulomb barrier and the entire second ridge lies essentially below the Coulomb barrier. Its appearance in our calculation is merely due to the "Gaussian shadow" of the

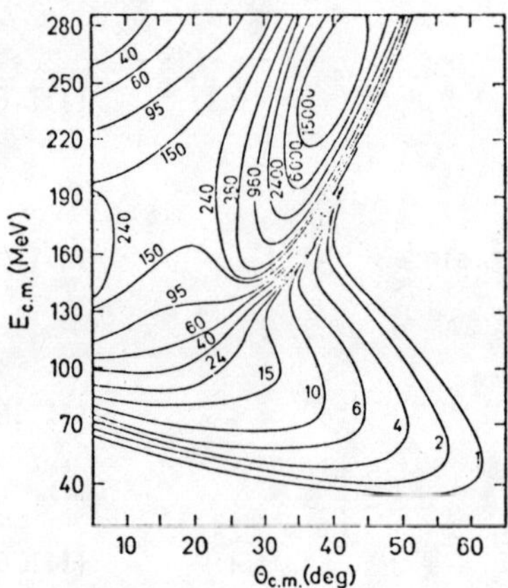

Fig. III.1. Wilzynski plot calculated from the statistical transport theory for the ion K from the reaction ^{40}Ar + ^{223}Th at 388 MeV [89].

first ridge. To reproduce the lower ridge requires collective modes not contained in our approach. In Fig. III.2, we show the charge-integrated angular distribution for various final energies from the reaction ^{136}Xe + ^{209}Bi at an incident energy of 1130 MeV [91]. The eight dashed curves in the figure, obtained from calculations, are in one-to-one correspondence with the eight experimental curves. While the disagreement for the first three curves never exceeds a factor of 2, increasing discrepancies build up as the energy of the reaction products decreases. This is again due to the neglect of other collective modes such as the deformation. In Ref. [92], the transport theory is generalized to include phenomenologically nuclear deformations. Calculated cross sections for the reaction ^{136}Xe + ^{209}Bi show substantial improvements over previous results obtained without deformations [93]. One of the results is given in Fig. III.3 which shows that even the low energy experimental cross sections

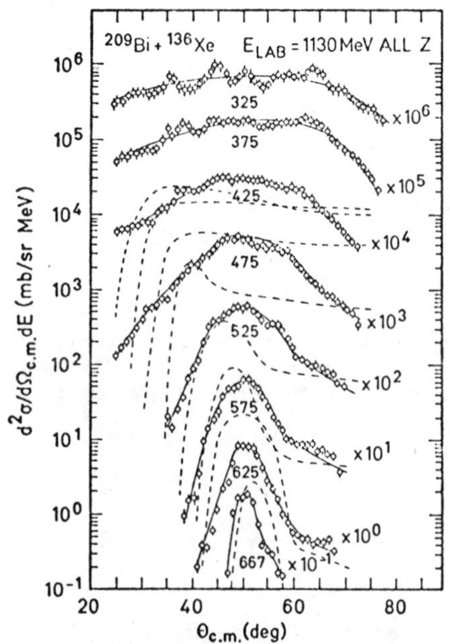

Fig. III.2. Charge-integrated angular distributions for different final kinetic energy bins. Solid curves are from the experiment [94], dashed curves are from calculations of Ref. [91].

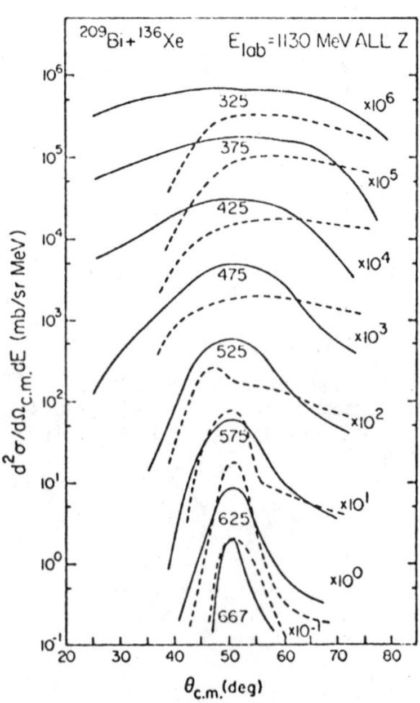

Fig. III.3. Same as Fig. II.2 with inclusion of deformation in theoretical calculations [93].

are reasonably reproduced. The lack of small-angle cross sections at large energy losses is due to the repulsive Coulomb barrier and can be removed if rotation of the intermediate complex were taken into account. Another example of the effect of nuclear deformation on the energy loss in heavy-ion collisions is shown in Fig. III.4. Here, we show the angular integrated energy loss spectrum in the same reaction ^{136}Xe + ^{209}Bi. The experimental spectrum, shown by the solid curve, has a peak at large energy loss. This peak is not obtained from a calculation without nuclear deformation as shown by the dashed curve. When nuclear deformations are included, one obtains the long-dashed curve which indeed exhibit a peak at low relative kinetic energy of the two outgoing nuclei.

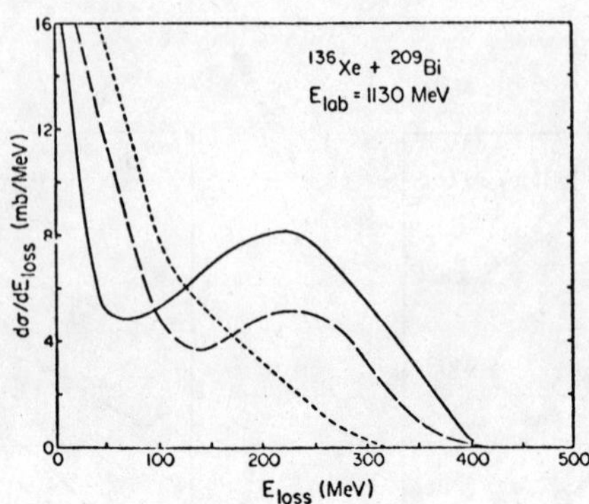

Fig. III.4. Energy-loss spectrum integrated over final angles. Solid curve is from the experiment of Ref. [94], short-dashed curve is from the calculation in Ref. [91], and long-dahsed curve is from calculation with deformation degree of freedom [93].

IV. HIGH ENERGY THEORY

IV.1 Introduction

In high energy heavy-ion collisions, the relative nucleon velocities exceed the sound velocities $\approx c/3$ in nuclear matter, and one expects the pile-ups of density. In order to achieve such a high density nuclear matter, high excitation energies $\approx 50-100$ MeV/nucleon have to be also deposited in the nuclear matter. Under these extreme conditions, one hopes that novel states of nuclear matter or unusual collective phenomena as described in Chapter I will then occur.

The effects of such collective phenomena on the experimental observables are not clear. One usually only measures a very small part of all final states. For example, only one or two particles are observed in the reactions without any attention to the other particles in the final state. It is therefore desirable to carry out model calculations of the collision processes to test the sensitivity of experimental results. In this way, one may obtain some idea of whether exotic phenomena do happen in the course of the collision.

Since both the linear and angular momentum involved in heavy ion collisions at high energies are very large, it is therefore conceivable that classical descriptions of the motion of all the nucleons may be appropriate [95,96]. If this is the case, then various approximations can be obtained depending on the time and length scales in the problem.

For high energy heavy ion collisions, the relevant time scales are

1) τ_{int} = duration of individual NN collision

          = force range/c $\approx O(\hbar/m_\pi c) \approx (1 - 2)$ fm/c

2) τ_{rel} = relaxation time between successive NN collisions

$$= \frac{\lambda}{v} = \frac{2\ \text{fm}}{(1/2\text{-}3/4)c} \approx (3 - 4)\ \text{fm/c}$$

3) τ_{col} = total collision

$$= \frac{L}{v} = \frac{10\ \text{fm}}{(1/2\text{-}3/4)c} \approx (10 - 20)\ \text{fm/c}$$

We therefore see that the following inequality holds:

$$\tau_{int} < \tau_{rel} < \tau_{col} \qquad . \tag{IV.1}$$

If $\tau_{int} \ll \tau_{rel}$, two-body collisions are isolated and no collective effect is involved. We therefore have the dilute gas limit. In this case, the collision process can be described by the Boltzmann equation or the so-called intranuclear cascade [97-102]. In such an approach, the essential physical input is the free space NN cross sections. If $\tau_{rel} \ll \tau_{col}$, many two-body collisions occur during the total collision time, we therefore expect the validity of local thermal equilibrium. In such a case, we have the hydrodynamic description [103-105]. The input is then the nuclear equation of state $W(\rho,T)$, which takes into account all the nuclear interactions, including binding and collective effects. On the other hand, if $\tau_{int} \ll \tau_{rel} \ll \tau_{col}$, then we can describe the two colliding heavy ions by hydrodynamical equations with an equation of state of an ideal gas.

In this chapter, we shall derive, based on the Glauber model, a Boltzmann equation for describing high energy heavy-ion collisions [106]. This offers a theoretical justification of the linear cascade model, which is a special case of the more general intranuclear cascade model. Then, we shall solve the linear cascade model using the rows on rows picture of Knoll and Hüfner [61]. Finally, the phase space model of Knoll [108] will be discussed.

IV.2 The Classical Transport Equation

Let us consider the collision between two heavy nuclei. Initially the projectile and the target are in their ground states $|0_P\rangle$ and $|0_T\rangle$, respectively. After the collision, they are scattered into the final states $|\alpha_P\rangle$ and $|\beta_T\rangle$, respectively. If the momentum exchanged between the nuclei is $\vec{k}$, then the cross section in the Glauber's theory is

$$\frac{d^2\sigma}{d\vec{k}}(0_P, 0_T \to \alpha_P, \beta_T) = \left| \frac{1}{2\pi} \int d^2b\, e^{-i\vec{k}\cdot\vec{b}} \langle \alpha_P \beta_T | 1 - \prod_{\substack{i\in T \\ j\in T}} (1 - \Gamma(\vec{b}+\vec{s}_i - \vec{s}_j)) | 0_P 0_T \rangle \right|^2 \quad (IV.2)$$

It is obvious that antisymmetrization between target and projectile is neglected in the above equation. The profile function Γ depends on the impact parameter $\vec{b}$, and the locations $\vec{s}_i$, $\vec{s}_j$ of the projectile and the target nucleons in the impact plane. It is related to the nucleon-nucleon cross section by

$$\frac{d^2\sigma_{NN}}{d\vec{q}} = \left| \frac{1}{2\pi} \int d^2b\, e^{-i\vec{q}\cdot\vec{b}} \, \Gamma(\vec{b}) \right|^2 \quad (IV.3)$$

We shall be interested in the one-nucleon inclusive cross section. In particular, we would like to calculate the cross section for observing one of the projectile nucleon with momentum $\vec{q}$. After summing over all projectile and target final states, it was shown [106] that this cross section can be written as

$$\frac{d^2\sigma_1}{d\vec{q}} = A_P \int d^2\vec{b} \, \langle 0_P 0_T | \left[\prod_{j\in T} (1-\Gamma_{j i_0}) \right]^+ P_{i_0}(\vec{q}) \quad (IV.4)$$

$$* \left(1 - |0_P\rangle\langle 0_T| \right) \prod_{i\in T} (1-\Gamma_{i i_0}) | 0_P 0_T \rangle$$

The operator $P(\vec{q})$ projects on a state with momentum $\vec{q}$, i.e.

$$\langle \vec{s}' \, | P(\vec{q}) | \, \vec{s} \rangle = \frac{1}{(2\pi)^2} \, e^{-i\vec{q} \cdot (\vec{s}-\vec{s}')} \tag{IV.5}$$

Since each projectile nucleon may be the observed nucleon, we have in Eq. (IV.4) the factor A_p, the number of projectile nucleons.

We introduce the definitions

$$\langle \vec{b}' \, | \, R \, | \, \vec{b} \rangle = \langle 0_T \, | [\prod_j (1-\Gamma(\vec{b}'-\vec{s}_j))]^+ \, [\prod_k (1-\Gamma(\vec{b}-\vec{s}_k))] | \, 0_T \rangle \tag{IV.6.a}$$

$$\langle \vec{b}' \, | \, R_0 \, | \, \vec{b} \rangle = \langle 0_T \, | \prod_j (1-\Gamma(\vec{b}'-\vec{s}_j)) | 0_T \rangle \langle 0_T \, | \prod_j (1-\Gamma(\vec{b}-\vec{s}_j)) | \, 0_T \rangle \tag{IV.6.b}$$

and the one-nucleon density matrix of the projectile

$$\langle \vec{s}' \, | \rho_P | \, \vec{s} \rangle = \int dz\, d^3\vec{x}_2 \, \cdots \, d^3\vec{x}_A \, \langle \vec{s}', z; \vec{x}_2 \cdots \vec{x}_A \, | \, 0_P \rangle \tag{IV.7}$$
$$* \, \langle 0_P \, | \, \vec{x}_A, \, \cdots \, \vec{x}_2; \vec{s}, z \rangle$$

then the one-nucleon inclusive cross section of Eq. (IV.4) can be written as

$$\frac{d^2\sigma_1}{d\vec{q}} = A_p \int d^2\vec{b} \int d^2\vec{s} \int d^2\vec{s}' \, \frac{e^{-i\vec{q} \cdot (\vec{s}-\vec{s}')}}{(2\pi)^2} \tag{IV.8}$$
$$* \, \{ \langle \vec{b}+\vec{s}' \, |R| \, \vec{b}+\vec{s} \rangle - \langle \vec{b}+\vec{s}' \, |R_0| \, \vec{b}+\vec{s} \rangle \} \, \langle \vec{s}' \, |\rho_P| \, \vec{s} \rangle \quad .$$

Making a change of variables and integrating over $\vec{b}$, we have

$$\frac{d^2\sigma_1}{d\vec{q}} = A_p \int d^2\vec{s} \int d^2\vec{s}' \, \frac{e^{-i\vec{q} \cdot (\vec{s}-\vec{s}')}}{(2\pi)^2} * \, \{ \langle \vec{s}' \, |R| \, \vec{s} \rangle - \langle \vec{s}' \, |R_0| \, \vec{s} \rangle \} \, \rho_P(s'-s) \tag{IV.9}$$

In the above, we have introduced the definition

$$\rho_P(\vec{s}'-\vec{s}) = \int d^2b \, \langle \vec{s}'+\vec{b} \, |\rho_P| \, \vec{s}+\vec{b} \rangle \quad . \tag{IV.10}$$

Introducing the transformation

$$W(\vec{S},\vec{q}) = \int d^2\vec{s} \, \frac{e^{-i\vec{q}\cdot\vec{s}}}{(2\pi)^2} \, \langle \vec{S}-\frac{1}{2}\vec{s} \, |R| \, \vec{S}+\frac{1}{2}\vec{s} \rangle \, \rho_P(\vec{s}) \tag{IV.11}$$

then the one-nucleon inclusive cross section is given by

$$\frac{d^2\sigma_1}{d\vec{q}} = A_p \int d^2\vec{S} \, \{ W(\vec{S},\vec{q}) - W_0(\vec{S},\vec{q}) \} \tag{IV.12}$$

Here, $W_0(\vec{S},\vec{q})$ is obtained from Eq. (IV.11) by replacing R with R_0.

According to Glauber [108], the optical phase shift function $\chi_{opt}(\vec{b})$ is related to the profile function

$$e^{i\chi_{opt}(\vec{b})} = \langle 0_T \mid \prod_{j \in T} (1-\Gamma(b-s_j)) \mid 0_T \rangle \qquad (IV.13)$$

In the absence of nucleon-nucleon correlations

$$\chi_{opt}(\vec{b}) \approx -A_T \int_\infty^\infty dz \; \rho_T(\vec{b},z) \int d^2\vec{s} \; \Gamma(\vec{b}-\vec{s}) \qquad (IV.14)$$

$$= -A_T \int^\infty dz \; \rho_T(\vec{b},z) \frac{1}{2} \sigma_{NN}^{tot} \left[i + \frac{Re \; f_{NN}(0)}{Im \; f_{NN}(0)} \right]$$

Here σ_{NN}^{tot} is the nucleon-nucleon total cross secstion, and $f_{NN}(0)$ is the nucleon-nucleon forward scattering amplitude.

In terms of $\chi_{opt}(\vec{b})$, we have

$$\langle \vec{s}' \mid R_0 \mid \vec{s} \rangle = e^{-i\chi_{opt}^*(\vec{s}')} \; e^{i\chi_{opt}(\vec{s})} \qquad (IV.15)$$

The operator R can be expressed as

$$\langle \vec{s}' \mid R \mid \vec{s} \rangle = e^{-i\chi_{opt}^*(\vec{s}')} \; e^{i\chi_{opt}(\vec{s})} \; e^{J(\vec{s}',\vec{s})} \qquad (IV.16)$$

Because of unitarity, we have

$$\langle \vec{s} \mid R \mid \vec{s} \rangle = 1 \qquad (IV.17)$$

which requires that

$$J(\vec{s},\vec{s}) = 2 \; Im \; \chi_{opt}(\vec{s}) \qquad (IV.18)$$

Eq. (IV.16) also has the correct asymptotic behavior, i.e.

$$\lim_{|\vec{s}| \to \infty} \langle \vec{s}' \mid R \mid \vec{s} \rangle = e^{-i\chi_{opt}^*(\vec{s}')} \qquad (IV.19.a)$$

$$\lim_{|\vec{s}'| \to \infty} \langle \vec{s}' \mid R \mid \vec{s} \rangle = e^{i\chi_{opt}(\vec{s})} \qquad (IV.19.b)$$

because the profile function $\Gamma(b)$ has a short range of ≈ 1 fm. In the absence of nucleon-nucleon correlation, one has

$$J(\vec{s},\vec{s}') \approx A_T \int_{-\infty}^\infty dz \; \rho_T(\frac{\vec{s}+\vec{s}'}{2},z) \int d^2b \; \Gamma^*(\vec{s}'-\vec{b}) \; \Gamma(\vec{s}-\vec{b}) \qquad (IV.20)$$

$$= A_T \int_{-\infty}^{\infty} dz \; \rho_T\left(\frac{\vec{s}+\vec{s}'}{2},z\right) \int d^2\vec{q} \; e^{-i\vec{q}\cdot(\vec{s}-\vec{s}')} \frac{d^2\sigma_{NN}(\vec{q})}{d\vec{q}}$$

At high energies, the nucleon-nucleon differential cross section is forward peaked with a characteristic slope of $\langle\vec{q}^2\rangle^{1/2} \simeq 400$ MeV/c. Therefore, the function $J(\vec{s},\vec{s}')$ goes to zero quickly for $|\vec{s}-\vec{s}'| > 1$ fm. The same is true for the difference of the operators $R - R_0$. As a result, we can expand the optical phase shift function χ_{opt}^{*} $(\vec{s}')$ and $\chi_{opt}(\vec{s})$ around $1/2$ $(\vec{s}+\vec{s}')$ and keep only the leading two terms,

$$i\left(\chi_{opt}(\vec{s})-\chi_{opt}(\vec{s}')\right) \approx -2 \; \mathrm{Im} \; \chi_{opt}^{*}\left(\frac{s+s'}{2}\right)+(\vec{s}-\vec{s}') \cdot \nabla_\sigma \; \mathrm{Re} \; \chi_{opt}(\vec{\sigma})\Big|_{\sigma \, = \, \frac{\vec{s}+\vec{s}'}{2}} \qquad \text{(IV.21)}$$

Substituting the above equation into Eqs. (IV.15) and (IV.16) we obtain

$$W_0(\vec{S},\vec{q}) = \int \frac{d^2\sigma}{(2\pi)^2} \; e^{-\vec{\sigma}\cdot[\vec{q}-\vec{\nabla} \; \mathrm{Re} \; \chi_{opt}(\vec{S})]-2 \; \mathrm{Im} \; \chi_{opt}(\vec{S})} \qquad \text{(IV.22.a)}$$

and

$$W(\vec{S},\vec{q}) = \int \frac{d^2\sigma}{(2\pi)^2} \; e^{-\vec{\sigma}\cdot[\vec{q}-\vec{\nabla} \; \mathrm{Re} \; \chi_{opt}(\vec{S})]-2 \; \mathrm{Im} \; \chi_{opt}(\vec{S})} * e^{J\left(\vec{S} - \frac{1}{2}\vec{\sigma},\vec{S} + \frac{1}{2}\vec{\sigma}\right)} \qquad \text{(IV.22.b)}$$

We are interested in deriving differential equations for W_0 and W. For this purpose, we introduce the variable z along the beam direction by defining

$$\Gamma(\vec{b}-\vec{s}_j,z-z_j) = \Gamma(\vec{b}-\vec{s}_j)\theta(z-z_j) \qquad \text{(IV.23)}$$

where $\theta(x)$ is the step function. Then the z-dependence can be introduced into the optical phase shift function. In terms of the optical potential $U_{opt}(b,z)$, defined by

$$U_{opt}(\vec{b},z) = -v \; A_T \; \rho_T(\vec{b},z) \frac{1}{2} \sigma_{NN}^{tot}\left[i + \frac{\mathrm{Re} \; f_{NN}(0)}{\mathrm{Im} \; f_{NN}(0)}\right], \qquad \text{(IV.24)}$$

with v the projectile velocity, the optical phase shift function is given by

$$\chi_{opt}(\vec{b},z) = - \frac{1}{v} \int_{-\infty}^{z} dz' \; U_{opt}(\vec{b},z') \qquad \text{(IV.25)}$$

Similarly, the z-dependence can also be introduced into the J function, i.e.

$$J(\vec{b},\vec{b}',z) = \frac{-2}{v\sigma_{NN}^{tot}} \int d^2\vec{q} \; e^{-i\vec{q}\cdot(\vec{b}-\vec{b}')} \frac{d^2\sigma_{NN}(\vec{q})}{d\vec{q}} * \int_{-\infty}^{z} dz' \; \mathrm{Im} \; U_{opt}\left(\frac{\vec{b}+\vec{b}'}{2},z'\right) \qquad \text{(IV.26)}$$

For later use, we introduce the definition

$$K(\vec{b},\vec{b}',z) = J(\vec{b},\vec{b}',z)/J(\vec{b},\vec{b},z) = \frac{1}{\sigma_{NN}^{tot}} \int d^2\vec{q} \; e^{-i\vec{q}\cdot(\vec{b}-\vec{b}')} \frac{d^2\sigma_{NN}}{d\vec{q}} \qquad (IV.27)$$

The z-dependence for the functions W_0 and W is obtained by introducing the z-dependent quantities $\chi_{opt}(\vec{s},z)$ and $J(\vec{s},\vec{s}',z)$ into Eq. (IV.22). Differentiating $W_0(\vec{S},z;\vec{q})$ and $W(\vec{S},z;q)$ with respect to z one finds

$$v \frac{d}{dz} W_0(\vec{S},z;\vec{q}) + \vec{F}(\vec{S},z) \cdot \vec{\nabla}_q W_0(\vec{S},z;q) = - \frac{v}{\lambda(\vec{S},z)} W_0(\vec{S},z;q) \qquad (IV.28.a)$$

and

$$v\frac{d}{dz} W_0(\vec{S},z;\vec{q}) + \vec{F}(\vec{S},z) \cdot \vec{\nabla}_q W(\vec{S},z;q) = - \frac{v}{\lambda(\vec{S},z)} W(\vec{S},z;\vec{q})$$
$$+ \frac{v}{\lambda(\vec{S},z)} \int d^2\vec{q}' \; \tilde{K}(\vec{q}-\vec{q}';\vec{S},z) \, W(\vec{S},z;\vec{q}') \qquad (IV.28.b)$$

In the above, the following definitions have been introduced,

$$F(\vec{S},z) = v \frac{d}{dz} \vec{\nabla}_S \, Re \, \chi_{opt}(\vec{S},z) = -\vec{\nabla}_S \, Re \, U_{opt}(\vec{S},z) \quad , \qquad (IV.29.a)$$

$$\frac{1}{\lambda(\vec{S},z)} = \frac{d}{dz} \, 2Im \, \chi_{opt}(\vec{S},z) = - \frac{2}{v} \, Im \, U_{opt}(\vec{S},z) \qquad (IV.29.b)$$

and

$$\tilde{K}(\vec{q};\vec{S},z) = \int \frac{d^2\sigma}{(2\pi)^2} K(\vec{S} - \frac{1}{2}\vec{\sigma},\vec{S} + \frac{1}{2}\vec{\sigma};z)e^{-i\vec{q}\cdot\vec{\sigma}} \qquad (IV.29.c)$$

$$= \frac{d^2\sigma_{NN}}{d\vec{q}} / \sigma_{NN}^{tot}$$

Eqs.(28) have the form of a stationary transport equation. Before the collision when the two nuclei are infinitely separated, the functions W and W_0 are equal and describe the transversal momentum distribution of a particle in the projectile nucleus, i.e.

$$W_0(\vec{S},-\infty;\vec{q}) = W(\vec{S},-\infty;\vec{q}) = \int \frac{d^2\vec{\sigma}}{(2\pi)^2} e^{-i\vec{q}\cdot\sigma} \rho_P(\vec{\sigma}) \qquad (IV.30)$$

As the projectile nucleons start to interact with the target, their momentum distributions are modified by a conservative force $\vec{F}$, which is related to the real part of the optical potential. Furthermore, the projectile nucleon collides with the target nucleons with a mean free path λ, determined by the imaginary part of the optical

potential. This results in a reduction of the strength of the momentum distributions. For the distribution W, however, there is a gain in this strength due to the
second term on the right-hand side of Eq. (IV.28.b). Since the integral of the
right hand side of Eq. (IV.28.b) over $\vec{q}$ is zero, the total probability of the distribution function W is conserved. This is not the case for the distribution function W_0, which always diminishes because of the collisions between the projectile
nucleons with the target nucleons. The one-nucleon inclusive cross section due to
the impact parameter $\vec{S}$ is obtained by subtracting from the asymptotic distribution
$W(S,+\infty;\vec{q})$ the distribution $W_0(S,-\infty;\vec{q})$, which describes those projectile nucleons
suffering no collisions with the target.

IV.3 The Rows on Rows Model

Eqs.(IV.28), describing the evolution of the momentum distribution of the
projectile nucleons, can be generalized to include that of the target nucleons as
well. The solution to such equations has been suggested by Knoll and Hüfner [61]
using simple geometrical and dynamical considerations. In their model, both the
projectile and the target are decomposed into rows of nucleons along the beam direction. It is assumed that only rows on the same straight line scatter with each
other and that interactions among projectile nucleons and among target nucleons are
neglected.

Let $\vec{S}_P$ be the position of a projectile row in the transverse plane, then the
mean number of collisions a nucleon would suffer if it were impinging on P with the
impact parameter $\vec{S}_P$ is

$$\alpha(\vec{S}_P) = \sigma_{NN} \int \rho_P(\vec{S}_P,z)dz \qquad (IV.31)$$

Here $\rho_P(\vec{r})$ is the matter density distribution of the projectile nucleus P and σ_{NN} is
the total nucleon-nucleon cross section, which is approximately 40 mb at high energies. The actual number of collisions suffered is a stochastic variable governed by
a Poisson distribution so that the probability for suffering exactly M collisions is
given by

$$P_M(\vec{S}_P) = \frac{1}{M!} \alpha(\vec{S}_P)^M e^{-\alpha(\vec{S}_P)} \qquad (IV.32)$$

The total cross section for the nucleon to suffer exactly M collisions when bombarded on to P is obtained by integrating $P(\vec{S}_P)$ over $\vec{S}_P$, i.e.

$$\sigma_P(M) = \int d\vec{S}_P \, P_M(\vec{S}_P) \qquad (IV.33)$$

If similar probability function $P_N(\vec{S}_T)$ is introduced to the target, the one-nucleon inclusive cross section can be written as

$$\frac{d^2\sigma}{d\vec{p}} = \frac{1}{\sigma_{NN}} \int d\vec{S} \int d\vec{S}_P \int d\vec{S}_T \; \delta(\vec{S}-\vec{S}_P-\vec{S}_T) * \sum_{M=1}^{\infty} P_M(\vec{S}_P) \sum_{N=1}^{\infty} P_N(\vec{S}_T) \; F_{MN}(\vec{P}) \qquad (IV.34)$$

Here the δ-function ensures that $\vec{S}_A$ and $\vec{S}_B$, the positions of the two colliding rows, refer to the same position in the transverse plane. The function $F_{MN}(P)$ gives the final momentum distribution of the nucleons after M projectile nucleons collide with N target nucleons. Using Eq. (IV.33), one can rewrite Eq. (IV.34) as

$$\frac{d^2\sigma}{d\vec{P}} = \frac{1}{\sigma_{NN}} \sum_{M=1}^{\infty} \sum_{N=1}^{\infty} \sigma_P(M) \; \sigma_T(N) \; F_{MN}(\vec{P}) \qquad (IV.35)$$

The function $F_{MN}(\vec{P})$ can be written as a sum of contributions from each of the participating nucleons,

$$F_{MN}(\vec{P}) = \sum_{m=1}^{M} f_{mN}^{P}(\vec{P}) + \sum_{n=1}^{N} f_{Mn}^{T}(\vec{P}) \quad . \qquad (IV.36)$$

Here $f_{mN}^{P}(\vec{P})$ is the momentum distribution of the mth projectile-row nucleon after it has collided with all N nucleons in the target row. Similarly, $f_{Mn}^{T}(\vec{P})$ is the momentum distribution of the nth target-row nucleon after the entire row of M projectile nucleons has struck it. Since each of these distributions is normalized to unity, the total one-nucleon inclusive cross section is

$$\sigma_N = \int \frac{d^2\sigma}{d\vec{p}} \, d^3p = \frac{1}{\sigma_{NN}} \sum_{M=1}^{\infty} \sum_{N=1}^{\infty} \sigma_P(M) \; \sigma_T(N) * (M+N) = A_P \Sigma_T + A_T \Sigma_P \qquad (IV.37)$$

In the above, we have used that

$$\sum_{M=1}^{\infty} \sigma_P(M) = \Sigma_P \approx \pi R_P^{\,2} \quad , \qquad (IV.38)$$

the total reaction cross section for nucleus P, and

$$\frac{1}{\sigma_{NN}} \sum_{M=1}^{\infty} M\sigma_P(M) = A_P \quad , \qquad (IV.39)$$

the total number of nucleons in the nucleus P.

Initially both nuclei are in their ground states and the momentum distribution of the nucleons can be approximated by a non-interacting Fermi gas, i.e.

$$f_{mo}^{p}(\vec{p}) = \theta\left(\left|\vec{p}_F-\vec{p}-\vec{p}_o\right|\right) \Big/ \frac{4\pi}{3} \, p_F^{\,3} \qquad (IV.40.a)$$

$$f^T_{on}(\vec{p}) = \theta(|\vec{p}_F - \vec{p}|) / \frac{4\pi}{3} p_F^3 \qquad (IV.40.b)$$

where $\vec{p}_F$ is the Fermi momentum and $\vec{p}_0$ is the incident momentum per nucleon.

In the original paper by Knoll and Hüfner, the linear cascade between the projectile and the target rows of nucleons is assumed to follow a Markoffian chain, i.e.

$$f_{mn}(\vec{p}'_P, \vec{p}'_T) = \int d^3\vec{p}_P \, d^3\vec{p}_T \, f^P_{m\,n-1}(\vec{p}_P) \, f^T_{m-1\,n}(\vec{p}_T) \times M(\vec{p}'_P, \vec{p}'_T \leftarrow \vec{p}_P, \vec{p}_T) \qquad (IV.41)$$

Here, $f_{mn}(p'_P, p'_T)$ denotes two-particle distribution function for the mth projectile nucleon and the nth target nucleon after their collision. The one-particle distribution functions can be obtained from f_{mn} by

$$f^P_{mn}(\vec{p}_P) = \int d^3\vec{p}_T \, f_{mn}(\vec{p}_P, \vec{p}_T) \qquad (IV.42.a)$$

$$f^T_{mn}(\vec{p}_T) = \int d^3\vec{p}_P \, f_{mn}(\vec{p}_P, \vec{p}_T) \qquad (IV.42.b)$$

The transition probability M is proportional to the nucleon-nucleon differential cross section, i.e.

$$M(\vec{p}'_1, \vec{p}'_2 \leftarrow \vec{p}_1, \vec{p}_2) = \frac{1}{N} \delta^3(\vec{P} - \vec{P}') \delta(|\vec{p}| - |\vec{p}'|) \frac{d\sigma(\vec{q})}{d\Omega}\Big|_{c.m.} \qquad (IV.43)$$

Here, $\vec{P}$ and $\vec{P}'$ are the initial and final total momenta, while $\vec{p}$ and $\vec{p}'$ are the initial and final relative momenta. The normalization factor N follows from the condition

$$\int d^3\vec{p}'_1 \int d^3\vec{p}'_2 \, M(\vec{p}'_1, \vec{p}'_2 \leftarrow \vec{p}_1, \vec{p}_2) = 1 \qquad (IV.44)$$

i.e. $N = P^2 \, \sigma^{tot}_{NN}$.

They consider the moments of the momentum distributions. For component p_α of $\vec{p}$, they are defined by

$$\langle (p_\alpha)^K \rangle^i_{mn} = \int d^3\vec{p} \, (p_\alpha)^K \, f^i_{mn}(\vec{p}) \qquad (IV.45)$$

In particular, only the first and the second moments are taken into account. The recursion relations for these moments can be derived from Eq. (IV.41). For the mean momenta, one obtains

$$\langle p_{||} \rangle^P_{mn} = (1-\alpha) \langle p_{||} \rangle^P_{mn-1} + \alpha \langle p_{||} \rangle^T_{m-1n} \qquad (IV.46.a)$$

$$\langle p_\parallel \rangle^T_{mn} = \alpha \langle p_\parallel \rangle^P_{mn-1} + (1-\alpha) \langle p_\parallel \rangle^T_{m-1n} \qquad\qquad (IV.46.b)$$

The mean values for the transverse components vanish due to symmetry. Eqs. (46) are in agreement with the conservation of momentum as each nucleon looses a fraction α of its momentum and gains a fraction α of the momentum of the other nucleon. The factor α is defined by the average of $\sin^2 \frac{\theta}{2}$ over the nucleon-nucleon differential cross section i.e.

$$\alpha = \langle \sin^2 \frac{\theta}{2} \rangle_{NN} \qquad\qquad (IV.47)$$

The initial conditions for the mean momenta are

$$\langle p_\parallel \rangle^P_{m0} = p_0, \quad \langle p_\parallel \rangle^T_{0n} = 0 \qquad\qquad (IV.48)$$

The variables, which are related to the second moments, are the longitudinal and transverse dispersion in momentum,

$$\langle \sigma_\parallel^2 \rangle^i_{mn} = \langle p_\parallel^2 \rangle^i_{mn} - (\langle p_\parallel \rangle^i_{mn})^2 \qquad\qquad (IV.49.a)$$

$$\langle \sigma_\perp^2 \rangle^i = \langle p_\perp^2 \rangle^i_{mn} - (\langle p_\perp \rangle^i_{mn})^2 \qquad\qquad (IV.49.b)$$

The recursion relations for these dispersions are

$$\langle \sigma_\parallel^2 \rangle^P_{mn} = (1-\alpha-\beta)\langle \sigma_\parallel^2 \rangle^P_{m\,n-1} + (\alpha-\beta)\langle \sigma_\parallel^2 \rangle^T_{m-1\,n} + \beta[\langle \sigma_\perp^2 \rangle^P_{m\,n-1} + \langle \sigma_\perp^2 \rangle^T_{m-1\,n}] \quad (IV.50.a)$$

$$+ (\alpha-\alpha^2-\beta)[\langle p_\parallel \rangle^P_{m\,n-1} - \langle p_\parallel \rangle^T_{m-1\,n}]^2$$

$$\langle \sigma_\perp^2 \rangle^P_{mn} = (1-\alpha-\tfrac{1}{2}\beta)\langle \sigma_\perp^2 \rangle^P_{m\,n-1} + (\alpha-\beta)\langle \sigma_\perp^2 \rangle^T_{m-1\,n} \qquad\qquad (IV.50.b)$$

$$+ \tfrac{1}{2}\beta[\langle \sigma_\parallel^2 \rangle^P_{m\,n-1} + \langle \sigma_\parallel^2 \rangle^T_{m-1\,n}] + \tfrac{1}{2}\beta[\langle p_\parallel \rangle^P_{m\,n-1} - \langle p_\parallel \rangle^T_{m-1\,n}]^2$$

where

$$\beta = \frac{1}{2}\langle \sin^2\theta \rangle_{NN} \qquad\qquad . \qquad\qquad (IV.51)$$

The dispersion in longitudinal momentum changes for three reasons. The first two terms in Eq. (IV.50.a) are due to scattering to different angles in the collision. The third term gives the dispersion in longitundinal momentum as a result of the dispersion in transverse momentum. In the last term, the longitundinal dispersion increases as the momentum is transferred. Similar recursion relations hold for the dispersions of the momentum distribution of the target nucleons. The initial dispersions due to the Fermi motion are

$$\langle\sigma_{\|}^2\rangle_{m0}^{P} = \langle\sigma_{\perp}^2\rangle_{m0}^{P} = \langle\sigma_{\|}^2\rangle_{0n}^{T} = \langle\sigma_{\perp}^2\rangle_{0n}^{T} = \frac{1}{5} P_F^2 \qquad (IV.52)$$

The final momentum distribution of the nucleons is assumed to have a Gaussian form with the mean values and dispersions determined by Eqs. (48) and (50),

$$f_{mn}^i(\vec{p}) = \frac{1}{[(2\pi)^3\langle\sigma_{\|}^2\rangle_{mn}^i(\langle\sigma_{\perp}^2\rangle_{mn}^i)^2]^{1/2}} * \exp\left\{-\frac{(p_{\|}-\langle p_{\|}\rangle_{mn}^i)^2}{2\langle\sigma_{\|}^2\rangle_{mn}^i} - \frac{p_{\perp}^2}{2\langle\sigma_{\perp}^2\rangle_{mn}^i}\right\} \qquad (IV.53)$$

This model can be generalized to include the possibility that one nucleon is excited to the Δ resonance when colliding with another nucleon [61]. Numerical calculations using this model have been carried out for various reactions. In Fig. IV.1, the invariant one-proton inclusive cross section as a function of the momentum

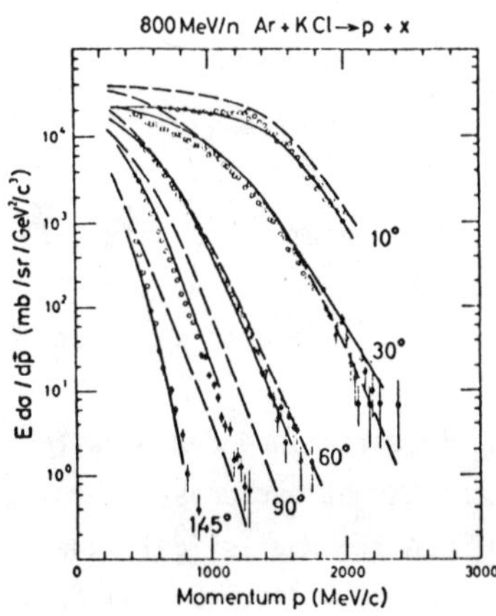

Fig. IV.1. The invariant proton inclusive cross section as a function of the momentum. Data from Ref. [109]. Solid curves from numerical solution of the linear cascade model [110]. Dashed curves from the moment approximation of Ref. [61].

is shown for the reaction of 800 MeV/N Ar on KCl. The experimental data are from [109]. The dashed curves are from the linear cascade model with Gaussian approximations to the momentum distribution.

On the other hand, the linear cascade maodel has been solved by Knoll and Randrup [110] using the method of numerical simulation. In this method, the initial momenta of the nucleons in the two rows are picked randomly out of the respective Fermi spheres. The sequence of binary collisions between the nucleons in the projectile row and the nucleons in the target row are carried through, the result of each collision being determined randomly in accordance with the appropriate differential cross sections. For given numbers of projectile nucleons M and target nucleons N, a large number of such linear cascades are performed. The momentum distribution $F_{MN}(\vec{p})$ is then obtained by appropriate averaging. The results of such a calculation are shown in the previous figure Fig. IV.1 by the solid lines. The experimental data are very well reproduced by the theoretical calculations. The model has also been used to study K^+ production in relativistic heavy ion collisions [111].

IV.4 The Statistical Model

After the collison of M projectile nucleons with N target nucleons, the final one-nucleon momentum distribution function $F_{MN}(\vec{p})$ of these nucleons can be written in the general form

$$F_{MN}(\vec{p}) = S_{MN}(\vec{p}) \; \phi_{M+N}(\vec{p}) \tag{IV.53}$$

Here, S_{MN} contains the dynamical information of the collision process while ϕ_{M+N} is the phase space function given by

$$\phi_{M+N}(\vec{p}) = \int \frac{d^3\vec{p}_2}{\varepsilon_2} \cdots \frac{d^3\vec{p}_{M+N}}{\varepsilon_{M+N}} \; \delta^3\left(\vec{P}_{MN} - \sum_{i=1}^{M+N} \vec{p}_i\right) * \delta\left(E_{MN} - \sum_{i=1}^{M+N} \varepsilon_i\right) / I_{M+N}(s) \tag{IV.54}$$

In the above, E_{MN} and P_{MN} are the total energy and momentum, respectively. The quantity s is the square of the invariant mass $s = E_{MN}^2 - P_{MN}^2$. The total energy of individual nucleons including the rest mass is ε_i. The phase space function $\phi_{M+N}(\vec{p})$ is normalized to one by the phase space integral $I_{M+N}(s)$.

In the rows on rows model, the function S_{MN} is determined dynamically from the collisions between the projectile nucleons and the target nucleons according to the free nucleon-nucleon differential cross section. In the statistical model, Knoll [107] assumes that S_{MN} is a constant so that the momentum distribution function $F_{MN}(\vec{p})$ is determined by the behavior of the phase space distributions. This is a good approximation when the number of observed particles is much smaller than the number of participating nucleons. In this case, $S_{MN}(\vec{p})$ is determined by the normalization of the momentum distribution, i.e. $\int \frac{d^3\vec{p}}{\varepsilon} F_{MN}(\vec{p}) = M + N$.

In Fig. IV.2 we show the phase space distribution function $\phi_K(\vec{p}_{CM})$ as a function of the center-of-mass momentum of the statistical ensemble of K = 2, 3, 5, 10, and ∞. The center-of-mass energy per particle is 100 MeV. This statistical model has been applied to calculate the proton inclusive spectra in heavy-ion collisions. In Fig. IV.3, we show the proton inclusive spectrum from the reaction of 800 MeV/N Ar on KCl for various laboratory angles as a function of the momentum of the observed proton. In comparison with the calculated results using the method of moment expansion and the numerical simulation, the statistical model gives essentially similar results. One therefore tends to conclude that the one-nucleon inclusive cross section is dominated by the phase space.

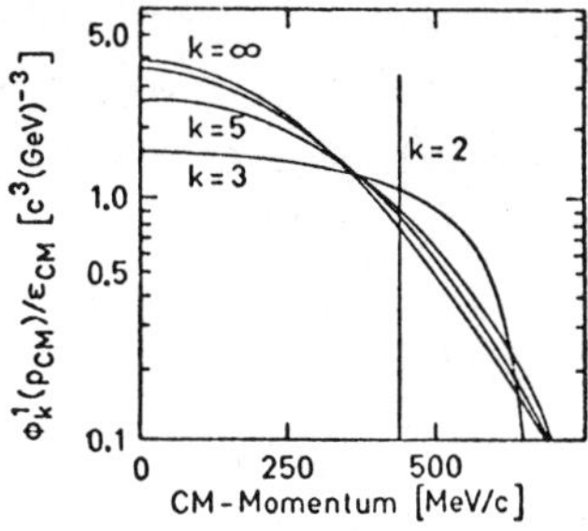

Fig. IV.2. The one-particle spectrum as a function of the c.m. momentum of the statistical ensemble of K = 2, 3, 5, 10, and ∞. The c.m. energy per particle is 100 MeV. (From Ref. [107].)

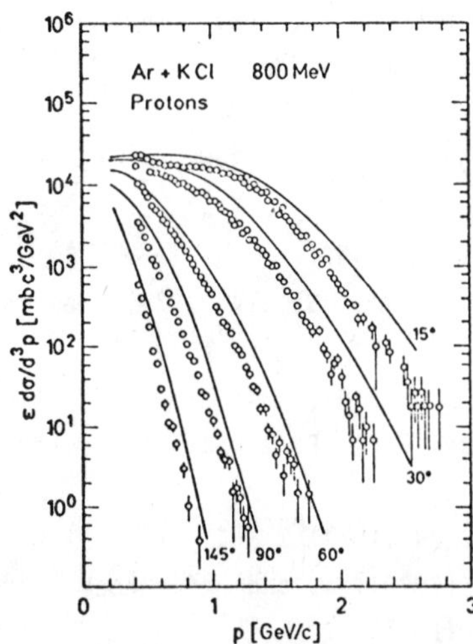

Fig. IV.3. Same as Fig. IV.1. Solid curves are from the statistical model [107].

The statistical model has also been extended to include pion production [112].
For M+N >> 1 and in the non-relativistic limit, Eq. (IV.54) reduces to the thermal distribution

$$\phi_{M+N}(\vec{p}) \xrightarrow[M+N\gg1]{} \frac{1}{(2\pi mT)^{3/2}} e^{-(\vec{p}-\vec{p}_{MN})^2/2mT} \qquad (IV.55)$$

with the temperature $3/2T = E^*_{MN}$. Here $\vec{P}_{MN}$ and E^*_{MN} are, respectively, the center of mass momentum and energy per nucleon. This limit then corresponds to the fire streak model [113]. If one further assumes that for each impact parameter complete thermalization between adjacent rows is reached as well, then the fireball model is obtained [54].

REFERENCES

[1]   W. U. Schröder and J. R. Huizenga, Ann. Rev. Nucl. Sci. 27 (1977), 465.

[2]   H.A. Weidenmüller, Prog. Part. Nucl. Phys. 3 (1980), 49.

[3]   A. S. Goldhaber and H. H. Heckman, Ann. Rev. Nucl. Part. Sci. 28 (1978), 161.

[4]   J. R. Nix, Prog. Part. Nucl. Phys. 2 (1979), 237.

[5]   Proc. Topical Conf. on Heavy Ion Collisions, Fall Creek Falls, ORNL Report Conf-770602 (1977).

[6]   "Theoretical Methods in Medium-Energy and Heavy-Ion Physics", ed. by K. W. McVoy and W. A. Friedman, Plenum Press, New York (1978).

[7]   Proc. 4th High Energy Heavy Ion Summer Study, Berkeley, California LBL-7766 (1978).

[8]   "Heavy-Ion Collision," ed. by R. Bock, North-Holland Press, Amsterdam (1981).

[9]   Proc. Symposium on Heavy-Ion Physics from 10 to 200 MeV/u, Brookhaven Report BNL-51115 (1979).

[10] I. Ragnarsson, S. G. Nilsson, and R. K. Sheline, Phys. Rep. C45 (1978) 1, and references therein.

[11] M. Schädel et al., Phys. Rev. Lett. 41 (1978), 469.

[12] D. K. Scott, in Inter. School of Nuclear Physics, Sicily (1979), Lawrence Berkely Preprint LBL-8931.

[13] T. J. L. Symons et al., Phys. Rev. Lett. 42 (1979), 40.

[14] R. M. Diamond and F. S. Stephens, Ann. Rev. Nucl. Part. Sci. (in press), Lawrence Berkeley Preprint LBL-10325.

[15] D. H. Youngblood et al., Phys. Rev. Lett. 39 (1977) 1188.

[16] G. E. Brown and W. Weise, Phys. Rep. C27 (1976), 1, and references therein.

[17] T. D. Lee and G. C. Wick, Phys. Rev. D 9 (1974), 2291.

[18] G. Baym and S. Chin, Nucl. Phys. A262 (1976), 527.

[19] S. A. Chin and A. K. Kerman, Phys. Rev. Lett. 43 (1979), 1292.

[20] J. Cugnon, T. Mizutani, and J. Vandermeulen, Nucl. Phys. A352 (1981), 505.

[21] M. Gyulassy and W. Greiner, Ann. Phys. 109 (1977), 485.

[22] M. Lefort, J. Phys. C5 (1976), 37.

[23] A. Artukh, G. Gridnev, V. Mikkev, V. Volkov, and J. Wilczynski, Nucl. Phys. A215 (1973), 91.

[24] J. Wilczynski, Phys. Lett. 47B (1973) 484.

[25] D. H. E. Gross and K. Kalinowski, Phys. Lett. 48B (1974), 302.

[26] J. Blocki, J. Randrup, W. J. Swiatecki, and C. F. Tsang, Ann. Phys. 105 (1977), 427.

[27] C. Ngô, B. Tamain, J. Galin, M. Beiner, and R. J. Lombard, Nucl. Phys. A240 (1975), 353.

[28] D. M. Brink and N. Rowley, Nucl. Phys. A219 (1974), 79.

[29] H. H. Deubler and K. Dietrich, Phys.Lett. 56B (1975), 241.

[30] D. H. E. Gross, R. C. Nayak, and L. Satpathy, Z. Phys. A299 (1981), 63.

[31] W. Nörenberg, Phys. Lett. 53B (1974), 289.

[32] M. Berlanger, A. Gobbi, F. Hanappe, U. Lynen, C. Ngô, A. Olmi, H. Sann, H. Stelzer, H. Richel, and M. F. Rivet, Z. Phys. A291 (1979), 133.

[33] W. Scobel, H. H. Gutbrod, M. Blann, and A. Mignerey, Phys. Rev. C 14 (1976), 1808.

[34] P. Dand, J. Bisplinghoff, M. Blann, T. Mayer-Kuckud, and A. Mignerey, Nucl. Phys. A287 (1977), 179.

[35] J. Dank, K. P. Lieb, and A. M. Kleinfeld, Nucl. Phys. A241 (1975), 170.

[36] B. B. Back, R. B. Betts, C. Gaarde, J. S. Larsen, E. Michelsen, and Tai Kuong-Hsi, Nucl. PHys. A285 (1977), 317.

[37] R. L. Kozub, N. H. Lu, J. M. Miller, D. Logan, T. W. Debrak, and L. Kowalski, Phys. Rev. C 11 (1975), 1497.

[38] Y. Eisen, I. Tserruya, Y. Eyal, Z. Fraenkel, and M. Hillman, Nucl. Phys. A241 (1977), 459.

[39] B. Borderie, M. Berlanger, R. Bimbot, C. Cabot, D. Gardes, C. Gregoire, F. Hanappe, C. Ngô, L. Nowicki, and B. Tamain, Z. Phys. A298 (1980), 235.

[40] C. Ngô, Lectures given at the PREDEAL Inter. Summer School, Rumania (1980).

[41] J. R. Birkeleund, J. R. Huizenga, J. N. De, and D. Sperber, Phys. Rev. Lett. 40 (1978), 1123.

[42] C. F. Tsang, Phys. Scripta 10A (1974), 90.

[43] J. Randrup, Ann. Phys. 112 (1978), 356.

[44] P. Gonthier, H. Ho, M. N. Namboodiri, L. Adler, J. B. Natowitz, S. Simon, K. Hagel, R. Terry, and A. Khodai, Phys. Rev. Lett. 44 (1980), 1387.

[45] C. K. Gelbke, C. Olmer, M. Buenerd, D. L. Hendrie, J. Mahoney, M. C. Mermaz, and D. K. Scott, Phys. Rep. 42 (1978), 312.

[46] D. E. Greiner, P. E. Lindstrom, H. H. Heckman, B. Cork, and F. Bieser, Phys. Rev. Lett. 35 (1975), 152.

[47] A. S. Goldhaber, Phys. Lett. 53B (1974), 306.

[48] G. F. Bertsch, Phys. Rev. Lett. 46 (1981), 472.

[49] K. Van Bibber, D. L. Hendrie, D. K. Scott, H. H. Wieman, L. S. Schroeder, J. V. Geaga, S. A. Chessin, R. Trenhaft, J. Y. Grossiard, J. O. Rasmussen, and C. Y. Wong, Phys. Rev. Lett. 43 (1979), 840.

[50] C. M. Ko, Phys. Rev. C 21 (1980), 2672.

[51] U. Udagawa, T. Tamura, T. Shimoda, H. Froehlich, M. Ishihara, and K. Nagatani, Phys. Rev. C 20 (1979), 1949.

[52] J. Hüfner, K. Schäfer, and B. Schürmann, Phys. Rev. C 12 (1975), 1888.

[53] T. Fujita and J. Hüfner Nucl. Phys. A343 (1980), 493.

[54] J. Gosset, H. H. Gutbrod, W. G. Meyer, A. M. Poskanzer, a. Sandoval, R. Stock, and G. D. Westfall, Phys. Rev. C 16 (1977), 629.

[55] J. I. Kapusta, Phys. Rev. C 16 (1977), 1493.

[56] F. Asai, H. Sato, and M. Sano, Phys. Lett. 98B (1981), 19.

[57] P. J. Siemens and J. O. Rasmussen, Phys. Rev. Lett. 42 (1979), 880.

[58] A. Mekjian, Phys. Rev. Lett. 38 (1977), 640; Phys. Rev. C 17 (1978), 1051; Nucl. Phys. A312 (1978), 491.

[59] P. J. Siemens and J. I. Kapusta, Phys. Rev. Lett. 43 (1979), 1486.

[60] I. Mishustin, F. Myhrer, and P. J. Siemens, Phys Lett. 95B (1980), 361.

[61] J. Hüfner and J. Knoll, Nucl. Phys. A290 (1977), 460.

[62] D. Vautherin and D. M. Brink, Phys. Rev. C 5 (1972), 626.

[63] P. Bonche, S. E. Koonin, and J. W. Negele, Phys. Rev. C 13 (1976), 1226.

[64] S. E. Koonin, K. T. R. Davies, V. Maruhn-Rezwani, H. Feldmeier, S. J. Krieger, and J. W. Negele, Phys. Rev. C 15 (1977), 1359.

[65] K. T. R. Davies, V. Maruhn-Rezwani, S. E. Koonin, and J. W. Negele, Phys. Rev. Lett. 41 (1978), 632.

[66] R. Vandenbosh, M. P. Webb, and T. D. Thomas, Phys. Rev. C 14 (1976), 143; and Phys. Rev. Lett. 36 (1976), 459.

[67] S. E. Koonin, B. Flanders, H. Flocard, and M. S. Weiss, Phys. Lett. 77B (1978), 13.

[68] K. R. Sandhya Devi and M. R. Strayer, Phys. Lett. 77B (1978), 135.

[69] R. Y. Cusson, R. K. Smith, and J. Maruhn, Phys. Rev. Lett. 36 (1976), 1166.

[70] R. Y. Cusson, J. A. Maruhn, and H. W. Meldner, Phys. Rev. C 18 (1978), 2589.

[71] H. Flocard, S. E. Koonin, and M. S. Weiss, Phys. Rev. C 17 (1978), 1682.

[72] P. Bonche, B. Grammaticos, and S. E. Koonin, Phys. Rev. C 17 (1978), 1700.

[73] P. Bonche, K. T. R. Davis, B. Flanders, H. Flocard, B. Rammaticos, S. E. Koonin, S. J. Krieger, and M. S. Weiss, Phys. Rev. C 20 (1979), 641.

[74] S. E. Vigdor, D. G. Kovar, P. Speer, J. Mahoney, A. Menchaca-Rocha, C. Olmer, and M. S. Zisman (unpublished).

[75] H. Blocki, Y. Boneh, J. R. Nix, J. Randrup, M. Robel, A. J.Sierk, and W. J. Swiatecki, Ann. Phys. 113 (1978), 330.

[76] J. Randrup, Nucl. Phys. A237 (1979), 490.

[77] H. Hofmann and P. J. Siemens, Nucl. Phys. A257 (1976), 165.

[78] H. Hofmann and P. J. Siemens, Nucl. Phys. A275 (1977), 464.

[79] W. Nörenberg, Z. Phys. A274 (1975), 241.

[80] S. Ayik, B. Schürmann, and W. Nörenberg, Z. Phys. A277 (1976), 299; A279 (1976), 145.

[81] S. Ayik, G. Wolschin, and W. Nörenberg, Z. Phys. A286 (1978), 271.

[82] S. Ayik and W. Nörenberg, Z. Phys. A288 (1978), 401.

[83] C. M. Ko, H. J. Pirner and H. A. Weidenmüller, Phys. Lett. 62B (1976), 248.

[84] D. Agassi, C. M. Ko, and H. A. Weidenmüller, Ann. Phys. 107 (1977), 140.

[85] B. R. Barrett, S. Shlomo, and H. A. Weidenmüller, Phys. Rev. C 17 (1978), 544.

[86] C. M. Ko, D. Agassi, and H. A. Weidenmüller, Ann. Phys. 117 (1979), 237.

[87] D. Saloner and H. A. Weidenmüller, MPI preprint MPI H-1979-V23, 1979.

[88] S. Shlomo, B. R. Barrett, and H. A. Weidenmuller, Phys. Rev. C 20 (1979), 1.

[89]   D. Agassi, C. M. Ko, and H. A. Weidenmüller, Ann. Phys. 117 (1978), 407.

[90]   D. Agassi, H. A. Weidenmüller, and C. M. Ko, Phys. Lett.73B (1978), 284.

[91]   D. Agassi, C. M. Ko, and H. A. Weidenmüller, Phys. Rev. C 18 (1978), 223.

[92]   C. M. Ko, Z. Phys. A286 (1978), 405.

[93]   C. M. Ko, Phys. Lett. 81B (1979), 299.

[94]   W. U. Schröder, J. R. Birkelulnd, J. R. Huizenga, K. L. Wolf, J. P. Unik, and V. E. Viola, Phys. Rev. Lett. 36 (1976), 514.

[95]   A. R. Bodmer, C. N. Panos and A. D. Mckeller, Phys. Rev. C 22 (1980), 1025.

[96]   D. J. E. Callaway, L. Wilets, and Y. Yariv, Nucl. Phys. A327 (1979), 250.

[97]   J. P. Bondorf, H. T. Feldmeire, S. Garpman, and E. C. Halbert, Phys. Lett. 65B (1976), 217.

[98]   E. C. Halbert, Phys. Rev. C 23 (1981), 295.

[99]   R. K. Smith and M. Danos, p. 363 of Ref. 5.

[100]   J. D. Stevenson, Phys. Rev. Lett. 41 (1978), 1702.

[101]   Y. Yariv and Z. Fraenkel, Phys. Rev. C 20 (1979) 2227.

[102]   J.Cugnon, Cal. Tech. prepring MAP-9, 1980.

[103]   A. A. Amsden, F. H. Harlow, and J. R. NIx, Phys. Rev. C 15 (1977), 2059.

[104]   H. Stöcker, J. A. Maruhn, and W. Greiner, Z. Phys. A293 (1979), 173.

[105]   H. H. K. Tang and C. W. Wong, Phys. Rev. C 21 (1980), 1848.

[106]   J. Hüfner, Ann. Phys. 115 (1978), 43.

[107]   J. Knoll, Phys. Rev. C 20 (1979), 773.

[108]   R. J. Glauber, in "Lectures in Theoretical Physics" (W. E. Brittin et al., Eds.), Vol. I., p. 315, Interscience, N.Y., 1959.

[109]   S. Nagamiya, L.Anderson, W. Bruckner, O. Chamberlain, M. C. Lemaire, S. Schnetzer, G. Shapiro, H. Steiner, and I. Tanikata, Phys. Lett. 81B (1979), 147.

[110]   J. Knoll and J. Randrup, Nucl. Phys. A324 (1979), 445.

[111]   J. Randrup and C. M. Ko, Nucl. Phys. A343 (1980), 519.

[112]   S. Bohrmann and J. Knoll, Nucl. Phys. A356 (1981), 498.

[113]   W. D. Myers, Nucl. Phys. A296 (1978), 177.

CHAPTER VIII

INTERACTING BOSON MODEL

Akito Arima

Department of Physics, Faculty of Science
University of Tokyo
Tokyo, Japan

Abstract

Truncation into the SD subspace is shown to be good approximation. In order to simulate shell model calculations in this subspace, s and d bosons are introduced. This boson model is called the Interacting Boson Model. Some results given by this model are discussed.

I.   INTRODUCTION

Nuclei consist of many strongly interacting particles whose number is neither very large nor very small. Therefore the nuclear systems are not amenable to statistical treatment while exact calculations are practically out of question. Nevertheless, the low-lying levels of many even-even nuclei show a relatively simple structure. The structure has been interpreted as arising from collective quadrupole motion of nuclei.[1] The Bohr-Mottelson model has been very successful in describing the collective motion; both rotational and vibrational.

The nuclear shell model on the other hand has succeeded in explaining the low-lying spectra of many light nuclei including their collective features.[2] Thus it is tempting to use the shell model to interpret the low-lying spectra of medium and heavy nuclei. However, counting the number of shell model states, one immediately sees that shell model calculations are almost impossible without drastic truncation. As an example, let us take ^{154}Sm which has twelve valence protons occupying all orbits in the 50-82 shell and ten valence neutrons occupying those in the 82-126 shell. There are 41, 654, 193, 516, 797 positive parity states with $J = 0^+$, 346, 132, 052, 934, 889 states with $J = 2^+$ and 530, 897, 397, 260, 575 states with $J = 4^+$. These numbers are just astronomical. Such large secular equations cannot be solved. Even if they are solved, the results are not particularly interesting, because we cannot examine all of them in our life time.

2. SD SUBSPACE

Iachello and I together with Talmi and Otsuka proposed a truncation scheme in which we assumed that collective states are approximately expressed in terms of L = 0 and L = 2 pairs of protons and neutrons;[3),4)]

$$S^+ = \frac{1}{2}\sum_j \alpha_j \, [a_j^+ \, a_j^+]_0^{(0)}$$

$$D_\mu^+ = P \sum \beta_{j_1 j_2} \, [a_{j_1}^+ \, a_{j_2}^+]_\mu^{(2)}$$

where P projects out seniority eigenstates. A subspace constructed from those pairs will be called the SD-subspace. The validity of this truncation was examined in several systems; large single j-shells[5)] (see fig. 1) degenerate ($g_{7/2}$, $d_{5/2}$, $d_{3/2}$, $s_{1/2}$) shell[6)] and the Ginocchio model.[7)]

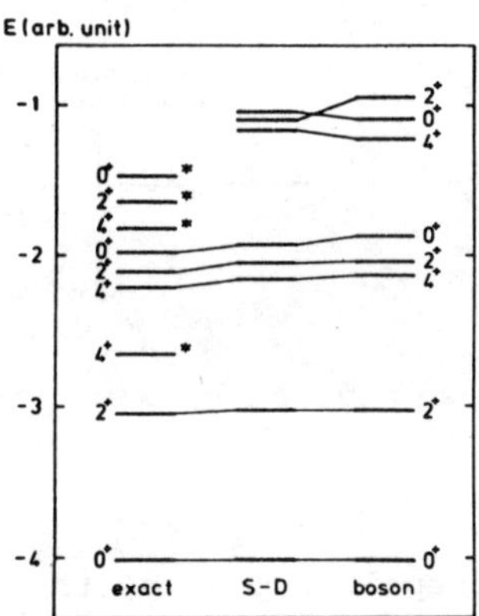

Fig. 1: Comparison between the calculated energies; (I) Exact shell model calculations $(\frac{23}{2})^8$ (II) Truncation in the SD subspace (III) Corresponding boson space.

One might be afraid that the truncation into the SD subspace is not good for well deformed nuclei,[8)] even if it could be good for spherical nuclei. There have been two calculations which prove that this truncation is good even for well deformed nuclei so far as low-spin states are concerned. The calculation by Dukelsky, Dussel and Sofia[9] based on the Principal Series Approximation shows that the SD subspace takes care of nearly 90 % of the ground state wave functions of well deformed nuclei (in probability). The other calculation by Otsuka[10] gives a similar conclusion. Otsuka assumed that both proton and neutrons occupy degenerate $0g_{7/2}$, $1d_{5/2}$, $1d_{3/2}$ and $2s_{1/2}$ shell. The interaction between protons is assumed to be a surface delta interaction. The same is assumed for neutrons. A quadrupole-quadrupole interaction is taken as a proton-neutron interaction. A result of the shell model calculation is shown in fig. 2, where six protons and six neutrons are put in the shell. We see that the level structure is almost rotational.

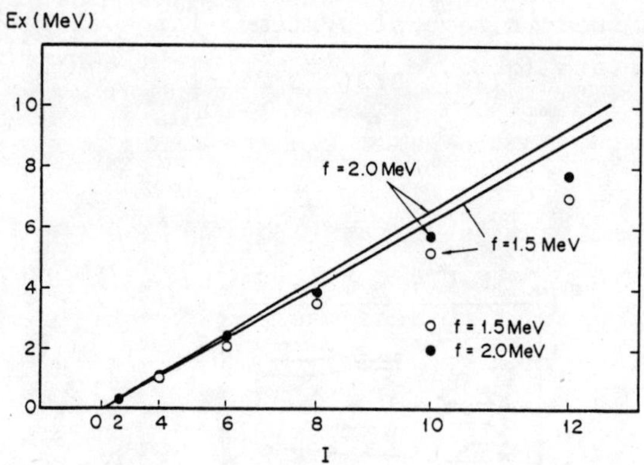

Fig. 2: Excitation energies produced by shell model calculations.
          f is the strength of a QQ interaction.

Otsuka found indeed the quadrupole moment of the first 2^+ state is almost equal to that given by the Nilsson model with $\delta = 0.25$. The probability of finding nucleons in the SD subspace is shown for each spin in fig. 3. One sees that the probabilities are large up to 4^+ state. Thus we can say that, even in well deformed nuclei, the truncation into the SD subspace is good for low-spin.

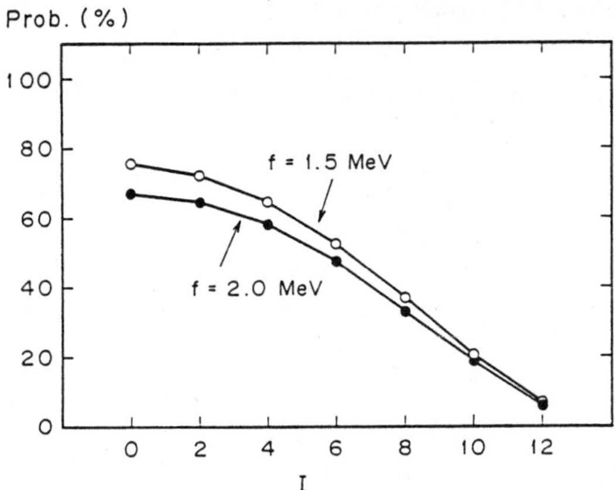

Fig. 3: Probabilities of the shell model eigenstates in the SD
subspace.

3. INTERACTING BOSON MODEL

We proceed to map fermion operators defined in the SD subspace
onto boson operators defined in a boson space consisting of s and d
bosons. One method for this purpose was proposed by Otsuka, Arima
and Iachello.[5] This method was shown to work well in many cases.

Now we have s and d bosons which are interacting with each other.
Then we call this model the interacting boson model[11] (IBM). The similar
model was introduced already in 1966 by Taruishi[12] and by myself.[12]
There were, however, not so many data in those days. In 1974, Iachello
found that the interacting boson model works very well in many nuclei.[13]
The name of the Interacting Boson Model or Approximation was given by
Iachello. Since s and d have altogether 6 degrees of freedom, they
can be taken as basis states of the SU(6) group. Then the model can be
called the SU(6) model. Janssen, Jolos and Dönau quite independently
invented the SU(6) model though they did not introduce s bosons.[14]

Let me now proceed to phenomenological analyses of low-lying
collective states of even-even nuclei using the Interacting Boson
Model. Here we ignore the difference between proton and neutron
degrees of freedom. The general Hamiltonian can be very complicated.
We can, however, set up three simplifying cases as described below.

The SU(6) groups has subgroups SU(5), SU(4), SU(3) and SU(2).
All physical states should have good angular momenta. Then the
subgroup SU(2) must be in any chain. Then we can have three different

chains of subgroups.[15]   They are[16),17),18)]

1      $SU(6) > SU(5) > SO(5) > SU(2)$

2      $SU(6) > SU(3) > SU(2)$

3      $SU(6) > SU(4) \equiv O(6) > SU(2)$.

Corresponding to each chain, we can express a Hamiltonian in terms of the Casimir operators C of those subgroups;

1) $H = \epsilon \, C_1(U_5) + \alpha \, C_2(SU_5) + \beta \, C_2(SO_5) + \gamma \, C_2(SU_2)$

2) $H = \alpha_3 C_2(SU_3) + \gamma_3 C_2(SU_2)$

3) $H = \alpha_2 C_2(O_6) + \beta_2 C_2(SO_5) + \gamma_2 C_2(SU_2)$

where $C_1(U_5)$ denotes the linear Casimir operator of the U_5 group and others are quadratic Casimir operators. $(C_2(SU_2) = L^2)$. Expectation values of these Hamiltonians are easily expressed as

1) $E = \epsilon \, n_d + \alpha \, \dfrac{n_d(n_d-1)}{2} + \beta(n_d - 2n_\beta)(n_d - 2n_\beta + 3) + \gamma \, J(J+1)$

where n_d is the number of d-bosons and n_β is the number of zero pairs of d bosons,

2) $E = \alpha_3\{\lambda^2 + \mu^2 + \lambda\mu + 3(\lambda + \mu)\} + \gamma_3 L(L + 1)$

where (λ, μ) are the labels of the SU(3) groups, and

3) $E = 2\alpha_2\sigma(\sigma + 4) + \beta_2 \, \dfrac{1}{6} \, \tau(\tau + 3) + \gamma_2 L(L + 1)$

where σ, and τ are labels of the SO(6) and SO(5) subgroups.

Some typical examples are shown in figs. 4, 5 and 6. One would see from those figures that some nuclei indeed show some characteristic patterns of symmetry. These three limiting cases have corresponding to Bohr-Mottelson pictures;

SU(5) - unharmonic vibrator

SU(3) - β stable deformed nuclei

SO(6) - γ unstable deformed nuclei.

Dynamical Symmetry. I

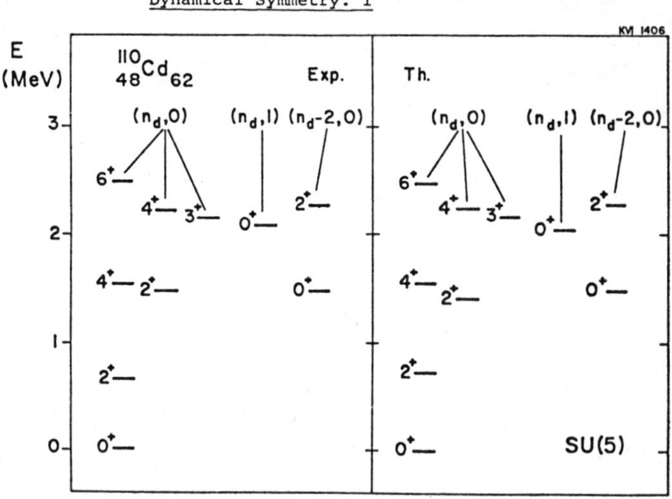

Fig. 4: An example of a spectrum with SU(5) symmetry: $^{110}_{48}Cd_{62}$.

Dynamical Symmetry. II

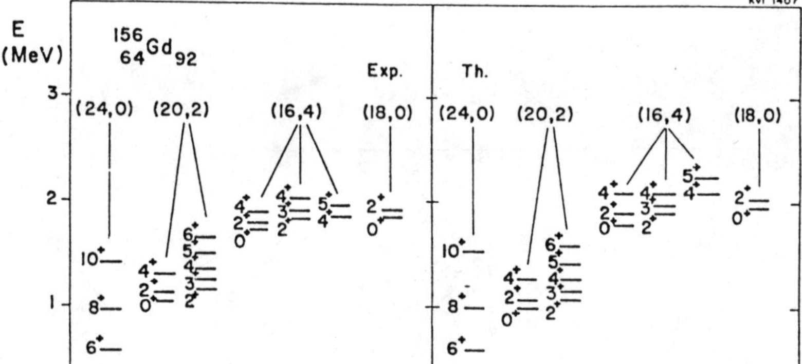

Fig. 5: An example of a spectrum with SU(3) symmetry: $^{156}_{64}Gd_{92}$.

<u>Dynamical Symmetry. III</u>

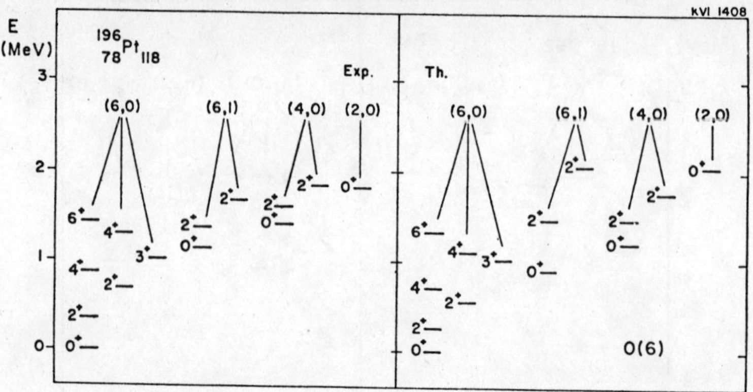

Fig. 6: An example of a spectrum with O(6) symmetry: $^{196}_{78}\text{Pt}_{118}$.

Real nuclei deviate more or less from those ideal cases. We need to
introduce perturbation to those simple Hamiltonians. For example[19]
the following Hamiltonian can be used to study phase transition from
SU(5) to SU(3) nuclei;

$$H = (\varepsilon_0 - \theta\hat{N})\hat{N}_d + \kappa C(SU_3) + \kappa'L^2,$$

where ε_0, θ, κ and κ' are constant. Some results are shown in
figs. 7 and 8.

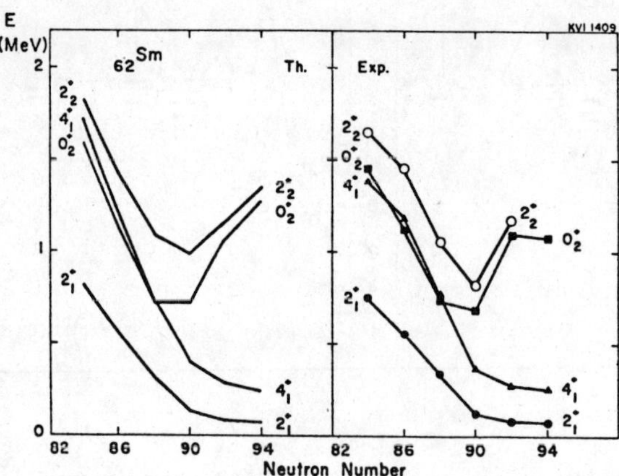

Fig. 7: Typical features of the transition from SU(5) to SU(3).
Electromagnetic transition rates.

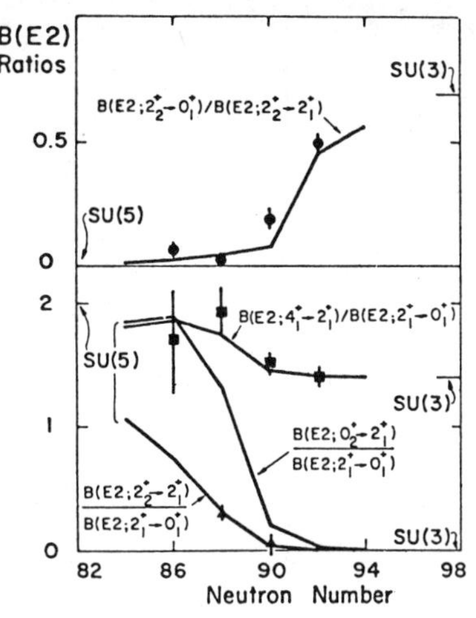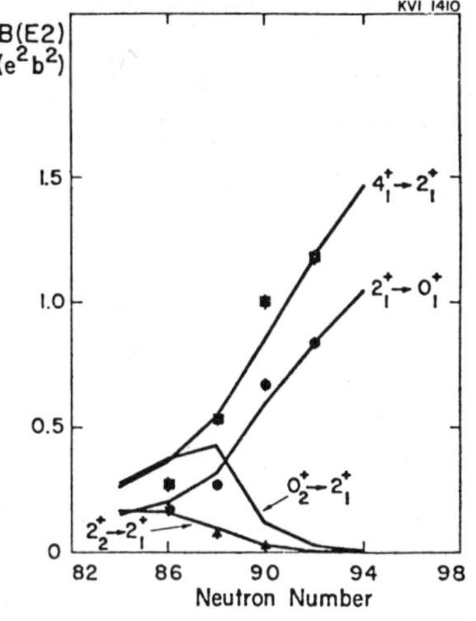

Fig. 8:
Typical features of the transition from SU(5) to SU(3).

One sees from those figures that the transition from SU(5) type to SU(3) type is well explained. The other transition from SO(6) to SU(3) was studied very thoroughly by Casten and Cizewski.[20]

An extremely interesting example was recently studied by Warner, Casten and Davidson[21] for the low-lying levels of ^{168}Er. The experimentally obtained levels have been analyzed by the same authors using the Interacting Boson Model. They concluded from their study that the levels up to 2 Mev are well described by the model. See figs. 8 and 9.

4. DISCUSSIONS

It is very natural from the shell model point of view to distinguish the proton degrees of freedom and those of neutrons.[3),4)] This leads us to introduce the Interacting Boson Model 2. There have been already several papers on it.[22] An advantage of the model 2 is that parameters used in this model can be related more directly with those provided by the shell model. The model 2 further more produces better results for SO(6) nuclei than the model 1.

The Interacting Boson Model is still young and premature.[23] I would like to invite many researchers to be interested in the model and tries to develop it in any new directions.

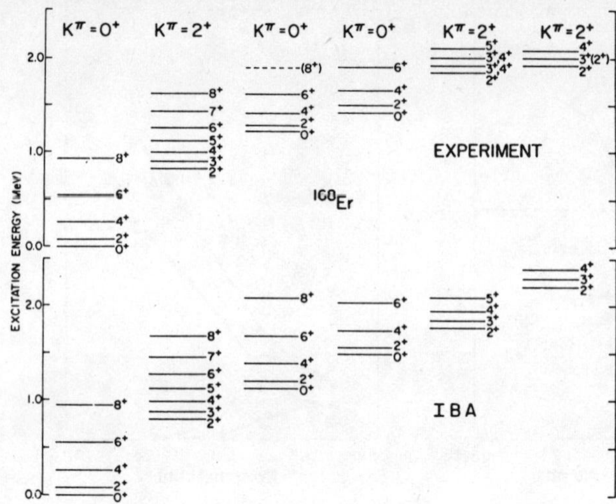

Fig. 9: Levels in ^{168}Er compared with the results of an IBA calculation for 16 bosons. Only experimental bands below the pairing gap, estimated as $\sim$2 MeV, have been included. (From ref. 21.)

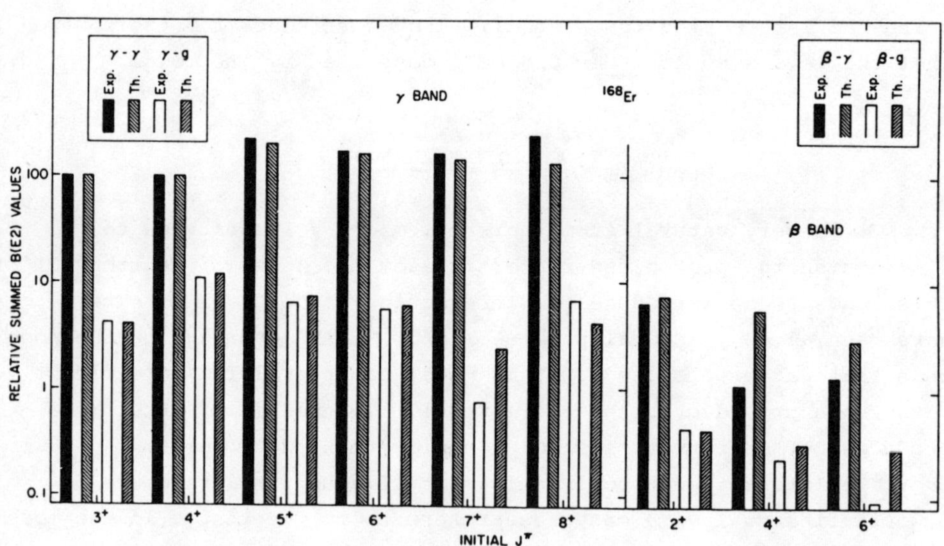

Fig. 10: Comparison of calculated and experimental values of the summed B(E2) strengths from states in the γ and β bands to the γ and ground bands. For each initial state the bars represent the sum of all observed or calculated transitions to a given final band. The B(E2) values have been normalized to 100 for an intraband transition each case. To keep the figure uncluttered for the β band, only the interband transitions are plotted. (From ref. 21.)

867

REFERENCES

1. A. Bohr, Mat. Fys. Medd. Dan. Vid. Selsk. <u>26</u> (1952) no. 14
   A. Bohr and B. R. Mottelson, Mat. Fys. Medd. Dan. Vid. Selsk.
   <u>27</u> (1953) no. 16.
2. J. P. Elliott, Proc. Roy, Soc, <u>A245</u> (1958) 128; 245 (1958) 562.
3. A. Arima, T. Otsuka, F. Iachello and I. Talmi, Phys. Lett. <u>66B</u>
   (1977) 205.
4. A. Arima, T. Otsuka, F. Iachello and I. Talmi, Phys. Lett. <u>76B</u>
   (1978) 139.
5. T. Otsuka, A. Arima and F. Iachello, Nucl. Phys. <u>A309</u> (1978) 1.
6. T. Otsuka, preprint and submitted to Phys. Rev. Lett.
7. A. Arima, Nucl. Phys. <u>A347</u> (1980) 339.
   A. Arima, N. Yoshida and J. N. Ginocchio, to be published in
   Phys. Lett.
8. A. Bohr and B. R. Mottelson, Phys. Scripta <u>22</u> (1980) 468.
9. J. Dukelsky, G. G. Dussel and H. M. Sofia, preprint.
10. T. Otsuka, private communication
11. A. Arima and F. Iachello, Phys. Rev. Lett. <u>35</u> (1975) 1069.
12. K. Taruishi, Prog. Theor. Phys. <u>39</u> (1968) 53.
    A. Arima, Soryushiron-Kenkyu (Japanese) <u>35</u> (1967) E47
13. F. Iachello, Proc. Intern. Conf. on Nuclear Structure and
    Spectroscopy, Amsterdam (1974), ed. by H. P. Blok and A. E. L.
    Dieperink.
14. D. Janssen, R. V. Jolos and F. Dönau, Nucl. Phys. <u>A224</u> (1974) 93.
15. F. Iachello, Lecture Notes in Physics 119, Nuclear Spectroscopy,
    "Lecture Notes of the Workshop, Gull Lake, Michigan, August
    27 - Sept. 7, 1979," ed. by G. F. Bertsch and D. Kurath,
    (Springer, 1980).
16. A. Arima and F. Iachello, Ann. Phys. (N. Y.) <u>99</u> (1976) 253.
17. A. Arima and F. Iachello, Ann. Phys. (N. Y.) <u>111</u> (1978) 201.
18. A. Arima and F. Iachello, Ann. Phys. (N. Y.) <u>123</u> (1979) 468.
19. O. Scholten, F. Iachello and A. Arima, Ann. Phys. <u>115</u> (1978) 325.
20. R. F. Casten and J. A. Cizewski, Nucl. Phys. <u>A300</u> (1978) 477.
21. D. D. Warner, R. F. Casten and W. F. Davidson, Phys. Rev. Lett.
    <u>45</u> (1980) 1761.
22. For examples
    U. Kaup and A. Gelberg, Z. Phys. <u>A293</u> (1973) 311.
    P. Van Isacker and G. Puddu, Nucl. Phys. <u>A348</u> (1980) 125.
    G. Puddu, O. Scholten and T. Otsuka, Nucl. Phys. <u>A348</u> (1980) 109.
    R. Bijker, A. E. L. Dieperink, O. Scholten and R. Spankof,
    Nucl. Phys. <u>A344</u> (1980) 207.
23. Interacting Bosons in Nuclear Physics, ed. by F. Iachello,
    (Plenum, New York, 1979).
    Interacting Bose - Fermi System in Nuclei, ed. by F. Iachello,
    (Plenum, New York, 1981).

CHAPTER IX

THE OSCILLATING BEHAVIOUR OF BACKBENDING[+]

Amand Faessler
Universität Tübingen
Institut für Theoretische Physik
D-7400 Tübingen, W.-Germany

Abstract:

The anomaly of the moment of inertia usually called backbending varies in its strength regularly as a function of the neutron number for the rare earth nuclei. This oscillating behaviour of backbending due to the rotational alignment of two $i_{13/2}$ neutrons is explained in a particle number and angular momentum conserving model. The neutrons in the $i_{13/2}$-shell are described by shell model wave functions and are weakly coupled to the rotation of the core. It is shown that the alignment is arranged by the nucleus by scattering two $i_{13/2}$ neutrons mainly from the $\Omega = \frac{1}{2}$ and $\Omega = \frac{3}{2}$ $i_{13/2}$ orbits into the core. The aligned band has therefore in average two neutrons less in the $i_{13/2}$-shell. Interaction between the valence nucleon and the core is due to the pairing force. It is shown that this interaction has $j - \frac{3}{2}$ zeros. No interaction between the ground band and the aligned band means strong backbending while a strong interaction means no backbending. In the second part of this talk we explain the behaviour of the second anomaly of the moment of inertia by the alignment of two $h_{11/2}$ protons corresponding to the same oscillating behaviour as discussed before. In this way we can explain that for the N=90 isotones $^{156}_{66}$Dy shows no backbending, $^{158}_{68}$Er shows upbending and $^{170}_{70}$Yb shows a strong second anomaly of the moment of inertia.

[+]Part of Lectures given at the Chinese Winter School on Nuclear Physics in Peking, China, December 22, 1980 - January 9, 1981.

1. Introduction

Today practically everyone working in the field of High Spin States agrees that the anomaly of the moment of inertia normally called backbending is due to the intersection of two rotational bands (see Fig. 1). It has first been proposed by Stephens and Simon[1] that the upper band is the aligned two quasi-particle $i_{13/2}$ neutron band. The large effective moment of inertia is due to the fact that by alignment one can obtain easily a large amount of angular momentum without needing too much energy. The alignment is due to the Coriolis force

$$V_c(i) = - \frac{\hbar^2}{\Theta} \vec{J} \cdot \vec{j}(i) \tag{1}$$

which tries to align the single particle angular momenta j. Parallel to the total angular momentum J. If the direction of the total angular momentum J defines the intrinsic x-axis then the Coriolis matrix element is proportional to

$$\langle V_c(i) \rangle \propto \sqrt{j(j+1) - \Omega(\Omega \pm 1)} \tag{2}$$

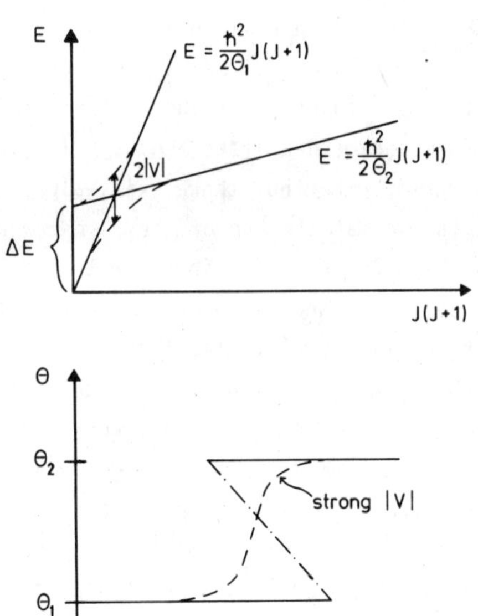

Fig.1

Fig. 1:

The upper part of this figure shows the intersection of two rotational bands in a diagram energy against angular momentum squared. The two bands interact with each other by the matrix element V. The distance between the perturbed band near the intersection point is given by $2|v|$. The lower part of the Fig. shows a schematic backbending plot. Without interaction between the two bands one gets the dashed dotted line while a strong interaction prevents backbending and one obtains an upbending as indicated by the dashed line.

The symbol Ω represents the angular momentum projection of the $i_{13/2}$ neutron state onto the intrinsic symmetry axis. Since the values of Ω near the Fermi surface increase with the increasing neutron number, one expects strong backbending at the beginning and no backbending at the end of the rare earth nuclei. The experimental situation is quite different[2]. A look to the data suggests that backbending oscillates over the rare earth nuclei with increasing neutron and proton numbers. The fact that one has strong backbending in ^{166}Yb no such backbending in ^{168}Yb and backbending again in ^{170}Yb and also no backbending in ^{180}Os is well described within the Hartree-Fock-Bogoliubov (HFB) Cranking model[4-6]. This description has been applied estensively by the groups in München, Jülich-Tübingen and Pittsburgh-New Orleans[4-7].

From the strong backbending found in some of the Os isotopes and the neighbouring odd mass nuclei experimental groups[8] deduced that the $h_{9/2}$ proton should be responsible for backbending in this region. But Faessler, Ploszajczak and Sandhya Devi have shown[6] that even in those nuclei backbending is due to the alignment of two $i_{13/2}$ neutrons. The simple notion that backbending should be proportional to the Coriolis matrix element (2) can therefore not be correct. As mentioned already above cranked HFB calculations[3-7] could explain the variations of backbending over the rare earth region. A first systematic study of the variation of the interaction between the ground state band and the aligned band was performed by Bengtsson, Hamamoto and Mottelson[9] in a simple HFB calculation built only on the neutron $i_{13/2}$ basis as a function of the neutron Fermi surface. They showed that the absolute value of this interaction oscillates as a function of the neutron surface and has five zeros at which one expects strong backbending (see Fig. 2). The interaction is increasing with increasing neutron number but there exist always zeros at special isotopes for which one expects even at the end of the rare earth nuclei strong backbending. Such a zero is for example responsible for the strong backbending in ^{182}Os, ^{184}Os and ^{186}Os. Bengtsson and Frauendorf[10] analysed the experimental data across the rare earth nuclei and extracted the absolute value of the interaction between two bands and compared it with the theoretical calculations. There is a rough qualitative agreement between the cranked HFB theory and the experimental data (see fig. 3). In detail the agreement is not too good since the interaction has been extracted from a cranked HFB theory using a constant deformation β and constant pairing gaps Δ_P and Δ_N. But there seems to be no doubt that the oscillations of the interaction between the ground band and the aligned band are qualitatively the same in experiment and theory. Nevertheless there remain some important open questions:

1. Is there a simple explanation for these oscillations? Bengtsson, Hamamoto and Mottelson[9] explained it as an interference effect. But it is not a convincing simple explanation. Why does one find $j-\frac{3}{2} = 5$ zeros for the $i_{13/2}$-shell?

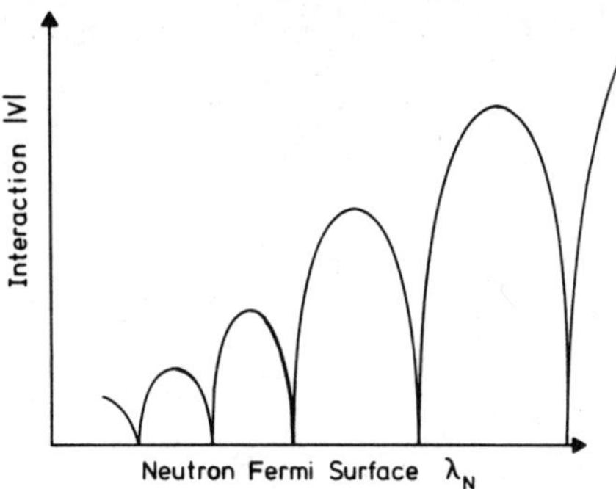

Fig. 2

Fig. 2:

Interaction between the ground band and the aligned band at the backbending point as a function of the neutron Fermi surface for an $i_{13/2}$ neutron shell qualitatively according to Hartree-Fock-Bogoliubov calculations.

2. The original calculation of Bengtsson, Hamamoto and Mottelson was restricted to the $i_{13/2}$-shell. Is this not a too small basis? Could the small size of the basis not introduce spurious effects?

3. All these calculations do not conserve the total proton and neutron number. If one describes an effect which varies strongly with the neutron number one should have a good total neutron number in the theory.

4. The cranking approach conserves angular momentum only in average. It was pointed out especially by Hamomoto[11] that this may lead to difficulties in the backbending region. Therefore one would like to have a theory with good total angular momentum.

Such a theory which answers the above questions will be developed in chapter 2. In detail it will give a simple explanation of the oscillating behaviour; it will include not only the $i_{13/2}$-shell but also the core, it will have total good particle number and conserve the angular momentum exactly.

In chapter 3 we shall then concentrate on the second anomaly of the moment of intertia around angular momentum J=28. We shall see that the behaviour of this second anomaly as a function of the proton number can be explained by the oscillating behaviour of the interaction between the aligned $i_{13/2}$ neutron two quasi-particle band and the $h_{11/2}$ proton two quasi-particle band. In chapter 4 I will summarize the results of this talk. The work which I am presenting here has been done in collaboration with F. Grümmer, M. Ploszajzcak and K.W. Schmid.

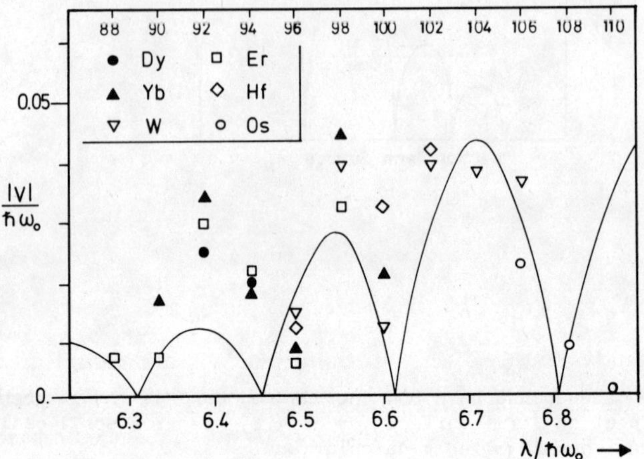

Fig. 3:

Interaction between the ground band and the aligned $i_{13/2}$ two quasi-particle neutron band as a function of the neutron Fermi surface in units of the oscillator energy $\hbar\omega_O$. The solid line represents the theoretical result of Hartree-Fock-Bogoliubov calculations. The points correspond to experimental interactions extracted from the data by Bengtsson and Frauendorf[10].

2. Core-Cluster Coupling Model for Backbending

To answer the questions 1-4 in the introduction we present here a core-cluster
coupling model[12]. This model should give a simple explanation of the oscillating
behaviour of backbending. It will not only include the $i_{13/2}$-shell but also describe
the coupling of this shell to the core, it will conserve the total particle number
and describe the nuclear states with good total angular momentum J.

We devide the nucleus in two parts, into the $i_{13/2}$ neutron shell and the rest.

$$H = H_{core} + H_{shell} + H'$$ (3)

The core is described as a simple axially symmetric rotor.

$$H_{core} = E(N_C)_{core} + \frac{\hat{R}^2}{2\Theta_0}$$ (4)

E_{core} depends on the number of nucleons in the core (N_C). The moment of inertia
Θ_0 is in principle also a function of N_C. But to simplify the calculations we shall
assume that the moment of inertia has the same value for all cores needed. The neutrons in the $i_{13/2}$-shell are described by a spherical shell model Hamiltonian.

$$H_{shell} = \sum_m \tilde{\varepsilon}_{i_{13/2}} \, C^+_{13/2;m} \, C_{13/2;m} + \frac{1}{2} \sum_{l,k} V(l,k)$$ (5)

As the residual interaction between the neutrons $V(l,k)$ we use the surface delta-
interaction.

$$H' = -\frac{1}{2} \lambda \, \hat{Q}_{core} \cdot \hat{Q}_{i_{13/2}}$$

$$-G \sum_{i,k} \left\{ \left[C^+_i C^+_i \right]^{core}_0 \left[C_i C_i \right]^{i_{13/2}}_0 + h.c. \right\}$$ (6)

The interaction between the core and the
valence nucleons consists of two parts. The quadrupole-quadrupole interaction between the core and the neutron $i_{13/2}$-shell introduces into the motion of the $i_{13/2}$
neutrons the fact that the potential in which they are moving is deformed. But
opposite to the strong coupling picture we use neutron wave functions in the laboratory frame. The second part is that part of the pairing force which scatters pairs
of particles coupled to angular momentum zero from the $i_{13/2}$ valence nucleons into
the core and vice versa. By this last interaction one has the possibility to exchange nucleons between the core and the $i_{13/2}$-shell. The pairing forces is here
the main contribution since we assume that all the nucleons in the core are paired
to angular momentum zero and we get only alignment in the $i_{13/2}$-shell. An exchange
of one nucleon alone is practically not possible since the $i_{13/2}$ is a positive parity

intruder into a region of only negative partiy states. Due to parity conservation
we can therefore only exchange two nucleons from the valence shell into the core
or the other way around. The Hamiltonian (3) has the following general solution.

$$|JM;A> = \sum_{RN_C,N_V,v\alpha I} C_{RN_C;v\alpha IN_V} \left\{ |RN_C> \quad v\alpha IN_V> \right\}_J$$

(7)

The mixing coefficients C have to be found by diagonalisation. The sum has to be
performed under the constraint N_C+N_V = A. This restriction ensures the conservation of the total particle number A but allows a variation of the nucleons in the
core N_C and the nucleons in the $i_{13/2}$-shell N_V. Further quantum numbers are the
angular momentum of the core R, the seniority v of the $i_{13/2}$ neutrons, an additional
quantum number α to describe different states with otherwise the same quantum numbers and the angular momentum I of the $i_{13/2}$ neutrons.

One could have the idea that a truncation of the sum in eq. (7) may be possible
and may reduce the numerical work. One could for example allow only a definite number of nucleons in the core and in the valence shell and restrict the states in the
valence shell to a limited seniority number v. If one takes into account the symmetry between particle and hole states the maximum seniority for the $i_{13/2}$-shell is
v=6. Fig. 4 shows the probability for the angular momentum distribution in the
valence shell if one allows for six $i_{13/2}$ neutrons (N_V=6) and restricts the maximum
seniority to 2,4 or 6, respectively. One sees that with the restriction to seniority
V=2 and smaller the nucleons show no alignment. The probability distribution for the
total valence angular momentum I is concentrated around I=0 for the total angular
momentum J=0 and J=24. This means that the total angular momentum J is essentially
due to the core angular momentum. If one allows for seniority $v \leqslant 4$ one finds no
alignment for the total angular momenta J=0 60 to J=18. But for J=20 and higher one
pair is fully aligned to angular momentum 12. One finds therefore a very rapid
alignment between total angular momentum J=18 and J=20. If one includes all possible
seniority states ($V \leqslant 6$) one finds a relative smooth alignment of the total valence
angular momentum I around total angular momentum J=16. A conclusion from Fig. 4 is
therefore that it is not allowed to restrict the seniority of the states in the
$i_{13/2}$-shell.

Let us now test if it is allowed to fix the distribution of particles between
the valence shell and the core. Fig. 5 gives the backbending plot for zero to 14
particles in the $i_{13/2}$ neutron valence shell. We find the result expected from the
Coriolis force (2). If we have only two particles in the $i_{13/2}$-shell and therefore
a Fermi surface at the beginning of the shell (small Ω) then we find strong backbending. If we increase the particle number backbending disappears. (0 and 14

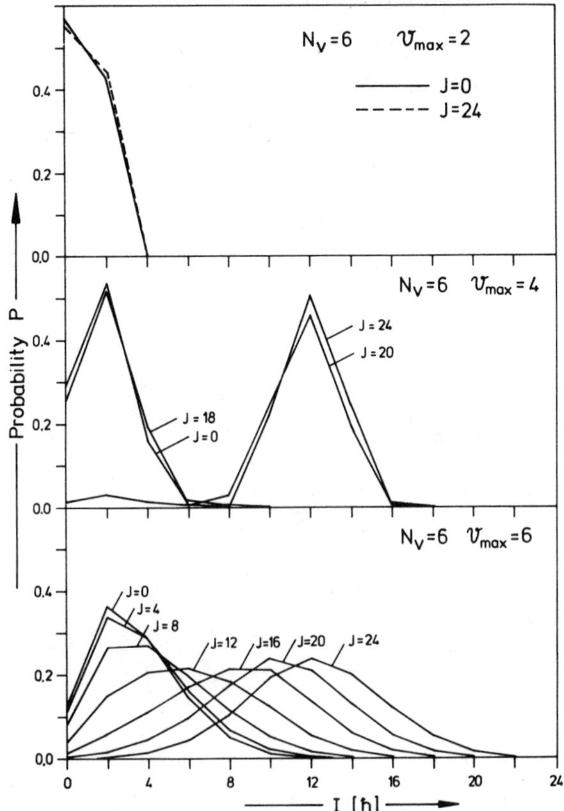

Fig. 4:

The probability of finding the $N_V=6$ nucleons of the $i_{13/2}$ neutron shell in a given yrast wave function with angular momentum J coupled to the different possible valence spins I. The seniority has been truncated in the upper and the middle part of the figure to a maximum value v=2 and v=4, respectively.

neutrons in the valence shell are pathological cases since in a totally empty or in a totally filled shell all the angular momentum has to come from the core). If we request therefore that there is no exchange between neutrons in the core and in the valence shell we are not getting the oscillating behaviour of backbending found in experiment. This seems to be strange in view of the fact that the oscillating behaviour as a function of the neutron number is seen in the cranked HFB model even if one restricts oneself to the $i_{13/2}$ basis. We shall see later on that this oscillating behaviour is seen instead of this restriction and is connected with a deficiency of the HFB approach, the non-conservation of the particle number.

We conclude therefore that we have to allow that neutron pairs are scattered between the core and the valence shell as requested by the pairing part of the interaction H' in eq. (6). To be able to calculate results with scattering of neutron pairs between the core and $i_{13/2}$-shell we have to know how the ground state

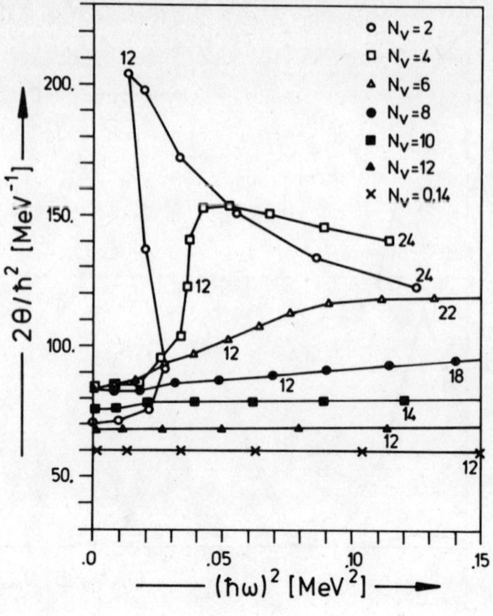

$2\theta/\hbar^2 \; [\mathrm{MeV}^{-1}]$

$(\hbar\omega)^2 \; [\mathrm{MeV}^2]$

Fig. 5:

The backbending plots obtained for different fixed particle numbers N_V in the $i_{13/2}$ valence shell. Some of the points are labeled with the values of the corresponding total angular momentum J.

energies for the core E_{core} (N_C) depends on the number of the nucleons in the core. If we assume a linear dependence

$$E_{core}(N_C) = aN_C = aA - aN_V \tag{8}$$

we can describe the core energy as a function of the neutrons in the valence shell N_V using the conservation of the total nucleon number $N_C + N_V = A$. A similar ex - pression proportional to the particle number in the valence shell we get from the independent shell model Hamiltonian contained in eq. (5). The terms proportional to N_V yield a contribution to the diagonal matrix elements of the form:

$$\Delta E(diag, N_V) = N_V \left[\tilde{\varepsilon}_{i_{13/2}} - a \right] = N_V \varepsilon \tag{9}$$

The new parameter ε regulates the average distribution of the particles between
the valence shell and the core. A large ε favours a small number of nucleons in
the valence shell N_V while a small or even negative ε prefers a large number of
valence neutrons. The quantity $-\varepsilon$ is therefore something like a Fermi surface.
But I want to stress that we are conserving here the total particle number A exactly. Only for the distribution of the nucleons to the valence shell and to the core
we can not give an exact number but only an average expectation value since neutrons pairs are always scattered from the core into the valence shell and back
again.

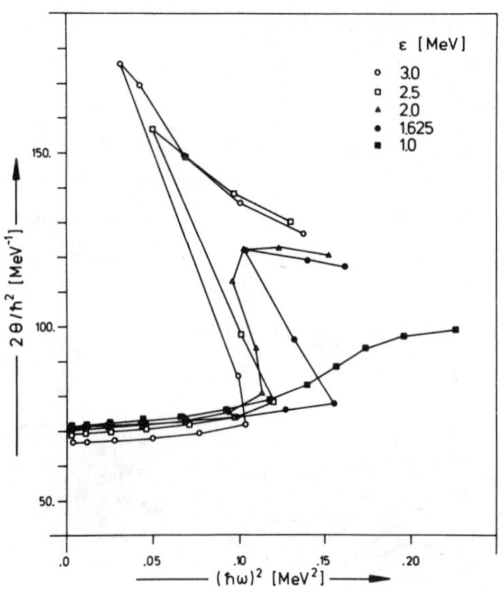

Fig. 6:

The backbending plots as obtained for different values of the particle number parameter ε, which describes the filling of the $i_{13/2}$-shell. The calculation includes
the scattering of neutron pairs from the valence shell into the core by the pairing
force. One sees the oscillations of backbending as a function of the particle num-
ber parameter ε. (Large value of ε corresponds to an almost empty and a small value
of ε to an almost filled shell).

Fig. 6 shows the backbending plot for different parameters ε and therefore for
different neutrons in the valence shell. One sees that backbending is now oscilla-

ting as a function of ε and therefore of the average particle number in the valence shell. How can this oscillation of backbending with the particle number now be explained?

An analysis of the result shows that the upper or aligned band has a tendency to have less nucleons than the ground state band. Two of the $i_{13/2}$ neutrons are aligned with their angular momentum along the rotational axis. This means the corresponding two quasi-particles have to sit in the low $|\Omega|$ states. Those states are below the Fermi surface and we have therefore two aligned holes. If we would only have the $i_{13/2}$-shell the two particles would have to be moved higher up in the sam shell into empty levels. But the nucleus tries to save energy. Between the last occupied $i_{13/2}$ level and the first empty $i_{13/2}$ are many core levels as can be seen from any Nilsson diagram for the neutron state in the rare earth nuclei. Due to the pairing force the two neutrons removed from the $i_{13/2}$-shell scatter into the core. If this interaction is very strong we find no backbending. If this interaction is weak one finds strong backbending. The interaction between a configuration with N and N+2 particles in the valence shell due to the pairing interaction (6) is shown in fig. 7 as a function of the total angular momentum J. We see that those interactions have all a zero as a function of the total angular momentum apart of the interaction with the total empty ($W_{0,2}$) and the total filled shell ($W_{12,14}$). Fig. 8 shows the total energies as a function of the total angular momentum J for a fixed number of neutrons in the $i_{13/2}$-shell N_V.

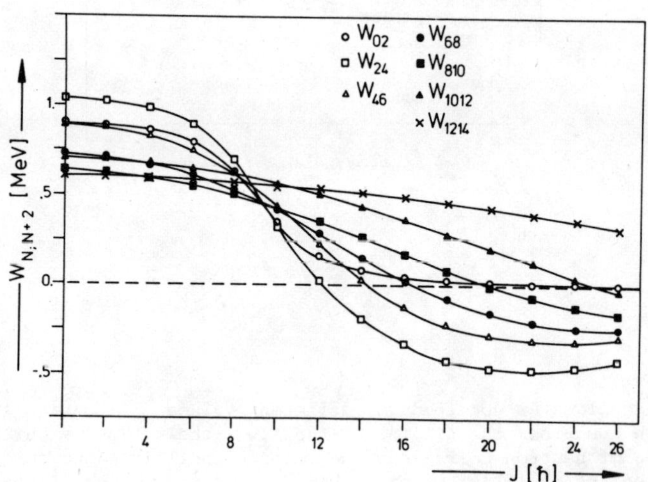

Fig. 7:

The residual pairing interaction $W_{N,N+2}$ between the 8 fixed number yrast solutions for the $i_{13/2}$ valence shell as a function of the total angular momentum J. One notes that the interaction with the total empty shell $W_{0,2}$ and with the total filled shell $W_{12,14}$ have no zeros as a function of the total angular J.

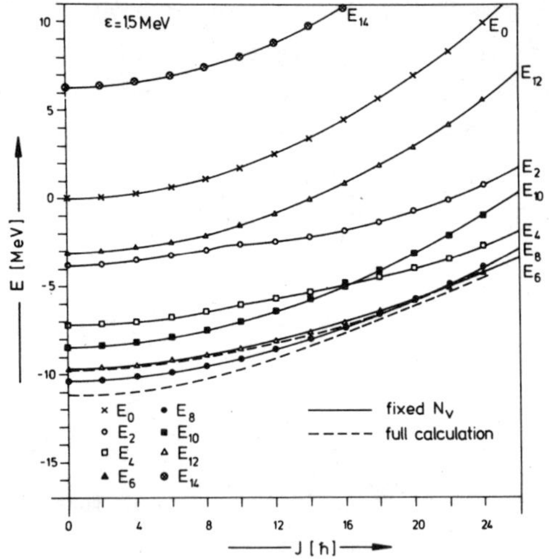

Fig. 8:

The yrast energies $E_N(J)$ as obtained conserving the particle number N_V in the $i_{13/2}$-shell for $\varepsilon=1.5$ MeV. The dashed curves represent the two energetically lowest solutions of the complete diagonalisation of the Hamiltonian (3) for the same particle number parameter ε (9).

The rotational band with 8 neutrons in the valence shell intersects with the one with 6 neutrons in the valence shell around angular momentum 20. If the particle number is decreased the energy E_6 moves down and the energy E_8 moves up. Then it happens that the two band intersect around angular momentum J=16. But at this energy the interaction $W_{6,8}$ has a zero and we find strong backbending. If we now increase or decrease the particle number in the valence shell the bands in Fig. 8 move up and down and always two different bands play the role of the ground state and the aligned band. From Fig. 7 one counts easily that we have five interactions with zeros as a function of the total angular momentum J. Increasing therefore the particle number in the valence shell from zero to 14 we expect also five zeros in the interaction between the two bands as a function of the neutron number. In general if we have a shell with single particle angular momentum j we expect j-3/2 zeros. The question for a simple explanation of the oscillating behaviour of backbending therefore boils down to the fact: Why do we have zeros in the interaction between the bands? To explain this we make an extremely naive and simplyfied model. We assume that the wave function for a definite particle number in the valence shell consists only of two contributions: the total paired wave function $\Phi_{I=0}$ and the wave function with one fully aligned neutron pair $\Phi_{I=12}$. The total wave functions for N_V-2 and N_V nucleons in the valence shell are now written in the following way:

$$\Psi_J(N-2) = a_J^{N-2} \, \Phi_{I=0}(N-2) + b_J^{N-2} \, \Phi_{I=12}(N-2)$$

$$\Psi_J(N) = a_J^N \, \Phi_{I=0}(N) + b_J^N \, \Phi_{I=12}(N)$$

$$(10)$$

For a fixed particle number in the $i_{13/2}$-shell the Coriolis force decreases with increasing this fixed particle number. Thus the critical angular momentum at which the two bands intersect moves up. This is now shown in Fig. 9. The intersection between the bands with N_V and N_V-2 id due to the pairing force. But the pairing force can not connect the two solutions in the angular momentum aerea where one solution is not yet aligned and the other is already aligned. Thus we expect a zero of this interaction. The range in which this zero is expected is indicated by an arrow in Fig. 9.

Fig. 10 shows now the final result of the absolute value of the interaction between the Yrast and the Yrare band as a function of the shell filling parameter ϵ. The solid lines are results of a calculation including only the lowest solution for each neutron number in the valence shell. It corresponds to a diagonalisation of a 8x8 matrix. The circles give results of the full calculation which corresponds to a diagonalisation of matrices with a dimension more than 2000. The lower part of the figure shows the critical angular momentum for backbending.

Backbending is thus explained as an interplay between the Coriolis and the pairing force. The Coriolis force aligns two holes preferentially in the occupied levels of the $i_{13/2}$-shell. But the nucleus wants to save energy. Thus the two nucleons are not moved into higher levels of the $i_{13/2}$-shell but are scattered as a pair coupled to angular momentum zero in the core. This pairing interaction which scatters pairs of particles form the valence shell into the core has zeros due to the fact that the critical angular momentum at which the $i_{13/2}$ pair aligns varies with the shell filling.

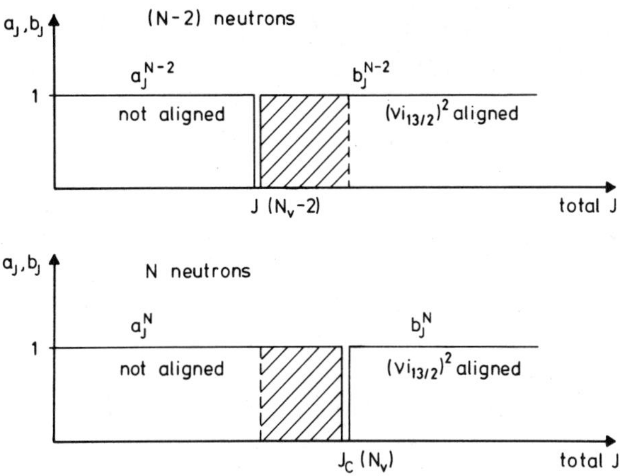

Fig.9

Fig. 9:

Sketch of the variation of the mixing parameters a_J and b_J for the extremely simplified consideration (10) to explain the zeros in the interaction $W_{N,N+2}$. The wave
functions for N_V-2 and N_V neutrons in the valence shell is represented as a function of the total angular momentum J. For larger particle number the Coriolis force
is weaker and therefore the critical angular momentum is increased. Since the
pairing force can only scatter pairs of particles coupled to angular momentum zero
it can not yield an interaction between the two states with different particle
numbers in the region between J_C (N_V-2) and J_C (N_V). We expect therefore always a
zero of the interaction between the ground band and the aligned band as a function
of increasing total angular momentum. The only exemptions are the interactions
with the totally empty shell and the totally filled shell where in both cases no
nucleon can align.

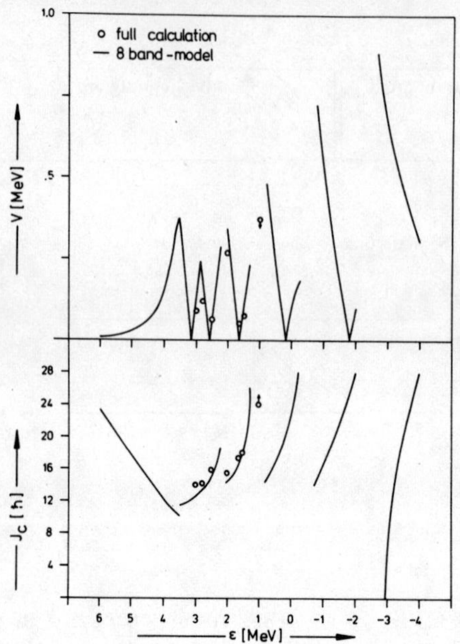

Fig. 10:

The yrast-yrare interaction and critical spin for the $i_{13/2}$ + rotor case as functions of the particle number parameter ε. The full lines correspond to the simple 8-band diagonalisation using the residual interaction from Fig. 8. The open dotts represent the results of the complete diagonalisation of the Hamiltonian (3). The lower part shows the critical angular momentum for the crossing of the two bands, if no interaction would be present.

3. Alignment of $h_{11/2}$ proton pair

This oscillating behaviour of the interaction between the 0 and 2 q.p. bands plays an essential role in the quantitative understanding of a second anomaly of the moment of inertia which has been detected[13] in 1977 and at the same time explained[14] as the alignment of an $h_{11/2}$ proton pair. Recently, the isotones $^{156}_{66}Dy_{90}$ and $^{160}_{70}Yb_{90}$ have been measured up to the angular momentum region of the second anomaly[15-17]. In the sequence of the isotones ^{156}Dy, ^{158}Er and ^{160}Yb (see Fig. 11), one finds around angular momentum 28 no anomaly of the moment of inertia, upbending and bb, respectively.

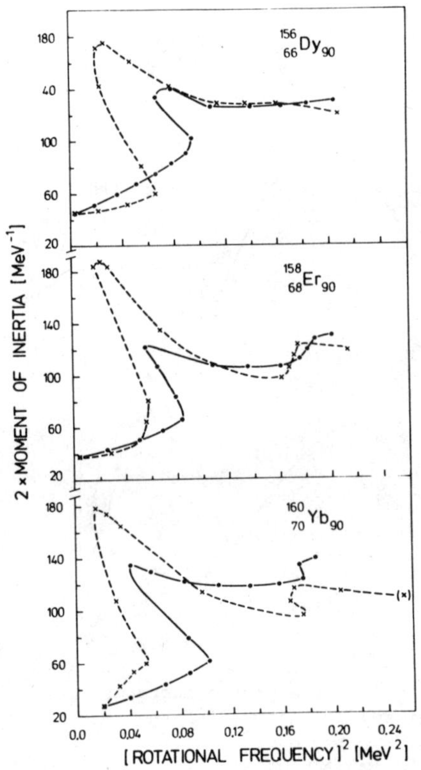

Fig. 11:

Backbending plots for the N=90 isotones $^{156}_{66}Dy_{90}$, $^{158}_{68}Er_{90}$ and $^{160}_{70}Yb_{90}$. The solid line gives the experiment[13,15-17] and the dashed line the theory. Theoretical results are calculated with the Hartree-Fock-Bogoliubov approach[4,18] and the Hamiltonian of Kumar and Baranger[21] (A·G_P=25 MeV; A·G_N=20 MeV; χ=72·A$^{-1.4}$ MeV). The proton gap parameter has been varied to minimize the total energy for each angular momentum. The other shape and pairing parameters are kept fixed (β=0.26; γ=0; β_4=0.08; Δ_N=0.9 MeV). If one shifts the proton $h_{11/2}$ level to the experimental value[24] then one gets the fit to the data with the experimental hexadecapole deformtion β_4=0.05.

Here we want to discuss the explanation of this strange behaviour. We use again the cranked Hartree-Fock-Bogoliubov theory[4-6,18]. The parameters of the Hamiltonian are given in the figure caption of Fig. 12. The expectation value of the total many-body Hamiltonian is in principle minimized as a function of the shape parameters β, γ and β_4 and the pairing gaps Δ_P and Δ_N.

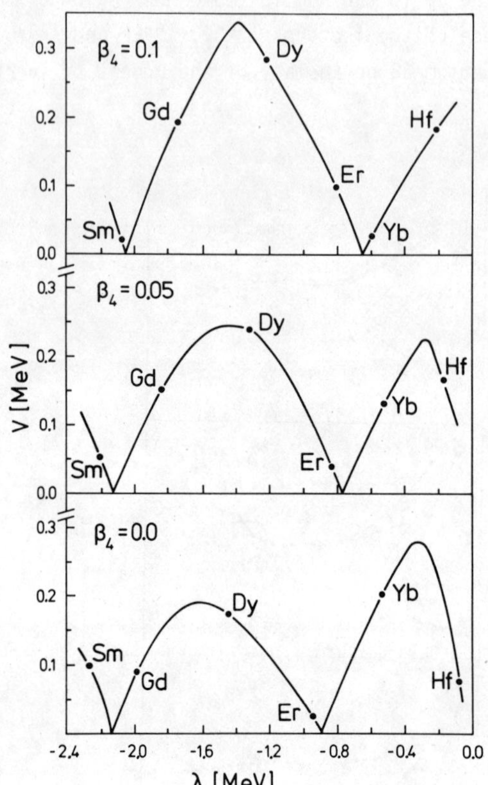

Fig. 12:

Interaction V at the crossing with the proton two quasi-particle $h_{11/2}$ band as a function of the proton Fermi surface λ for different β_4 values. The position of the Fermi surface for different N=90 isotones is indicated. A small absolute $|V|$ yields a large second anomaly for the moment of inertia. A shift of the proton $h_{11/2}$ level to the experimental value yields the best agreement with the data at the experimental value $\beta_4 = 0.05$.

The calculation in which all these parameters are varied to find the minimum of the total energy is numerically not feasible. But extensive numerical studies[4,5,6] showed that it is allowed to keep the shape parameters fixed with increasing angular momentum and to choose a constant value of the proton gap Δ_P if one studies the first bb and a constant value of the neutron gap Δ_N if one is interested in the second anomaly. The parameters chosen independently of the total angular momentum in the wave function have been chosen to minimize the energy in the intrinsic system[20]: β=0.26, γ=0, Δ_p=0.9 MeV. The choice of the hexadecapole deformation β_4=0.08 will be discussed below. The proton gap parameter Δ_p is varied to yield the minimum of the total energy for each average angular momentum $\langle I \rangle$.

Fig. 11 shows the results for the three isotones in the bb plot. The variation of the second anomaly of the moment of inertia is nicely reproduced. This is essentially due to the choice of the β_4 deformation (β_4=0.08).

Fig. 12 displays the interaction[19,20] V in MeV between the aligned $h_{11/2}$ protons. The interaction is defined as half the energy distance at the "crossing point" as a function of the cranking frequency ω. The interaction with the aligned $h_{11/2}$ proton pair shows the familiar oscillations first discussed by Bengtsson, Hamamoto and Mottelson[9] for the $i_{13/2}$-shell. For β_4=0 one obtains a strong second bb for ^{158}Er but none for ^{156}Dy and ^{160}Yb. The β_4 has to be chosen to be larger than or equal to β_4=0.08 to find for a spherical single particle energy of Kumar and Baranger[21] a strong second bb in ^{160}Yb, upbending in ^{158}Er and no anomaly in ^{156}Dy. The analysis of alpha scattering data by Hendrie et al.[22] gives around Z=62, N=90 the value β_4=0.05$\pm$0.01. The theoretical calculations tend to give larger values. Nilsson and coworkers[25] obtain with the Strutinsky method for N=90 isotones β_4=0.075 (Sm), 0.07 (Gd), 0.075 (Dy), 0.08 (Er). (These values are partially extrapolated). The value β_4=0.08 needed to obtain agreement with the data seems therefore slightly high but is still within the range of the theoretical results. One could obtain also the correct quantitative behaviour of the second bb for β_4=0.05 if one increases the $h_{11/2}$ proton single particle energy by 900 keV. Such an increase by the same amount has also recently been suggested by Chasman[24]. Such a shift is also favoured by the suggestion of Kleinheinz[25] that Z=64 should be a closed shell.

Fig. 12 shows that one expects (for a hexadecapole deformation larger than or equal to β_4=0.08 and no energy shift) a very small interaction between the intersecting bands at the second anomaly for Yb, a larger interaction for Er, and a very large interaction for Dy. According to this we find in these three isotones strong backbending, upbending and no anomaly at all.

Fig. 13 shows the alignment plot for ^{158}Er. It demonstrates that the first bb is due to the alignment of an $i_{13/2}$ neutron pair while the second anomaly is caused by the alignment of an $h_{11/2}$ proton pair.

Since we are able to reproduce the second anomaly for various proton numbers, this supports strongly the explanation of the second anomaly as the alignment of two $h_{11/2}$ protons.

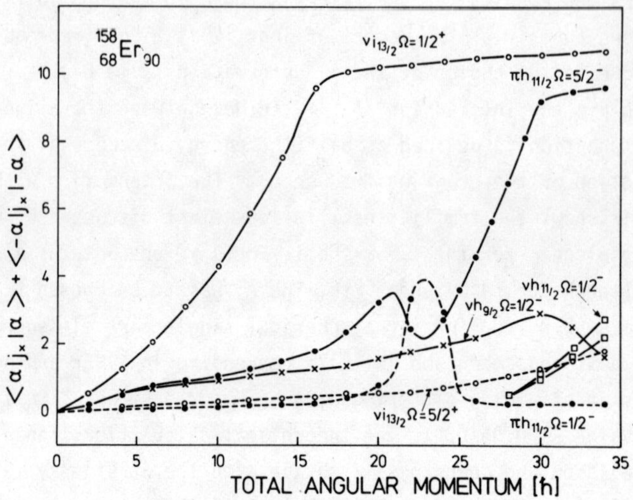

Fig. 13:

Alignment of the single particle angular momenta of a conjugate pair $|\alpha\rangle$ and $|-\alpha\rangle$ of nucleons along the rotational axis as a function of the total angular momentum for ^{158}Er. The quantum numbers assigned are only good for small total angular momentum I. At high I the main amplitude of $|\alpha\rangle$ may be characterized by a different single particle angular momentum projection Ω to the symmetry axis. The conjugate states are defined in the canonical representation as the quasi-particle states with the same occupation probability $v_\alpha^2 = v_{-\alpha}^2$.

4. Summary

In the first part of this talk we looked for a simple explanation of the oscillating behaviour of backbending as a function of the neutron and the proton number. To understand this effect we introduced a model which devides the nucleus in valence nucleons of the $i_{13/2}$ neutron shell and the core and conserves the total particle number and the total angular momentum. We found that it is essential not to restrict the wave functions in the valence shell to lower seniorities. The cnetral point was the possibility to exchange pairs of particles between the valence shell and the core. From this work emerges the following understanding of backbending:

The upper band has a tendency to have less nucleons in the $i_{13/2}$ than the ground state band. The two hole states are aligned by teh Coriolis force but not scattered into higher $i_{13/2}$ states. The nucleus saves energy. It therefore moves the two $i_{13/2}$ neutrons with the help of the pairing force into core states which lie between the last occupied and the first empty $i_{13/2}$ level. The pairing force connects therefore the ground band and the aligned band. Since it can only scatter pairs of particles coupled to angular momentum zero, this interaction has always a zero since the alignment for the $i_{13/2}$ pairs occurs at different critical angular momentum if we change the number of neutrons in the valence shell.

In the second half we discussed the second anomaly of the moment of inertia at around angular momentum J=28. We concentrated on the N=90 isotones, ^{156}Dy, ^{158}Er and ^{160}Yb. They show no second anomaly, an upbending and a strong backbending, respectively. This variation can be explained if one assumes that this anomaly is due to the alignment of two $h_{11/2}$ protons as originally proposed in Ref. 14. The variation of the backbending is nicely explained by the oscillating behaviour of the pairing interaction between the two intersecting bands. To reproduce the correct behaviour a hexadecapole deformation $\beta_4 = 0.05$ is needed in agreement with the measurements of this deformation.

References

1. F.S. Stephens, R. Simon, Nucl. Phys. $\underline{A138}$ (1972) 257.
2. R.M. Lieder, H. Ryde, Adv. in Nucl. Phys. 10 (1978) 1.
3. P. Ring, H.J. Mang, B. Banerjee, Nucl. Phys. $\underline{A225}$ (1974) 141.
4. A. Faessler, K.R. Sandhya Devi, F. Grümmer, K.W. Schmid, R.R. Hilton, Nucl. Phys. A256 (1976) 106.
5. A. Faessler, K.R. Sandhya Devi, A. Barroso, Nucl. Phys. $\underline{A286}$ (1977) 101.
6. A. Faessler, M. Ploszajczak, K.R. Sandhya Devi, Nucl. Phys. $\underline{A301}$ (1978) 529.
7. A.L. Goodman, Nucl. Phys. A265 (1976) 113.
8. A. Neskakis, R.M. Lieder, et al. Nucl. Phys. $\underline{A261}$ (1976) 189.
9. R. Bengtsson, I. Hamamoto, B. Mottelson, Phys. Lett. 73B (1978) 259.
10. R. Bengtsson, S. Frauendorf, Nucl. Phys. $\underline{A314}$ (1979) 27.
11. I. Hamamoto, Nucl. Phys. A271 (1976) 15; Phys. Lett. 66B (1977) 222.
12. F. Grümmer, K.W. Schmid, A. Faessler, Nucl. Phys. A326 (1979) 1.
13. I.Y. Lee et al., Phys. Rev. Lett. $\underline{38}$ (1977) 1454.
14. A. Faessler, M. Ploszajczak, Phys. Lett. $\underline{76B}$ (1978) 1.
15. F.A. Beck, E. Bozek, T. Byrski, C. Gehringer, J.C. Merdinger, Y. Schutz, J. Styczen, J.P. Vivien, Phys. Rev. Lett. $\underline{42}$ (1979) 493.
16. L.L. Riedinger et al. to be published.
17. D. Ward et al., Proc. of Int. conf. Nucl. Phys., Canberra, Australia, September 1978.
18. A. Faessler, M. Ploszajczak, K.W. Schmid, to be published.
19. A. Faessler, M. Ploszajczak, Phys. Lett. 76B (1978) 1.
20. F. Grümmer, K.W. Schmid, A. Faessler, Nucl. Phys. A308 (1978) 77.
21. M. Baranger, K. Kumar, Nucl. Phys. A110 (1968) 490 and 529.
22. D.L. Hendrie et al., Phys. Lett. 26B (1968) 127.
23. S.G. Nilsson et al., Nucl. Phys. A131 (1969) 1.
24. R.R. Chasman, Phys. Rev. C21 (1980) 456.
25. P. Kleinheinz et al., Z. Physik $\underline{A290}$ (1979) 279.

Chapter X

EXCITATION AND DECAY OF THE NEW GIANT MULTIPOLE RESONANCES

C. C. Chang
Department of Physics and Astronomy
University of Maryland
College Park, Maryland 20742/USA

I. Introduction

The giant multipole resonances (GMR) are highly collective modes of excitation in the nuclear continuum in which an appreciable fraction of the nucleons in a nucleus move together. These collective excitations are characterized as shape oscillation of various multipolarities. They are termed "giant" in the sense of containing a large fraction ($\geq 30\%$) of their respective sum rule limits (see Sec. II.3), the theoretical limits for the excitation of a given multipole strength in a given nucleus.

In 1947, a strong resonance behavior was observed in photonuclear reactions.[1] The existence of such a resonance was predicted theoretically by Migdal in 1944.[2] This resonance was later interpreted as due to dipole absorption. It was found that this giant resonance is a general feature of all nuclei, and that the centroid energy and the width of this resonance change smoothly from nucleus to nucleus.

Two models were proposed to explain the dipole vibration. In the Goldhaber-Teller (GT) model,[3] the proton sphere as a whole moves against the neutron sphere under the influence of the incident photon field. The restoring potential against the separation of protons and neutrons is the symmetry energy. On the other hand, in the Steinwedel-Jensen (SJ) model,[4] the proton and neutron fluids are two compressible interpenetrating fluids moving against each other within the rigid surface of the initial nucleus.

In this talk, I will not discuss the well-known electric giant dipole resonance (GDR). Instead, I will concentrate on the new non-dipole giant resonances.

In the early 1970's, a new giant resonance was discovered to locate at $E_x \sim 63 A^{-1/3}$ MeV.[5-7] This resonance was later identified as isoscalar electric giant quadrupole resonance (GQR). This discovery led to a large amount of activity throughout the world in trying to establish the main features of this new giant resonance, and in searching for more new giant resonances. In Sec. III, the experimental investigations of these new giant resonances using various singles experiments will be presented. Section IV discusses the general features of the new collective modes of excitation.

Various macroscopic[8,9] and microscopic[10-13] theories have been advanced to describe the systematics of these giant resonances. The calculations reproduced the resonance energy as a function of mass number and multipole degree correctly.

However, the experimentally determined resonance widths are generally larger than any theory can predict. Since the width of any excited state is the sum of the partial widths for its various decay channels, it is necessary to study the decay properties of the giant resonances to learn more about their structures and to understand the origin of their widths. In Sec. V, the various experiments which study the decay modes of the giant resonances are discussed, and in Sec. VI, the decay property as a function of nucleus mass in terms of statistical or non-statistical processes will be discussed.

II. A Brief Theoretical Framework

II.1   The Collective Model Description of Giant Resonances

By the very definition of a giant resonance, the coherent, as well as the specific motions of all nucleons in a nucleus must be considered. It comes as no surprise that collective model should provide a good description for some of the experimental features of the GMR. In the framework of the liquid drop model, one can classify the various GMR's according to the basic oscillations of a nucleus. Unlike an ordinary liquid drop, a nucleus fluid has four components; i.e., protons and neutrons with spin up, and protons and neutrons with spin down. Therefore, for each multipole there are four combinations of oscillation. The mode of oscillation where the protons and neutrons move in phase without any differentiation of their spin, is called an isoscalar ($\Delta T=0$) electric ($\Delta S=0$) resonance. When the protons of any spin move against neutrons of any spin, like in the electric dipole case, this mode of oscillation is called an isovector ($\Delta T=1$) electric ($\Delta S=0$) resonance.

The magnetic mode of oscillation is characterized by spin oscillation ($\Delta S=1$) rather than charge oscillation. In the isoscalar magnetic mode ($\Delta T=0, \Delta S=1$), the protons and neutrons with spin up move against protons and neutrons with spin down. In the case of isovector magnetic mode ($\Delta T=1, \Delta S=1$), the protons with spin up and neutrons with spin down move against neutrons with spin up and protons with spin down.

The multipolarity of a giant resonance is determined by its mode of oscillation. For instance, the monopole oscillation ($L=0$) is a compressional mode without change of shape, i.e., the so-called breathing mode. The study of isoscalar monopole resonance is of special interest because its energy is directly related to the compressibility of the nucleus. The other higher multipoles are primarily surface oscillations. For dipole ($L=1$), the oscillation is axially symmetric. For quadrupole, the oscillation is bi-axially symmetric. These various modes of oscillation are sketched in Fig. 1.

II.2   The Microscopic Model Description of Giant Resonances

The collective oscillations described in the previous section can also be described microscopically in terms of a shell model.[14] Figure 2 shows single-particle transitions that take place between major oscillator shells. Since the energy difference between major shells is approximately $1\hbar\omega$ or $\sim 41A^{-1/3}$ MeV, taking the $\Delta T = \Delta S = 0$ operator $r^L Y_L^M$ as an example, this operator can only excite a nucleon by $L\hbar\omega$, i.e., through L major shells. We would then expect these single-particle transitions to have excitation energies of $0\hbar\omega$, $1\hbar\omega$, $2\hbar\omega$, ... The even number excitations correspond to even parity states while the odd number excitations

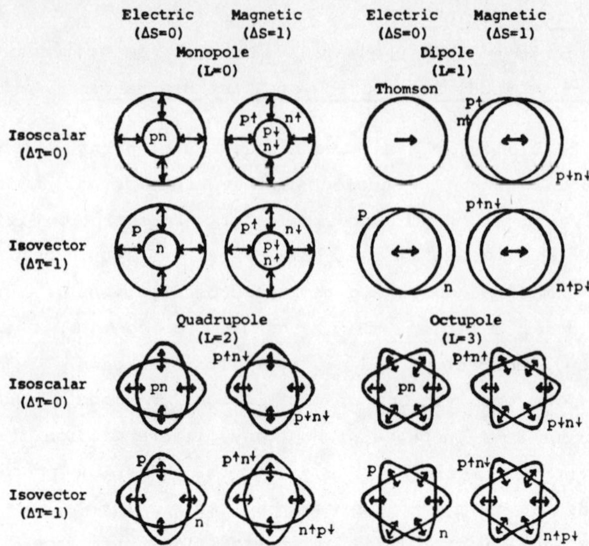

Fig. 1.   Various modes of oscillations of a nucleus.

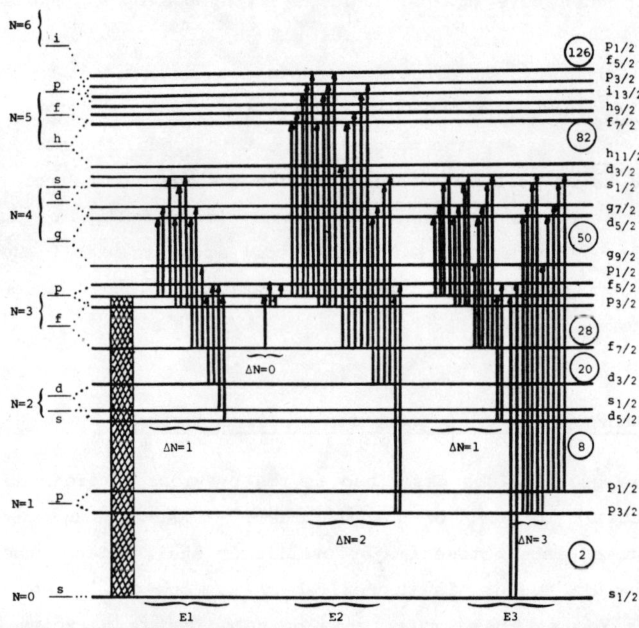

Fig. 2.   Elementary single-particle transitions of E1, E2, E3, and M1.[15]   The sub-shells are assumed to be filled up to the $f_{5/2}$ level ($f_{5/2}$ is assumed to be partially filled).

correspond to odd parity states.

The electric giant dipole resonance, E1, is built up of transitions from the last occupied shell to the next unoccupied shell above the Fermi surface, i.e., transitions due to $1\hbar\omega$ excitation, or at an excitation energy of $\sim 41A^{-1/3}$ MeV. However, due to the residual particle-hole interaction which is repulsive for isovector case, the excitation energy is shifted up, in better agreement with the experimentally observed value of $\sim 77A^{-1/3}$ MeV.

The E2 transitions could result from transitions within a major shell ($0\hbar\omega$) or between two major shells ($2\hbar\omega$). Of course, the $0\hbar\omega$ transitions will not occur if the shell is fully occupied ("closed shell"). For non-closed shell nuclei, the $0\hbar\omega$ transitions correspond to the familiar low-lying 2^+ states. The strength of these states varies from nucleus to nucleus, but typically exhausts $\sim 30\%$ of the energy-weighted sum rule (see Sec. II.3). The remaining energy-weighted sum rule strength is to be found in the $2\hbar\omega$ transitions which comprise the new GQR. Due to the residual particle-hole interaction, these $2\hbar\omega \simeq 80A^{-1/3}$ MeV transitions split into two giant resonances, the isoscalar and the isovector GQR. The energy of the isoscalar GQR is pushed down to $\sim 63A^{-1/3}$ MeV, while the isovector GQR is pushed up in energy to $\sim 120A^{-1/3}$ MeV.

Similary, the E3 excitations correspond to the $1\hbar\omega$ and $3\hbar\omega$ transitions, while the E4 excitations are formed by the $0\hbar\omega$, $2\hbar\omega$, and the $4\hbar\omega$ transitions.

II.3 <u>The Sum Rule Limits</u>

An electric or magnetic multipole resonance is termed giant if its transition rate for the excitation (or de-excitation) is much larger than some single-particle transition estimate, which represents the effect of a single-nucleon jump between two shell orbitals. The single-particle transition rates are usually expressed as the Weisskopf units:

$$T_{sp}^{(EL)} = \frac{2(L+2)}{L[(2L+1)!!]^2} \left(\frac{3}{L+3}\right)^2 \frac{e^2}{\hbar c} \left(\frac{\omega R}{c}\right)^{2L} \omega \ \sec^{-1} \qquad (II.3-1)$$

$$T_{sp}^{(ML)} = \frac{2(L+1)}{L[(2L+1)!!]^2} \left(\frac{3}{L+3}\right)^2 \frac{e^2}{\hbar c} \left(\frac{\hbar}{mcR}\right)^2 \cdot \left(\frac{\omega R}{c}\right)^{2L} \omega \ \sec^{-1} \ , \qquad (II.3-2)$$

where R is the nucleus radius.

The other useful criterion for recognizing a collective excitation is to compare the observed transition strength with the anticipated sum rules. Among the various sum rules, the energy weighted sum rule (EWSR) is the most useful one because it is the most model independent sum rule, and depends only on the properties of the ground state. If one neglects the effects of exchange interactions and sums the final states only over excitation energies well below the meson threshold, the EWSR for electric transition with multipolarity $L \geqslant 1$ can be written as[15]

$$S(EL) = \sum_f B(EL;i \to f)(E_f - E_i)$$

$$= \frac{L(2L+1)^2}{4\pi} \frac{\hbar^2}{2m} Z <R^{2L-2}>, \tag{II.3-3}$$

where $B(EL;i \to f)$ is the reduced transition probability, $<R^{2L-2}>$ is the expectation value of R^{2L-2} in the ground state of the nucleus, and m is the nucleon mass. The sum in Eq. (II.3-3) includes both $\Delta T=1$ and $\Delta T=0$ terms. The $\Delta T=0$ part of the full EWSR is obtained by multiplying Eq. (II.3-3) by (Z/A):

$$S(EL,\Delta T=0) = \frac{L(2L+1)^2}{4\pi} \frac{\hbar^2}{2m} \frac{Z^2}{A} <R^{2L-2}> . \tag{II.3-4}$$

For a uniform distribution of radius R, one has

$$<R^{2L-2}> = \frac{3}{2L+1} R^{2L-2} , \tag{II.3-5}$$

and

$$S(EL,\Delta T=0) = \frac{3\hbar^2}{8\pi m} \frac{Z^2}{A} L(2L+1)R^{2L-2} . \tag{II.3-6}$$

The $B(EL;i \to f)$ value, and therefore the sum rule strength, can easily be determined in the electromagnetic interaction (see Sec. III.1). However, the $B(EL;i \to f)$ is not directly measured in the inelastic hadron scattering. The quantity that one measures in the inelastic hadron scattering is a deformation parameter, β_L, which is determined by comparing the measured cross section to that predicted by a distorted wave Born approximation (DWBA) calculation (see Sec. III.2):

$$\beta_L^2 = \left(\frac{d\sigma}{d\Omega}\right)_{exp} / \left(\frac{d\sigma}{d\Omega}\right)_{DWBA} . \tag{II.3-7}$$

If one assumes that β_L is proportional to the mass multipoles $\sum_i r_i^L Y_L^M (\theta_i, \phi_i)$, one has for a uniform mass distribution:

$$B(EL;i \to f) = \left(\frac{3ZR^L \beta_L}{4\pi}\right)^2 , \tag{II.3-8}$$

and the EWSR $(\Delta T=\Delta S=0, L>1)$ may then be written as

$$S(EL,\Delta T=0) = \sum_f \beta_L^2 (E_f - E_i)$$

$$= L(2L+1) \frac{4\pi}{3A} \frac{\hbar^2}{2mR^2} . \tag{II.3-9}$$

If a single state with excitation energy E_x exhausts 100% of the EWSR, one would expect

$$\beta_L^2 R^2 = L(2L+1) \frac{4\pi}{3A} \frac{\hbar^2}{2mE_x} \, . \qquad (II.3-10)$$

For an isoscalar monopole excitation (breathing mode), the EWSR is given by[16]

$$S(E0,\Delta T=0) = \sum_f \beta_0^2 (E_f - E_i)$$

$$= 5 \frac{4\pi}{3A} \frac{\hbar^2}{2mR^2} . \qquad (II.3-11)$$

Again, for a single state exhausting 100% of the monopole EWSR, one has

$$\beta_0^2 R^2 = 5 \frac{4\pi}{3A} \frac{\hbar^2}{2mE_x} \, . \qquad (II.3-12)$$

In the case of electric dipole resonance, with the absence of exchange forces, the EWSR is given as[17]

$$S(E1,\Delta T=1) = \frac{\hbar^2}{2m} \frac{NZ}{A} \, . \qquad (II.3-13)$$

The sum rules for magnetic transitions are much more model dependent. However, they can be evaluated within the context of the shell model.[18]

III. Experimental Investigations – Singles Experiments

Unlike the giant dipole resonance where most of the experimental information was obtained from photonuclear reactions, the non–dipole giant resonances, which are the subject of this talk, have been studied mostly in the inelastic electron and hadron scattering experiments.[5-7] In the case of GDR study, a virtually background free gamma absorption cross section was observed. Unfortunately, the photonuclear reactions are found not to be a good tool in the study of GMR other than GDR.

Although inelastic scattering experiments are known to be particularly suited in exciting the collective states of nuclei, one no longer has a background free situation. In fact, the GMR appear as broad bumps superimposed on a very large continuum. Furthermore, there are many resonances with different modes of oscillation (see Sec. II.1, for instance), and most of these resonances are broad, thus, overlapping. To sort out and to study each resonance separately, one would have to choose a certain nuclear reaction which selectively excites only a few (but not all) modes of oscillation. The choice of a particular projectile and reaction is important because the nature and strength of its interaction with the target nucleus will determine the cross section of excitation of the various modes. For instance, because of the isoscalar nature of alpha particles and deuterons, only transitions with $\Delta T=0$ will be excited. In this section, I will discuss the advantages and the disadvantages of various singles experiments. Each experiment complements the others. The hope is that by using a variety of reactions, one may be able to unravel the complicated GMR spectra.

III.1 Inelastic Electron Scattering

Since the interaction between an electron and the target nucleus is electro-magnetic in nature, the reaction mechanism is well known, and the results of an experiment can be directly related to nuclear properties. In photonuclear reactions involving real phone, such as γ–decay, photoexcitation, photodisintegration, and radiative capture, etc., the 3–momentum transfer q is fixed for a fixed energy transfer ω, $q=\omega$. For $\omega \sim 20$ MeV, for instance, the photonuclear reactions are dominated by electric dipole transitions. Therefore, photonuclear reactions are not suitable for studying the higher multipole giant resonances. One advantage of electron scattering is that the 3–momentum transfer q can be varied at will for any value of the energy transfer ω, provided $q \geq \omega$. Thus, inelastic electron scattering is very useful in exciting resonances with higher multipolarity.

Let's consider an electron with initial 4–momentum $k_\mu = (\vec{k}, E_e)$ is scattered from a nucleus through an angle θ to a final state with 4–momentum $k'_\mu = (\vec{k'}, E'_e)$. In this process, a single virtual photon of 4–momentum $q_\mu = (\vec{q}, \omega)$ is exchanged with the nucleus, and the nucleus goes from its ground state $|i\rangle$ to a final state $|f\rangle$ (see Fig.

3). The energy transfer ω is equal to the excitation energy of $|f\rangle$ plus the recoil energy.

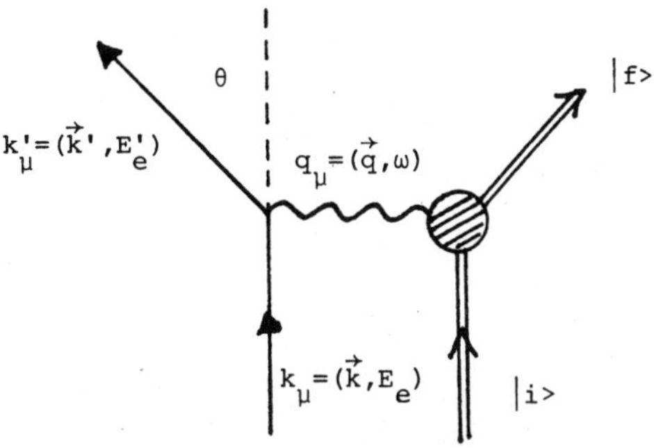

Fig. 3. Electron scattering with one-photon exchange.

In the plane wave Born approximation (PWBA), the (e,e') cross section may be written as

$$\frac{d^2\sigma}{d\Omega dE_{e'}} = \frac{\sigma_{Mott}}{\eta} \, |W(q,\omega)|^2 \quad , \qquad (III.1-1)$$

where $\eta=1+2E_e\sin^2(\theta/2)/M_T$ is a recoil factor with M_T being the target mass and $\sigma_{Mott} = (Z\alpha/2E_e)^2 \cdot \cos^2(\theta/2)/\sin^4(\theta/2)$ is the Mott cross section, where Z is the atomic number of the target and $\alpha \approx 1/137$ is the fine structure constant. $|W(q,\omega)|^2$ is the total differential form factor, which can be decomposed into a Coulomb or longitudinal form factor $|W^L(q,\omega)|^2$ and the transverse form factor $|W^T(q,\omega)|^2$ as follows:

$$|W(q,\omega)|^2 = \left(\frac{q_\mu^2}{q^2}\right)^2 |W^L(q,\omega)|^2 + \left[\frac{1}{2}\frac{q_\mu^2}{q^2} + \tan^2(\frac{\theta}{2})\right] |W^T(q,\omega)|^2 \quad . \qquad (III.1-2)$$

Since $|W^L(q,\omega)|^2$ and $|W^T(q,\omega)|^2$ depend only on q and ω but not on θ, they may be separated either by making a plot of the cross section against $\tan^2(\theta/2)$ for fixed value of q and ω (Rosenbluth plot) or by working at $\theta \approx 180^\circ$ where only $|W^T(q,\omega)|^2$ contributes. The form factors $|F(q)|^2$ (and thus $|F^L(q)|^2$ and $|F^T(q)|^2$) are related to $|W(q,\omega)|^2$ by

$$|F(q)|^2 = \int |W(q,\omega)|^2 d\omega . \qquad (III.1-3)$$

For medium and heavy mass nuclei, the PWBA formalism is severely limited due to the distortions of the incoming and outgoing electron plane waves by the Coulomb field of the nucleus. The effect of the Coulomb distortions can be partly taken into account by replacing q with an effective 3-momentum transfer:

$$q_{eff} = q \left(1 + \frac{3}{2} \frac{Z\alpha}{E_e R}\right) , \qquad\qquad (III.1-4)$$

where R is the uniform nucleus radius. For the usual calculation of the inelastic electron scattering, however, the more realistic distorted wave Born approximation (DWBA) is used.[19]

Consider, for example, a transition from an 0^+ ground state to an L^π final state. The longitudinal form factor can be written in PWBA as

$$|F^L(q)|^2 = \frac{4\pi}{Z^2} \left| \int \rho_L^{tr}(r) j_L(qr) r^2 dr \right|^2 , \qquad\qquad (III.1-5)$$

where the transition charge density $\rho_L^{tr}(r)$ for multipolarity L is defined as

$$\rho_L^{tr}(r) = <J_f| |\hat{\rho}_L| |J_i> . \qquad\qquad (III.1-6)$$

$\hat{\rho}_L$ is the charge density operator.

The reduced transition probabilities and the monopole matrix elements for monopole transitions can be written as

$$B(EL;i \to f) = \left| \int \rho_L^{tr}(r) r^{L+2} dr \right|^2 \qquad\qquad (III.1-7)$$

and

$$|<r^2>|^2 = 4\pi \left| \int \rho_{L=0}^{tr}(r) r^4 dr \right|^2 . \qquad\qquad (III.1-8)$$

Nuclear model is required in order to generate the transition charge density, thus the form factor. For isoscalar giant resonances, it is common practice to use the Tassie model of an oscillating charged, unmagnetized liquid drop.[20] In this model, the transition charge density for multipolarity L is given in terms of the ground state charge distribution $\rho_0(r)$, by

$$\rho_L^{tr}(r) = C\, r^{L-1} \frac{d\rho_0(r)}{dr} . \qquad\qquad (III.1-9)$$

For monopole transitions, the transition charge density of a breathing mode is used[21]:

$$\rho_{L=0}^{tr}(r) = C \left[3\rho(r) + r \frac{d\rho_0(r)}{dr} \right] . \qquad\qquad (III.1-10)$$

Except for the normalization constant C, the transition charge densities are completely determined if the ground state charge density $\rho_0(r)$ is specified. For all but the lightest nuclei, $\rho_0(r)$ is reasonably well given by the Fermi distribution:

$$\rho_0(r) = \rho_0 \{1 + \exp(r-c)/z\}^{-1} . \qquad\qquad (III.1-11)$$

For giant dipole and other isovector resonances, the Goldhaber-Teller model is often used, and the transition charge density is given as[3]:

$$\rho_{GT,L}^{tr}(r) = N_{GT}\, r^{L-1}\, \frac{d\rho_0(r)}{dr}\; . \qquad\qquad (III.1\text{-}12)$$

It should be noted that the transition charge densities in Tassie and GT models have precisely the same shape.

Another useful model for the isovector resonances is the Steinwedel–Jensen model.[4,22] In this case,

$$\rho_{SJ,L}^{tr}(r) = N_{SJ}\, j_L(k_L r), \quad r < R \quad , \qquad\qquad (III.1\text{-}13)$$

where k_L is determined from the boundary condition of zero normal velocity across the spherical boundary, i.e.,

$$\frac{d}{dr}\,[\,j_L(k_L R)\,] = 0 \; . \qquad\qquad (III.1\text{-}14)$$

Figure 4 shows the E0, E1, and E2 form factors for various models, including the modified SJ model $\rho_L^{tr} \propto r^L \rho_0(r)$, calculated in DWBA.[23] Firstly, it is interesting to note that the isoscalar E0 and E2 form factors are identical. Secondly, the form factors predicted by GT and SJ models are quite different. This model dependence of the form factor leads to differences in the strengths of the overlapping broad resonances extracted from the data.[23]

Experimentally, the giant multipole resonances are never truely dominant in the inelastic electron scattering spectrum. Before the experimental form factor can be extracted by dividing the cross section by the Mott cross section (see Eq. (III.1-1)), a number of contributions to the underlying continuum spectrum should be

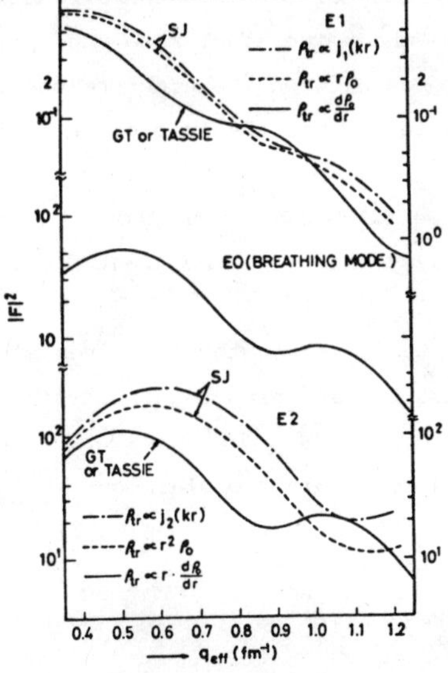

Fig. 4.　E0, E1, and E2 form factors for Tassie, GT, SJ, and modified SJ models, calculated in DWBA. See Ref. 23.

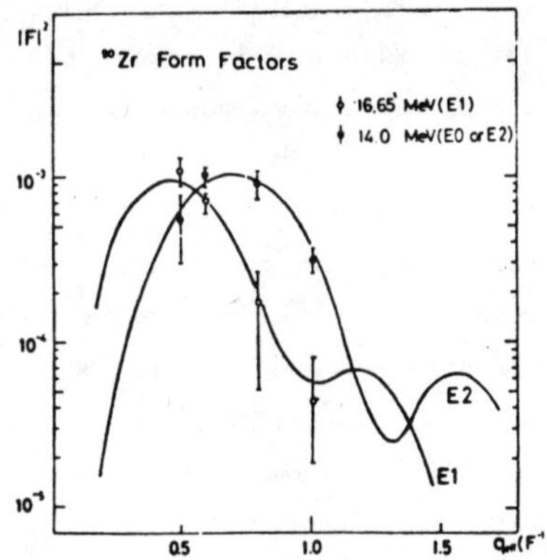

Fig. 5.　The experimental form factors for the 14.0- and 16.65-MeV peaks in ^{90}Zr. The solid curves are the DWBA E1 and E2 form factors. See Ref. 7.

subtracted. At low momentum transfer, the radiative tail due to elastic scattering, which amounts to 80-90% of the observed cross section, dominant. In principle, the continuum due to radiative tail is understood. In practice, empirically adjusted radiative tail calculations are used to perform the continuum subtraction. At high momentum transfer, quasi-free scattering can also contribute to the continuum, and should also be subtracted.

As an example, Fig. 5 shows the experimental form factors for the 14.0- and 16.5-MeV states in ^{90}Zr.[7] The form factor for the 16.5-MeV state, which is the well-known GDR, is well reproduced by the GT model calculated in DWBA. As can be seen, the form factor for the 14.0-MeV resonance follows the theoretical E2 curve, calculated with the Tassie model. However, since an E2 inelastic form factor cannot be distinguished from an E0 form factor, a monopole assignment for the 14.0-MeV resonance cannot be ruled out.

III.2 Inelastic Light Ion Scattering

The inelastic scattering of light ions (p, d, ^{3}He, and α particle) at intermediate energy has played a major role in establishing the existence of the new giant resonances in the past ten years.[24-26] One major drawback of these studies is that the new giant resonances are observed as broad bumps superimposed on a very large nuclear continuum. Furthermore, the nuclear processes which give rise to this continuum are not yet fully understood, although considerable theoretical[27-31] and experimental[32,33] efforts have been spent in trying to understand this continuum. Due to the lack of a reliable theory to predict the shape of continuum as a function of angle, it is a common practice to subtract from under the giant resonance peaks an arbitrary background shape (assuming no interference between the giant resonances and the continuum). This procedure usually resulted in a large absolute error in the extracted EWSR strength ($\sim$30%).

Inspite of this drawback, a wealth of experimental information about the new giant resonances has been obtained, especially the systematics of the giant monopole and quadrupole resonances (see Secs. IV.1 and IV.2).

The angular momentum transfer, L, and the excitation strength of a given state are determined by comparing the experimental inelastic scattering angular distribution with the theoretical predictions, calculated in the distorted wave Born approximation. The DWBA has been shown to be very successful in explaining the low-lying collective states. It has been assumed that the DWBA is still applicable even for particle-unstable high-lying GMR.

The DWBA is an extension of the optical model to inelastic scattering. The differential cross section in DWBA is given by

$$\frac{d\sigma}{d\Omega} = \left(\frac{\mu}{2\pi\hbar^2}\right)^2 \left(\frac{k_f}{k_i}\right) \Sigma |T_{fi}|^2 , \qquad\qquad (\text{III.2-1})$$

where μ is the reduced mass, k_i and k_f are the wave numbers of the initial and final states, Σ indicates a sum over the final and average over the initial spin orientations, and T_{fi} is the transition amplitude and is given by

$$T_{fi} = \int d\vec{r}\ \chi_f^{(-)*}(\vec{k}_f,\vec{r})\ \langle\psi_f|V|\psi_i\rangle\ \chi_i^{(+)}(\vec{k}_i,\vec{r})\ , \qquad (\text{III.2-2})$$

where ψ_i and ψ_f are the nuclear wave functions of the initial and final states, and $V(|\vec{r}-\vec{x}|)$ is the projectile-target interaction with $\vec{r}$ being the coordinate of the projectile and $\vec{x}$, the internal coordinates of the target. The $\chi(\vec{k},\vec{r})$ are the distorted waves which describe the elastic scattering of the projectile by the target nucleus before and after the inelastic scattering.

To obtain the angular distribution, a nuclear model must be used. The interaction V can be expanded into multipoles:

$$V(|\vec{r}-\vec{x}|) = \sum_{L,M} V_{LM}(r,x) Y_L^{M*}(\Omega) Y_L^{M}(\Omega_x)\ . \qquad (\text{III.2-3})$$

By use of the Wigner-Eckart theorem:

$$\langle\psi_f|V|\psi_i\rangle = \sum_{L,M} \langle J_f M_f|J_i LM_i M\rangle\ \langle J_f||V_{LM}(r,x)Y_L(\Omega_x)||J_i\rangle\ Y_L^{M*}(\Omega)\ . \qquad (\text{III.2-4})$$

The reduced matrix element in Eq. (III.2-4) is defined as the nuclear form factor:

$$F_L(r) = \langle J_f||V_{LM}(r,x)Y_L(\Omega_x)||J_i\rangle \qquad (\text{III.2-5})$$

and all the model dependence is in $F_L(r)$.

It has been customary to use the collective model to calculate the form factor $F_L(r)$. The collective model of the nucleus attributes many low-lying excited states either to oscillation in shape about a spherical mean or to the rotations of a deformed shape. This leads naturally to extending the optical model to include nonspherical optical potential. In the collective model excitation of the nucleus, it is then assumed that the nonspherical parts of the optical potential induce the inelastic scattering to these collective states.

By Taylor series expanding the optical potential $U(r-R)$ about $R=R_0$ and retaining only to first order in the deformations, we can write $F_L(r)$, for isoscalar excitations with multipolarity $L \geq 2$:

$$F_L(r) = (2L+1)^{-\frac{1}{2}}\ \beta_L R_0\ \frac{dU(r-R_0)}{dr}\ , \qquad (\text{III.2-6})$$

where β_L is the deformation parameter. If we take the Woods-Saxon form for the central optical potential $U(r-R_0)$, then

$$F_L(r) = (2L+1)^{-\frac{1}{2}}\ \frac{\beta_L R_0}{a}\ (V+iW)\ \frac{df(x)}{dx}\ , \qquad (\text{III.2-7})$$

where $x=(r-R_0)/a$ and $f(x)=(1+e^x)^{-1}$.

The model for the monopole vibration that is commonly used is the radial density oscillation of Satchler (Version 1).[16] The form factor is:

$$F_0(r) = \beta_0 (3U+r \frac{dU}{dr}) . \qquad\qquad (III.2-8)$$

The values of the deformation parameters, β_L, are then determined by the ratio of the experimental to the calculated differential cross sections, $\beta_L^2 = (\frac{d\sigma}{d\Omega})_{exp} / (\frac{d\sigma}{d\Omega})_{DWBA}$.

The excitation of the isovector GDR by inelastic hadron scattering has been discussed in great detail by Satchler in terms of the GT model or the SJ model.[34] The possibility of exciting the isovector GDR by T=0 projectiles, such as deuterons and alpha particles, due to the difference in shape of the neutron and proton density distributions of the target nucleus, was also discussed. The form factor for the GT type GDR excitation by the T=0 projectiles, assuming the central neutron and proton densities to be equal, is given by[34]:

$$F_1(r) = d_1 \ (\frac{\pi}{3})^{\frac{1}{2}} \ (\frac{N-Z}{A}) \left(\frac{dU}{dr} + \frac{R}{3} \frac{d^2U}{dr^2} \right) , \qquad\qquad (III.2-9)$$

where d_1 is proportional to the distance of separation of the proton and neutron distributions, and is determined by the EWSR. If a single state located at excitation energy E_x exhausts the dipole EWSR (Eq. II.3-13), then

$$d_1^2 = (\frac{\hbar^2}{2mE_x}) \ (\frac{NZ}{A}) . \qquad\qquad (III.2-10)$$

III.3 Inelastic Heavy Ion Scattering

The interest in the excitation of giant resonances in inelastic heavy ion scattering stems from a suggestion made by Broglia et al.,[35] that giant resonances might play a role in the energy dissipation process in deeply inelastic heavy ion reactions. The first experimental observation of giant resonance structures in the inelastic heavy ion scattering spectrum was reported by Betts et al. in $^{12}C+^{27}Al$ at 82 MeV.[36] Because of the angular momentum mis-match condition in the incoming and outgoing channels in heavy ion scattering, the excitation of giant resonances with higher multipolarities might be favored. In fact, broad peaks were observed in $^{16}O+^{208}Pb$ at 315 MeV.[37] In addition to the well-studied GQR, a peak at $\sim$19 MeV excitation energy has been suggested as a possible candidate for 3^- and 5^- giant resonances, which are predicted to be located at about this excitation region by RPA calculations.[30,38]

Unfortunately, the existence of these structures in ^{208}Pb at $E_x \sim$19 MeV has not been confirmed by the reaction $^{208}Pb(^{12}C,^{12}C')$ at 200 MeV.[39] One of the possible explanations is that these structures could be produced in a two-step mechanism such as pickup-breakup process via an intermediate state in ^{17}F at 3.1 MeV.

Other advantages and disadvantages exist in the heavy ion inelastic scattering experiment. It has been demonstrated that the continuum part of the inelastic heavy ion scattering spectra is much smaller than that of other light ion spectra (see Fig. 6). This, of course, is desirable, if one hopes to search for other broad resonances high up in excitation. One of the reasons for the reduction of continuum could be due to the quasi-free contributions which are less important for heavy ion scattering.

On the other hand, unlike the light ion experiments, the excitation of giant resonances by heavy ion inelastic scattering exhibits little or no sensitivity to the multipolarity of the transition. Figure 7 shows DWBA calculations for inelastic ^{16}O scattering on ^{208}Pb at 312 MeV for various different multipolarities.[40] The angular distributions are dominated by "Fresnel"-type diffraction pattern, due to interference between Coulomb and nuclear excitations. The angular distributions, therefore, do not exhibit the characteristic dependence on the angular momentum transfer that is the case in light ion inelastic scattering.

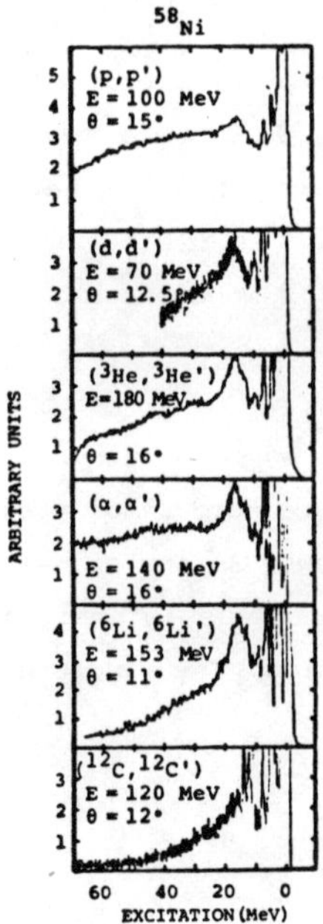

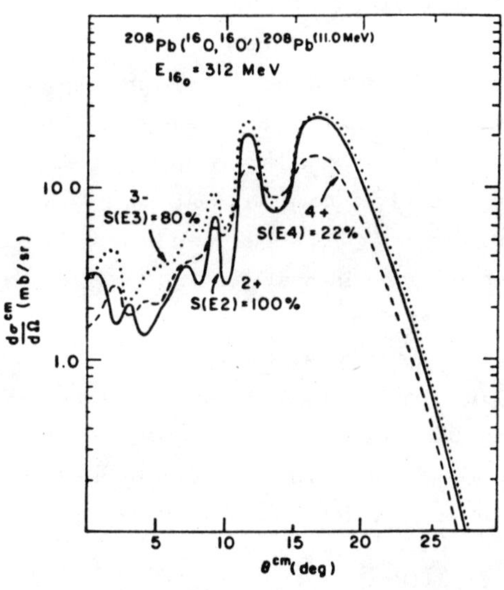

Fig. 6. Inelastic proton, deuteron, ^{3}He, α particle, ^{6}Li and ^{12}C scattering spectra on ^{58}Ni. The incident energies and angles are indicated.

Fig. 7. DWBA calculations for inelastic ^{16}O scattering on ^{208}Pb at 312 MeV for L=2, 3, and 4. See Ref. 40.

In addition to the structures produced by multi-step processes mentioned earlier, projectile excitation, which is significant for heavy ion, could also give rise to spurious structures in the inelastic heavy ion scattering spectra. This problem can be partially overcome by using projectiles with low particle emission thresholds, such as ^{6}Li.[41-43]

III.4 <u>Inelastic Pion Scattering</u>

The pion, being the lightest of hadrons, has a mass of ~140 MeV and quantum number $(J^{\pi},T)=(1^-,1)$. At low energy, the pion-nucleus interaction is weak, and the nucleus appears transparent to the incoming pions. At energy near the (3,3) resonance, the mean free path of pion in nuclear matter becomes small (~0.5 F) and the pions are strongly absorbed by nucleus, thus, likely to excite the various GMR. This strong absorption property leads then to diffraction processes in pion- nuclear interaction, and the angular distribution becomes sensitive to the angular momentum transfer. Unlike the heavier hadrons, however, the diffraction pattern of the pion inelastic scattering is spread out over a much larger angular range with the same range of momentum transfer than that of heavier hadrons. This feature makes the small momentum transfer measurement, like the study of giant monopole resonance (see Sec. IV.1), easier. Figure 8 shows the giant monopole and quadrupole excitations in ^{208}Pb for ^{3}He and pions.[44] In each case, a 100% energy weighted sum rule strength is assumed. It should be noted that for a momentum transfer of ~ 140 MeV/c (or ~ 0.7 F^{-1}), the angular range covered is ~ 30° for pions, while for ^{3}He, it is ~ 10°. The first deep minimum, which is the characteristic of monopole excitation, occurs at ~4° for ^{3}He, while for pions, it occurs at ~15°, which is more accessible experimentally.

Another feature of pion induced reactions is that the isovector non spin-flip ($\Delta T=1,\Delta S=0$) and isoscalar spin-flip ($\Delta T=0,\Delta S=1$) transitions are more readily excited than in proton induced reactions. This can be seen as follows. As the pion-nucleon interaction governs the pion-nucleus interaction, the pion-nucleus interaction can be expressed as a sum over single nucleon transition operation. The single nucleon transition operator is given in the form[45]:

$$t(\vec{k},\vec{k}') = t_{00} + t_{01}\vec{\tau}\cdot\vec{I} + t_{10}\vec{\sigma}\cdot\vec{L}_\pi + t_{11}(\vec{\tau}\cdot\vec{I})(\vec{\sigma}\cdot\vec{L}_\pi) , \qquad (III.4-1)$$

where $t_{\Delta S,\Delta T}$ are functions of the pion-nucleon partial wave scattering matrix elements, $\vec{\sigma}(\vec{\tau})$ represents the nucleon Pauli spin (isospin) operator, $\vec{I}$ is the pion isospin operator, and $\vec{L}_\pi$ is the pion angular momentum operator. Equation (III.4-1) gives the spin and isospin dependence of the pion-nucleon transition operator and enables one to understand the spin and isospin character of the nuclear particle-hole states excited. It was predicted by Walker[46] that $t_{00}:t_{01}:t_{10}:t_{11}$ are in the ratios 4:-2:2.88:1.4. The corresponding ratios in proton nucleus effective interaction are 3:-1:-1:-1. This suggests that with respect to the isocalar non spin-flip transition ($\Delta T=\Delta S=0$), the $\Delta T=1,\Delta S=0$ and $\Delta T=0,\Delta S=1$ transitions may indeed be stronger in pion induced reactions than in proton induced reactions.

The first experimental observaton of the GQR in inelastic pion scattering was reported by J. Arvieux et al. in 1979.[47] The GQR in ^{40}Ca at $E_x \sim 18.2$ MeV was clearly seen. It is interesting to note that the continuum part of the energy spectrum resembles that of the heavy ion inelastic scattering, i.e., the continuum drops off as the scattered pion energy decreases.

Figure 9 shows three inelastic pion scattering spectra on ^{89}Y with π^- at 163 MeV and π^+ at 163 and 240 MeV.[48] A peak located at the expected location for the GQR ($E_x \sim 14$ MeV) was seen. However, the observed full width at the half maximum (FWHM) of this peak (~ 5 MeV) is larger than that observed in the (α,α') experiment (~ 3.2 MeV). It was suggested[44] that some GDR might also be excited, since pions are expected to excite, in addition to the isoscalar GQR the isovector giant resonances, as mentioned previously.

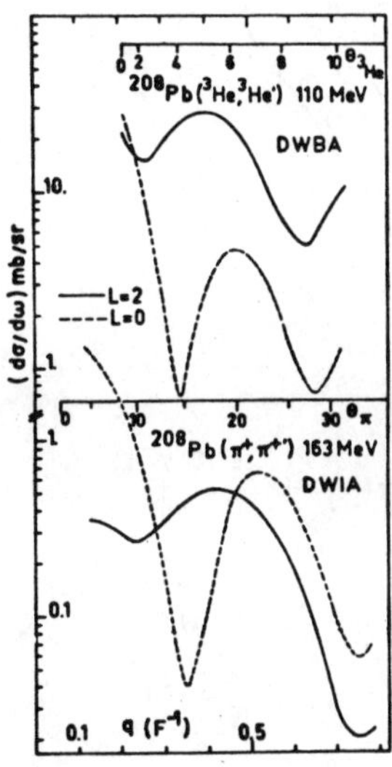

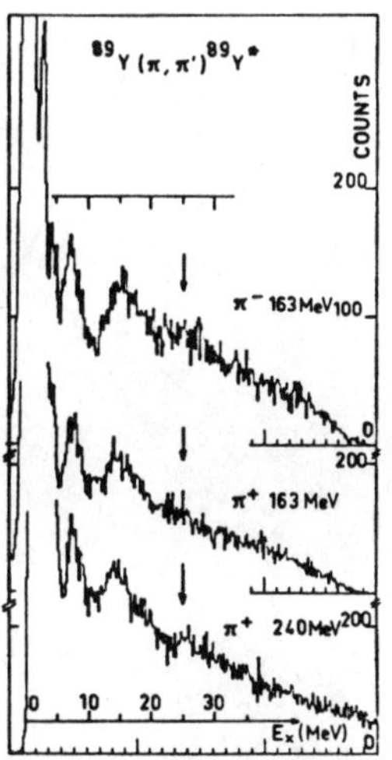

Fig. 8. Calculated giant monopole and quadrupole excitations in ^{208}Pb for ^{3}He and pions as a function of momentum transfer. A 100% EWSR strength is assumed. See Ref. 44.

Fig. 9. Inelastic pion scattering on ^{89}Y at $\theta_L=29°$. The arrows indicate an enhancement observed at $E_x \sim 25$ MeV. See Ref. 48.

III.5 Charge Exchange Reactions

Although, in principle, both the isoscalar and the isovector giant resonances can be excited by inelastic electron and hadron scattering (except for the T=0 projectiles where only the isoscalar modes are excited), in reality, only the isoscalar giant resonances are strongly excited. For example, in (e,e') experiments, the isoscalar and isovector states are excited with approximately equal strength, while protons and ^{3}He will preferentially excite the isoscalar states with cross sections roughly a factor of ~ 9 and ~ 30, respectively, over the isovector states. For this reason, hadron inelastic scattering is found not to be a good tool for studying the isovector giant resonances.

Charge exchange reactions at intermediate energy may offer a new approach to the study of isovector giant resonances because they will not induce isoscalar excitations. Examples of charge exchange reactions are $(\pi^-,\pi^0),(\pi^+,\pi^0)$; $(n,p),(p,n)$; $(d,^2\mathrm{He})$; $(t,^3\mathrm{He}),(^3\mathrm{He},t)$; and $(^6\mathrm{Li},^6\mathrm{He})$. Figure 10 shows the isospin relationships among the states reached by charge exchange reactions.

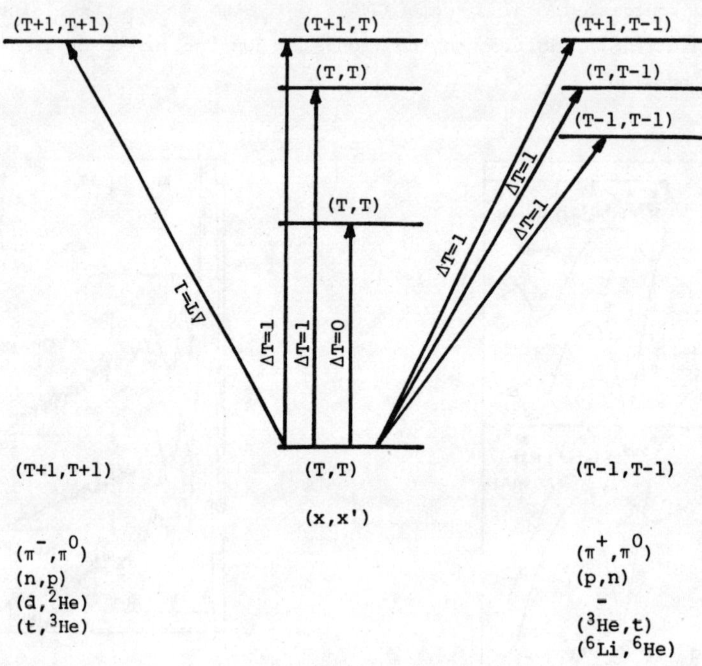

Fig. 10.   Isospin relationship among states reached by charge exchange reactions.

A convincing evidence for the excitation of isovector giant resonance in charge exchange reaction is shown in Fig. 11 for the $^{27}\mathrm{Al}(n,p)^{27}\mathrm{Mg}$ reaction.[49] The (n,p) spectrum at 15°, after subtracting a continuum background, appears to be correlated to the GDR in $^{27}\mathrm{Al}$ observed in photonuclear reactions. Further evidence comes from the angular distribution for the peak at $E_x \sim 14.5$ MeV which is consistent with an $L=1$ transfer. It is interesting to note that no appreciable cross section was observed for the isovector GDR in $^{27}\mathrm{Al}$ by the inelastic proton scattering.

Recently, considerable progress has been made in understanding the Gamow-Teller giant resonance with the (p,n)[50-52] and (^{3}He,t)[53] reactions. It has been shown that the (p,n) reaction is quite suitable in mapping out the Gamow-Teller strength at higher excitation energy. A more detailed discussion about the giant Gamow-Teller resonance will be given in Sec. IV.6.

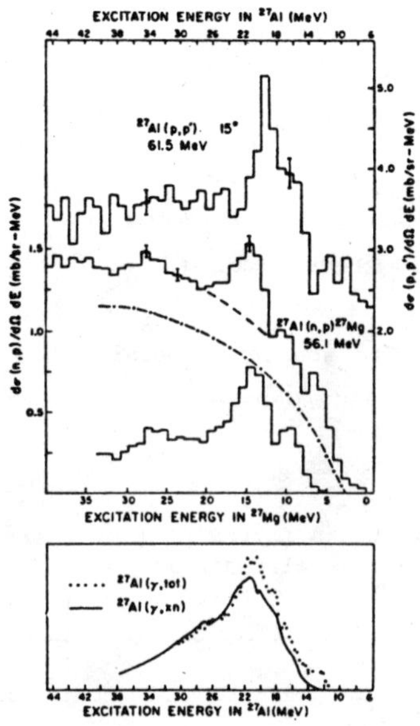

Fig. 11.   Comparison between the ^{27}Al(n,p) reaction and the ^{27}Al(p,p') and the gamma absorption cross sections.   The background subtracted (n,p) spectrum correlates nicely with the GDR in ^{27}Al.   See Ref. 49.

IV. Systematics of the Giant Multipole Resonances

Ten years have passed since the discovery of the isoscalar GQR in 1971. Many review articles[24,25] and conference proceedings[26,54-58] on giant multipole resonances have appeared since then. For more detailed discussion, I would refer the readers to these review articles and conference proceedings.

Table 1 gives the theoretical predictions for the excitation energy E_x (in units of $A^{-1/3}$ MeV) and the EWSR strength S for the isoscalar and isovector components of various electric multipoles.[59] Among these modes of oscillation (see Sec. II.1 and Fig. 1), the isoscalar giant monopole and quadrupole resonances are the two that have been studied the most, both theoretically and experimentally. Information on other higher multipole resonances, isovector giant resonances, and the giant Gamow-Teller resonance have also been studied extensively in the past few years. It would be impossible to discuss all of these at great lengths in this short course. In the following, I will briefly summarize mainly the systematic of the various electric giant resonances. The giant Gamow-Teller resonance will also be discussed briefly.

TABLE 1. Theoretical predications for excitation energy E_x and the EWSR strength S for the isoscalar and isovector electric giant multipole resonances.

L	$\hbar\omega$	$\Delta T=0$		$\Delta T=1$	
		$E_x A^{1/3}$	S (%)	$E_x A^{1/3}$	S (%)
2	2	58	100	135	110
3	1	25	28	53	2
	3	107	72	197	98
4	2	62	51	107	3
	4	152	49	275	97

IV.1 Isoscalar Giant Monopole Resonance

The observation of the isoscalar giant monopole resonance, a compressional mode without change of shape, is of special importance because its energy is related to the incompressibility of a finite nucleus

$$K_A = R_0^2 \frac{d^2(E/A)}{dR_0^2} \qquad (IV.1-1)$$

by[60]:

$$E_0 = \frac{\hbar}{R_0} \sqrt{K_A/m} \quad , \tag{IV.1-2}$$

where m is the nucleon mass, and R_0 is the root-mean-square radius. It should be noted, however, that K_A is different from the incompressibility of a nuclear matter K_{nm} in that volume, surface, Coulomb, and symmetry effects are all contributing to K_A, while only volume term contributes to K_{nm}. By examining the mass formula

$$E/A = -a_{vol} + a_{surf} A^{-1/3} + a_{sym} \left(\frac{N-Z}{A}\right)^2 + E_{coul}/A \quad , \tag{IV.1-3}$$

the relationship between K_A and K_{nm} can then be parameterized as[60]:

$$K_A = K_{nm} + K_{surf} A^{-1/3} + K_{sym} \left(\frac{N-Z}{A}\right)^2 + K_{coul} \quad , \tag{IV.1-4}$$

where K_{nm}, K_{surf}, K_{sym}, and K_{coul} can be viewed as the second derivatives of the co-efficients $-a_{vol}$, a_{surf}, a_{sym}, and E_{coul}/A with respect to the radius of the nucleus.

The compressibility χ of nuclear matter is related to K_{nm} by[60]

$$\chi = \frac{9}{\rho} \frac{1}{K_{nm}} \quad , \tag{IV.1-5}$$

where ρ is the density of nuclear matter.

The nuclear matter incompressibility, or the compressibility, is one of the three important quantities which characterize the nuclear matter. Unfortunately, the theoretical values of K_{nm} range from ~ 100 to ~ 400 MeV, depending upon the nucleon-nucleon potentials and effective interactions used in the calcula-tions.[61] Experimentally, there are many quantities, such as the isotope shifts, the surface thickness, the shape of the density inside the nucleus, and the energy of the giant monopole vibrations which are expected to be related to K_{nm}. Of all of these quantities, only the energy of the giant monopole resonance can be well determined by the value of K_{nm}, thus making it possible to limit the range of K_{nm}.

The possible existence of the giant monopole resonance was first suggested in 1975 by Marty et al. in their inelastic deuteron scattering data.[62] By comparing their (d,d') spectra with those of (α,α'), they concluded that additional cross sec-tions were observed at the higher excitation energy side of the GQR. The centroid of these yields happen to coincide closely with that of the GDR, i.e., at $E_x \sim 80A^{-1/3}$ MeV for heavy nuclei. Since the GDR is not expected to be excited strongly by the inelastic deuteron scattering, a possible explanation for these excess yields was the excitation of the giant monopole resonance.

Additional evidences for the excitation of the giant monopole resonance in ^{90}Zr and ^{208}Pb were reported in inelastic electron scattering by the Sendai group.[23,63] However, due to the dependence of the GDR form factor on the models used (see Sec. III.1), it was concluded that there was no need to assume the monopole excitation if the SJ model was used for the GDR. On the contrary, a giant monopole resonance exhausting $\sim 100\%$ of EWSR strength was needed if the GT model was used for the GDR. Because of this ambiguity, the existence of the giant monopole resonance is uncertain.

In 1977, Harakeh et al.[64] reported the observation of a second isoscalar resonance located at $E_x \sim 80A^{-1/3}$ MeV in addition to the GQR at $E_x \sim 63A^{-1/3}$ MeV (see Sec. IV.2) in their inelastic alpha particle scattering at 120 MeV on a number of targets. The angular distribution of this second resonance, though not inconsistent with an L=4 transfer, is better described by a monopole or a quadrupole resonance. This is because over the angular range where data were available, the calculated angular distributions for L=0, 2, and 4 transfers are similar in shape.

An unambiguous evidence for the existence of the giant monopole resonance was first established by the Texas A&M group[65] in their inelastic alpha particle scattering taken at very small angles. The success of their works can be seen from DWBA calculations of the angular distributions for L=0, 1, 2, and 4 transfer, as shown in Fig. 12. Although for angles greater than $\sim 10^{\circ}$, one cannot distinguish between L=0, 2, and 4 transfer, there is a characteristic pronounced minimum for an L=0 transfer at $\theta \sim 4^{\circ}$. The cross section at $\theta = 0^{\circ}$ for an L=0 transfer also seems to be enhanced over that of the GQR. Since the GDR is not excited strongly by alpha particles, one would expect the resonance shape to change somewhat at small angles if both the monopole and the quadrupole are present. This is indeed the case as can be seen in Fig. 13 for the ^{144}Sm(α,α') spectra taken at $E_\alpha = 96$ MeV.[65] A more drastic case is shown in Fig. 14 for the ^{208}Pb(α,α') spectra taken at $E_\alpha = 127$ MeV.[66] (In all of these spectra shown, a continuum has been subtracted from each spectrum.) By decomposing the composite spectrum shape into two components, it was found that the angular distribution for the upper one of the two peaks agrees quite well with an L=0 transfer, as shown in Fig. 15.

In addition to the works at Texas A&M, small angle inelastic scattering data with a variety of projectiles have also been reported. Marty et al.,[67] at Orsay reported the identification of the giant monopole resonance in many target nuclei in 108 MeV deuteron inelastic scattering. M. Buenerd et al.[68] at Grenoble, have also reported the study of the giant monopole resonance in several nuclei using 109 MeV ^{3}He as projectile.

The excitation of the giant monopole resonance has also been reported in (p,p') experiment by Bertrand et al.[69] In their previous analysis of the (p,p') data, the cross section at $E_x \sim 80A^{-1/3}$ MeV has been attributed entirely to the GDR.[70] Based on the new information on the isovector part of the proton-nucleus interaction,[71] they have reanalyzed their (p,p') data and have shown that the data are in better agreement with the excitation of a giant monopole resonance at approximately the same position as that of the GDR.

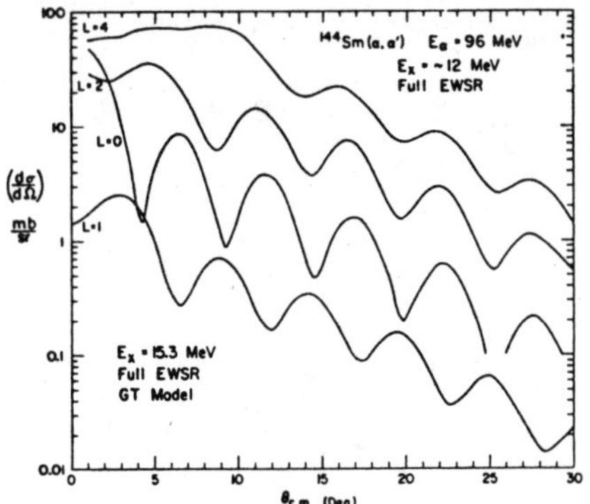

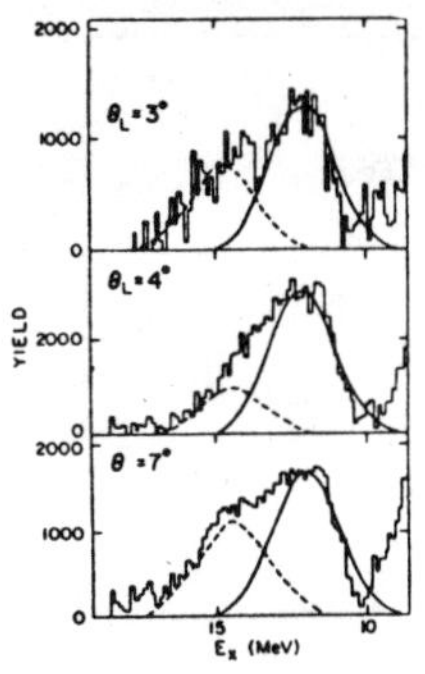

Fig. 12. DWBA predications of the angular distributions for L=0, 1, 2, and 4 transfers. See Ref. 65.

Fig. 13. ^{144}Sm(α,α') spectra taken with 96 MeV α particles at the angles indicated. The continuum has been subtracted. Two peak fits are shown superimposed. See Ref. 65.

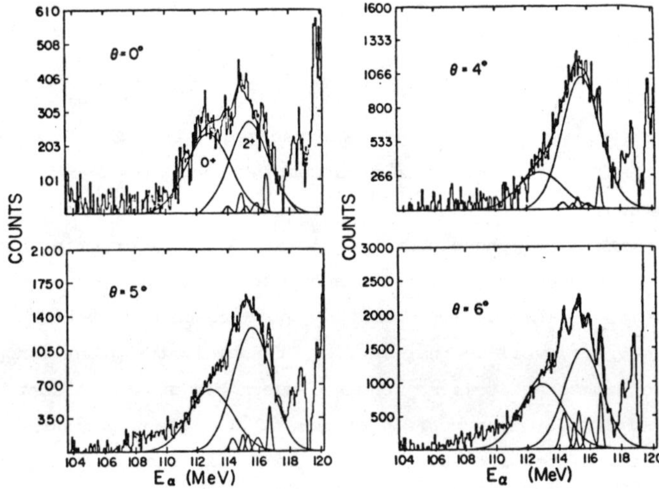

Fig. 14. ^{208}Pb(α,α') spectra taken with 127 MeV α particles at the angles indicated. The continuum has been subtracted. See Ref. 66.

By now, it is quite convincing that the existence of the giant monopole resonance has been well established in a wide range of target nuclei. Figure 16 shows the systematics of the location, the width, and the EWSR strength for the isoscalar giant monopole resonance. As can be seen from this figure, the giant monopole resonance is located at $E_x \sim 80A^{-1/3}$ MeV for heavier mass targets, and for lighter nuclei, it moves to lower excitation energy. This trend has also been seen for the GDR, as well as the GQR (see Sec. IV.2). The width of the giant monopole resonance shows a slight increase with decreasing target mass. For heavier mass targets, the

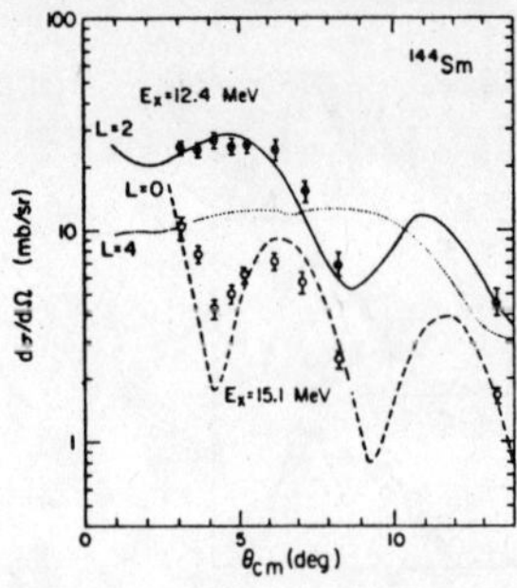

Fig. 15. Angular distributions for the ^{144}Sm(α,α') reaction at $E_\alpha = 96$ MeV. DWBA calculations are shown for several L transfers. See Ref. 65.

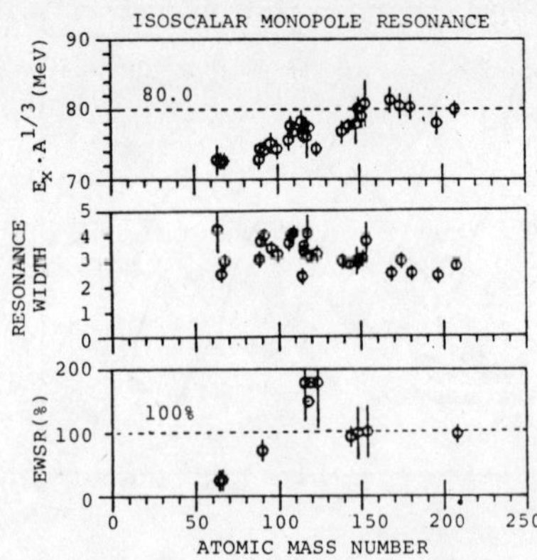

Fig. 16. Systematics of the locations, the width, and the EWSR strength for the isoscalar giant monopole resonance.

full isoscalar EWSR strength for monopole is exhausted, although for light nuclei, only about half of the full EWSR strength is accounted for. This trend is again true for the GDR and the GQR.

One thing is worth noting here. For nuclei with $A \lesssim 50$, no convincing evidences for the existence of the giant monopole resonance were reported. There are no apparent theoretical reasons as to why the giant monopole resonance should disappear in lighter nuclei. It may very well be that in lighter nuclei, the giant monopole resonance is becoming so wide that it is no longer discernible above the continuum.

From the experimentally measured locations of the giant monopole resonance, values of K_A can be calculated from Eq. (IV.1-2) if R_0 is taken to be equal to $\sqrt{3/5}R \simeq 0.97A^{1/3}$. By performing a least square search of Eq. (IV.1-4) to these K_A's (assuming $K_{coul} = (6/5)\ Z^2 e^2/(1.24A^{4/3})$), a value of $K_{nm} \sim 236$ MeV was obtained.[72] This result is in a good agreement with the predictions of Pandharipande.[73]

IV.2 Isoscalar Giant Quadrupole Resonance

Historically, the GQR was the first of the new giant resonances to be discovered in electron and proton inelastic scattering experiments in 1971–1972.[5-7] Subsequent works have been carried out with (p,p') reaction,[74] with (d,d') reaction,[75] with (^{3}He, ^{3}He') reaction,[76] and with (α,α') reaction.[77,78] Information up to 1976 has been summarized in two review articles.[24,25] More recent works, including the pion and heavy ion reactions can be found in various conference proceedings.[26,55-57]

The GQR was observed mainly in inelastic scattering experiments as a broad bump superimposed on a large nuclear continuum (see Fig. 17). The excitation energy of this broad peak changes smoothly with the target mass and is located at $E_x \sim 63A^{-1/3}$ MeV. A systematic study for the GQR has been carried out by the Texas A&M group using 96- and 115-MeV alpha particles.[78] Figure 18 shows the (α,α') spectra from the Texas A&M work. The arrows on the figure mark the locations $E_x \sim 63 A^{-1/3}$ MeV. Angular distributions obtained were generally best described by an L=2 DWBA calculation. Although monopole excitation has similar angular distribution as that of E2 in this angular range (see Sec. IV.1), the E0 assignment was ruled out because several hundred percent of the E0 EWSR would have to be required to explain the observed cross section.

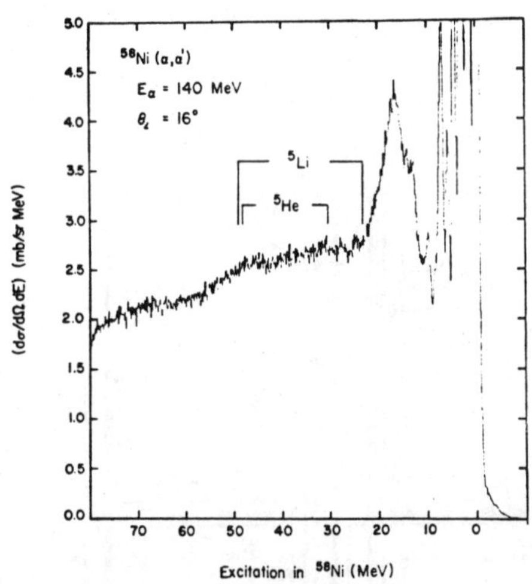

Fig. 17. The spectrum of 140 MeV α particles on ^{58}Ni at $\theta_L=16°$. The kinematic limits for the processes ^{58}Ni(α,^{5}Li $\to$ α+p)^{57}Co and ^{58}Ni(α,^{5}He$\to$x+n)^{57}Ni are as indicated.

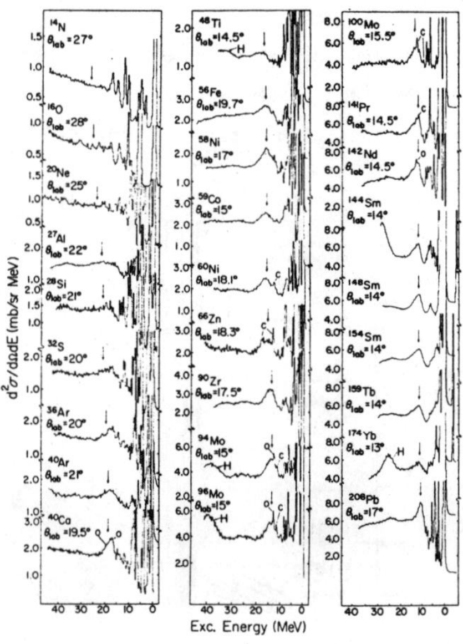

Fig. 18. Inelastic α particle spectra at E_α = 96 MeV (for targets with A $\leq$ 56 and ^{90}Zr) and 115 MeV (all others). The arrows indicate the location $E_x \sim 63A^{-1/3}$ MeV. See Reg. 78.

As was mentioned earlier, one no longer has the background free situation in inelastic hadron scattering spectra. A few possible contributions to the continuum part of the spectra have been discussed[79]: (1) the collective excitation of higher multipoles, (2) the quasi-free scattering contribution between the projectile and a target nucleon or cluster, and (3) the sequential process contribution. One well-known example of the sequential process is the broad bumps seen in the (α,α') spectra, especially in heavy target nuclei (see Fig. 19).[78] These bumps are due to the formation of the short lived ^{5}He or ^{5}Li, the so-called messa-resonance,[80] which subsequently breaks apart into an alpha particle and a nucleon. At low incident alpha particle energies and at low angles, this "resonance" could mask the effect of the GQR.

The GQR is quite compact for targets with $A \geq 40$ (see Fig. 18). As target mass decreases, the GQR becomes broader and even disappears for lighter nuclei. This seems to be in consistent with the radiative alpha particle capture reactions (see also Sec. V.1), where the E2 strength was found to be fragmented over a large range of excitation energy.[81]

In contrast to the Texas A&M work and the radiative alpha capture reaction, the inelastic alpha particle scattering at higher energies showed a compact resonance-like bump at $E_x \sim 63A^{-1/3}$ MeV even for sd-shell nuclei.[82-84] Figure 20 shows the (α,α') spectra for ^{24}Mg and ^{27}Al at three alpha particle incident energies.[82] At lower energy, no enhancement was observed near $E_x \sim 63 A^{-1/3}$ MeV, a result consistent with the Texas A&M group. As incident energy increases, a compact peak begins to appear. The observation of the GQR in light nuclei persisted even for the p-shell nuclei.

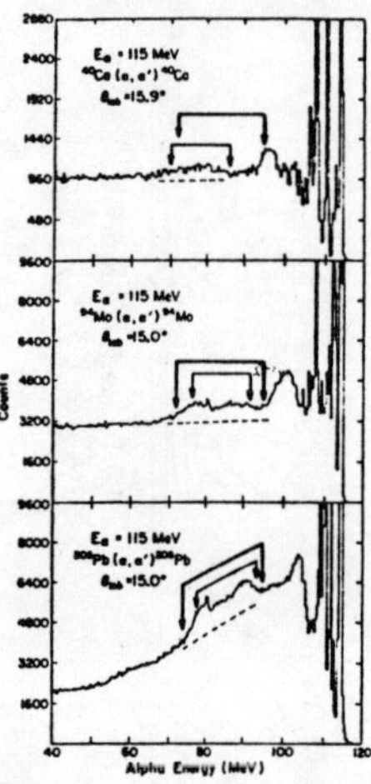

Fig. 19. (α,α') spectra from ^{40}Ca, ^{94}Mo, and ^{208}Pb at $E_\alpha = 115$ MeV. The kinematic limits for the decay of short-lived ^{5}He and ^{5}Li are indicated by narrow and broad lines, respectively. See Ref. 78.

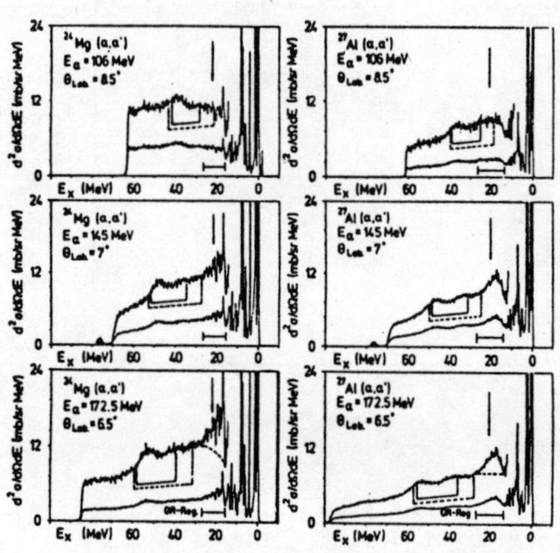

Fig. 20. (α,α') spectra on ^{24}Mg and ^{27}Al at $E_\alpha = 106$, 145, and 172.5 MeV. The arrows indicate $E_x \sim 63A^{-1/3}$ MeV. See Ref. 82.

There are at least two reasons for the success to observe a compact GQR in light nuclei. Firstly, the messa resonance mentioned previously may have merged into the GQR at lower alpha particle bombarding energy, thus obscuring the GQR. As incident energy increases, the messa resonance peaks move away from the GQR to higher excitation energy, and make the GQR more readily observed. Secondly, according to the DWBA prediction, the cross section for the GQR increases as the incident energy increases.[82] Although the total continuum yield may also increase with incident energy, it spreads over a much larger energy range, and as a result, the continuum cross section per unit energy may remain the same.

^{12}C seems to be the lightest nucleus studied so far where the resonance-like bumps were seen.[85] These bumps have been identified as quadrupole resonance from their angular distribution. However, no more than 20% of the E2 EWSR was exhausted, the so-called "pygmy-resonance". This may suggest the disappearance of any collective phenomenon in very light nuclei.

It is well established that the GDR splits into $K^{\pi}=0^-$ and 1^- components for deformed nuclei.[86] The amount of splitting depends on the deformation of the ground state. Similar effect is expected for the GQR.[87] As long as the nuclei preserve rotational symmetry, the GQR should split into $K^{\pi}=0^+$, 1^+, and 2^+ components.

Experimentally, no splitting of the GQR was observed. Instead, only a small broadening of the GQR due to deformation was reported.[69,78,88,89] Figure 21 shows the 115 MeV alpha particle inelastic scattering spectra on ^{144,148,154}Sm taken at $\theta_{Lab}=14^\circ$.[78] One sees that the width increases in going from the spherical ($\Gamma=3.89$ MeV in ^{144}Sm) to the deformed ($\Gamma=4.72$ MeV in ^{154}Sm) isotopes.

Few theoretical predictions have been made on the effect of deformation on the width (or splitting) of the GQR.[89-91] Using the ordinary quadrupole-quadrupole (Q-Q) interaction, a total splitting of the $K^{\pi}=0^+$, 1^+, and 2^+ components of about 6 MeV is predicted for ^{154}Sm, which gives a width much wider than the experimental observation. By modifying the Q-Q interaction,[89] the splitting reduced to only ~ 2 MeV, which fits the experimental width rather well, as can be seen from Fig. 22.

It is generally believed that the giant monopole resonance will not be affected by the ground state deformation. However, Zawischa et al.[91] predicted that the giant monopole resonance for deformed nuclei also splits into two components. One of these components is predicted to locate just below the GQR while the other component lies ~ 8 MeV higher in excitation energy. This splitting is caused by the mixing of $K^{\pi}=0^+$ component of the GQR with the giant monopole resonance. As a result of this mixing, two $K^{\pi}=0^+$ states are formed: the lower of the two is a mixture of both the monopole and quadrupole strengths, while the upper of the two is predominantly monopole. The experimental evidence for the splitting of the giant monopole resonance has been reported by Garg et al.[92] Figure 23 shows the decompositions of the resonance structure in ^{144}Sm and ^{154}Sm into various components. For ^{154}Sm, the giant monopole resonance splits into two components while the GQR is only broadened, as expected.

By now, it is well established that the isoscalar GQR exists in all nuclei with $A \geq 12$. Figure 24 shows the systematics of the location, the width, and the E2 EWSR strength depleted. As can be seen from this figure, the GQR is located at $E_x \sim 63A^{-1/3}$ MeV (a value of $64.7A^{-1/3}$ MeV was predicted in Ref. 9) for heavier mass nuclei. For lighter nuclei, the GQR moves to lower excitation energy than that expected from $63A^{-1/3}$ MeV. This behavior seems to be a universal feature for all giant multipole resonances. The width of the GQR is about 3 MeV for heavy mass nuclei, and it increases with decreasing target mass (~10 MeV for p-shell nuclei). This behavior is again similar to those observed for other giant resonances. An

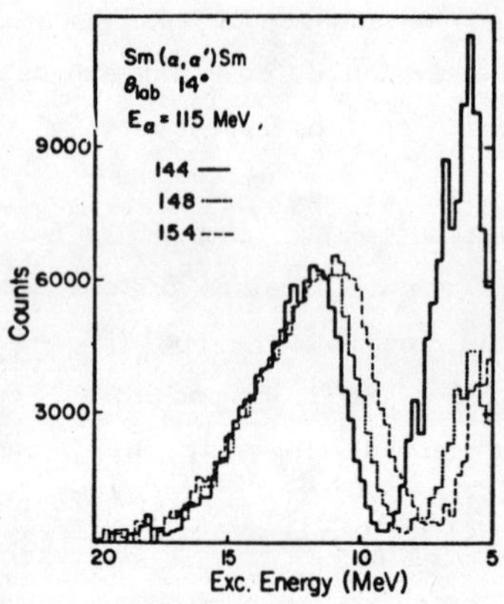

Fig. 21. Background subtracted (α,α') spectrum from ^{144,148,154}Sm at $E_\alpha =$ 115 MeV. The broadening of the GQR peak due to deformation is obvious. See Ref. 78.

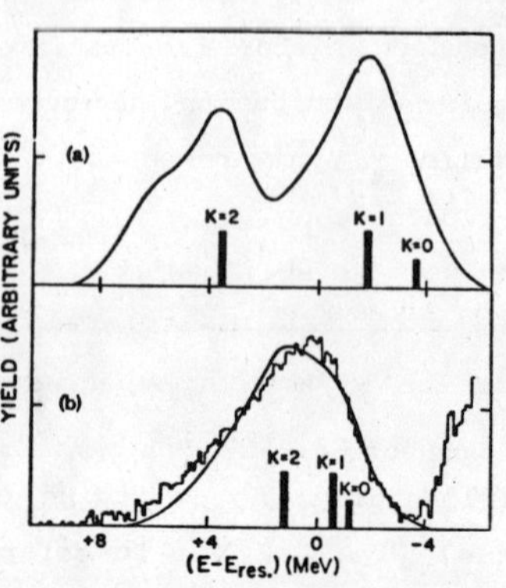

Fig. 22. The predicted strength distributions of the GQR for ^{154}Sm using the ordinary quadrupole-quadrupole interaction (upper curve) or the modified quadrupole-quadrupole interaction (lower curve). The histrogram is the data. See Ref. 89.

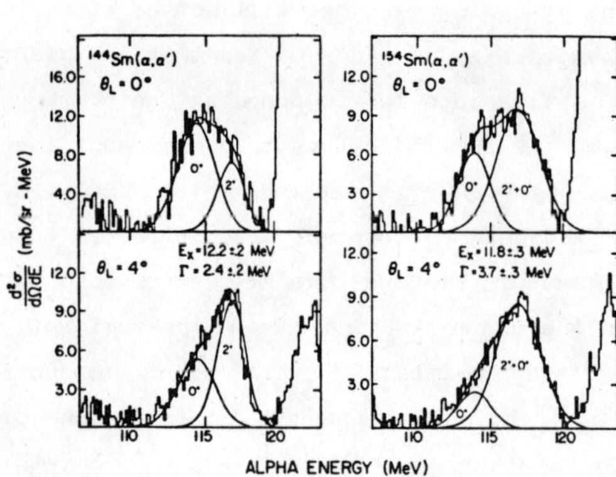

Fig. 23. Background subtracted (α,α') spectra at $E_\alpha =$ 129.4 MeV. The solid curves are the results of two peak fits. See Ref. 92.

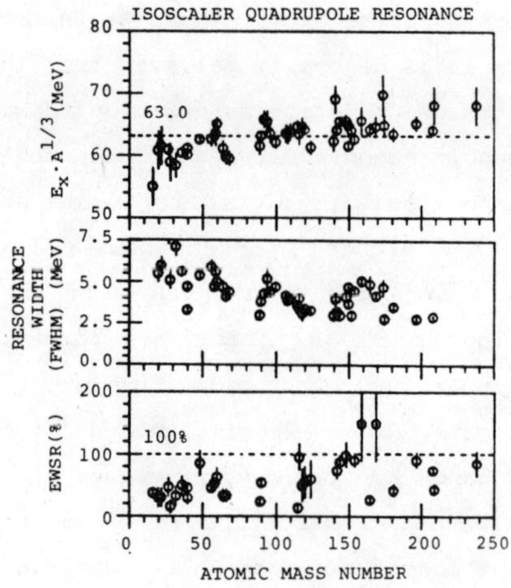

Fig. 24. Systematics of the location, the width, and the E2 EWSR strength depleted for the isoscalar GQR.

$A^{-2/3}$-dependence of the GQR width has been predicted by Nix et al.[9] This $A^{-2/3}$-dependence is in reasonable agreement with the experimental observation, although the predicted widths underestimate the observed widths.

For nuclei with $A \geq 100$, the full E2 EWSR strength is depleted by the $2\hbar\omega$ excitations. However, less than half of the full E2 EWSR strength is depleted for lighter nuclei by the $2\hbar\omega$ excitations. If one includes the E2 strengths exhausted by the low-lying 2^+ states (the $0\hbar\omega$ excitations), a near 100% of the E2 EWSR can then be accounted for.

IV.3 <u>Isoscalar Giant Octupole Resonance</u>

As mentioned in Sec. II.2 (see also Fig. 2), there are two types of E3 transitions, one corresponding to $1\hbar\omega$, the other $3\hbar\omega$. The $1\hbar\omega$ transitions should be at lower excitation than those of $3\hbar\omega$.

In inelastic alpha particle scattering experiment carried out at Texas A&M, resonance-like structure centered at $E_x \sim 32A^{-1/3}$ MeV was observed in many nuclei except in ^{40}Ca and ^{208}Pb.[93] Figure 25 shows results of this measurement. The angular distributions are well fitted by L=3 DWBA calculations, suggesting that this structure corresponds to the $1\hbar\omega$ low energy octupole resonance (LEOR). The fraction of E3 EWSR exhausted by this LEOR, together with the known low-lying 3^- states, is about 25-30%.

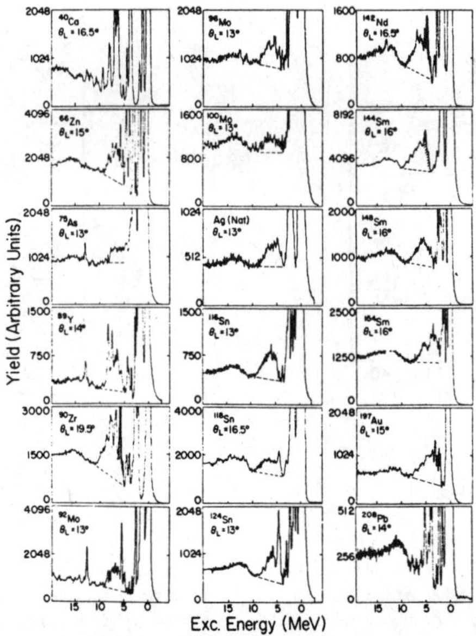

Fig. 25. (α,α') spectra from the targets indicated at $E_\alpha = 115$ MeV. Resonance-like structures are seen at $E_x \sim 32A^{-1/3}$ MeV. See Ref. 93.

The observation of $3\hbar\omega$ high energy octupole resonance (HEOR) has recently been reported in inelastic scattering of 800 MeV protons on ^{40}Ca, ^{116}Sn, and ^{208}Pb.[94] Figure 26 shows the (p,p') spectra taken at angles corresponding to a maximum and a minimum, respectively, of the L=3 angular distribution. In addition to the GQR and the LEOR, broad structure at $E_x \sim 110A^{-1/3}$ MeV was seen in all three nuclei. The angular distributions of this structure, shown in Fig. 27 are consistent with L=3 calculations, suggesting that the observed structure is octupole in nature. However, the deduced E3 EWSR of $\sim 20\%$ is only about one half that predicted by random-phase-approximation calculations.[95]

Additional evidences for the observation of the HEOR have also been reported in 172 MeV alpha particle[96] and 120 MeV ^{3}He[97] inelastic scattering experiments. In all three experiments, the HEOR is found to be located at $E_x \sim 110A^{-1/3}$ MeV. However, the E3 EWSR deduced from the (^{3}He,^{3}He') and (α,α') experiments are larger than that deduced from the (p,p') experiment. Obviously, more data are needed before any systematics can be made on the HEOR.

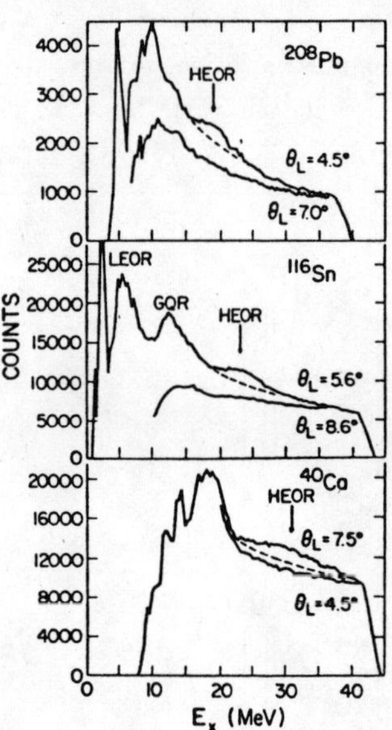

Fig. 26. Inelastic scattering of 800 MeV protons on ^{40}Ca, ^{116}Sn, and ^{208}Pb, taken at angles corresponding to a maximum (upper spectra) and a minimum (lower spectra) of the L=3 angular distribution. See Ref. 94.

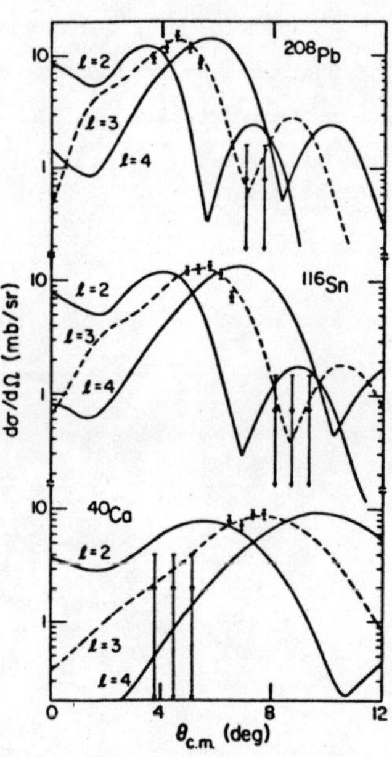

Fig. 27. Angular distributions for the resonance-like structures seen in Fig. 26. The solid curves are DWBA predictions for L=2, 3, and 4. The L=3 curve is normalized to the data. See Ref. 94.

IV.4   <u>Other Isoscalar Higher Multipole Resonances</u>

For E4 giant resonance, the $4\hbar\omega$ transitions are predicted to lie around $E_x \sim 150A^{-1/3}$ MeV, while the $2\hbar\omega$ transitions are predicted to coincide with the GQR.[98] Up to now, there is no direct experimental evidence for any E4 strength in a spectrum. The first indirect evidence was obtained when the structure seen in the (p,p') spectrum on ^{208}Pb at E_x=9-15 MeV was fitted to microscopic DWBA calculations.[99] It was concluded that the spectral shape as well as the angular distribution of this structure were best described if L=0, 2, and 4 are all present. The strength extracted for E4 is about 38% of the E4 EWSR.

Other indirect evidences for the E4 strength based on the fit to the total resonance cross section have also been reported.[100] In view of the fact that the HEOR is becoming wider than the width of other lower multipole resonances, it may be that the E4 giant resonance would be too wide to be detectable in inelastic hadronic scattering experiments. Experiments like (e,e'x) at higher electron energy might be able to demonstrate convincingly the presence of E4 strength in nuclei.

IV.5   <u>Isovector Giant Quadrupole Resonance</u>

The isovector GQR is predicted to be located at $E_x \sim 130A^{-1/3}$ MeV.[87] So far, most of the data on the isovector GQR were obtained from the photonuclear reactions. This is because the hadronic probes are found not to be good tools for exciting the isovector modes of oscillation (see discussion in Sec. III.5).

From the measured angular distributions for the ^{15}N(p,γ_0)^{16}O reaction,[101] it was concluded that the data are consistent with considerable E2 strength in the 30 to 50 MeV excitation region. However, this interpretation is uncertain due to the fact that direct capture and resonant capture cannot be separated. Since the recoil effective quadrupole charge for neutrons is very small, the direct neutron capture cross section should be very small also, and any E2 strength should be attributed to the resonant capture.

In fact, ^{16}O(γ,n_0)^{15}O reaction has been measured over the excitation energy range of 25 to 45 MeV.[102] By fitting the measured angular distributions to a fourth-order Lengendre polynomial expansion (see Sec. V.1), the E2 cross section was extracted, and is shown in Fig. 28. The solid curves indicate the upper and lower limits of the experimental E2 cross section. The data points correspond to $\sim$ 68% of the isovector E2 EWSR.

Other proton capture reactions have also been reported. In the ^{89}Y(p,γ_0)^{90}Zr reaction,[103] the author concluded that an isovector GQR at 26-28 MeV with a strength of 40-80% of the isovector EWSR is necessary in order to explain their data. Likewise, in the ^{208}Pb(p,γ_0)^{209}Bi reaction,[104] an isovector GQR at $E_x \sim 23.7$ MeV with a width of $\sim$ 3.5 MeV is observed from the asymmetry caused by the E1 and E2 interference. This resonance in ^{209}Bi exhausts ~ 60% of the isovector E2 EWSR if the branching ratio for the ground state proton decay from the isovector GQR is assumed to be the same as that from the GDR.

Other evidences for the observation of isovector GQR have been reported mainly in the inelastic electron scattering experiments. The targets studied include ^{28}Si,[105] ^{58}Ni and ^{60}Ni,[106] ^{89}Y,[107] ^{90}Zr,[7] ^{140}Ce,[108] ^{165}Ho,[109] ^{181}Ta,[110] ^{197}Au,[111] ^{208}Pb,[111] and ^{238}U.[112] Figure 29 shows typical inelastic electron scattering spectra in ^{58}Ni and ^{60}Ni. Yields at $E_x \sim 32$ MeV, corresponding the isovector GQR, increase with angles. The strength of the isovector E2 EWSR varies between 50–100%, depending on the models used in the calculation. The observed width varies from ~ 5 MeV for ^{238}U to ~ 10 MeV for ^{58}Ni. The increasing width with decreasing mass is similar to that observed in other giant multipoles, and is a general feature of all giant multipole resonances.

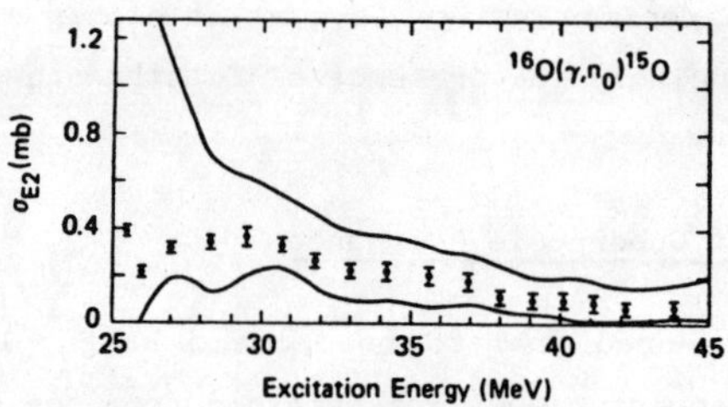

Fig. 28. Extracted E2 cross section for the $^{16}O(\gamma,n)^{15}O$ reaction. The curves indicate the upper and lower limits of the experimental E2 cross section. The data points represent ~ 68% of the isovector E2 EWSR. See Ref. 102.

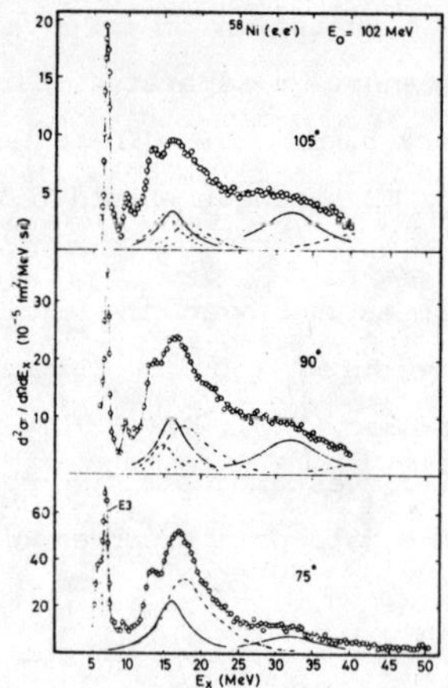

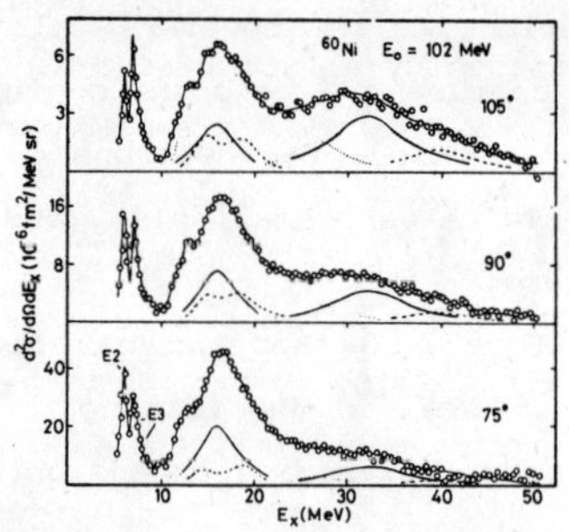

Fig. 29. Inelastic electron scattering spectra from ^{58}Ni and ^{60}Ni at $E_e = 120$ MeV. See Ref. 106.

IV.6 Giant Gamow-Teller Resonance

Soon after the discovery of isobaric analog states (IAS)[113] (equivalent to superallowed or giant Fermi (non-spin-flip) transitions in β decay) in (p,n) reactions by Anderson and Wong, Ikeda and co-workers[114] predicted the possible existence of a giant Gamow-Teller resonance in charge exchange reactions. Figure 30 shows the Fermi and Gamow-Teller states that should be strongly populated in the (p,n) reaction, together with the M1 states in the target nucleus.[115]

Since the nulcear matrix element for a $(\sigma \cdot \sigma)(\tau \cdot \tau)$ interaction is the same as that which appears in Gamow-Teller β decay, one would expect a simple relationship between the (p,n) cross sections and the Gamow-Teller β decay matrix elements.

Knowledge of the distribution of Gamow-Teller strength is very important in understanding nuclear structure and the spin dependence of nuclear forces. Most of the Gamow-Teller strength has not been located (e.g., only $\sim 12\%$ of Gamow-Teller strength is found in ^{48}Sc, where mapping was done up to 6.4 MeV).[116] The missing Gamow-Teller strength is probably concentrated at higher excitation energy. This may account for the observed hindrance of the allowed Gamow-Teller β decay observed for $N > Z$ nuclei when compared to calculated single particle transition rates.

The first experimental evidence for the observation of Gamow-Teller resonance was reported in the ^{90}Zr(p,n)^{90}Nb reaction at $E_p = 45$ MeV.[117] The Gamow - Teller resonance was observed to be centered at $E_x \sim 8.4$ MeV. At higher incident proton energies, the Gamow-Teller resonance is even more pronounced. To some extent, this can be understood in terms of a changing ratio of $V_{\sigma\tau}$ to V_τ (the strength parameters of the central part of spin-isospin and isospin terms, respectively, of the effective nucleon-nucleon interaction) as a function of incident proton energy. At $E_p = 120$ MeV, $(V_{\sigma\tau}/V_\tau)^2 \simeq 4.5$,[114] i.e., the spin-flip transition enhances over the non-spin flip transition by a factor of ~ 4.5.

Because direct observation of the Gamow-Teller β decay strength is limited by energy conservation to decay to nuclear states with lower energy than the parent (minus 511 keV), the (p,n) reaction has been proven to be very useful in mapping out the high-lying Gamow-Teller strength. It was experimentally observed that the Gamow-Teller resonance seen in (p,n) reaction is well localized (see Fig. 31). This is in disagreement with the suggestion made by Brown et al.[118] that the Gamow-Teller strength should be highly fragmented in heavy nuclei. The square of Gamow-Teller matrix element summed over all final states, $\Sigma <GT>^2$, has a value of ~ 11[52] and ~ 48,[52] respectively, as deduced from the ^{90}Zr(p,n)^{90}Nb and ^{208}Pb(p,n)^{208}Bi reactions. These values correspond to $\sim 30\%$ of the sum rule strength, i.e., $\sim 3(N-Z)$.

The Gamow-Teller strength has also been looked for with the ^{90}Zr(^{3}He,t)^{90}Nb reaction at 80 MeV at Grenoble[53] and at 130 MeV at Jülich.[53] The Grenoble experiment further separated the 8.4 MeV peak, seen in the (p,n) reaction,[117] into two groups, one at 7.2 MeV and the other at 9.7 MeV. The 7.2 MeV peak was believed to be a Gamow-Teller state while the multipolarity of the 9.7 MeV peak is unknown. There is a severe experimental difficulty in the (^{3}He,t) reaction. The continuum spectrum underlying the Gamow-Teller resonance is very large, and is rising as the excitation energy increases. This is because both the $\Delta S=0$ and $\Delta S=1$ transitions contribute to the continuum. The peak to continuum ratio may be improved if higher energy ^{3}He is used. This work is currently underway.[119]

Another useful tool for mapping out the Gamow-Teller strength in nuclei at high excitation energy is the (^{6}Li,^{6}He) reaction. It has been demonstrated[116] that an excellent correlation exists between the known Gamow-Teller strengths and the L=0 components of the (^{6}Li,^{6}He) cross sections. A second advantage of the (^{6}Li,^{6}He) reaction is that this reaction selects only the $\Delta T=\Delta S=1$ transitions, thus eliminating the large part of the continuum due to $\Delta T=1,\Delta S=0$ transitions, which contribute to the (^{3}He,t) and (a lesser degree to) (p,n) reactions. Search for the Gamow-Teller strengths using the (^{6}Li,^{6}He) reaction at $E_{^6Li}$ =90 MeV has begun at Maryland.[120] The preliminary results indicate that the continuum is indeed small.

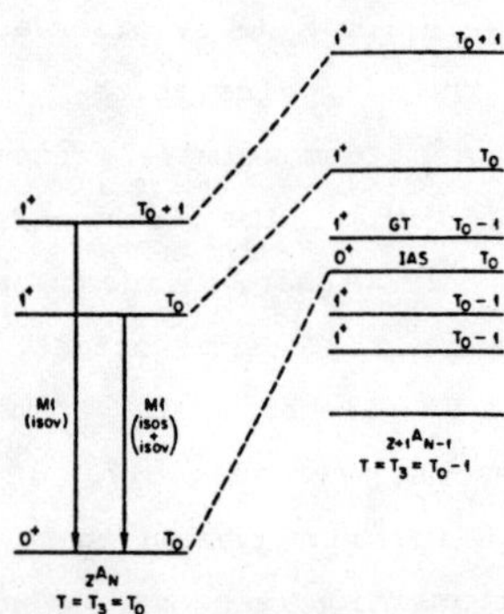

Fig. 30. Fermi and Gamow-Teller states that can be strongly populated in the (p,n) reaction. See Ref. 115.

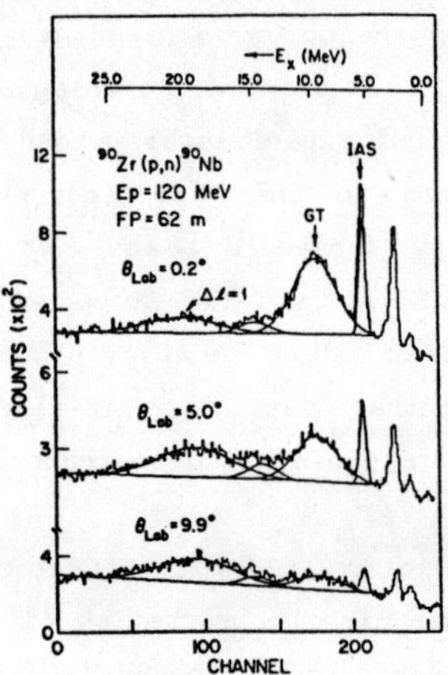

Fig. 31. ^{90}Zr(p,n)^{90}Nb time-of-flight spectra. See Ref. 115.

V.   Experimental Investigations of the Decay Modes of Giant Resonances

The systematics of the giant multipole resonances, especially the location, the width, and the exhaustion of the energy weighted sum rule strength have been well established experimentally, as described in Sec. IV. Various macroscopic and microscopic theories have been advanced to describe these systematics. (For an extensive list see Ref. 9). For example, a recent macroscopic theory describes the isoscalar giant resonances as the motion of a viscous fluid within deformable boundaries.[9] Good fits are obtained for the giant resonance energies as a function of mass number. However, the widths are generally underestimated. The microscopic theories of giant resonances have also had some difficulty in accounting for the resonance widths. Calculations for spherically symmetric nuclei, such as ^{16}O, ^{40}Ca, and ^{208}Pb, have shown that the large resonance widths cannot be predicted unless multi-particle-multi-hole states are included along with the simple one-particle-one-hole giant resonance configurations.[10-13]

In an independent particle harmonic oscillator picture (see Sec. II.2), the giant resonance is initially excited as a superposition of many one-particle-one-hole states. If the giant resonance immediately decays by emitting a particle, this process can be characterized by an "escaping width", and the decay properties will reflect the microscopic structure of the giant resonance. On the other hand, if the giant resonance proceeds to more complex many-particle-many-hole states before decaying, the process can be described by a "spreading width", and the decay may be statistical in nature. Since the total width of a state is the sum of all partial widths of its various decay modes, it is, therefore, necessary to study the decay properties of the giant resonance in order to learn more about the origin of their observed widths.

In the past four to five years, there have been some experimental efforts in trying to measure and understand the decay properties of the giant resonances, especially the GQR on a number of target nuclei. These studies were made mainly in coincidence experiments where the decay particles were detected in coincidence with the inelastically scattered hadrons, mostly alpha particle. Other reactions, such as the radiative capture reactions, the electroproduction experiments, and the coincidence experiments induced by high duty-cycle electron beam have also been used for these studies. In this secction, I will briefly review the various techniques used.

V.1   Radiative Capture Reactions

The radiative capture reactions have been used extensively for many years to study the giant resonances. Although most of these studies are on the GDR, attempts to extract the E2 strength which shows up in interference with the dominant E1 radiation have also been made. Two types of experiments are usually used: (i) the polarized proton or neutron capture reactions, and (ii) the alpha particle capture reactions.

It should be pointed out, however, that the information obtained in these studies is limited to the ground state decay channel only. For most of the nuclei studied so far, since the ground state decay branch is a small fraction of the total decay width (see Sec. VI), the extracted E2 strength from radiative capture reactions can not be compared directly with that obtained from the singles inelastic scattering experiments. If it is assumed that the capture reactions excite only the compound nuclear part of the GDR or the GQR, which in turn decays into various channels in a purely statistical way, then the total E1 or E2 absorption cross section can be obtained from the measured (γ, x_0) [obtained from capture (x, γ_0) reaction by the detailed balance] by use of Hauser–Feshbach theory:

$$\sigma_{tot}^{CN} = (\gamma, x_0) \frac{\sum\limits_{n,p,\alpha} T_i}{T_{x_0}} \, , \quad x_0 = p_0 \text{ or } \alpha_0 \, , \tag{V.1-1}$$

where T_i's are the transmission coefficients for decay into various p, n, or alpha particle channels.

The comparison between σ_{tot}^{CN} and the cross sections obtained from singles inelastic scattering experiments might be valid for medium and heavy mass nuclei, but not for light nuclei. This is because the GQR for light nuclei may not decay in a statistical way, while for medium and heavy mass nuclei, there are good evidences that the GQR decays statistically (see Sec. VI).

The unpolarized differential cross section $\sigma(\theta)$ and the analyzing power $A(\theta)$ for the capture of polarized protons or neutrons may be written as

$$\sigma(\theta) = [\sigma^\uparrow(\theta) + \sigma^\downarrow(\theta)]/2$$

$$= A_0[1 + \sum_{k=1}^{2L_{max}} a_k P_k(\cos\theta)] \, , \tag{V.1-2}$$

and

$$\sigma(\theta)A(\theta) = [\sigma^\uparrow(\theta) - \sigma^\downarrow(\theta)]/2P$$

$$= A_0 \sum_{k=1}^{2L_{max}} b_k P_k^1(\cos\theta) \, , \tag{V.1-3}$$

where $P_k(\cos\theta)$ and $P_k^1(\cos\theta)$ are the Legendre and associated Legendre polynomial functions, respectively, $\sigma^\uparrow(\theta)$ and $\sigma^\downarrow(\theta)$ are the cross sections for an incident polarized beam with spin parallel ($\uparrow$) or anti-parallel ($\downarrow$) to the normal $\hat{n} = \vec{k}_{in} \times \vec{k}_{out}$ to the reaction plane, and P is the beam polarization. L_{max} is the maximum multipole which contributes. For example, $L_{max} = 2$ if both dipole and quadrupole contribute.

By fitting the experimentally measured $\sigma(\theta)$ and $\sigma(\theta)A(\theta)$ to Eqs. (V.1-2) and

(V.1-3), the coefficients a_k and b_k can be extracted. These coefficients are related to the reaction amplitudes for various multipoles involved. Taking the $^{15}N(\vec{p},\gamma_0)^{16}O$ reaction as an example, i.e., a $J^\pi = \frac{1}{2}^-$ target and a final state of $J = 0^+$. In this case, only two complex reaction amplitudes contribute for each multipole:

$$E1: \quad |s_{1/2}|e^{i\phi_s}, \quad |d_{3/2}|d^{i\phi_d}$$

$$E2: \quad |p_{3/2}|e^{i\phi_p}, \quad |f_{5/2}|e^{i\phi_f}$$

$$M1: \quad |p_{1/2}|e^{i\phi_p'}, \quad |p'_{3/2}|e^{i\phi_p''} \quad .$$

The expressions for a_k and b_k in terms of E1 and E2 amplitudes and phases are given by[121]:

$$1 = 0.750\,(s_{1/2}{}^2 + d_{3/2}{}^2) + 1.250(p_{3/2}{}^2 + f_{5/2}{}^2)$$

$$a_1 = 2.372|s_{1/2}||p_{3/2}|\cos\phi_{sp} - 0.335|d_{3/2}||p_{3/2}|\cos\phi_{dp}$$
$$+ 2.465|d_{3/2}||f_{5/2}|\cos\phi_{df}$$

$$a_2 = 1.061|s_{1/2}||d_{3/2}|\cos\phi_{sd} - 0.375\,d_{3/2}{}^2 + 0.625\,p_{3/2}{}^2$$
$$- 0.437|p_{3/2}||f_{5/2}|\cos\phi_{pf} + 0.714\,f_{5/2}{}^2$$

$$a_3 = 1.936|s_{1/2}||f_{5/2}|\cos\phi_{sf} + 2.012|d_{3/2}||p_{3/2}|\cos\phi_{dp}$$
$$- 1.095|d_{3/2}||f_{5/2}|\cos\phi_{df}$$

$$a_4 = 3.499|p_{3/2}||f_{5/2}|\cos\phi_{pf} - 0.714\,f_{5/2}{}^2$$

$$b_1 = 1.186|s_{1/2}||p_{3/2}|\sin\phi_{sp} - 0.671|d_{3/2}||p_{3/2}|\sin\phi_{dp}$$
$$- 1.232|d_{3/2}||f_{5/2}|\sin\phi_{df}$$

$$b_2 = -0.530|s_{1/2}||d_{3/2}|\sin\phi_{sd} + 0.365|p_{3/2}||f_{5/2}|\sin\phi_{pf}$$

$$b_3 = -0.646|s_{1/2}||f_{5/2}|\sin\phi_{sf} + 0.671|d_{3/2}||p_{3/2}|\sin\phi_{dp}$$
$$+ 0.0913|d_{3/2}||f_{5/2}|\sin\phi_{df}$$

$$b_4 = -0.875|p_{3/2}||f_{5/2}|\sin\phi_{pf} \quad . \tag{V.1-4}$$

where $\phi_{ij} = \phi_i - \phi_j$ is the phase difference. With the measurement of 9 independent coefficients for 7 unknowns (4 amplitudes and 3 relative phases), the problem is over-determined.

If M1 radiation also contributes, there will in general be 11 unknowns, and the problem becomes underdetermined. Fortunately, M1 is small in the region of giant resonances in ^{16}O, and can be neglected.

Figure 32 shows the $\sigma(\theta)$ and $\sigma(\theta)A(\theta)$ for the $^{15}N(\vec{p},\gamma_0)^{16}O$ capture reaction at two incident energies.[81,101] The solid curves are the best fits to Eqs. (V.1-2) and (V.1-3) through k=4 (or $L_{max}=2$). The coefficients a_k and b_k are also indicated in the figure. On the basis of these measurements, the $p_{3/2}$ and $f_{5/2}$ amplitudes and phases were extracted from Eq. (V.1-4) and the total E2 cross sections, $\sigma_{E2} = p_{3/2}^2 + f_{5/2}^2$, are shown in Fig. 33.

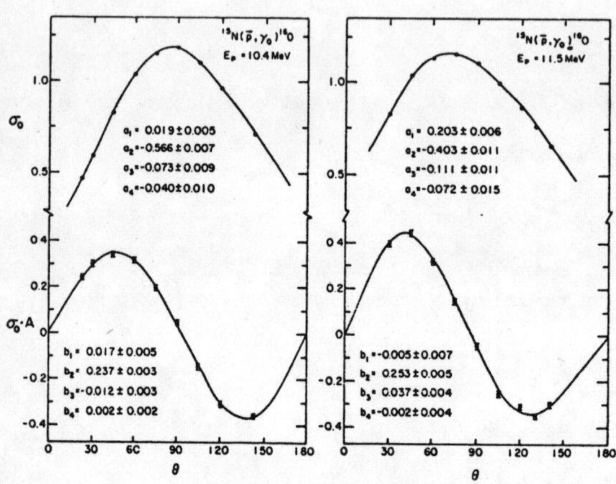

Fig. 32. $\sigma(\theta)$ and $\sigma(\theta)A(\theta)$ for the $^{15}N(\vec{p},\gamma_0)^{16}O$ capture reaction at $E_p = 10.4$ and 11.5 MeV. See Ref. 81.

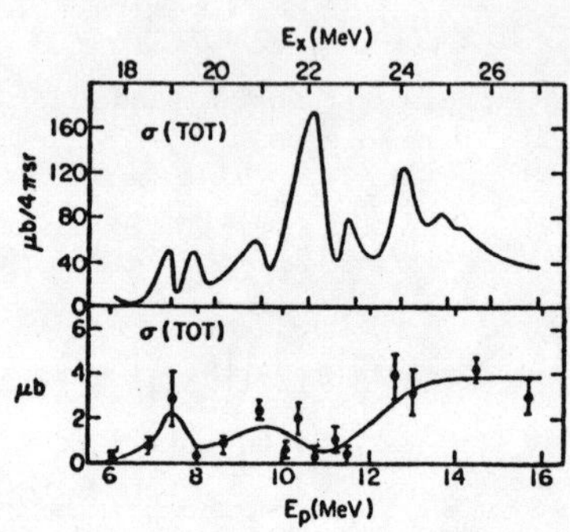

Fig. 33. Total and extracted E2 cross sections for the $^{15}N(\vec{p},\gamma_0)^{16}O$ capture reation. See Ref. 81.

The $^{15}N(\vec{p},\gamma_0)^{16}O$ over a larger energy range has also been performed at Seattle.[122] It was concluded that the structure seen in the excitation region E_x=23-27 MeV above the smooth "background" roughly accounts for 5-10% of the isoscalar E2 EWSR strength, in qualitative agreement with Stanford results.[81,101] For the excitation region E_x=17.9-27.3 MeV, the $^{15}N(\vec{p},\gamma_0)^{16}O$ experiment finds 12-22% of the E2 EWSR strength. This is to be compared with the $^{16}O(\alpha,\alpha'p_0)$ coincidence experiment[123] (see Secs. V.3 and VI.1) where 9% of the EWSR strength was found in p_0 channel over the same excitation energy region.

E2 cross sections in the $^{11}B(\vec{p},\gamma_0)$,[124] $^{12}C(\vec{p},\gamma_0)$,[125] $^{13}C(\vec{p},\gamma_1)$,[121] and $^{14}C(\vec{p},\gamma_0)$[126] reactions have also been extracted. For the $^{14}C(\vec{p},\gamma_0)$ and $^{13}C(\vec{p},\gamma_1)$ reactions, there is little evidence for significant E2 strength in excess of calculated direct E2 capture.[127]

Let's now turn to the radiative alpha particle capture reactions. It has been demonstrated that the alpha particle capture reaction to the ground state, (α,γ_0), on even-even target nuclei is a sensitive tool for obtaining the E2 strength. This is because the angular distribution is uniquely determined by the multipolarity of the radiation. Furthermore, for self-conjugate nuclei, the E1 radiation is supressed because only T=0 states can be formed directly in alpha particle capture reaction and ΔT=0 transition is forbidden by E1 radiation. Any E1 transition must then occur by isospin mixing. Therefore, the very small amounts of E2 radiation can be detected using the (α,γ_0) radiative reaction.

Consider a simple case where an alpha particle is captured into a $J^{\pi}=0^+$ target nucleus leading to a $J^{\pi}=0^+$ residual state by gamma emission. In this case, only electric multipoles contribute. Assuming only E1 and E2 contributions (E3 and higher multipoles are in general small and may be neglected), the angular distribution for the captured gamma ray can be written as:

$$\sigma(\theta) = \frac{1}{4\pi}\left\{(\sigma_{E1}+\sigma_{E2})-(\sigma_{E1}-0.71\ \sigma_{E2})P_2(\cos\theta)\right.$$
$$- 1.71\ \sigma_{E2}P_4(\cos\theta)$$
$$\left.- 2.68(\sigma_{E1}\sigma_{E2})^{\frac{1}{2}}\cos\phi_{12}[P_1(\cos\theta)-P_3(\cos\theta)]\right\}\ , \qquad (V.1-5)$$

where σ_{E1} and σ_{E2} are the partial cross sections for capture into 1^- and 2^+ resonances, respectively, and ϕ_{12} is the phase difference between the two modes of excitation.

By fitting the experimentally measured angular distribution to Eq. (V.1-5), the E2 partial cross section can be deduced. Figure 34 shows the measured angular distributions at three incident alpha particle energies for the $^{54}\text{Fe}(\alpha,\gamma_0)^{58}\text{Ni}$ reaction.[128] The solid curves are fits to Eq. (V.1-5). The extracted E2 cross sections for this reaction together with the $^{24}\text{Mg}(\alpha,\gamma_0)^{28}\text{Si}$ reaction[129] are shown in Fig. 35. By integrating over the structure for ^{58}Ni, it was found that about 4.3% of the total isoscalar E2 strength is observed in α_0 channel alone. This is to be compared with an upper limit of 3% of the E2 EWSR for the ground state alpha

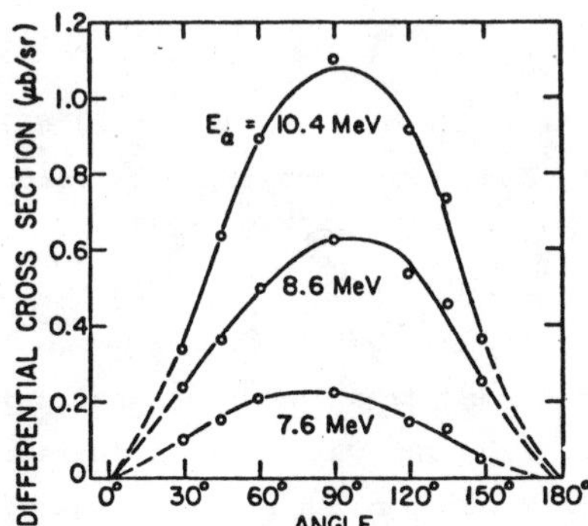

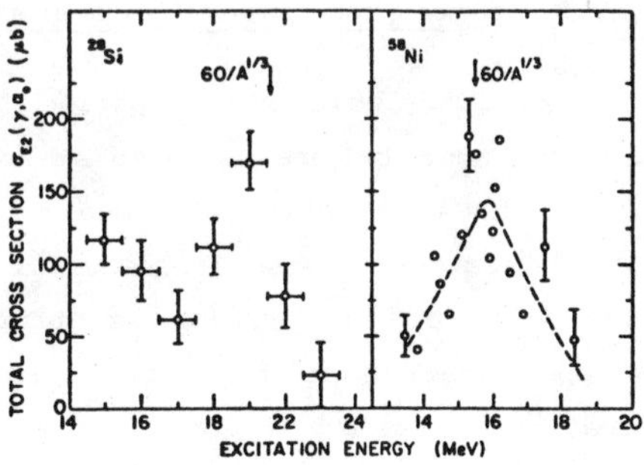

Fig. 34. Measured angular distributions for the $^{54}\text{Fe}(\alpha,\gamma_0)^{58}\text{Ni}$ reaction at three incident α particle energies. See Ref. 128.

Fig. 35. Extracted E2 cross sections for the $^{24}\text{Mg}(\alpha,\gamma_0)^{28}\text{Si}$ and $^{54}\text{Fe}(\alpha,\gamma_0)^{58}\text{Ni}$ reactions. See Ref. 128.

particle decay deduced from the ^{58}Ni$(\alpha,\alpha'\alpha_0)$ coincidence experiment[130] (see Sec. VI.2). In a recent alpha particle capture reaction by the Seattle group,[131] the E2 strength in the ^{58}Ni GQR region was found to be smaller by a factor of ~ 2 compared to that of Ref. 128.

An extensive work has been carried out at Stanford to extract the E2 strengths for nuclei in the sd-shell using the radiative alpha particle capture reactions.[81] Figure 36 summarizes the (α,γ_0) results, including the strengths measured in the low-lying discrete states. The arrows in the figure indicate the energy $63A^{-1/3}$ MeV where one would expect to see the GQR. It is obvious that there is little indication of such a resonance in these data, in drastic disagreement with the inelastic scattering experiments. Perhaps this discrepancy may be due to the small α_0 decay width from the GQR.

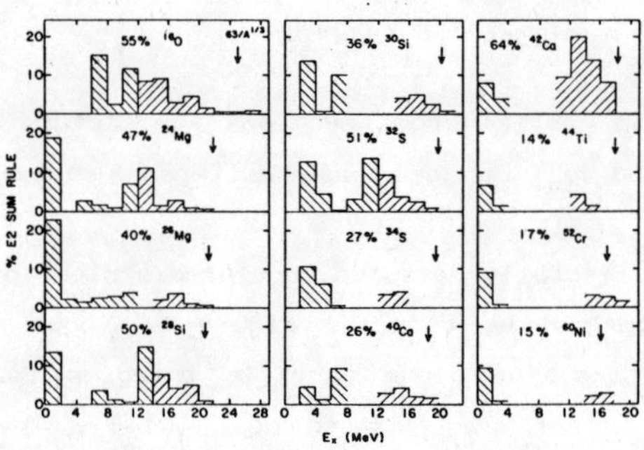

Fig. 36. Extracted E2 strength distributions from (α,γ_0) capture reactions and those of low-lying discrete states. The arrows indicate the energy $63A^{-1/3}$ MeV. See Ref. 81.

V.2 Electroproduction Experiments

In electron scattering experiment, the spectra of scattered electrons, which excite the nucleus, are measured at some angle θ. These data are then analyzed in terms of the momentum transfer to the target nucleus (see Sec. III.1). Since the electron scattering is inclusive in nature, no information concerning the de-excitation of the nucleus is obtained in this study. On the other hand, the electroproduction experiment is complementary to the electron inealstic scattering in that the emission of some kind of particles following the electron inelastic scattering is measured. In this case, the experiment integrates over the momentum transfer of the outgoing electrons.

In a conventional photoproduction experiment, the incident electrons with energy E_e are allowed to strike a radiator of high atomic number. The real photons generated are then incident on the target nucleus and are absorbed by the target which emits particles of type x. Since real photons can excite only the first few multipoles, the photoproduction experiment has been employed mainly in the study of the GDR, as well as the important M1 transitions in the p- and sd-shell nuclei.

In an electroproduction experiment, however, the electrons strike the target nucleus directly, and the target absorbs the radiation emitted by the electron in a single interaction (see Fig. 3). Unlike the real photon spectrum which has all multipole components present in equal amounts, the radiation (or the virtual photons with momentum greater than energy) seen by the target depends on the multipolarity L of the nuclear transition.

The usefulness of the electroproduction experiment (the virtual photon technique) is to relate the cross section of electroproduction experiment to that of photoproduction experiment. The relationship between the cross sections for electro- and photo-production with both leading to the emission of particle x may be written as

$$\sigma_{e,x}(E_e) = \int_0^{E_e - m} \sum_{\lambda L} \sigma_{\gamma,x}^{\lambda L}(E_x) N^{\lambda L}(E_e, E_x, Z) dE_x / E_x , \qquad (V.2-1)$$

where $\sigma_{e,x}(E_e)$ is the electroproduction cross section induced by an electron with energy E_e, $\sigma_{\gamma,x}^{\lambda L}(E_x)$ is the photoproduction cross section associated with the absorption of real photons of multipolarity λL(EL or ML), and $N^{\lambda L}(E_e, E_x, Z)$ represents the virtual photon spectrum of multipolarity λL generated when an electron of energy E_e interacting with a target nucleus of atomic number Z.

The virtual photon spectra in the plane wave Born approximation (PWBA) have been given in Ref. 132. They are quite adequate for very light nuclei. For heavier targets, due to the distortion of the incoming and outgoing electron waves in the Coulomb field of the target nucleus, the virtual photon spectra have to be calculated in the DWBA. This was done by Gargaro and Onley[133] who included the effect of Coulomb distortion when integrating the Möller interaction cross section over the scattered electron's angular distribution.

Figure 37 shows the calculated E1 and E2 virtual photon spectra in PWBA and DWBA for a 9.5 MeV electron inelastically scattered by a uranium nucleus.[132] As can be seen from this figure, the E2 virtual photon spectra are enhanced over those of E1 both in PWBA and DWBA calculations. The enhancement is further increased when the distortion is taken into account. Another example of the calculated E1 and E2 virtual photon spectra when 50 MeV electrons are inelastically scattered by a nickel nucleus is shown in Fig. 38.[134] It is observed that near the giant resonance region ($E_x \sim 16$ MeV) the E2 virtual photon spectrum is enhanced over that of E1 by at least a factor of 4. This enhancement makes the electroproduction experiment (involving virtual photons) more suitable for studying the GQR than the photoproduction experiment. In fact the virtual photon technique has been used by many groups to study the decay properties of the GQR for fp-shell nuclei and transuranium target. These results will be discussed in more details in Secs. VI.3 and VI.4.

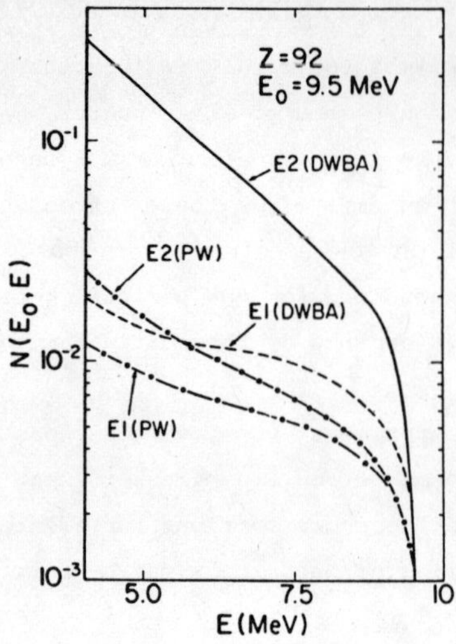

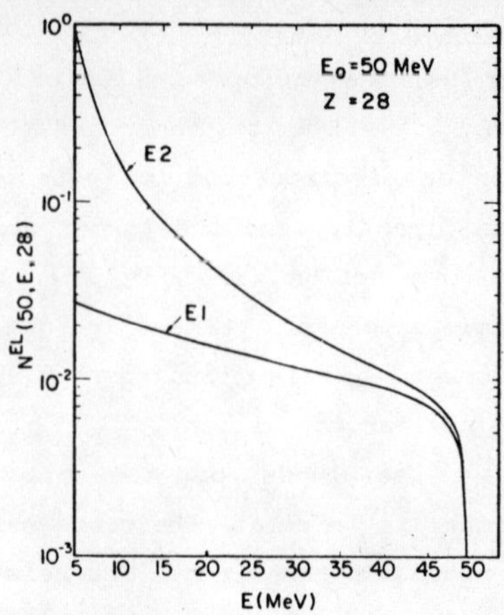

Fig. 38. Calculated E1 and E2 virtual photon spectra when 50 MeV electrons are inelastically scattered by a nickel nucleus. See Ref. 134.

Fig. 37. Calculated E1 and E2 virtual photon spectra in PWBA and DWBA for a 9.5 MeV electron inelastically scattered by uranium. See Ref. 132.

Experimentally, it is a common practice to place a radiator in the electron beam ahead of the target so that the observed outgoing particles are produced by both the electroproduction and the photoproduction (by the bremsstrahlung from the radiator). The usefulness of this practice is that with a radiator in place, the multipole composition of the photons seen by the target can then be varied. The yield of particle x when a radiator is in can then be written as

$$Y_{e,x}(E_e) = \sigma_{e,x}(E_e) + N_r \int_0^{E_e-m} \sum_{\lambda L} \sigma_{\gamma,x}^{\lambda L}(E_x) K(E_e,E_x,Z) dE_x/E_x \ , \qquad (V.2-2)$$

where N_r is the number of radiator per unit area (nuclei/cm^2) and $K(E_e,E_x,Z)$ is the Schiff bremsstrahlung spectrum.[135] $\sigma_{e,x}(E_e)$ is, of course, given by Eq. (V.2-1).

As an example of this technique, we discuss the electroproduction of protons and alpha particles emitted at $48°$, $90°$, and $132°$ with electron bombarding energies in the range 16-50 MeV.[134] To obtain the total production cross sections $\sigma_{e,p}(E_e)$ and $\sigma_{e,\alpha}(E_e)$, one would have to integrate the proton or alpha-particle energy spectra over energy and angle. Figures 39 and 40 show the experimental $\sigma_{e,p}(E_e)$ and $\sigma_{e,\alpha}(E_e)$ cross sections as function of the incident electron energy E_e. These data have been fitted to Eqs. (V.2-1) and (V.2-2) using E1 and E2 virtual photon spectra and experimental (when available) or hypothetical photonuclear cross sections. It was found that $\sigma_{e,p}(E_e)$ can be fitted rather nicely if only E1 is assumed (see Fig. 39). For $\sigma_{e,\alpha}(E_e)$, neither E1 nor E2 alone can fit the radiator in and out data simultaneously (see Fig. 40). The best fit was obtained if the absorption cross section has both E1 and E2 components. From this analysis, the E2 contribution to the (e,α) cross section thus the alpha particle decay of the GQR can then be deduced. This would have been very difficult if not impossible to obtain in photoproduction experiment.

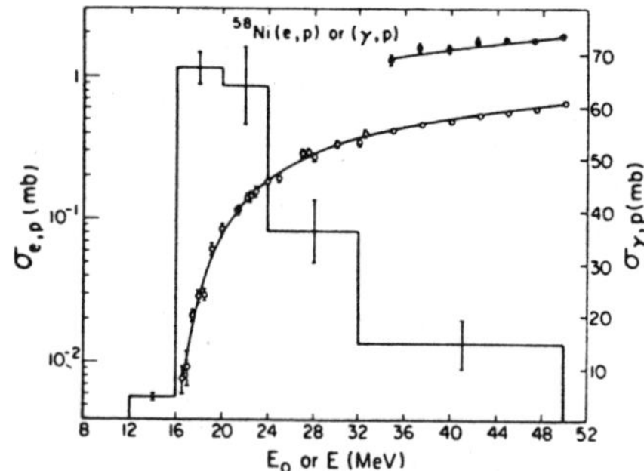

Fig. 39. Electroproduction cross sections for protons with (closed symbols) and without (open symbols) a radiator in position. The lines are calculated results by folding the (γ,p) cross section, represented by histogram, with the E1 virtual photon spectrum. See Ref. 134.

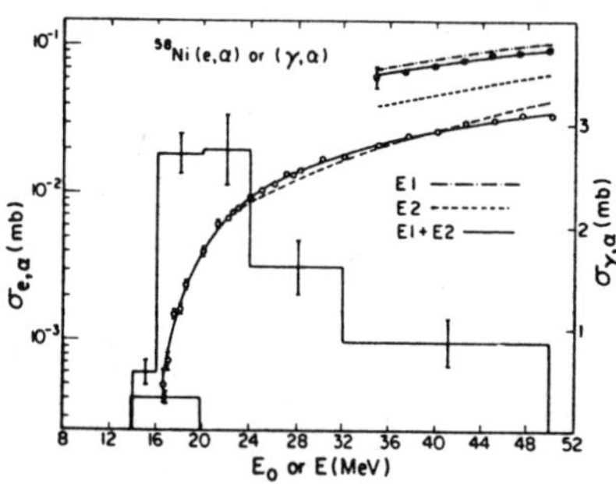

Fig. 40. Electroproduction cross sections for α particles with (closed symbols) and without (open symbols) a radiator in position. (γ,p) and (γ,α) cross sections are shown as histograms. Best fit is obtained when both E1 and E2 are included. See Ref. 134.

V.3 Hadron Induced Particle-Particle Coincidence Experiments

The strong excitation of the giant resonances in inelastic hadron scattering has made direct particle-particle coincidence measurements of the giant resonance decay feasible. Such experiments have now been performed on several nuclei from ^{12}C to ^{62}Ni.[43,85,123,130,136–138] Some coincidence works have also been done to look for giant resonance neutron and/or fission decay in the mass region of ^{208}Pb and above.[42,139–141]

The experimental arrangement for the giant resonance decay measurement is shown schematically in Fig. 41. The reaction is assumed to proceed in two steps. The giant resonance is first excited by the incoming projectile. The subsequent decayed particles are then detected in coincidence with the inelastically scattered particles.

Before discussing the particle-particle angular correlation function, the basic three-body kinematics will first be discussed briefly.

Let's consider the decay of a giant resonance following the inelastic scattering:

$$m_1 + m_T \rightarrow m_1 + m_T^*$$
$$\hookrightarrow m_2 + m_R^* \ .$$

Particles m_1 and m_2 are detected in coincidence while the unobserved recoiling particle m_R^* (residual nucleus) is left in an excited state E_x. For a given value of E_x, all events should lie on a kinematic locus in an E_1 vs. E_2 plot (see Fig. 42). Different loci correspond to different excited states of the recoiling nucleus. Nuclear structure information and reaction mechanism information are contained in the distribution of events along these loci. For instance, sequential decay (giant resonance decay) would show up as enhancements along the loci as shown in Fig. 42, and the intensity of events on different loci gives the branching ratio.

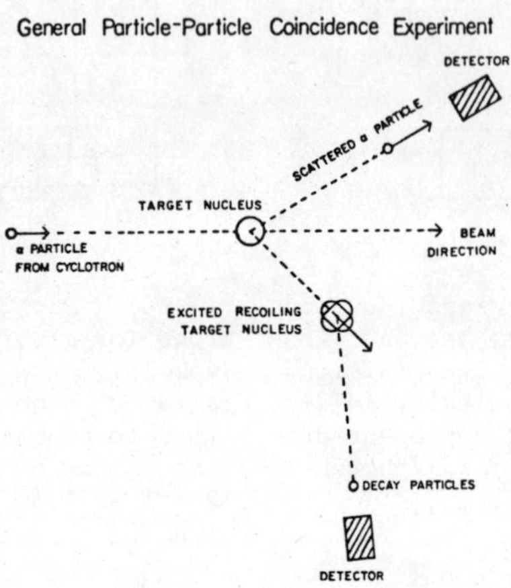

Fig. 41. A schematic experimental arrangement for the GQR particle decay experiments.

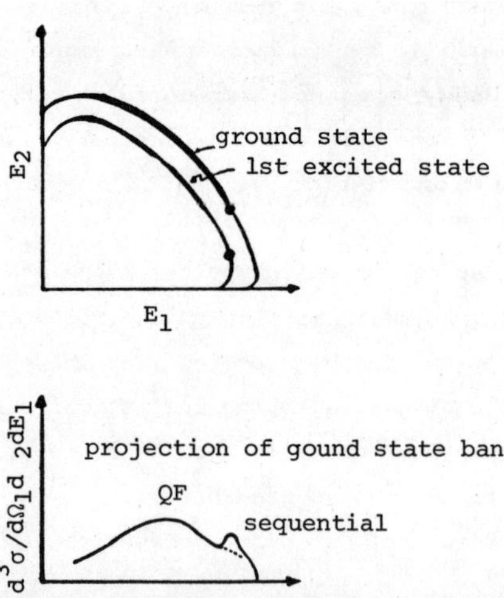

Fig. 42. A sketch of an E_1 vs. E_2 plot, showing various kinematic loci, and the projection of one of the loci onto the E_1 energy axis. Sequential processes are indicated as dots on the loci.

For a fixed excitation energy of the giant resonance, m_T^*, and fixed θ_1, E_1 is fixed also. E_2 would then depend on the excitation energy of the residual nucleus, m_R^*, i.e., E_2 is maximum when decaying to the ground state of m_R. As was mentioned in the last paragraph, the sequential decay would show up as enhancements on the kinematic loci. The projection of events along any of the loci onto the E_1 axis is sketched in Fig. 42. From this, the decay yield can be obtained. In general, data taken at various decay angles, θ_2, for a fixed scattered angle, θ_1, are transformed to the recoiling c.m. system before they can be angle-integrated to obtain the branching ratio or to compare with the calculated angular correlation function.

For light nuclei, various decay branches have been obtained using the procedure described above.[85,123] For medium and heavy mass nuclei, there are serious problems.[130] Figure 43 shows a typical E_1 vs. E_2 plot for the ^{58}Ni(p,p'p) reaction. In addition to the kinematics loci corresponding to leaving ^{57}Co in its low-lying states, an intense band is observed. This band corresponds to m_2 being an evaporation proton. Similar result is observed for the ^{58}Ni(α,α'p) reaction.

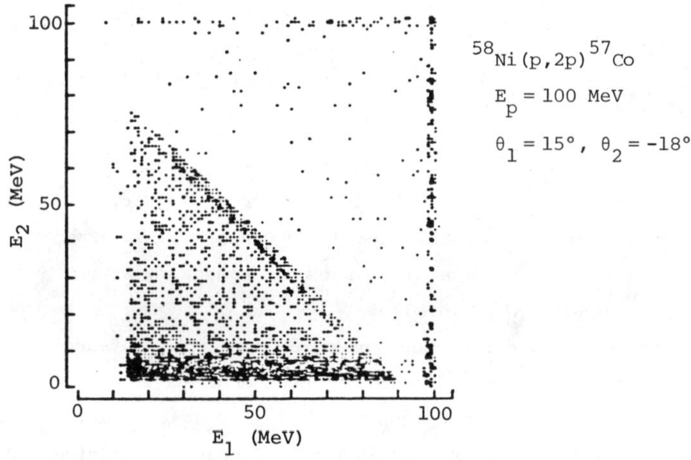

Fig. 43. E_1 vs. E_2 plot for the ^{58}Ni(p,p'p)^{57}Co reaction at $E_p = 100$ MeV. Two kinematic loci and a horizontal band, called the "evaporation band" are shown, together with two accidential bands (one vertical line in the far right and one horizontal line at the top).

Because the "evaporation" band dominates the entire E_1 vs. E_2 plot, it complicates the determination of the GQR decay branches to individual states of the residual nuclei for medium and heavy mass nuclei. Figure 44 shows the various projected E_1 spectra for proton decay in the ^{58}Ni(α,α'p)^{57}Co reaction at $E_\alpha = 140$ MeV with windows set on the low-lying states of ^{57}Co as indicated.[130] As can be seen, these projections result in a narrow peak which "walks" as the final state window is moved. Because the narrow peak is no wider than the GQR and "walks" smoothly through the GQR region, it is very difficult to determine what part of the peak results from GQR decay.

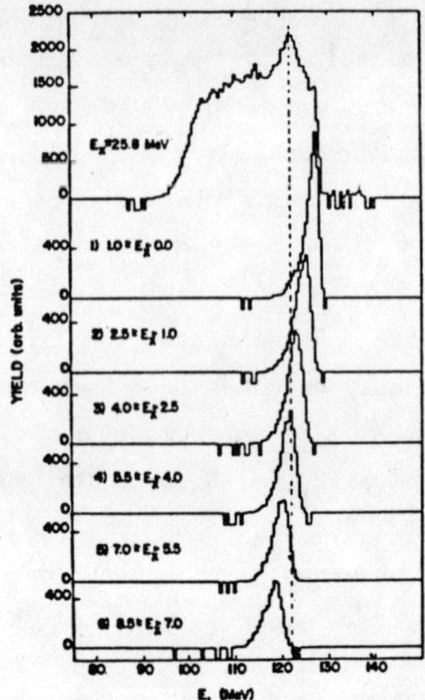

Fig. 44. The projections onto the inelastic alpha particle energy axis for the
^{58}Ni$(\alpha,\alpha'p)^{57}$Co reaction at $E_\alpha = 140$ MeV with gates set on seven excitation energy
regions of ^{57}Co.   See Ref. 130.

Because of this "walking", the decay branches to various final states are quite
difficult to measure for medium and heavy mass nuclei. However, the sum of decay
branches to all final states for a certain decay channel (such as n, p, or α
particle) can be obtained as follows. Instead of trying to extract the decay yield
from each individual coincidence sprectrum and then integrating over the angular
correlation function, the coincidence spectra at different angle pair (subject to
the condition that a certain type of decay particle of any energy is detected in
coincidence) are first angle integrated assuming cylindrical symmetry about the
recoil direction. An estimated continuum under the GQR peak is then subtracted from
both the angle-integrated coincidence spectrum and the single spectrum. The ratio of
these two estimated areas gives the total branching ratio for a particular channel.
This method not only eliminates the problem of "walking", but also·increases the
statistics in the coincidence spectra. Of course, the price paid is that the angular
correlation information is lost.

The general expressions for angular correlation function are rather complex.[142] For an interesting case where all the particles have zero spin except the
intermediate state, i.e., $X(0^+)(\alpha,\alpha')X^*(\alpha)Y(0^+)$, the angular correlation function
can be written as[142]:

$$W_L(\theta) = \sum_{km} P_m^J (2L+1)(LL00|k0)$$
$$\cdot (JJm-m|k0)P_k(\cos\theta), \qquad L=J . \qquad (V.3\text{-}1)$$

To calculate $W_L(\theta)$, one needs to know the magnetic substate populations, P_m^J, which are usually calculated in DWBA.[143] A simple case arises if the scattered alpha particles are detected along the beam axis. In this case, only the m=0 substate contributes in the sum of Eq. (V.3-1), and $W_L(\theta)$ can be simplified. Detecting alpha particles along the beam axis is experimentally rather difficult. It has been demonstrated that[144] if the scattered alpha particles are detected at a forward maximum of the GQR's angular distribution, the PWBA should be adequate. Choosing the direction of recoiling nucleus as the quantization axis, one has $P_m^J=1$ for m=0 and $P_m^J=0$ for m$\neq$0. In this case, using the additional theorem associated with Legendre polynomial, Eq. (V.3-1) reduces to

$$.W_L(\theta) \propto [P_L(\cos\theta)]^2 \ . \tag{V.3-2}$$

Figure 45 shows the angular correlation functions, $W_L(\theta)$, for a few L values.

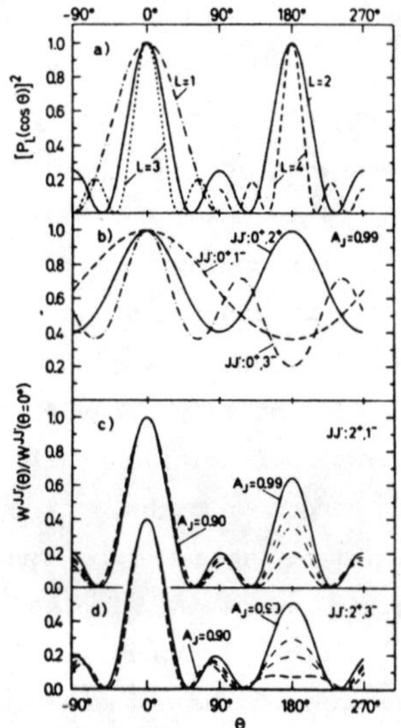

Fig. 45. Angular correlation functions for the sequential process $X(0^+)(\alpha,\alpha')X^*(\alpha)Y(0^+)$ calculated in plane wave: (a) for isolated resonance, (b)-(d) for overlapping resonances as indicated. See Ref. 144.

If there are two overlapping resonances with opposite parity, then the angular correlation function can be written as[144]:

$$W_{LL'}(\theta) \propto |A|^2(2L+1)[P_L(\cos\theta)]^2 + |B|^2(2L'+1)[P_{L'}(\cos\theta)]^2$$

$$+2|AB|\cos\delta \ \sqrt{(2L+1)(2L'+1)} \ P_L(\cos\theta)P_{L'}(\cos\theta) \ , \tag{V.3-3}$$

and $\qquad |A|^2+|B|^2 = 1 \ .$

$W_{LL'}(\theta)$ is also shown in Fig. 45. It is observed that by mixing in a few percent of L=3 into L=2 decay results in the forward-backward asymmetry. This asymmetry is indeed observed experimentally (see Sec. VI.1).

V.4 (e,e'x) Coincidence Experiments

With the high beam intensity and high duty cycle superconducting electron accelerators available at Stanford and at the University of Illinois, it is now feasible to perform the (e,e'x) coincidence experiment for studying the decay properties of the giant resonances. High beam intensity and high duty cycle accelerators are essential because the (e,e'x) experiment has very small cross sections, thus low coincidence count rates.

The (e,e'x) experiment has the advantages of the photonuclear reaction (γ,x) and the inelastic electron scattering combined. Like the (γ,x) reaction, the (e,e'x) coincidence experiment permits the simultaneous measurements of all the allowed decay channels. Furthermore, because the virtual photon is involved in the (e,e'x) experiment, the 3-momentum transfer, q, can be varied while keeping the energy transfer, ω, fixed (see Sec. III.1). This makes the measurement of the complete form factor for the nuclear excitation corresponding to a given final state possible.

One major advantage of the (e,e'x) experiment is that the continuum background is greatly reduced. The large continuum underlying the giant resonances seen in the singles (e,e') experiment is due essentially to elastically scattered electrons which have lost energy through radiation (radiation tail). Since this continuum does not involve the excitation of target nucleus, it should not give any true coincidences, and therefore, should be absent from the (e,e'x) spectra. This is indeed nicely demonstrated in Fig. 46, where the singles $^{12}C(e,e')$ spectrum taken at E_e=86 MeV is compared with the $^{12}C(e,e'p_0)$ experiment.[145] It should be emphasized that no background subtraction was made on the coincidence spectrum! The coincidence spectrum can now be compared directly to that for the $^{11}B(p,\gamma_0)^{12}C$ capture reaction.[146] As can be seen from Figs. 46 and 47, the agreement between the two is excellent.

It will be discussed in the next section that the decay branching ratios obtained from the electroproduction experiments are, in general, much larger than those obtained from the hadron-induced particle-particle coincidence experiments, especially the fission width of the ^{238}U GQR. Since (e,e'x) experiment has many nice features over the other two experiments, it is hoped that the (e,e'f) experiment might be able to resolve some of this discrepancy.

In a preliminary experiment at E_e=86 MeV, the ^{238}U fission fragments are detected in coincidence with the scattered electrons. Figure 48 shows the result from Stanford.[145] For comparison, the results for (e,e')[112] and (γ,f)[147] experiments are shown as solid curves in Fig. 48. The overall fission cross section is ~25%

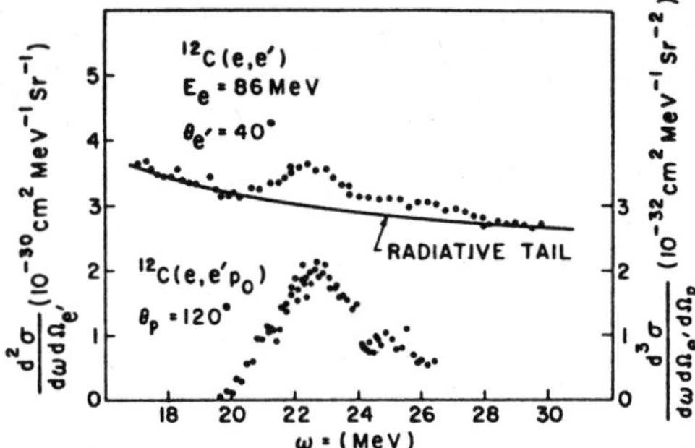

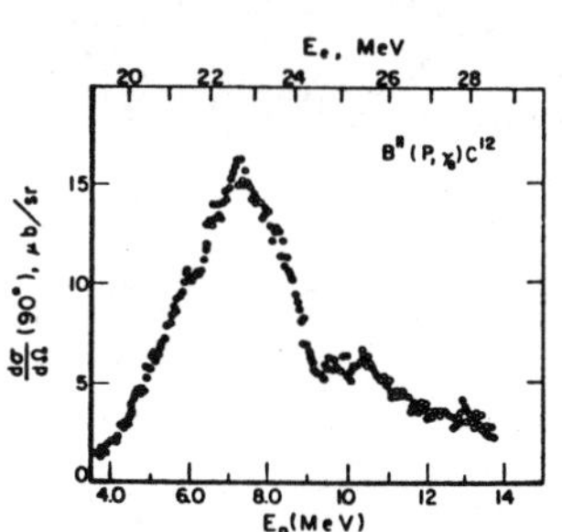

Fig. 46. Singles and coincidence (e,e') spectra at $E_e = 86$ MeV. No background has been subtracted from the coincidence spectrum. See Ref. 145.

Fig. 47. Proton radiative capture experiment on ^{11}B populating the ground state of ^{12}C. See Ref. 146.

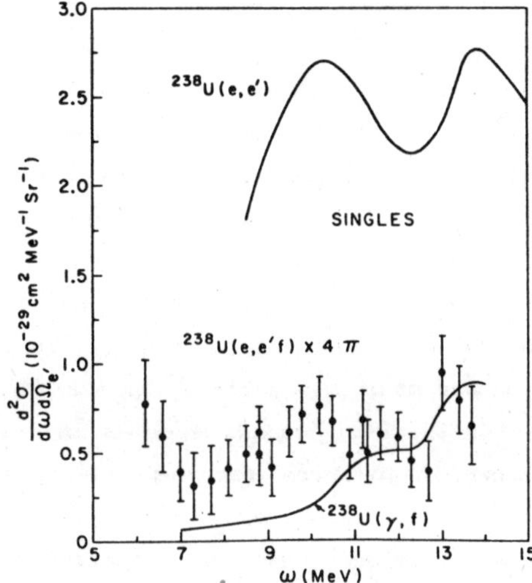

Fig. 48. Coincidence ^{238}U(e,e'f) spectrum multiplied by 4π. The solid curves are the cross sections for the ^{238}U(e,e') and ^{238}U(γ,f) reactions. See Ref. 145.

that of the (e,e'). By subtracting a normalized (γ,f) yield from the coincidence data, the excess yield seems to peak at $\omega=10$ MeV, corresponding to the expected excitation energy of the GQR. The fission branching ratio for the GQR is found to be $\sim 30\%$. More detailed discussion on this subject can be found in Sec. VI.3.

VI. Systematics of the Decay Properties

VI.1 Light Nuclei

The information concerning the decay of the isoscalar GQR in light nuclei has been obtained mainly through alpha particle[81,148,149] capture reaction and the particle-particle angular correlation experiments induced by alpha particles.[85,123] In the following, I will concentrate my discussion mainly on the results obtained from the coincidence experiments.

Up to now, the proton and alpha particle decay branches have been measured in coincidence experiments on ^{12}C,[85] ^{16}O,[123] ^{20}Ne,[144] ^{24}Mg,[150] and ^{28}Si.[144] Figure 49 shows typical ^{16}O(α,α') singles spectrum, together with the (α,α') spectra in coincidence with decayed protons and alpha particles detected at $\theta_{Lab}=-65^{\circ}$ (the direction of the recoiling ^{16}O nucleus for a Q-value of -20 MeV). It is obvious from this figure that the GQR in ^{16}O decays predominantly by alpha particle emission to ground and first excited states of ^{12}C. The (α,α') spectrum in coincidence with proton decay is very small and is non-resonant.

The large alpha particle decay width from the GQR seems to be a general feature of the decay mode in light nuclei. As the nuclear mass increases, the proton decay branches become more important. In fact, the proton decay branches are comparable to those of alpha particle for ^{28}Si.

At first sight, the large alpha particle decay widths may be difficult to understand because the GQR may be pictured as built up by the coherent superposition of the 1p-1h states. For doubly magic ^{16}O nucleus, the GQR can be thought of as the particles being in the 1f or 2p shell while the holes in the 1p shells (see Sec. II.2). In fact, it was shown in continuum random-phase-approximation calculations that the GQR in ^{16}O has predominantly $1f_{7/2}1p_{3/2}^{-1}$ configuration.[151]

This apparent discrepancy has been resolved in a calculation within the framework of SU(3) including predominantly the 1p-1h excitation.[152] In this calculation, Faessler et al. have shown that the large alpha particle decay widths stem from the fact that (i) the spectroscopic factors for protons and alpha particles are roughly of equal magnitude, and transmission factors favor the alpha particle emission, and that (ii) there is large overlap in light nuclei between 1p-1h shell model configurations and cluster wave function after antisymmetrization.

In fact, the calculation predicts the preferential emission of alpha particles from the ^{16}O GQR to the first excited 2^{+} state in ^{12}C with an L=4 partial wave, which is in good agreement with the observation.

The measured angular correlations of p_0, α_0, and α_1 decay from various regions of excitation energy in ^{16}O are shown in Fig. 50. One interesting feature worth noting is that the measured angular correlations for the GQR region exhibit a strong forward-backward asymmetry in the α_0 channel and to a lesser degree also in the

α_1 channel. By assuming a dominant E2 strength interferring with a few percent of E3 amplitudes, the experimental angular correlations, including the forward-backward asymmetry, are well reproduced by calculations assuming a pure m=0 substate population with respect to the recoil axis (PW calculations). This calculation offers the first unambiguous spin determination of 2^+ for the GQR. The same forward-backward asymmetry has also been observed in other light nuclei.

Various decay branching ratios can be obtained by integrating the experimental angular correlations assuming axial symmetry relative to the symmetry axis (recoil axis) as expected from plane wave calculations. The results are listed in Table 2 in Sec. VI.4, together with results for other nuclei.

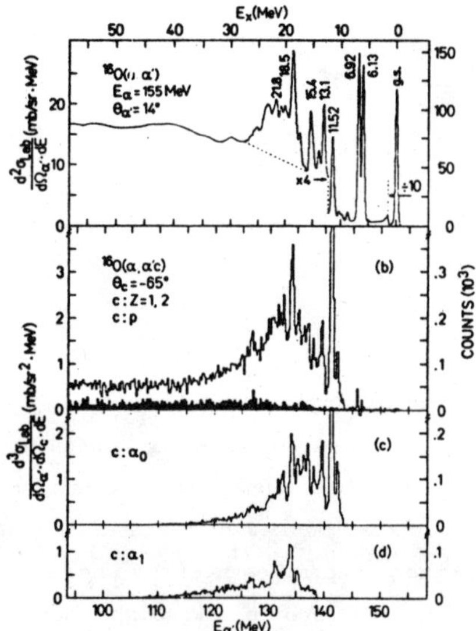

Fig. 49. (α,α') single spectrum at θ=14°. The same spectrum is also shown subject to the requirement of a proton (black area) or an alpha particle coincidence in a second detector at θ=-65°. See Ref. 123.

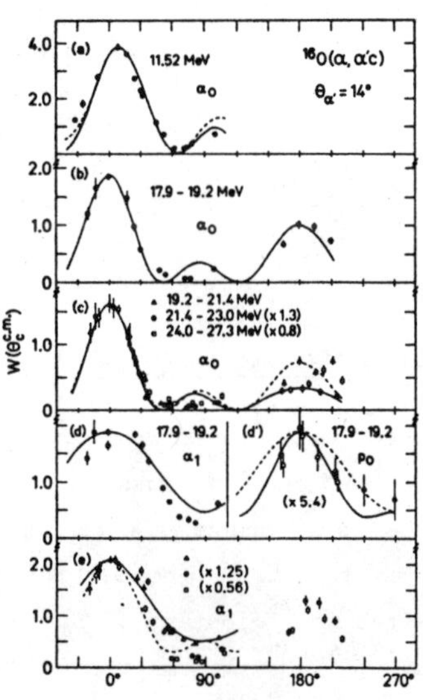

Fig. 50. Angular correlation functions of p_0, α_0, and α_1 decay from various regions of excitation energy. Angles are measured with respect to the recoiling ^{16}O nucleus. See Ref. 123.

The E2 strength inferred from the p_0 and α_0 channels can now be compared directly with the E2 strength deduced from proton and alpha particle capture reactions. These comparisons are not quite clear for sd-shell nuclei. In fact, ^{16}O is currently the best case for such a comparison. Figure 51 shows the extracted E2 strength distribution from the ^{12}C(α,γ_0)^{16}O capture reaction[148] and the ^{16}O($\alpha,\alpha'\alpha_0$) coincidence experiment.[123] It is encouraging to see that there is a close similarity in the structure. However, the E2 strength from the ($\alpha,\alpha'\alpha_0$) experiment is more than a factor of 2 greater than (α,γ_0) E2 strength. This difference has been attributed to the isospin mixing which reveals itself through a destructive interference of the isoscalar and isovector parts of the electromagnetic transition amplitudes.[123]

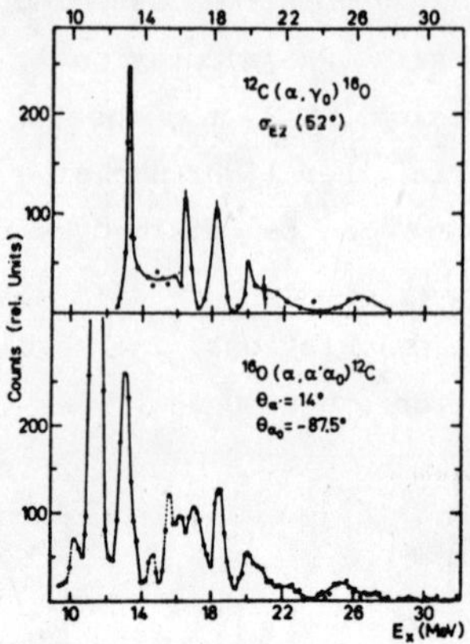

Fig. 51. Extracted E2 strength distribution from the $^{12}C(\alpha,\gamma_0)^{16}O$ capture reaction. See Ref. 148.

VI.2 <u>Medium Mass Nuclei</u>

The first coincidence experiment to measure the giant resonance decay modes was carried out on ^{40}Ca induced by 70 MeV ^{3}He.[136] In this experiment, large alpha particle decay branches were reported. In fact, the total alpha particle branch is equal to the total proton branch. However, this experiment was plagued by large quasi-free contributions. Subsequently, two other experiments have been performed.[137,138] Both of them tried to measure the GQR proton and alpha particle decays in ^{40}Ca. By now, it has generally been accepted that the total proton decay branch in ^{40}Ca dominates over the total alpha particle decay branch. The values reported are 70% branch for protons and 21% for alpha particles.[137]

Many experiments have been reported which use different techniques to measure the decay of the GQR in ^{58}Ni, with contradictory results. The first of these is an electroproduction experiment, and the data were analyzed using the DWBA E1 and E2 virtual photon specta (see Sec. V.2).[153] It was concluded that alpha particle emission results from a combination of E1 and E2 absorptions. The deduced E2 cross section from alpha particle channel alone was found to contain $56 \pm 4\%$ of the E2 EWSR. Since the GQR in ^{58}Ni has been found to exhaust about 50% of the E2 EWSR (see Sec. IV.2), this essentially implies a 100% GQR alpha particle decay branch. No proton decay strength was reported.

The dominant alpha particle decay branch in ^{58}Ni is surprising in view of the systematics of decreasing alpha particle branches with increasing A (see Sec. VI.1). In fact, this large alpha particle decay branch disagrees drastically with that obtained from the coincidence ^{58}Ni$(\alpha,\alpha'\alpha)$ experiments.[43,130] Figure 52 shows the

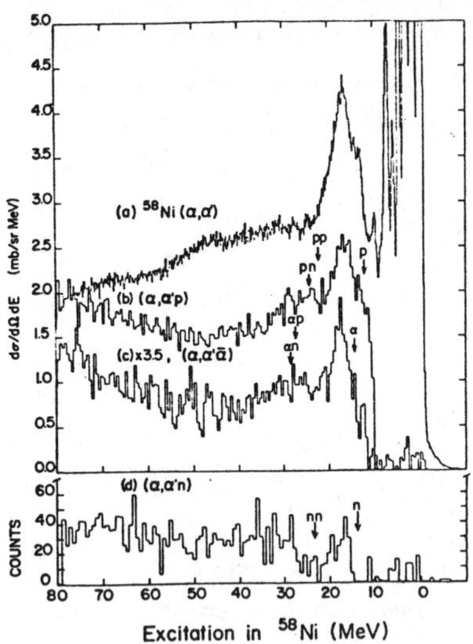

Fig. 52. (α,α') sprectrum at θ=16° for the reaction ^{58}Ni(α,α') at $E_\alpha = 140$ MeV. The same spectrum is also shown subject to the requirement of a proton, an alpha particle, or a neutron coincidence in the second detector.

^{58}Ni(α,α') singles as well as coincidence spectra.[130] Two of the coincidence spectra are angle-integrated (α,α') spectra subject to the condition that either a proton or an alpha particle of any energy was detected in coincidence. By subtracting a continuum spectrum from both the singles and coincidence spectra, the total proton and alpha particle decay branches are determined to be 59 ± 12% and 12 ± 4%, respectively, or equivalently, 30± 6% and 6 ±2% of the E2 EWSR.

It should be noted that the coincidence experiment is a direct measurement of the decay width, while the electroproduction experiment is a somewhat indirect singles measurement. An improved re-analysis of the electroproduction data has been reported by Wolynec et al.[134] and by Hayward.[132] The new result, although gives a much smaller alpha particle decay branch, still shows 15 ± 3% of the E2 EWSR (or ~ 30% decay branch) and no proton decay branch. A discrepancy of a factor of 2 still exists between the electroproduction and the coincidence experiments. It should be pointed out that a recent ^{58}Ni(e,α) experiment carried out by the Edinburg group[154] suggests that about 10% of the E2 EWSR was observed, in better agreement with the coincidence experiment.

The alpha particle decay of the GQR in ^{58}Ni has also been investigated through alpha particle capture reaction on ^{54}Fe.[128] This reaction was studied for various alpha particle energies between 7.6 and 12.8 MeV. The measured angular distributions were fitted with a theoretical expression in terms of Legendre polynomials (see Sec. V.1), assuming only E1 and E2 components. It was found that the E2 component was about 10% of the E2 component, and the extracted E2 cross section shows a 3 MeV wide peak located at about the GQR excitation energy region in ^{58}Ni (see Fig. 35). Integrating over this peak, the α_0 channel alone exhausts 4.3% of the E2 EWSR.

From the ^{58}Ni$(\alpha,\alpha'\alpha_0)$ coincidence experiment, an upper limit of 3% of the E2 EWSR is set on the α_0 channel by assuming zero contribution from the continuum.[130] A more realistic limit of 0.8% of the E2 EWSR was also set for the α_0 channel if the continuum contribution was subtracted.[130] This raises the possibility that the ground state alpha particle decay of the ^{58}Ni GQR may be greatly overestimated in the alpha particle capture reaction. In fact, the ^{54}Fe$(\alpha,\gamma_0)^{58}$Ni capture reaction has recently been repeated,[131] and it was found that the α_0 channel exhausts only about 2% of the E2 EWSR, a factor of ~ 2 less than that reported in Ref. 128. This brings the agreement between the alpha particle capture reaction and the coincidence experiment closer.

It is interesting to note that the energy of GQR ($\sim 63A^{-1/3}$ MeV) happens to lie in the vicinity of effective one-particle emission thresholds ($-Q+E_c$ for charged particle emission and $-Q+0.5$ for neutron emission) for nuclei with $40 \leq A \leq 90$. A natural consequence of this fact is that the various decay branches (n, p, and α particle branches) might be affected by the effective particle emission thresholds and the penetrabilities. The even zinc isotopes provide a good test of these effects because if these effects are large, one would expect a dramatic shift from charged particle to neutron decay strength with increasing neutron excess.

Coincidence measurements induced by 160 MeV alpha particles have been performed on ^{64}Zn, ^{66}Zn, and ^{68}Zn.[155] Indeed, appreciable branches were observed for both proton (38 ± 7%) and alpha particle (20 ± 4%) in ^{64}Zn (see Fig. 53). For ^{68}Zn, only negligible charged particle branches (0.8 ± 0.8% proton and 3.1 ± 0.8% alpha particle) were observed. The significance of these results in terms of the statistical decay of the GQR will be discussed in Sec. VI.4.

Neutron decay from the GQR in ^{119}Sn has recently been reported.[156] The decay neutrons were detected in coincidence with the inlastic alpha particle scattering at 109 MeV. Figure 54 shows the singles and coincidence (α,α') spectra. The coincidence spectrum is gated by "fast" neutrons ($E_n > 2.5$ MeV). Unlike the alpha particle decay from the GQR where the energy of the decaying alpha particles have almost the same energy as the evaporation alpha particles, even for decaying to the ground state of the residual nucleus, the neutron decay to low-lying states of the residual nucleus will have energies greater than the evaporation neutrons ($E_n \sim 1$ MeV). This makes it possible to investigate the direct neutron emission from the GQR. Figure 55 shows the angular correlations of the "fast" neutrons gated by the inelastic alpha particle exciting the GQR in ^{119}Sn. From these angular correlations, the total "fast" neutron branch from the GQR was found to be about 18%. It should be noted, however, that this 18% branch is an upper limit because the quasi-free process may contribute to the coincidence yields near the recoil axis ($\theta_n \sim -70°$), thus causing the large forward-backward asymmetry (see Fig. 55).

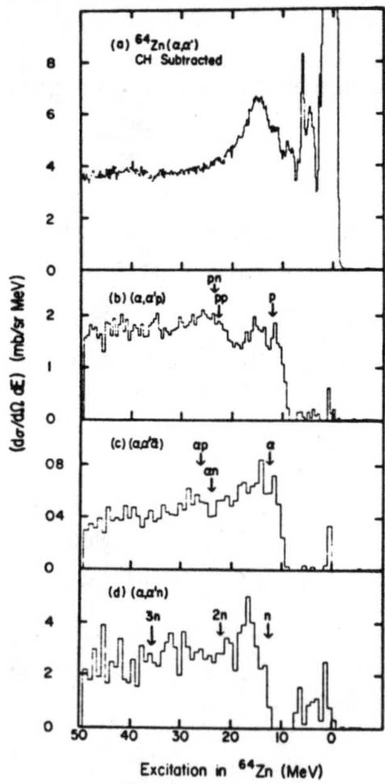

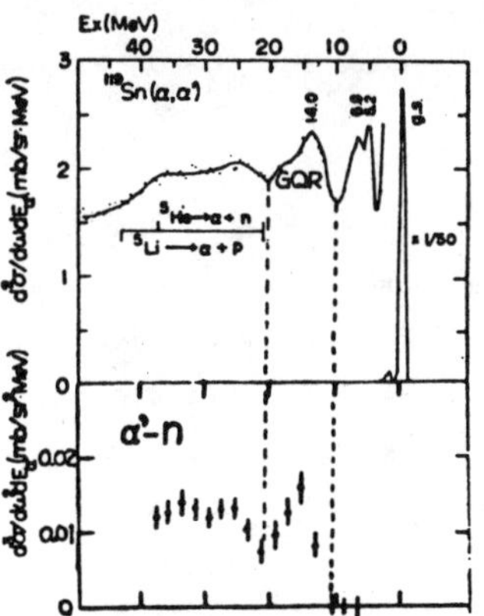

Fig. 54. Singles and neutron coincidence (α,α') spectra at $E_\alpha = 109$ MeV and $\theta_\alpha = 19.5°$. See Ref. 156.

Ref. 53. (α,α') singles spectrum at 14°. The same spectrum is also shown subject to the requirement of a proton, an alpha particle, or a neutron coincidence in a second detector.

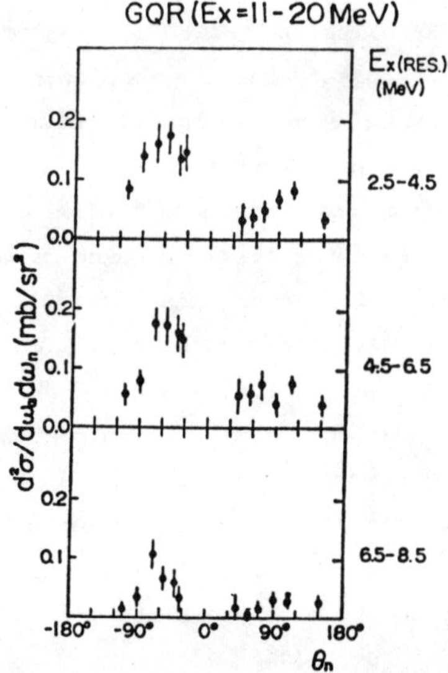

Fig. 55. Angular correlations of the "fast" neutrons gated by inelastic alpha particle exciting the GQR in ^{119}Sn. See Ref. 156.

VI.3 Heavy Mass Nuclei

The situation for the decay mode of the GQR in the actinide region is, to say the least, confusing. The first experiment was carried out using the vitual photon technique (see Sec. V.2) on ^{238}U. The cross sections for the ^{238}U$(e,\alpha)^{234}$Th reaction,[157] and later the ^{238}U$(e,n)^{237}$U reaction,[158] were measured by counting the residual gamma-ray activities. It was concluded that these activities were produced as a result of E2 absorption, and that the E2 resonance near 9 MeV decays predominantly by alpha particle emission (~50% of the E2 EWSR), while an upper limit of ~8% of E2 EWSR was set for neutron decay. However, the large alpha particle decay branch was not confirmed by subsequent ^{238}U(e,α) experiments.[159,161] In Ref. 160, alpha particle emission was measured directly and no alpha particles were observed that might result from the excitation of the isoscalar GQR near 9 MeV.

Electrofission measurements on ^{238}U have been undertaken by several investigators using a virtual photon technqiue. In these studies, a large fraction of the E2 EWSR was reported by Shotter et al. (27%)[162] and Arruda Neto et al. ($\geq$55%).[163] A 72% of E2 EWSR has also been reported for electrofission on ^{236}U.[164]

Contrary to the electrofission experiments, there is no evidence for a large fission width of the GQR in ^{232}Th and ^{238}U.[42,140,165] The first experiment using the coincidence ^{23}Th$(\alpha,\alpha'f)$ and ^{238}U$(\alpha,\alpha'f)$ experiments gave an upper limit of 1% for the fission probability of the GQR.[140] This value is at least a factor of 5 smaller than the fission decay of the underlying continuum ($5.5\pm1.5\%$). The authors of Ref. 140 suggest that the fission width of the GQR is inhibited as compared to the GDR fission decay, which decays statistically.

No inhibition of the GQR fission decay is reported in the ^{238}U$(^6$Li$,^6$Li$'f)$ experiment.[42] Shotter et al.[42] suggest that the fission probability of the GQR is no less than half that of the underlying continuum. Unfortunately, no values of the decay probability were reported. This is because no clear correspondence between the singles and coincidence spectra near the GQR region can be made. Part of this difficulty is caused by multiple-chance fissions which give rise to false peaks, and make the comparison difficult. Of course, this difficulty will be associated with every coincidence experiment involving detecting a fission product.

In a recent $(\alpha,\alpha'f)$ experiment at $E_\alpha=152$ MeV,[141] an enhancement is observed in the alpha-fission coincidence spectrum from ^{238}U. This enhancement is interpreted as arising from decay of the K=0 component of the GQR. A value of $\Gamma_f/\Gamma_t=0.25\pm0.10$ is reported for the GQR fission decay, while the continuum at 11.5° in the same excitation energy as the GQR has a value of $\Gamma_f/\Gamma_t=0.21\pm0.08$. This latest $(\alpha,\alpha'f)$ result agrees rather well with the $(e,e'f)$ experiment,[145] where the fission branching ratio for the GQR is reported to be ~ 30%.

In view of the situation among the various measurements, it would be too early to suggest that the GQR in the actinide region decays statistically. However, based on the systematics that have been seen so far for light and medium mass nuclei, I would suspect that a statistical equilibrium is reached for the GQR in heavy nuclei before the fission decay occurs.

VI.4 Statistical vs. Direct Decay of the Giant Quadrupole Resonance

A systematic picture of the GQR decay properties is beginning to emerge from the experiments that have been performed to date, at least for light and medium mass nuclei. Table 2 lists the experimental branching ratios for light and medium mass nuclei. The total proton and alpha particle branches are also plotted in Fig. 56. For the p- and sd-shell nuclei, a large alpha particle decay component is seen. On the other hand, the GQR decays predominantly by nucleon emission for nuclei with A ≥ 40. Single nucleon emission is found to dominate the GQR decay of ^{64}Zn and neutron decay is the overwhelming decay mode in ^{66}Zn and ^{68}Zn. It is also inferred from the lack of charged particle decay that neutron emission is the most important GQR decay channel in ^{62}Ni. As mentioned earlier (see Sec. VI.3), the results for very heavy nuclei are still unclear. For example, the coincidence experiments report that no GQR is seen in the fission decay channel of ^{238}U, while the electrofission experiments find an appreciable fission branch for ^{238}U.

It was first pointed out in the ^{58}Ni coincidence experiment[130] that there is a great similarity in the behavior of the GQR and the underlying continuum background in the charged particle decay channels. Both the angular correlations (see Fig. 57) and the total branching ratios for the GQR and the continuum (see Table 2) are essentially the same. Since one would expect the continuum to decay in a statistical way, the question was raised whether the particle decay of the GQR in medium mass nuclei may also be governed by statistical processes. One would expect that if the GQR decays statistically, its decay branches should be predictable by the Hauser-Feshbach calculation for $J^{\pi}=2^{+}$. The continuum decay branches should contain contributions from many different spins. They should be comparable with the predictions of the Hauser-Feshbach calculation summed over J. The Hauser-Feshbach predictions are also listed in Table 2 and are plotted in Fig. 56.

It was observed in the ^{16}O(α,α'x) coincidence experiment that the continuum underlying the GQR is reduced by more than a factor of five relative to the GQR.[123] This was taken to imply that the continuum and the GQR decay differently, namely the GQR in ^{16}O decays mostly by direct emission. This conclusion can also be reached by examining the content of Table 2. For ^{16}O, the ratio of the measured total alpha particle decay branch to the total proton decay branch of about 5/1 far exceeds the statistical prediction. Also, the α_0 and α_1 branching ratios are larger than those predicted by a factor of three. This trend is also observed in ^{28}Si.

Unlike the lighter mass nuclei, the ^{40}Ca total proton and alpha particle decay branches of 70% and 21% are in good agreement with the statistical values of 74% and 21%, respectively.[130] In fact, the measured branching ratios for all medium mass nu-

TABLE 2   Experimental Decay Branches and Statistical Model Predictions

Nucleus	Decay Particle	Experimental Γ_i/Γ (%)		Hauser–Feshbach Prediction Γ_i/Γ (%)		Ref.
		GQR	Underlying Continuum	$J^\pi=2^+$	All J^π	
^{12}C	p_0	<27*				85
	α_0	27*				85
	α_1	45*				85
^{16}O	p_0	<14		7		123
	α_0	20		6		123
	α_1	55		19		123
^{24}Mg	α_0	15		3		150
^{28}Si	p_0	22		6		144
	$p_{1,2}$	20		8		144
	Σ_p	56		62		144
	α_0	11		3		144
	α_1	28		10		144
	$\alpha_{2,3}$	12				144
	Σ_α	51		30		144
^{40}Ca	p_0	8^{+5}_{-3}				137
	p_1	22^{+5}_{-8}				137
	Σ_p	70^{+15}_{-20}		74	76	137
	α_0	< 6				137
	α_1	<10				137
	α_2	24^{+3}_{-6}				137
	Σ_α	21^{+5}_{-15}		21	18	137
^{56}Fe	Σ_α	24±6+				132
^{59}Co	Σ_α	22±4+				132
^{58}Ni	Σ_p	59±12	53±10	51	59	130
		65		51	59	43
	α_0	< 6				130
	Σ_α	30±6+		6	3	132
		20+		6	3	154
		12±4	7±1	6	3	130
		<30		6	3	43
^{60}Ni	Σ_α	30±8+				132
^{62}Ni	Σ_p	<20		2		43
	Σ_α	12±4+		4		132
		<15		4		43
^{64}Zn	Σ_n	59±27	58±23	28	34	155
	Σ_p	38±7	24±2	46	43	155
	Σ_α	20±4	9.3±1.2	26	23	155
		78±16+		26	23	132
^{66}Zn	Σ_n	113±45	88±34	86	91	155
	Σ_p	4.4±1.6	7.4±0.4	6	5	155
	Σ_α	3.7±0.9	3.7±0.4	8	4	155
^{68}Zn	Σ_n	124±51	111±43	96	97	155
	Σ_p	0.8±0.8	4.6±0.4	1	1	155
	Σ_α	3.1±0.8	2.6±0.2	3	2	155

*Assuming an E2 EWSR of 14.1% is exhausted.

+Assuming an E2 EWSR of 58% is exhausted.

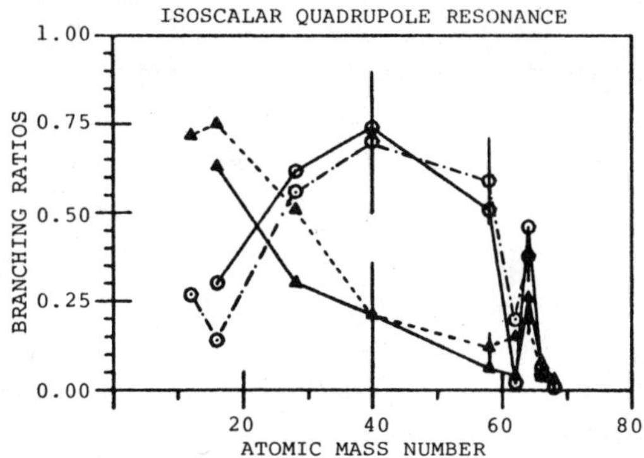

Fig. 56. Proton (circles) and α particle (triangles) decay branches as a function of nuclear mass. The experimental data points are connected by broken lines while the Hauser-Feshbach predictions are connected by solid lines.

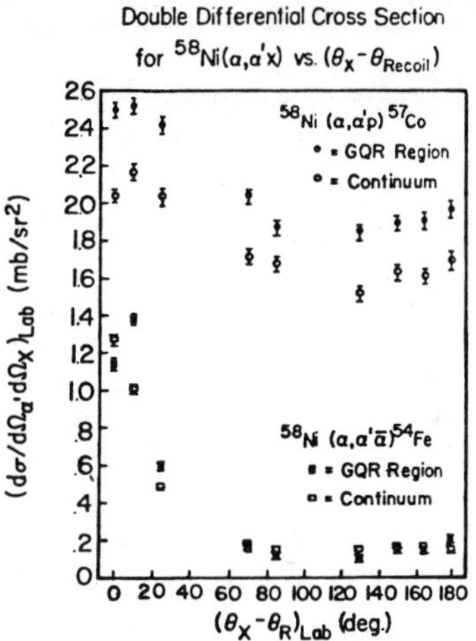

Fig. 57. Angular correlation functions for the ^{58}Ni(α,α'p) (circles) and ^{58}Ni(α,α'α) (squares) reactions. Solid symbols are yields for the GQR region (GQR plus background), while the open symbols are yields for the continuum region adjacent to the high excitation energy side of the GQR.

clei are in good agreement with the statistical predictions. On the basis of the good agreement between the statistical model predictions and experimental decay branches and the similarity between the branching ratios for GQR and underlying continuum, it was concluded that the GQR in the medium mass nuclei decay mainly by statistical process.[130]

Based on the systematics that have been discussed so far and the fact that the GDR in heavy mass nuclei decays mainly in a statistical way,[166,167] it may be tempting to conclude that the GQR for heavy mass nuclei might also decay by statistical process. Unfortunately, the data accumulated so far cannot substantiate this claim. The measurement of neutron decay branch on heavy mass nuclei in particle-particle angular correlation experiment, for instance, might be able to shed some light on this interesting question.

VII. Spreading and Escaping Widths

An extensive theoretical work on calculating the total width Γ of the giant resonances has been made both on a macroscopic[8,9] as well as a microscopic[168] approach. The simplest picture of the giant resonances is to describe them as a coherent mixture of one-particle, one-hole (1p-1h) states (see Sec. II.2), and the width of a giant resonance is determined by its decay properties in that it consists of a sum of the escaping width and the spreading (or damping) width. The escaping width, $\Gamma^{\uparrow}$, is due to the coupling of those 1p-1h states with the continuum states.[10-13] On the other hand, the spreading width, $\Gamma^{\downarrow}$, is due to the coupling of these 1p-1h states to more complicated multi-particle, multi-hole states.

If there is strong coupling between the doorway 1p-1h states and the continuum states, the giant resonance will decay by direct particle emission, and its total width will be determined mainly by the escaping width. However, if the coupling of the doorway 1p-1h states to the complex multi-particle, multi-hole states is strong, then the giant resonance will equilibrate and its decay will be characteristic of a compound nucleus, and its total width will be determined by the spreading width.

It has been known that the total width of the GDR in light nuclei is due mainly to th escaping width.[166,169] For heavy nuclei, however, the spreading width is dominant.[170] The situation for the GQR seems to be different. All theoretical calculations suggest that the spreading width is more important than the escaping width. For instance, the calculated one-particle escaping widths are only 0.5 MeV for ^{40}Ca and 1.2 MeV for ^{16}O, which are much smaller than the observed widths. On the other hand, spreading width of several MeV was obtained even in light nuclei.[13,171,172] Figure 58 shows the broadening of the GQR width in ^{16}O caused by the coupling of 1p-1h excitations with 2p-2h states, calculated in the core-coupling random-phase approximation.[172]

The relative importance of the spreading width over the escaping width as a function of the nuclear mass A can be inferred from the systematics of the GQR decay widths, as described in Sec. VI. It was observed that for light nuclei, alpha particle decay dominates over the nucleon decays, while this trend reverses as the nuclear mass increases beyond A $\sim$ 30 (see Fig. 56). Although Hauser-Feshbach calculation is quite successful in predicting the ratio of sum of alpha particle decay widths to proton widths, $\Sigma\Gamma_{\alpha}/\Sigma\Gamma_{p}$, it underestimates $\Sigma\Gamma_{\alpha}$ for light nuclei, suggesting a possible nonstatistical alpha particle decay component in light nuclei.

This large alpha particle decay width from the GQR in ^{16}O has been explained in the framework of a SU(3) calculation including predominantly 1p-1h excitation.[152] The same calculation predicts a ratio of alpha particle to proton width of $\sim$ 0.5, while the measured ratio is $\sim$ 0.3. In addition, the calculation also fails to predict the preferential population of relatively high-lying states in ^{39}K. This is expected because the experimental branching ratios for ^{40}Ca agree well with Hauser-Feshbach prediction, suggessting that multi-particle, multi-hole states will

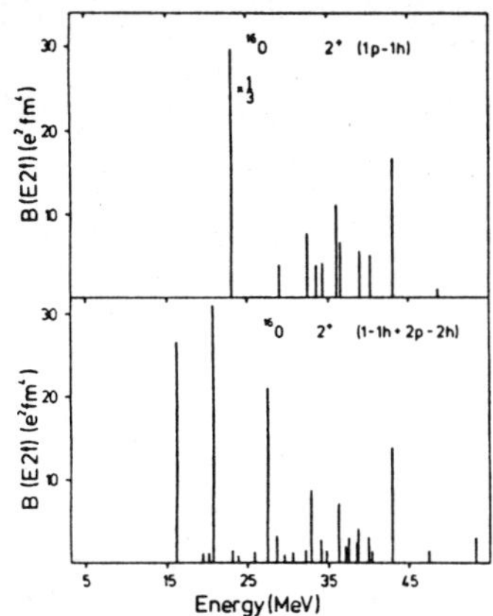

Fig. 58. The GQR in ^{16}O calculated in the 1p-1h and in the core-coupling (1h-1p + 2p-2h) random-phase-approximation. The broadening of the GQR due to the coupling is obvious. See Ref. 172.

have to be included in the calculation.

Without a detailed knowledge of the wave function for the GQR for most of the target nuclei studied, the interplay between the spreading and escaping widths may be partially understood in the context of a pre-equilibrium exciton model calculation.[173] In the following, the exciton model will be used to estimate the spreading and escaping widths as a function of nuclear mass.

The pre-equilibrium exciton model assumes that a composite system is formed in a specific initial particle-hole state after the projectile-target interaction. The system is then allowed to proceed to more complicated multi-particle, multi-hole states via energy conserving two-body residual interactions. Each step in the reaction is characterized by the exciton number (the number of particles and holes excited) and by the excitation energy, E. At each stage, the system is allowed to create or destroy a particle-hole pair or decay by particle emission. In the exciton model formalism, the widths for these three processes are Γ_+, Γ_-, and Γ_c, respectively. We identify the spreading width, $\Gamma^{\downarrow}$, with Γ_+, and the escaping width, $\Gamma^{\uparrow}$, with Γ_c. Making the "never-come-back" assumption, i.e., neglecting Γ_- (this is justified in the early stages of the equilibration process), the total width, Γ, is given by:

$$\Gamma = \Gamma_+ + \Gamma_c$$
$$= \Gamma^{\downarrow} + \Gamma^{\uparrow} \quad . \qquad\qquad\qquad (VII-1)$$

The pre-equilibrium exciton model estimate for the spreading width is given in terms of the particle number, p, the hole number, h, and the excitation energy, E, as:

$$\Gamma^{\downarrow}(p,h,E) = \pi|M|^2 \frac{g}{(p+h+1)} (gE-C_{p+1,h+1})^2 , \qquad\qquad (VII-2)$$

where

$$|M|^2 = \text{square of the average two-body transition matrix element}$$
$$= 200\ A^{-3}\ E^{-1}$$
$$g = \text{single-particle state density} = 6a/\pi^2$$
$$C_{p,h} = \tfrac{1}{2}(p^2+h^2)\ .$$

The total escaping width from the (p,h) state is given by:

$$\Gamma^\uparrow(p,h,E) = \hbar \sum_{x=p,n,\alpha} \int_0^{E-B_x} W_x(p,h,E,\varepsilon)d\varepsilon\ , \qquad \text{(VII-3)}$$

where B_x is the binding energy of particle x, and $W_x(p,h,E,\varepsilon)$ is the rate of emission of particle x with energy ε from a (p,h) state, and is given by:

$$W_x(p,h,E,\varepsilon)d\varepsilon = \frac{2S_x+1}{\pi^2\hbar^3}\ \mu_x\sigma_x(\varepsilon)\varepsilon$$

$$\cdot\ \left(\frac{\omega(p-p_x,h,U)}{\omega(p,h,E)}\ \frac{\omega(p_x,0,E-U)}{g_x}\right) R_x(p)\gamma_x d\varepsilon\ . \qquad \text{(VII-4)}$$

All the symbols have been defined in Ref. 173.

We take the initial configuration of the GQR in the exciton model to be a 1p-1h state of the target nucleus with an excitation energy equal to $63A^{-1/3}$ MeV. Pre-equilibrium exciton model estimates of the ratio of the GQR escaping width to the total width, $(\Gamma^\uparrow/\Gamma)$, for various nuclei from ^{12}C to ^{209}Bi are listed in Table 3. They were calculated with the program PREQEC.[174] The table shows the results for the initial 1p-1h state only, and the contributions summing up to the 4p-4h stage of the calculation. The calculated ratios for the 1p-1h through the 3p-3h step are essentially the same and are dominated by single nucleon emission. There is an increase in $(\Gamma^\uparrow/\Gamma)$ at the 4p-4h step, especially for the lighter nuclei, because alpha particle emission becomes possible when four or more particles are excited. There is then little change in $(\Gamma^\uparrow/\Gamma)$ beyond the 4p-4h stage.

The exciton model calculation predicts that escaping width will provide a 30% to 40% contribution to the total GQR width for nuclei of mass A < 28. This contribution decreases with increasing mass down to values of only a few percent for ^{90}Zr and ^{209}Bi. It is less than 9% for any of the nuclei with mass A > 40. Comparison of the 1p-1h results with the 4p-4h values shows that direct alpha particle emission is competitive with direct nucleon emission in the lighter nuclei and that it becomes insignificant for the heavier ones. This is consistent with the measured branching ratios (see. Sec. VI). In summary, the pre-equilibrium exciton model predicts that escaping width (or direct emission) and spreading width (or statistical emission) are comparable in light nuclei. However, in the mass range A > 40, escaping wdith is less than 10% that of total width.

It should be noted that the exciton model calculations described above use an unrealistic equal-spacing model for the nuclear level density. Also, the model uses a simple 1p-1h initial configuration for the GQR state even though the resonance is a coherent superposition of many 1p-1h states. It has been suggested by Luk'yanov[175] that the collective nature of the GQR could be treated by using an effective initial 1p-1h state density which is much larger than the normal values. This would tend to enhance direct particle emission.

TABLE 3   Pre-Equilibrium Exciton Model Estimates of the GQR Escaping Widths for Various Nuclei

Nucleus	GQR Excitation Energy (MeV)	$(\Gamma^{\uparrow}/\Gamma)_{1p-1h}$ (%)	$(\Gamma^{\uparrow}/\Gamma)_{\text{up to }4p-4h}$ (%)
^{12}C	27.5	21	43
^{16}O	21	17	33
^{20}Ne	23.2	15	41
^{24}Mg	21.8	13	20
^{26}Mg	21.3	20	32
^{27}Al	21	21	35
^{28}Si	20.7	9	12
^{32}S	19.8	13	20
^{40}Ca	18.4	9	12
^{58}Ni	16.3	6	7
^{62}Ni	15.9	7	8
^{64}Zn	15.75	6	7
^{66}Zn	15.59	6	7
^{68}Zn	15.43	7	8
^{90}Zr	14.1	1	1
^{209}Bi	10.6	3	3

VIII. Future Research

Although many aspects of the new giant resonances have been studied in the past ten years, there are still many new and exciting subjects or unsolved questions remaining. In the following, I will only briefly mention a few.

The observations of the giant monopole resonance in the vicinity of the GQR and the splitting of the giant monopole resonance in deformed nuclei (see Fig. 23) complicate the procedure for extracting the EWSR strength. In addition, the subtraction of underlying continuum has also contributed to a large extent to the uncertainty in the quoted EWSR. Up to now, it has been a common practice to assume an arbitrary continuum shape and to subtract it from the inelastic scattering spectrum. It would be very important to be able to understand, both theoretically and experimentally, the dominant nuclear reactions which are responsible for the continuum. This subject is very complex, and works along this line have just begun. By understanding the dominant reaction mechanisms responsible for the continuum, one can either calculate the continuum shape for background subtraction purpose, or choose a nuclear reaction which can minimize the continuum with respect to the giant resonance yields.

Besides the study of the Gamow-Teller resonance, no convincing evidences on the excitation of any new giant resonances have been reported in charge exchange reactions (see Fig. 11, however). Because of the isospin selectivity (and in other cases, the spin selectivity as well, such as the $(^6\mathrm{Li},^6\mathrm{He})$ reaction), the charge exchange reactions at higher incident energies may become quite useful tools in the study of isovector giant resonances. For instance, a tentative assignment for the observation of the analog GDR in $(^3\mathrm{He},t)$ reaction at $E_{^3\mathrm{He}}$ =130 MeV[119] has been reported.

The search for isoscalar giant resonances of higher multipolarities is also interesting. It remains to be seen to what extent the higher multipole resonances are broadened. As was discussed in Sec. IV.3, the HEOR is already quite broad. It may be that the higher multipole resonances become so broad that they constitute part of the continuum spectrum.

As for the decay modes of the giant multipole resonances, it would be important to determine what fraction of the decay branch is non-statistical for medium and heavy mass nuclei. This information is, unfortunately, difficult to obtain from the charged particle decays. A careful measurement of "fast" neutron decay from the GQR should be carried out, especially in the (e,e'n) coincidence experiment. This measurement not only enables one to learn about the non-statistical decay property of the GQR, it may help shed some light on the problem of fission decay probability for the GQR in heavy nuclei.

No information concerning the decay properties on giant multipole resonance other than the GDR and GQR has been obtained as yet. The systematics of the decay properties as functions of nuclear mass and multipolarity may help us understand the damping mechanisms in nuclei.

This work was supported in part by the National Science Foundation of the United States of America. I would also like to thank my colleagues, Dr. M. T. Collins and Dr. S. L. Tabor, for their contributions to part of the coincidence measurements which are reported here.

References

1. G. C. Baldwin and G. S. Klaiber, Phys. Rev $\underline{71}$, 3 (1947).
2. A. B. Migdal, J. Phys. USSR $\underline{8}$, 331 (1944).
3. M. Goldhaber and E. Teller, Phys. Rev. $\underline{74}$, 1046 (1948).
4. H. Steinwedel and J. H. D. Jensen, Z. Naturforsch. $\underline{5a}$, 413 (1950).
5. R. Pitthan and Th. Walcher, Phys. Lett. $\underline{B36}$, 563 (1971).
6. M. B. Lewis and F. E. Bertrand, Nucl. Phys. $\underline{A196}$, 337 (1972).
7. S. Fukuda and Y. Torizuka, Phys. Rev. Lett. $\underline{29}$, 1109 (1972).
8. N. Auerback and A. Yeverechyahu, Ann. Phys. (N.Y.) $\underline{95}$, 35 (1975).
9. J. R. Nix and A. J. Sierk, Phys. Rev. C $\underline{21}$, 396 (1980).
10. S. Krewald, J. Birkholz, A. Faessler, and J. Speth, Phys. Rev. Lett. $\underline{33}$, 1386 (1974).
11. K. F. Liu and N. van Giai, Phys. Lett. $\underline{65B}$, 23 (1976).
12. S. Shlomo and G. Bertsch, Nucl. Phys. $\underline{A243}$, 507 (1975).
13. T. Hoshino and A. Arima, Phys. Rev. Lett. $\underline{37}$, 266 (1976).
14. D. H. Wilkinson, Ann. Rev. Nucl. Sci. $\underline{9}$, 1 (1959).
15. O. Nathan and S. G. Nilsson, Alpha-, Beta- and Gamma-Ray Spectroscopy, Vol. 1, edited by K. Siegbahn (North-Holland, Amsterdam, 1966), p. 601.
16. G. R. Satchler, Particle and Nuclei $\underline{5}$, 105 (1973).
17. E. Hayward, Photonuclear Reactions, National Bureau of Standards Monograph 118 (1970).
18. D. Kurath, Phys. Rev. $\underline{130}$, 1525 (1963).
19. S. T. Tuan, K. E. Wright, and D. S. Onley, Nucl. Instrum. & Methods $\underline{60}$, 70 (1968).
20. L. J. Tassie, Aust. J. Phys. $\underline{9}$, 407 (1956).
21. H. Uberall, Electron Scattering from Complex Nuclei, (Academic, New York, 1971), Part B, Chap. 6.
22. H. Steinwedel, J. H. D. Jensen, and P. Jensen, Phys. Rev. $\underline{79}$, 1109 (1950).
23. M. Sasao and Y. Torizuka, Phys. Rev. C $\underline{15}$, 217 (1977).
24. G. R. Satchler, Phys. Rep. $\underline{14C}$, 97 (1974).
25. F. E. Bertrand, Annu. Rev. Nucl. Sci. $\underline{26}$, 457 (1976).
26. For a detailed account of the giant multipole resonances, see the Proceedings of the Giant Multipole Resonance Topical Conference, Oak Ridge, TN, Oct. 15, 1979, edited by F. E. Bertrand (Harwood Academic Publishers).
27. J. J. Griffin, Phys. Rev. Lett. $\underline{17}$, 478 (1966).
28. M. Blann, Phys. Rev. Lett. $\underline{21}$, 1357 (1968); Annu. Rev. Nucl. Sci. $\underline{25}$, 123 (1975).
29. K. Chen et al., Phys. Rev. $\underline{166}$, 949 (1968).
30. G. F. Bertsch and S. F. Tsai, Phys. Rep. $\underline{18C}$, 125 (1975); S. F. Tsai and G. F. Bertsch, Phys. Rev. C $\underline{11}$, 1634 (1975).
31. T. Tamura, T. Udagawa, D. H. Feng, and K. K. Kan, Phys. Lett. $\underline{66B}$, 109 (1977); T. Tamura and T. Udagawa, Phys. Lett. $\underline{78B}$, 189 (1978).
32. F. E. Bertrand and R. W. Peelle, Phys. Rev. C $\underline{8}$, 1045 (1973).
33. J. R. Wu, C. C. Chang, and H. D. Holmgren, Phys. Rev. C $\underline{19}$, 370 (1979); $\underline{19}$, 659 (1979); $\underline{19}$, 698 (1979).
34. G. R. Satchler, Nucl. Phys. $\underline{A195}$, 1 (1972).
35. R. A. Broglia, C. H. Dasso, and A. Winther, Phys. Lett. $\underline{61B}$, 113 (1976).
36. R. R. Betts, S. B. DiCenzo, M. H. Mortensen, and R. L. White, Phys. Rev. Lett. $\underline{39}$, 1183 (1977).
37. P. Doll et al., Phys. Rev. Lett. $\underline{42}$, 366 (1979).
38. J. Speth, E. Werner, and W. Wild, Phys. Rep. $\underline{33C}$, 127 (1977).
39. R. Kamermans et al., Phys. Lett. $\underline{82B}$, 221 (1979).

40. A. M. Sandorfi, in Proceedings of the Symposium on Heavy-Ion Physics from 10 to 200 MeV/amu, July 16-20, 1979 (Brookhaven National Laboratory, Upton, NY).

41. H. J. Gils, H. Rebel, J. Buschmann, and H. Klewe-Nebenius, Phys. Lett. 68B, 427 (1977).

42. A. C. Shotter et al., Phys. Rev. Lett. 43, 569 (1979).

43. K. T. Knöpfle et al., Lecture Notes in Physics 92, 445 (1979).

44. M. Buenerd and J. Arvieux, p. 381 of Ref. 26.

45. M. K. Gupta and G. E. Walker, Nucl. Phys. A256, 444 (1976).

46. G. E. Walker, 2nd International Conference on Meson Nuclear Physics, Houston, March 5-9, 1979.

47. J. Arvieux et al., Phys. Rev. Lett. 42, 753 (1979).

48. J. Arvieux et al., to be published.

49. N. S. P. King and J. L. Ullmann, in Proceedings of the (p,n) Reaction and the Nucleon-Nucleon Force, Telluride, CO, March 29-31, 1979, edited by C. D. Goodman et al., (Plenum Press).

50. R. R. Doering et al., Phys. Rev. Lett. 35, 1691 (1975).

51. C. D. Goodman et al., Phys. Rev. Lett. 44, 1755 (1980).

52. D. E. Bainum et al., Phys. Rev. Lett. 44, 1751 (1980).

53. A. Galonsky et al., Phys. Lett. 74B, 176 (1978); D. Ovazza et al., Phys. Rev. C 18, 2438 (1978).

54. International School on Electro- and Photonuclear Reactions, Lecture Notes in Physics (1976), Vol. 61.

55. International Conference on Nuclear Interactions, Canberra (1978), Lecture Notes in Physics, Vol. 92.

56. International Conference on Nuclear Physics with Electromagnetic Interactions, Mainz (1979), Lecture Notes in Physics, Vol. 108.

57. Proceedings of the International Symposium on Highly Excited States in Nuclear Reactions, Suita Osaka, Japan (1980).

58. Proceedings of the International Conference on Nuclear Structure Studies Using Electron Scattering and Photoreaction, Sendai (1972).

59. I. Hamamoto, Suppl. Res. Rep. Nucl. Sci., Tohoku Univ. 5, 205 (1972).

60. J. P. Blaizot et al., Nucl. Phys. A265, 315 (1976).

61. H. A. Bethe, Ann. Rev. Nucl. Sci. 21, 93 (1971); W. Myers and W. Swiatecki, Ann. Phys. 55, 395 (1969); D. Vautherin and D. Brink, Phys. Rev. C 5, 626 (1972); D. Gogny, Nuclear Self-Consistent Fields, edited by G. Ripka and M. Porneuf (North-Holland, Amsterdam, 1975) p. 333i; D. M. Brink and E. Boeker, Nucl. Phys. A91, 1 (1967); M. Beiner et al., Nucl. Phys. A238, 29 (1975).

62. N. Marty et al., in Proceedings of the International Symposium on Highly Excited States in Nuclei, Julich, 1975, p. 17.

63. S. Fukuda and Y. Torizuka, Phys. Lett. 62B, 146 (1976).

64. M. N. Harakeh et al., Phys. Rev. Lett. 38, 676 (1977); Nucl. Phys. A327, 373 (1979).

65. D. H. Youngblood et al., Phys. Rev. Lett. 39, 1188 (1977).

66. D. H. Youngblood, p. 113 of Ref. 26.

67. N. Marty et al., Bull. Am. Phys. Soc. 24, 844 (1979); Orsay Report IPNO-PhN-79-18 (1979).

68. M. Buenerd et al., Phys. Lett. 84B, 305 (1979).

69. F. E. Bertrand et al., Phys. Rev. C 18, 2788 (1978); F. E. Bertrand et al., Phys. Lett. 80B, 198 (1979).

70. D. J. Horen et al., Phys. Rev. C 9, 1607 (1974).

71. D. M. Patterson et al., Nucl. Phys. A263, 261 (1976).

72. C. M. Rozsa et al., preprint.

73. V. R. Pandharipande, Phys. Lett. 31B, 635 (1970).

74. N. Marty et al., Nucl. Phys. A238, 93 (1975).

75. C. C. Chang et al., Phys. Rev. Lett. 34, 221 (1975).

76. A. Moalem et al., Phys. Rev. Lett. 31, 482 (1973).

77. J. M. Moss et al., Phys. Rev. Lett. 34, 748 (1975).

78. D. H. Youngblood et al., Phys. Rev. C 13, 994 (1976).

79. C. C. Chang, p. 191 of Ref. 26.

80. G. Chenevert et al., Phys. Rev. Lett. 27, 434 (1971); see also N. S. Chant and E. M. Henley, Phys. Rev. Lett. 27, 1657 (1971).

81. S. S. Hanna, p. 275 of Ref. 54.

82. A. Kiss et al., Phys. Rev. Lett. 37, 1188 (1976).

83. K. T. Knöpfle et al., Phys. Lett. 69B, 263 (1975).

84. D. H. Youngblood et al., Phys. Rev. C 15, 1644 (1978).

85. H. Riedesel et al., Phys. Rev. Lett. 41, 377 (1978).

86. M. Danos, Nucl. Phys. 5, 23 (1958); K. Okamoto, Phys. Rev. 110, 143 (1958); E. G. Fuller and E. Hayward, Nucl. Phys. 30, 613 (1962); B. L. Berman and S. C. Fultz, Rev. Mod. Phys. 47, 713 (1975).

87. A. Bohr and B. Mottelson, Nuclear Structure, Vol. II, (Benjamin, Reading, Mass. 1975); R. Ligensa and W. Greiner, Nucl. Phys. A92, 673 (1967).

88. J. Moss et al., Phys. Lett. 53B, 51 (1974); D. J. Horen et al., Phys. Rev. C 9, 1607 (1974); D. J. Horen et al., Phys. Rev. C 11, 1247 (1975); A. Schwierc-zinski et al., Phys. Lett. 55B, 171 (1975).

89. T. Koshimoto et al., Phys. Rev. Lett. 35, 552 (1975).

90. D. Zawischa and J. Speth, Phys. Rev. Lett. 36, 843 (1976); T. Suzuki and D. J. Rowe, Nucl. Phys. A289, 461 (1977); Nucl. Phys. A292, 93 (1977).

91. D. Zawischa et al., Nucl. Phys. A311, 445 (1978).

92. U. Gary et al., preprint from Texas A&M.

93. J. M. Moss et al., Phys. Rev. Lett. 37, 816 (1976); J. M. Moss et al., Phys. Rev. C 18, 741 (1978).

94. T. A. Carey et al., Phys. Rev. Lett. 45, 239 (1980).

95. K. F. Kiu and G. E. Brown, Nucl. Phys. A265, 385 (1976).

96. H. P. Morsch et al., Phys. Rev. Lett. 45, 337 (1980).

97. T. Yamagata et al., p. 474 of Ref. 26.

98. P. Ring and J. Speth, Phys. Lett. B44, 477 (1973); Nucl. Phys. A235, 315 (1974).

99. E. C. Halbert et al., Nucl. Phys. A245, 189 (1975).

100. M. H. Harakeh et al., Phys. Rev. Lett. 38, 676 (1977); Nucl. Phys. A327, 373 (1979); J. Wambach et al., Nucl. Phys. A324, 77 (1979); J. P. Didelez et al., Nucl. Phys. A318, 205 (1979).

101. S. S. Hanna et al., Phys. Rev. Lett. 32, 114 (1974); P. Paul et al., in Proceedings of the International Symposium on Highly Excited States in Nuclei, Jülich, 1975, edited by A. Faessler et al., Julich, Germany, 1975, Vol. I, p. 2.

102. T. W. Phillips and R. G. Johnson, Phys. Rev. C 20, 1689 (1979).

103. F. S. Dietrich et al., Phys. Rev. Lett. 38, 156 (1977).

104. K. A. Snover et al., Phys. Rev. Lett. 32, 317 (1974).

105. R. Pitthan et al., Phys. Rev. C 19, 299 (1979).

106. R. Pitthan et al., Phys. Rev. C 21, 147 (1980).

107. R. Pitthan et al., Phys. Rev. C 16, 970 (1977).

108. R. Pitthan et al., Phys. Rev. C 19, 125 (1979).

109. G. L. Moore et al., Z. Naturforsch. 31a, 668 (1976).

110. R. S. Hicks et al., Nucl. Phys. A278, 261 (1977).

111. R. Pitthan et al., Phys. Rev. Lett. 33, 849 (1974); 34, 848 (1975).

112. R. Pitthan et al., Phys. Rev. C 21, 28 (1980).

113. J. D. Anderson and C. Wong, Phys. Rev. Lett. 7, 250 (1961).

114. K. Ikeda et al., Phys. Lett. 3, 771 (1963).

115. D. J. Horen, p. 223 of Ref. 57.

116. W. R. Wharton and P. T. Debevec, Phys. Rev. C 11, 1963 (1975).

117. R. R. Doering et al., Phys. Rev. Lett. 35, 1691 (1975).

118. G. E. Brown et al., Nucl. Phys. A350, 290 (1979).

119. S. L. Tabor, C. C. Chang, and M. T. Collins, to be published.

120. A. Guterman, D. L. Hendrie, C. C. Chang, and T. J. Symons, private communication.

121. R. W. Carr and J. E. E. Baglin, Nucl. Data Tables 10, 143 (1971); J. D. Turner et al., Phys. Rev. C 21, 525 (1980).

122. J. E. Bussoletti, Ph.D. Thesis, University of Washington, 1978; K. A. Snover et al., to be published.

123. K. T. Knöpfle et al., Phys. Lett. 74B, 191 (1978).

124. Progress Report, Nuclear Lab., Stanford University, pp. 31-64.

125. R. Helmer et al., Nucl. Phys. A336, 219 (1980).

126. K. A. Snover et al., Phys. Rev. Lett. 37, 273 (1976).

127. K. A. Snover, in Proceedings of the 3rd International Symposium on Neutron Capture Gamma Ray Spectroscopy and Related Topics, Brookhaven, 1978 (Penum Press, NY), p. 319.

128. L. Meyer-Schutzmeister et al., Phys. Rev. C 17, 56 (1978).

129. L. Meyer-Schutzmeister et al., Nucl. Phys. A108, 180 (1968).

130. M. T. Collins et al., Phys. Rev. Lett. 42, 1440 (1979).

131. K. Davis, to be published, and Annual Report, Nuclear Physics Lab., University of Washington, 1980.

132. E. Hayward, p. 275 of Ref. 26.

133. W. W. Gargaro and D. S. Onley, Phys. Rev. C 4, 1032 (1971).

134. E. Wolynec et al., Phys. Rev. C 22, 1012 (1980).

135. H. W. Koch and J. W. Motz, Rev. Mod. Phys. 31, 920 (1959).

136. A. Moalem et al., Phys. Lett. 61B, 167 (1976).

137. D. H. Youngblood et al., Phys. Rev. C 15, 246 (1977).

138. T. Yamagata et al., Phys. Rev. Lett. 40, 1628 (1978).

139. W. Eyrich et al., Phys. Rev. Lett. 43, 1369 (1979).

140. J. van der Plicht et al., Phys. Rev. Lett. 42, 112 (1979).

141. F. E. Bertrand et al., preprint.

142. J. G. Pronko and R. A. Lindgren, Nucl. Instrum. & Methods 98, 445 (1972).

143. G. R. Satcher, Nucl. Phys. 55, 1 (1964).

144. H. Ridesel, Ph.D. Thesis, Heidelberg (1979).

145. J. R. Calarco, p. 543 of Ref. 57.

146. R. G. Allas et al., Nucl. Phys. 58, 122 (1964).

147. J. T. Caldwell et al., Phys. Rev. C 21, 1215 (1980).

148. K. A. Snover et al., Phys. Rev. Lett. 32, 1061 (1974).

149. E. Kuhlmann et al., Phys. Rev. C 11, 1525 (1975).

150. A. Djaloeis et al., contributed paper 512 of Ref. 56; K. van der Borg et al., p. 468 of Ref. 26.

151. S. Krewald et al., Nucl. Phys. A281, 166 (1977).

152. A. Faessler et al., Nucl. Phys. A330, 333 (1979).

153. E. Wolynec et al., Phys. Rev. Lett. 42, 27 (1979).

154. J. C. McGeorge et al., to be published in J. Phys. G.

155. M. T. Collins, C. C. Chang, and S. L. Tabor, submitted to Phys. Rev. C.

156. K. Okada et al., p. 477 of Ref. 57.

157. E. Wolynec et al., Phys. Rev. Lett. 37, 585 (1976).

158. M. N. Martins et al., Phys. Rev. C 16, 613 (1977).

159. D. H. Dowell et al., Phys. Rev. C 18, 1550 (1978).

160. W. R. Dodge et al., Phys. Rev. C 18, 2435 (1978).

161. J. C. McGeorge et al., J. Phys. G 4, L145 (1978).

162. A. C. Shotter et al., Nucl. Phys. A290, 55 (1977).

163. J. D. T. Arruda Neto et al., Phys. Rev. C 18, 863 (1978); J. D. T. Arruda Neto and B. L. Berman, Nucl. Phys. (to be published).

164. J. D. T. Arruda Neto et al., Phys. Rev. C 22, 1996 (1980).

165. J. van der Plicht et al., Nucl. Phys. A346, 349 (1980).

166. C. Dover et al., Ann. Phys. 70, 458 (1972).

167. A. Veyssier et al., Nucl. Phys. A199, 45 (1973).

168. G. F. Bertsch et al., Phys. Lett. 80B, 161 (1979).

169. J. Raynal et al., Nucl. Phys. A101, 369 (1967).

170. M. Danos and W. Greiner, Phys. Rev. 138, B876 (1965).

171. W. Knupfer and M. G. Huber, Z. Phys. A276, 99 (1976).

172. J. S. Dehesa et al., Phys. Rev. C 15, 1858 (1977).

173. J. R. Wu and C. C. Chang, Phys. Lett. B60, 423 (1976); Phys. Rev. C 16, 1812 (1977).

174. J. R. Wu, Ph.D. Thesis, University of Maryland, 1977.

175. V. K. Luk'yanov et al., Nucl. Phys. 21, 508 (1975).

Chapter XI

Some Applications of Small Accelerators

Nelson Cue
Department of Physics
State University of New York at Albany
Albany, N.Y. 12222, USA

Abstract: With the frontiers of nuclear physics pushed toward
higher and higher energies, an increasing number of small accelerators
are being freed for other uses. A description of some of the areas
of small accelerator applications will therefore serve a useful purpose. Since the topics are both rich and varied, a reasonably detailed description must be confined to a selected few. These topics
are compositional studies of layered materials, channeling of heavy
charged particles, and characteristic radiation of channeled relativistic electrons. All interface closely with nuclear physics and
are aspects of particle-solid interactions which is a field receiving
increasing attention.

1.   Introduction and Scope

Small accelerators capable of producing charged particle beams of
<10 MeV traditionally play important roles in nuclear physics research.
Today, they still are used to address significant nuclear physics
problems such as, for example, in fast neutron induced reactions, in
thermonuclear reactions with charged particles, in the stability and
decay of proton-rich light nuclei, and in nuclear hyperfine structure.
As the frontiers of nuclear physics are pushed toward higher and
higher energies, however, small accelerators are increasingly applied
to areas other than nuclear physics. A useful purpose therefore will
be served by discussing some of these other areas both in order to
illustrate their connections to nuclear physics methods and the type
of problems that can be addressed by small accelerators.

As can be surmised from a number of monographs [Ma70, De73, Mo73,
Zi75, Ca76, To76, Le77, Th78, Ch78, Hi80, Pe80] and conference proceedings [Pi74, Ca75, Ch77, Du78, Gy78, Pr79, An80, Ja80, Be80] the
applications are both rich and varied; addressing problems in such
practical areas as materials characterization and modification,

radiation therapy, environmental pollution and dating of objects, to such fundamental areas as atomic and molecular structure and processes. For the present purpose three specific topics will be described in details and all are related to particle-solid interactions. The choices are necessarily biased by my own familiarity and interest, but they also interface closely with nuclear physics.

The first, discussed in Sec. 2., is the topic of "Compositional Studies of Layered Materials." The technique of materials characterization with MeV ion beams utilizes well-known close-encounter binary collision processes as characteristic signals in conjunction with our good understanding of energy loss phenomena for charged particles. The technique is widely accepted because it is one of the very few that can provide quick depth-profile information.

Section 3 treats the topic of "Channeling of Heavy Charged Particles." Channeling refers to the governed (or steering) motion brought on by a series of small-angle scattering when a particle is directed along one of the high symmetry axes or planes of a single crystal. The most dramatic consequence of channeling motion is the great reduction in violent collisions with lattice atoms for positively charged particles. Close-collision signals due to crystal imperfections are thereby easily detected and this forms a sensitive technique for investigating impurity atom locations and defects in single crystals. Moreover, the closely related phenomenon of "blocking" provides one of the few techniques for measuring nuclear lifetime in the 10^{-18} to 10^{-16} s range.

Finally, in Section 4, the recently predicted and observed radiations characteristic of emission from channeled relativistic charged particles are taken up and focussed on recent results using MeV electrons from small accelerators. These characteristic radiations not only provide fundamental information on particle-solid interactions but, also, and perhaps more exciting, form the basis for a versatile photon source with its promise of new areas of experimentations.

The area of materials modification by ion implantation has received widespread attention in recent years. In this case, the small accelerator, usually of the high current variety, serves to introduce in a controlled manner foreign atoms into a host material in concentration exceeding the solubility limit. The "new" materials thus formed possess many interesting properties which can have significant impact on technology. Amplification of this subject matter will not be attempted here. Interested readers may refer to the

various treatises [Ma70, De73, Ca76, To76, Hi80] and conference
proceedings [Pi74, Ca75, Ch77, Gy78, Pr79, Be80] for details.

REFERENCES

[An80] "IVth Intl. Conf. on Ion Beam Analysis", ed. by H.H. Andersen,
J. Bottiger and H. Knudsen, North-Holland, 1980.

[Be80] "Proceedings of the 2nd Intl. Conf. on Ion Beam Modification
of Materials, 14-18 July 1980, Albany, N.Y., USA", ed. by R.E. Benenson, E. Kaufmann, L. Miller, and W.W. Scholz (to be published in
Nuclear Instr. Methods).

[Ca75] "Applications of Ion Beams to Materials, 1975", ed. by G.
Carter, J.S. Colligon and W.A. Grant, The Institute of Physics, London,
1976.

[Ca76] "Ion Implantation of Semiconductors", G. Carter and W.A. Grant,
E. Arnold, London, 1976.

[Ch77] "Vth Intl. Conf. on Ion Implantation in Semiconductors, 1976",
ed. by F. Chernow, J.A. Borders and D.K. Brice, Plenum Press, N.Y.,
1977.

[Ch78] "Backscattering Spectrometry", Wei-Kan Chu, J.W. Mayer and M-A.
Nicolet, Academic Press, N.Y. 1978.

[De73] "Ion Implantation", G. Dearnaley, J.H. Freeman, R.S. Nelson
and J. Stephen, North-Holland, 1973.

[Du78] "1978 Conference on the Applications of Small Accelerators in
Research and Industry", ed. by J.L. Duggan and I.L. Morgan, IEEE Trans.
on Nuclear Science, Vol. NS-26, Feb. 1979.

[Gy78] "Proceedings of the 1st Intl. Conf. on Ion Beam Modification of
Materials, 4-8 September, 1978, Budapest, Hungary", ed. by J. Gyulai,
T. Lohner and E. Pasztor, Central Research Institute for Physics,
H-1525 Budapest 114, POB, Hungary (1978).

[Hi80] "Treatise on Materials Science and Technology - Volume 18 - Ion
Implantation", ed. by J.K. Hirvonen, Academic Press, N.Y., 1980.

[Ja80] "8th Intl. Conf. on Atomic Collisions in Solids", ed. by D.P.
Jackson, J.E. Robinson and D.A. Thompson, North-Holland, 1980.

[Le77] "Interactions of Radiation with Solids and Elementary Defect
Production", Chr. Lehmann; Vol. 10 of series in "Defects in Crystalline
Solids," North-Holland Publ. Co., Amsterdam, 1977.

[Ma70] "Ion Implantation in Semiconductors", J.W. Mayer, L. Eriksson, and J.A. Davies, Academic Press, N.Y., 1970.

[Mo73] "Channeling", ed. by D.V. Morgan, J. Wiley & Sons, N.Y., 1973.

[Pe80] "Site Characterization and Aggregation of Implanted Atoms in Materials", ed. by A. Perez and R. Caussement, Plenum Press, N.Y., 1980.

[Pi74] "Applications of Ion Beams to Metals", ed. by S.T. Picraux, E. P. EarNisse, and F.L. Vook, Plenum Press, N.Y., 1974.

[Pr79] "Ion Implantation Metallurgy", ed. by C.M. Preece and J.K. Hirvonen, The Metallurgical Society of AIME, 1979.

[Th78] "Material Characterization Using Ion Beams", ed. by J.P. Thomas and A. Cachard, Plenum Press, N.Y. 1978.

[To75] "Ion Implantation, Sputtering and Their Applications", P.D. Townsend, J.C. Kelly, and N.E.W. Hartley, Academic Press, N.Y., 1976.

[Zi75] "New Uses of Ion Accelerators", ed. by J.F. Ziegler, Plenum Press, N.Y., 1975.

2. Compositional Studies of Layered Materials

The richness of collisional phenomena associated with ion beam interactions with solids gives rise to a variety of techniques for compositional studies. These microanalytic techniques, which rely on the detection of characteristic signals generated in the collisions may be classified broadly as either "destructive" or "non-destructive" according to whether or not material is eroded from the sample to be analyzed. Of course the solid is altered to some degree even if no erosion occurs since the mere passage of the ions creates lattice vacancies and interstitials along their paths and these can later agglomerate to form extended defects. Moreover, if these ions are stopped inside the solid they constitute impurities if they differ from the host species. Thus the term "non-destructive" as used here is meant to imply that the level of impurities or defects produced is either inconsequential or can be controlled to a tolerable degree.

Because an ion beam has a finite range, the techniques to be described are most appropriate for the analysis of layered structures such as those encountered in thin films and thick solid surfaces. To the extent that the surface is flat and laterally uniform in composition across the area sampled by the probing beam, many of these techniques have the ability to determine the concentrations of atoms with high sensitivity as a function of depth and with a resolution approaching tens of angstroms and, consequently, provide invaluable tools for the investigation of solid state phenomena such as thin film reactions and impurity solubility and diffusion. In subsequent sections, the more established "non-destructive" techniques are described. Complications due to channeling in crystallographic axes and planes will be avoided here by focussing on amorphous or polycrystalline targets. When single crystal targets are discussed, the beam will be assumed to travel in a direction other than the symmetry axes or planes. Finally, we call attention to a recently published ion beam handbook for material analysis [Ma77] which contain convenient tables and information pertaining to the subjects discussed below.

2.1 Rutherford Backscattering Spectrometry (RBS)

Back angle elastic scattering is widely used particularly in the regime where the well known Rutherford cross section [Ru11] is applicable. For a given projectile incident on a target, this regime corresponds to a bombarding energy range such that at the distance of closest approach in the collision, the screening of the atomic electrons is ineffective on the one hand and the short range nuclear interactions are not significant on the other. In the case of light ions (atomic number $Z \lesssim 3$) incident on a target at rest, the bombarding energy range 100 keV/amu $\lesssim E_o \lesssim$ 1 MeV/amu generally fits into this regime for not too light a target ($Z_2 \gtrsim 20$), and is accessible by many small accelerators. The important point here is that, at these energies, an incident ion penetrating a solid loses energy quasicontinuously through the predominant interactions with the sea of valence or conduction electrons [Si74] and is thereby hardly deflected until a comparatively rare event of close impact encounter which causes the particle to backscatter and reemerge in a straight line path. The backscattered particles can thus be interpreted as single collision events and the simplicity of the two-body kinematics (with appropriate consideration of the energy losses) can be exploited. This analysis technique is known as Rutherford Backscattering Spectrometry (RBS).

Similar considerations apply to higher bombarding energies except that the elastic scattering cross section now includes the lesser known contributions from the nuclear interactions. This then restricts the method to special cases. In particular the cases of larger cross sections can be used to enhance the sensitivity for particular impurity species and the higher energy can be used to probe deeper layers. Further considerations of such cases and others involving the detection of nuclear reaction products are deferred until Section 2.2.

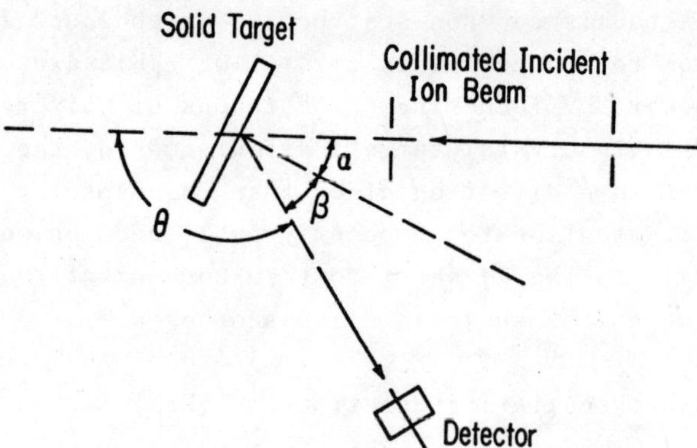

Fig. 2.1 - Schematic of experimental geometry for backscattering measurements.

2.1.1 Experimental Considerations

A schematic diagram of a typical experimental set up is shown in Fig. 2.1. Ion beams of 0.1-1 MeV H^+ and 1-4 MeV He^+ are commonly used because they are readily available, incur the least damage among heavy ions of comparable velocity and their stopping power (energy loss per path length) has been extensively investigated [Zi74, An77, Zi77]. Also the regime for which the Rutherford cross section is applicable is better known for these ions. The momentum analyzed and collimated beam from an accelerator is directed toward the target at angle α relative to the surface normal and the elastically scattered yield coming off at an angle β is recorded by a charged particle detector. Commonly used are solid state detectors which are relatively inexpensive, convenient to use because of their energy dispersive nature and reasonable energy resolution ($\delta E \overset{\sim}{=} 15$ keV). Improved energy resolution can be achieved by using the bulkier electrostatic [Fe76] or magnetic [Hi76] spectrometers but this must be weighed against the lengthier data accumulation time. Also the larger accumulated beam dose contributes to a larger radiation damage in the crystal. One other advantage of the spatial dispersive devices is the absence of electronic pile up distortion in the backscattering spectra which is often encountered with solid state detectors. However such distortions in the latter devices can be minimized with a lower counting rate or the use of standard pile-up rejection circuitry. Base pressure in the scattering chamber of 10^{-6} torr is adequate for most applications except in instances where the cleanliness of the sample surface needs to be preserved for other purposes. Surrounding the sample with a cold surface (cold can) at liquid nitrogen temperature is easily implemented and help to slow down the build up of surface impurities on the sample. Biasing the cold can to -300 V relative to the target is a convenient means of suppressing secondary electrons if the true beam current on the target is to be measured. For targets which are insulators, voltage built up due to the poor conduction of accumulated charges can be alleviated by depositing a thin conducting masking layer on the sample [Be72].

2.1.2 Basic concepts

In order to illustrate the principles and limitations of the technique, consider first the case of a homogenous monoisotopic solid target. For a fixed bombarding energy and geometry, the yield as a function of the energy of the scattered particle is depicted in Fig. 2.2. The edge occuring at an energy $E_1(0)$ corresponds to the scattering from the atoms at the surface layer and is governed by the two-body kinematics of elastic scattering [Ma68]. For the energies of interest here, relativistic effects may be neglected. In terms of the incident particle mass M_o and energy E_o, target mass M and laboratory scattering angle Θ,

$$E_1(0) = K^2 E_o, \qquad\qquad 2.1.1$$

where

$$K = [M_o \cos\Theta \pm (M - M_o^2 \sin^2\Theta)^{1/2}] / (M_o + M). \qquad 2.1.2$$

Here the plus sign is applicable for $M > M_o$. The yield at energies below the step therefore correspond to the scattering from deeper layers because the scattered particle suffers energy loss both in the inward and outward paths. It can also be seen that the presence of surface impurities will show up as a peak at an energy above the edge if their mass is heavier and below if lighter. At a depth z, the incident particle energy is reduced to a value E(z) before a large angle deflection occurs and which, according to Eq. 2.1.1, will result in a scattered energy of $K^2 E(z)$. This energy is further reduced to $E_1(z)$ as the scattered particle emerged from the surface towards the detector. In the inward and outward paths, the energy losses are related to the depth z, respectively, according to

$$\frac{z}{\cos\alpha} = -\int_{E_o}^{E(z)} \frac{dE}{S} \qquad\qquad 2.1.3$$

$$\frac{z}{\cos\beta} = -\int_{K^2 E(z)}^{E_1(z)} \frac{dE}{S}, \qquad\qquad 2.1.4$$

where $S = \dfrac{dE}{dz}$ \qquad\qquad 2.1.5

is known as the stopping power. Using tabulated values for S or their analytic approximations [An77, Zi77], the integrations of the above equations allow one to convert the measured energy E_1 into a depth scale z with the elimination of E(z).

For a thin surface layer in which the range of the incident beam far exceeds the layer thickness, the variation of S is small and the use of an average value is a reasonably good approximation. The integrations of 2.1.3 and 2.1.4 are then simple and result in the relation

$$E_1(0) - E_1(z) = [S] \, z, \qquad 2.1.6$$

where $[S] = \{ \dfrac{K^2}{\cos \alpha} \, S_{in} + \dfrac{1}{\cos \beta} \, S_{out} \}.$ $\qquad$ 2.1.7

The $[S]$ is commonly called the energy loss factor. From these equations, the overall energy resolution δE_1 with which the particles are detected is seen to determine the depth resolution for a fixed geometry. This δE_1 is usually governed by the detector resolution since the energy loss straggling is negligible and the kinematical energy spread due to the finite acceptance angle of the detector can be arranged to be small. For a given δE_1, the depth sensitivity is seen to increase as either α or β approaches 90°. In practice the maximum values for α and β are restricted by the experimental geometry and by the flatness of the scattering surface [Wi75a]. Moreover, if the surface layer contains scatterers of closely-spaced masses, the scattering angle $\Theta = \pi - (\alpha + \beta)$, according to Eqs. 2.1.1 and 2.1.2, must be chosen to be sufficiently large in order to separate out their respective contributions. The angles must thus be optimized by taking into account these experimental factors.

The above results can be applied to the case of a thick slab by considering it to be made up of many thin layers of thickness Δz. Using the notation given in Fig. 2.3, Eq. 2.1.6 becomes

$$\delta' = [S(E)] \Delta z, \qquad 2.1.8$$

where E is the reduced incident energy at z and $[S(E)]$, as given in Eq. 2.1.7, is to be evaluated with the corresponding energy E. Experimental spectra are usually recorded as counts per channel in which each channel corresponds to a fixed E_1 increment δ. The relation between δ and δ' can be obtained by noting that the outward path lengths of $K^2 E$ and $K^2 E - \delta'$ are the same (see Fig. 2.3), and thus

$$\int_{K^2 E}^{E_1} \frac{dE}{S} = - \int_{K^2 E - \delta'}^{E_1 - \delta} \frac{dE}{S}. \qquad 2.1.9$$

Since both δ and δ' are small compared to $K^2 E$ and E_1, they can be treated as differentials and Eq. 2.1.9 reduces to

$$\frac{\delta'}{\delta} = \frac{S(K^2 E)}{S(E_1)} \; .$$

$$2.1.10$$

The correspondence between E_1 and z may thus be established by an iterative procedure [Zi76] starting from the surface. With increasing depth, energy loss straggling increases and therefore the depth resolution becomes progressively worse.

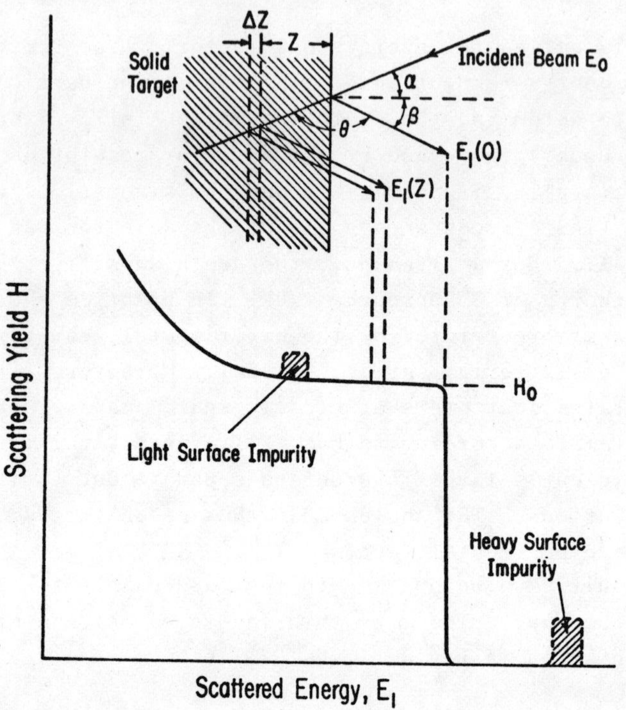

Fig. 2.2 - Schematic of RBS yield from a thick target with surface impurities.

Turning to the backscattered yield which corresponds to the height of the spectrum, the contribution of a layer Δz at a depth z can be expressed as

$$H(z) = \sigma \Omega N n \Delta z,$$

$$2.1.11$$

where σ is the differential scattering cross section evaluated at $E(z)$ and averaged over the solid angle Ω subtended by the detector, N is the total number of particles incident on the sample having an atomic volume density n for that layer. Note that the product $n\Delta z$ is the number of atoms per unit area, the areal density, and the yield is proportional to this. To make this relation explicit in the expression for H and therefore more convenient for comparison with measured backscattered spectra, the stopping cross section ε defined as

$$\varepsilon = \frac{1}{n} S \qquad\qquad 2.1.12$$

can be used in Eqs. 2.1.8, 2.1.10, and 2.1.11 to obtain

$$H(E_1) = \sigma(E)\Omega N \frac{\delta}{[\varepsilon(E)]} \frac{\varepsilon(K^2 E)}{\varepsilon(E_1)} \qquad\qquad 2.1.13$$

For the surface layer, $E = E_o$ and $E_1 = K^2 E_o$, the equation simplifies to

$$H_o = \sigma_o \Omega N \frac{\delta}{[\varepsilon_o]} . \qquad\qquad 2.1.14$$

As indicated in Fig. 2.2, this is the height of the edge.

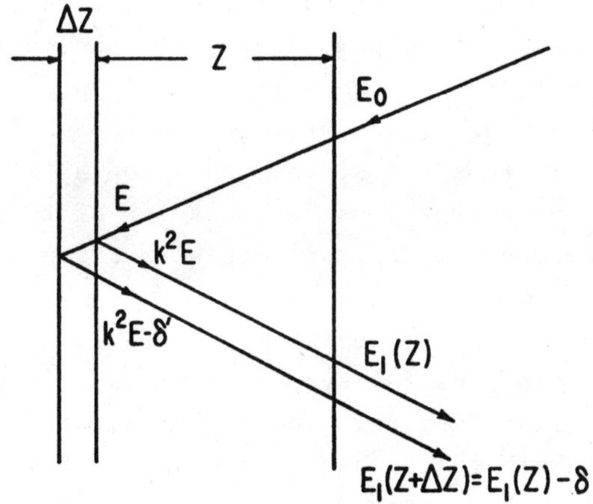

Fig. 2.3 - Energies associated with a particle scattered from a thin layer Δz at a depth z.

The case of a thin heavy impurity layer on top of a light substrate contains illuminating features. This heavy impurity corresponds to the peak above the edge in Fig. 2.2 for which Eq. 2.1.14 is applicable. If the overall energy resolution is small compared to the width of peak, this width is a direct measure of the film thickness if the stopping power is known. On the other hand, the yield represented by the total area under the peak is also a direct measure of the film thickness as implied, for example, by Eq. 2.1.14 if δ is interpreted as the total energy loss in the thin film. This then offers a self consistency check on the analysis. Because the Rutherford cross section for a fixed scattering angle is proportional to $[Z_1 Z_2 / E]^2$, where Z_1 and Z_2 are the atomic numbers of the incident and target nuclei, respectively, and E is the incident energy, the detection sensitivity increases with decreasing E and, for impurities, with increasing Z_2. Effects of the finite energy resolution are also manifested in the peak shape. The rounded edge and the corresponding knee on the high energy side reflect the detector resolution and finite acceptance angle, while on the low energy side the added contribution of straggling effects resulting from the penetration into and out of the film makes the round off and knee more pronounced [Br73a].

Areal density and not the volume density is also the determining factor in the energy loss since collisional (soft) processes are involved. Thus in the particle traversal through two thin samples of the same material, one of which is porous and the other not, a thicker layer must be crossed in the porous case in order to produce the same energy loss. If the volume densities are denoted respectively n' and n for the porous and non-porous cases, the condition $n'\Delta z' = n\Delta z$ implies $\Delta z' > \Delta z$ for $n' < n$. Thus the energy loss is a measure of the areal density; only when the volume density of the sample is given can a depth scale be established.

For homogeneous solids consisting of multiple elements such as chemical compounds and solutions (alloys), the backscattered spectra can be analyzed in a similar fashion. The essential difference is that the energy loss in both the inward and outward paths will now be governed by the S or ε of the compound medium. For a two element composite of the form $A^a B^b$ where a/b is the atomic ratio, the assumption of linear additivity of the elemental stopping cross sections results in $\varepsilon^{AB} = a\varepsilon^A + b\varepsilon^B$. This is the so called Bragg's rule [Br05] which has been widely used. The extension to more than two elements is straightforward. In addition, the backscattering kinematic factor and cross section will be different for the different elements and due account of these must be kept in reducing the spectra to depth profiles.

2.1.3 General Limitations

In deciding on what are the most suitable applications for this technique, it is worthwhile to describe briefly its general limitations. First of all it is a mass sensitive technique and according to Eq. 2.1.1, the scattering from neighboring masses becomes increasingly difficult to distinguish with increasing mass. Thus depth profiling is simple only if the sample contains a few elements of highly differing masses. In the case of impurities, a high detection sensitivity can be achieved only for those more massive than the substrate atoms. Secondly, the depth scale established on the basis of energy loss phenomena is meaningful only if the surface is flat and laterally uniform. Although the degree of flatness can be checked by analyzing the backscattering yield as a function of the target tilt angle α or detector angle β or both, the observed deviation from a flat surface can not be directly related to the nature of the irregularity. Other techniques such as x-ray diffraction and electron microscopy must be used to obtain the details of surface topography. It is also clear that the effects of any surface irregularity will be magnified in a glancing geometry (large α or β)which enhances the depth resolution. Thirdly the accuracy of the depth scale is governed by our knowledge of the stopping cross section ε. For H and He ions, ε has been extensively investigated both experimentally and theoretically and the values for nearly all elements are known with an accuracy of a few percent in the energy region of interest here. However several recent observations which are not completely understood must be taken into consideration. It has been reported that the ε for He ions stopping in elemental gas targets appear consistently higher than those for elemental solids [Zi76a] and that bonding effects in molecular gases not considered in the Bragg rule may be important [Lo74]. In any case the RBS itself can provide information on the stopping cross sections [Sc76]. Finally, the specific validity of the Rutherford cross section must be independently verified.

The limitations described above have not been restrictive as evidenced by the widespread applications of the technique to a large variety of problems, some of which are described in the following section. As can be surmised, the main advantage of the technique is that the analysis is fast, simple and quantitative.

2.1.4 Applications of RBS

Many technologically important aspects of material sciences involve phenomena occuring at or near the solid surface. It is thus

not surprising that the RBS "non-destructive" technique has been
extensively applied to a wide variety of problems since the composition
in depth of the materials is basic to the understanding of the underlying phenomena. The applications range from routine ones such as
quality control in the fabrication of semiconductor devices to specialized investigation such as interfacial reactions in thin films. Their
number and diversity can be seen in the proceedings of various conferences [Ma74, Me76, Wo77, An80] on ion beam analysis. A description of
a few representative ones will suffice to illustrate some of the more
salient features of the technique.

In the modification of materials properties by ion implantation,
the intent is to achieve a certain spatial distribution for the implanted species. Although general theories for the slowing down of
ions in amorphous solids [Li63, Fi57] give predictions on the specific
range distributions, deviations from these have been observed which
can be attributed to a number of factors. At high dose, target sputtering, changes in the stopping power of the target as the foreign
species builds up and radiation enhanced diffusion can contribute to
the final ion distribution. Accurate final profile must thus be
measured and the RBS technique is particularly suited to analyze the
cases of implants more massive than that of the substrate species.

Another area where the RBS technique has been extensively applied
is in the study of thin film interdiffusion [Po76]. The large surface-
to-volume ratio of materials in film form gives rise to many properties which differ markedly from those in the corresponding bulk form.
Again its non-destructive nature, relatively direct reflection of the
concentration profiles and short data accumulation time make the RBS
technique ideally suited for the study of diffusion in thin films.

In the thin film interdiffusion studies, the unambiguous determination of the diffusing species has been made in a number of cases
using an inert marker analogous to the Kirkendall marker [Ki42] for
diffusion in bulk materials. The ion implantation of inert gases as
markers was first used by Brown and Mackintosh [Br73] and applied by
other investigators [Ch74, Ch75]. Other kinds of thin inert markers
have also been used successfully [Gu76, Pr76]. The basic idea can be
seen from the RBS spectra in Fig. 2.4 taken from work of Van Gurp et
al. [Gu76] on cobalt silicide formation starting from a cobalt film
on a silicon substrate. In this case the marker consists of islands
of tungsten about 30Å thick deposited at the Si-Co interface. After
annealing, the backscattered tungsten peak has increased in energy
indicating that Co is the faster diffusing species. The surface peak

corresponding to Sn is the remnant associated with the making of the
tungsten marker. Shape changes in the backscattering from Si and Co
has been attributed to the formation of mainly Co_2Si.

Many other phenomena in layered structure have been investigated
by the RBS techniques, including thin film reactions, impurity solubility and diffusion in bulk materials, epitaxial layers, oxidation
and corrosion, superconducting or magnetic thin films. As we shall
see in Sec. 3, backscattering measurements in a channeling geometry
also give crystallographic information such as lattice site location
of impurities and lattice disorder.

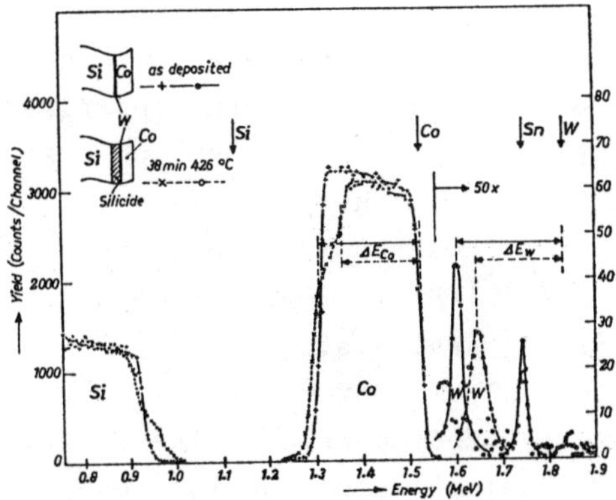

Fig. 2.4 - Spectra of 2.0-MeV He$^+$ backscattered
from a thick Si target deposited with islands of
W and a 0.18-μm layer of Co showing features
before and after silicide formation. The arrows
indicate the energy for scattering by Si, Co, Sn
and W on the surface. The movement of the W peak
(marker) gives information on the faster diffusion
species (From[Gu76]).

2.2 Nuclear Reactions Analysis

The great variety of nuclear reaction events offers a wide selection of highly specific techniques for the analysis of isotopic (and
thus elemental) compositions in a sample. Such events are of course
the consequence of close encounters of nuclear dimensions which, for
an incident ion of nuclear charge Z_1e directed at a target atom of Z_2e
with a relative kinetic energy E, are more probable if the repulsive
Coulomb barrier is surmounted. This condition can be expressed as

$$E \gtrsim 1.44 \frac{Z_1 Z_2}{r} \qquad\qquad 2.2.1$$

with r being the interaction radius in $F(10^{-15}m)$ and E in MeV. However, because of the finite probability for barrier penetration, nuclear reactions can proceed even if this condition is not satisfied; the reaction rates will generally be small except in instances where compound nuclear resonance conditions are satisfied or the target has a low lying level which connects to the ground state with a large Coulomb excitation amplitude.

With the more commonly available small accelerators producing ion beams of a few MeV, nuclear reaction analysis (NRA) is therefore usually practical only for the lighter elements. The relative insensitivity to heavy elements is a positive feature particularly in the analysis of light impurities in a heavy element substrate. Indeed it is precisely this area which the other "non-destructive" techniques do not adequately address. Thus NRA is complimentary to particle induced x-ray emission analysis (PIXE) in impurity concentration studies and to RBS in depth profiling analysis.

Nuclear reaction analysis may be broadly categorized under three headings for the purpose of discussion. These are: 1) Charged particle activation analysis (CPAA) in which the ion beam is merely used to produce specific radioactive nuclei whose characteristic decays are followed after the bombardment is stopped; 2) Prompt radiation analysis (PRA) in which the characteristic reaction products are detected during bombardment usually at a fixed energy; and 3) Resonant reaction analysis (RRA) in which the energy shift and width broadening of an otherwise isolated narrow compound nuclear resonance cross section measurement provides the depth profile information of the corresponding isotope. There are of course many variations within these groupings. Since an excellent detailed review has been given recently by Wolicki [Wo75], the discussion which follows will be brief and directed at the salient features. Literature on a variety of applications can be found in the proceedings of various conferences on ion beam analysis [Ma74, Me76, Wo77, An80].

2.2.1 Charged Particle Activation

Since its discovery more than eight decades ago, radioactivity has been investigated for a number of purposes and consequently many of the measurement techniques have become highly developed. For compositional study, neutron activition analysis (NAA) is perhaps the most familiar. The principles underlying CPAA differ from those of NAA only in the consequences of the use of an ion beam instead of neutrons to produce radionuclides. Thus the finite range R of an ion beam with an incident laboratory energy of E_0 can produce radioactive nuclei only within this range.

Applications of CPAA have not been confined to any accelerator type. However, the CPAA of heavy elements using high energy ion beams offers no general advantage over NAA. As already mentioned, the analysis of light elements using small accelerators, which is facilitated by a lesser interfering background, is particularly suited for surface layers. Moreover, since unambigous analysis requires the monitoring of the characteristic decay curves, the easily implementable feature of pulsed beam with variable pulse durations may prove advantageous in following decays as short as a few nanoseconds. Most applications to date have generally focussed on longer lived radionuclide produced in p,d,t and ^{3}He bombardments and on the detection of γ-rays including those from the annihilations of positrons from β^+ emitters.

2.2.2 Prompt Reaction Analysis

Analysis by the detection of prompt reaction signals generally relies on the simplicity of the two-body final state kinematics [Ma68] in which the application of the energy and momentum conservation laws to the observed energy of one of the two final products is sufficient to identify the reaction unambigously. Within this class of reactions the preferred ones are clearly those having a reasonably large cross section or for which the measurement is simple to implement and interpret. Bombardment with simple ions and detection of γ-rays or simple particles generally fit into this class. Typically, the energy spectrum will contain well separated sharp peaks corresponding to the discrete low lying states of the heavy final product. The simplicity alluded to lies in the simple detection scheme of one detector. Interferences from the more probable elastic scattering events are avoided by blocking out the charged particles in the case of γ or n measurement or, in the case of charged particle detection, by choosing a reaction with a positive Q-value which places the reaction peaks above the elastic peak.

In some instances several reactions can be used to analyze one particular species. The choice is clearly dictated by the available equipment, the nature of the material to be analyzed and degree of detailed information desired.

All PRA can potentially yield depth profile information because

the energy of the detected particle directly reflects the energy loss
phenomena associated with the beam particles and, if charged, the detected particles also. The depth resolution attainable, however, differs for the different reactions.

At a fixed bombarding energy, considerations entering into the
conversion of energy loss to the depth at which the reactions occur
are analogous to those for RBS analysis discussed in Sec. 2.1.1, but
due consideration must be given to the different nature of the particles detected. In the case of uncharged outgoing particles such as in
(p,γ) and (d,n) reactions, only the energy loss of the incident ion
need be considered.

In the charged particle case such as in (d,p) and $(^{3}He,\alpha)$ reactions, not only is the outgoing energy determined by a different kinematical relation, but, the different charged species also entail different sets of stopping powers. The depth resolution is similarly determined by the overall detection resolution which includes contributions from the incident energy, finite solid angle, detection system
and straggling effects.

The depth profile reflected by the peak shape in the energy spectrum is a convolution of the experimental factors and the reaction
cross section. Extracting such a profile can be accomplished either
by a deconvolution procedure [Hu76], by simulating the experimental
data with a model concentration profile, or by comparison with a standard in which the species of interest is uniformly distributed [Pa65].
Whatever the procedure, errors can be minimized if the cross section
varies smoothly with energy. Reactions with a sharp resonant cross
section can be more profitably utilized in a slightly different approach described in the next section.

A majority of PRA applications has centered on the profiling of
low-Z elements [Wo75, Pa76]. In general, the use of photonuclear reactions is hampered by the relatively small cross section while that
of neutron producing reactions suffers from the difficulties associated with detecting energetic neutrons with a reasonable energy resolution. These reactions also inherently yield a poorer depth resolution
than reactions yielding charged particles at the same penetration depth
because of the smaller overall energy loss associated with the reaction.
However, the smaller energy loss can be used advantageously to probe
deeper layers for which the increasing importance of energy straggling
phenomena places a lesser demand on the detector resolution. Finally,
although the heavy product left in an excited state may decay
promptly such as, for example, in $(p,p'\gamma)$ and $(p,\alpha\gamma)$, the detection of

such prompt decay at a fixed bombarding energy carries little depth information.

2.2.3 Resonant Reaction Analysis

An isolated compound nuclear resonance exhibited as a sudden enhancement in the cross section over a narrow interval of bombarding energy provides an alternative and sometimes superior method for depth profiling. The basic ideas underlying the technique can be seen by idealizing the resonance as a delta function $\delta(E-E_r)$ and neglecting energy loss straggling effects. The particular reaction yield will then result if the beam energy is exactly $E=E_r$ and with a strength proportional to the areal density of the relevant target atoms, as will be the case if these atoms all lie on the surface. For atoms in a deeper layer z, the corresponding yield will result only if the bombarding energy is raised to $E=\Delta E+E_r$, where ΔE is the energy loss suffered by the beam particles penetrating into the depth z. The situation is illustrated schematically by the middle diagram in Fig. 2.5 for a resonance in the ^{27}Al$(p,\gamma)^{28}$Si reaction for a target whose composition is indicated in the upper figure. Folding in the beam energy spread, finite resonance width and straggling effects will broaden the profile as depicted in the bottom diagram.

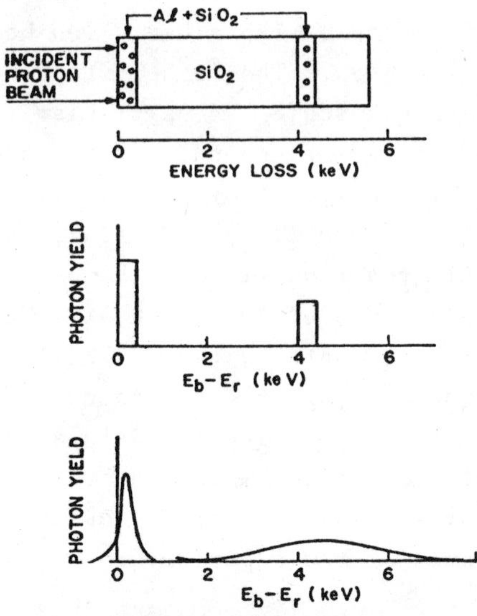

Fig. 2.5   Schematic diagrams of a SiO$_2$ sample with two thin layers of Al and the corresponding idealized and realistic curves of photon yield from the ^{27}Al$(p,\gamma)^{28}$Si resonance reaction. (From[Wo75]).

Unfolding techniques which determine the distribution profiles have been described by a number of authors [Wo75, La76]. In general the depth resolution is determined by the resonance width for the surface layer and by straggling effects for the deeper layers. The total probing depth for which the analysis remains simple is governed by the interval of bombarding energies in which the single resonance dominates the cross section.

Depth profiling with the resonance technique was first reported by Amsel and Samuel [Am62] using the $^{27}Al(p,\gamma)^{28}Si$ resonance at 992 keV and $^{18}O(p,\alpha)^{15}N$ resonance at 1167 keV in a study of anodic oxidation mechanisms. Since then many other resonances including those in the elastic channel such as in $^{16}O(\alpha,\alpha)^{16}O$ at 3048 keV [Me76a] have been used for a variety of applications. The technique is flexible in the sense that the signature may also be manifested in the secondary reaction products such as, for example, in the $^{19}F(p,\alpha\gamma)^{16}O$ resonance at 1375 keV where the detection of the secondary γ-rays of fixed energy is more convenient.

A notable of recent applications of RRA is in the elucidations of the role of hydrogen in various solid state phenomena [La77]. There exist very few techniques for the microscopic profiling of hydrogen and, within these limited choices, RRA can perhaps provide the most detailed information without destroying the sample. The high depth sensitivity of RRA in this case is achieved by reversing the usual procedure and bombard the sample with a heavy ion beam to produce the known proton induced resonances. The higher stopping power for heavy ion beams magnifies the depth scale. A particular example is the $^{1}H(^{15}N,\alpha\gamma)^{12}C$ reaction with a corresponding resonance at 6385 keV. In an interesting recent comparative study [Zi77a] of a number of ion beam techniques for hydrogen profiling in which identically prepared standards were used, this reaction yielded the highest depth resolution of ~40Å for hydrogen at a depth of 4000 Å in Si. The standard consists of a high purity Si wafer implanted with $10^{16}H/cm^2$ at 40 keV and the 4.43 MeV γ-ray yield observed [Zi77a] as a function of ^{15}N bombarding energy is reproduced in Fig. 2.6. The raw data as they stand already reflect the theoretically expected implanted hydrogen profile. Hydrogen contamination on the surface which is almost unavoidable is seen to pose no problem because of the superior depth resolution.

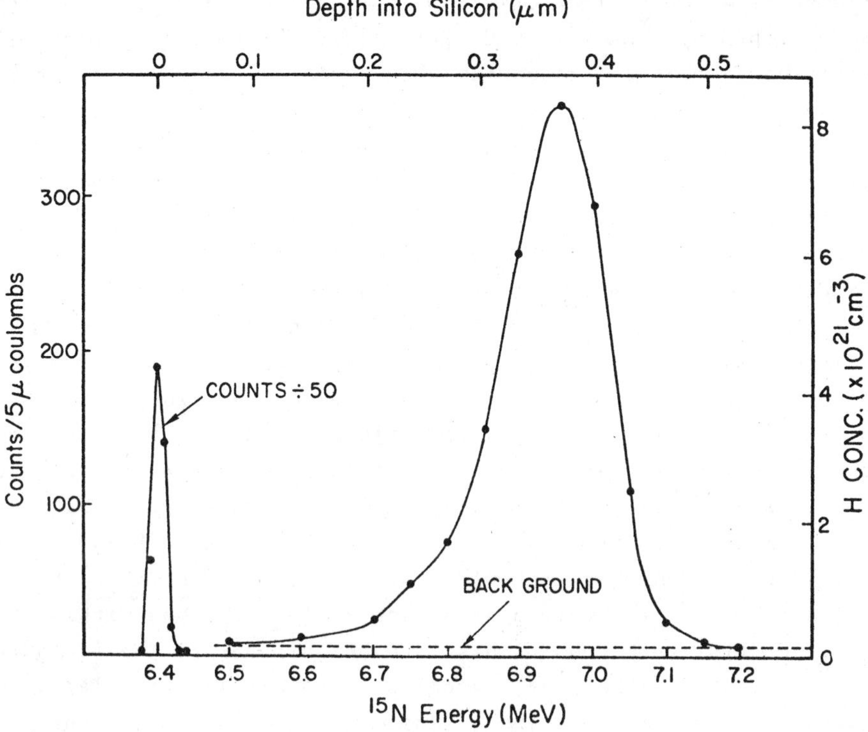

Fig. 2.6 - Depth profile of H in a silicon implanted with 10^{16}H/cm^2 at 40 keV obtained from the H(^{15}N,$\alpha\gamma$)^{12}C resonance reaction. The peak on the left is due to hydrogen contaminants on the surface.(From[La76]).

2.3 Particle Induced X-Ray Emission (PIXE)

In the collision of an ion with an atom, there is a significant probability for ejecting electrons from the inner shells of the colliding partners. The subsequent filling of such a vacancy can proceed by either x-ray emission or radiationless Auger transition, both of which are characteristic of the emitting ion. The characteristic emission is of course the basis for the better known techniques of x-ray analysis and Auger electron spectroscopy (AES) with electron and photon beams. The impetus for the analysis by Particle Induced X-ray Emission (PIXE) can be traced to the expectation of lesser background signals, higher characteristic x-ray cross sections and the ready availability of energy dispersive semiconductor x-ray detectors of sufficient resolution to resolve the K-x-rays of all but the lighter

elements, as well as the L-x-rays of the heavier elements. The recent
international conference at Lund [Pr77] devoted solely to the subject
of PIXE attests to its increasing acceptance as a sensitive analytical
technique particularly for the detection of trace amount of impurities.

For analysis, the use of simple ion like H^+ and He^+ to excite K
or L x-rays is favored for a number of reasons. Perhaps the most
important is the fact that the x-ray production mechanisms for such
ions in solids are reasonably well understood. The creation of a
vacancy in the inner shell of a target atom in this case is dominated
by the Coulomb interaction between the bare nucleus of the projectile
and the corresponding target electron, as evidenced by the remarkable
agreement over six orders of magnitude between the experimental K-
vacancy cross sections and the predictions based on direct Coulomb
ionization theories [Ma75]. Not only is the perturbation due to the
attached electron in the projectile small, but such projectiles are
stripped bare of electrons when traversing solids with $E \gtrsim 100$ keV/amu.
Additionally, the probability of creating multiple inner vacancies in
a single collision is small and is adequately described within the
context of multiple Coulomb ionization [Cu76]. Since the fluorescence
yield ω, which expresses the branching ratio for a particular x-ray
emission in the vacancy filling, depends on the number of inner shell
vacancies [Bh76], the simple vacancy produced by the simple ions means
that the uncertainties in the ω are minimized. Chemical bonding
effects are also insignificant except perhaps for the lighter elements.

At a fixed incident velocity, the direct Coulomb ionization cross
section for a given target atom is proportional to the square of the
projectile's nuclear charge Z_1^2 and this would suggest higher analytical sensitivities with the use of heavier ion beams. Moreover substantial enhancement in the cross sections over those predicted by the
direct ionization theories have been observed [Ma75] for incident
velocities below that of the corresponding target electron. Such
enhancements have been attributed to the capture of target electrons
into bound states of the projectile [Ha73] and to ionization through
electron promotion via the formation of transient molecular orbitals
[Ba72]. For compositional analysis, the larger cross section, must
however, be weighed against other factors. The enhancement is selective because it depends on the particular combination of colliding
species and on their electronic structure at the time of collision
which, for heavy ions moving inside a solid, is only known in a statistical sense due to the many prior collisions [Be72a]. This selectivity may be advantageous for the analysis of a particular element but

introduces complications when the same beam is used for multielement
analysis. Another complicating aspect can be seen in Fig. 2.7 where
the Al Kα x-ray region has been scanned with a high resolution curved
crystal spectrometer during bombardment with e^-, H_3^+ and Ne^+ beams.
One-electron transitions from an initial state having a single K- and
n L- vacancies are denoted by KL^n. Thus KL^o is the normal diagram
line and its dominance in the cases of e^- and H_3^+ reflects the fact
that the creation of only a single vacancy is likely. In the case of
Ne^+, the shift of the intensity maximum away from the diagram line is
a dramatic illustration of the comparatively large probability for
multiple inner shell ionization. The interpretation of the observed
satellite structure must in addition take into account the variation
in the fluorescence yields of the KL^n satellites and the vacancy
rearrangement processes prior to the x-ray emission [Cu76]. Observed
chemical bonding effects upon the satellite structure have been
attributed [Ho76, Wa75] to the latter type of processes. With the
poorer resolution semiconductor detectors, the satellite structure will
of course be merged into a single peak. Nevertheless, some sort of
average of the effects described above will enter into the conversion
of the x-ray intensity to the number of primary collision events. In
most instance the basic information from which such averages can be
obtained are lacking and thus hampers the analysis with heavy ion beams.

In PIXE as well as in most microanalysis techniques, the ultimate
trace-element detection sensitivity in the case of no interference of
characteristic lines is governed by the background radiation which
arise mainly from the interactions of the probing beam with the most
abundant elements in the sample matrix. The background processes in
PIXE have been identified [Fo76] as 1)bremsstrahlung of secondary
electrons (SEB), 2)projectile bremsstrahlung (PB), 3)Compton scattering
of γ-rays (CS), 4)radiative electron capture (REC) and 5)quasi-mole-
cular transition (MO).

Bremsstrahlung radiation arises from the deceleration of the
charged particle in the field of the nucleus and is more probable for
electrons than for other ions of the same velocity because the cross
section varies inversely with the square of the particle mass. Since
in the close encounters necessary to produce PB there is a much larger
probability for ionizing the target atoms in which electrons are
invariably ejected, SEB will dominate the low energy continuum background radiation. This is shown in Fig. 2.8 for 2-MeV proton bombardment of a carbon and an aluminum matrix. The upper cut off for SEB is
due to the fact that secondary electrons with energy exceeding the

value acquired by a free electron in a head-on collision with the
incident projectile are increasingly difficult to produce because the
ionization involves more tightly bound orbitals. Note that because
the secondary electron yield is proportional to the ionization cross
section, the larger characteristic x-ray yield with the heavier ion
beam will correspondingly be accompanied by a larger SEB background.

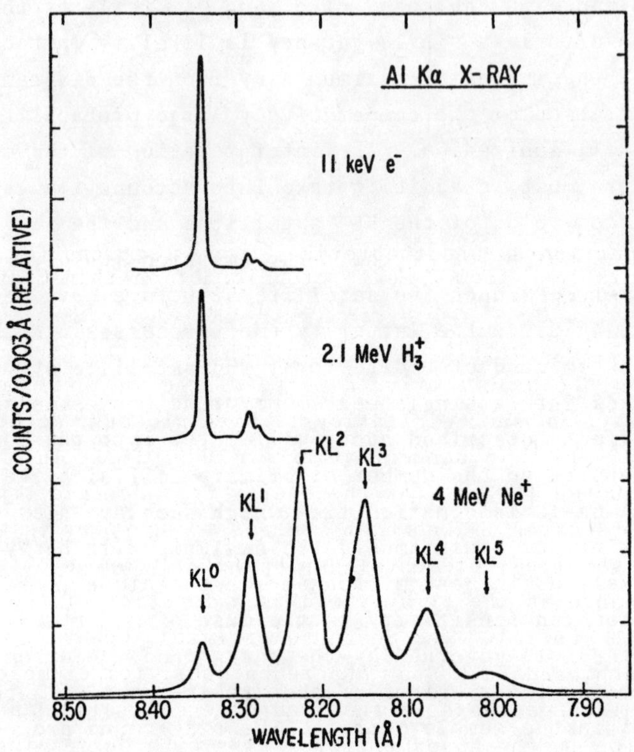

Fig. 2.7 - Aℓ Kα x-ray spectra recorded with a 4" curved-
crystal spectrometer in the impact of 11-keV e⁻, 2.1-MeV
H₃⁺, and 4-MeV on a thin Aℓ target.

The use of heavier ions also has the further disadvantage that
the x-ray region corresponding to the projectile emission will be
masked. The projectile's characteristic lines are intense because the
projectile can interact with all of the target constituents along its
path, and additional structure can be present as a result of specific
interactions. At high velocity, a vacancy in the projectile produced
in a prior collision can be filled by a radiative capture of a target
electron. Further complication may arise from quasimolecular (MO)
transitions [Li74, Ma74a].

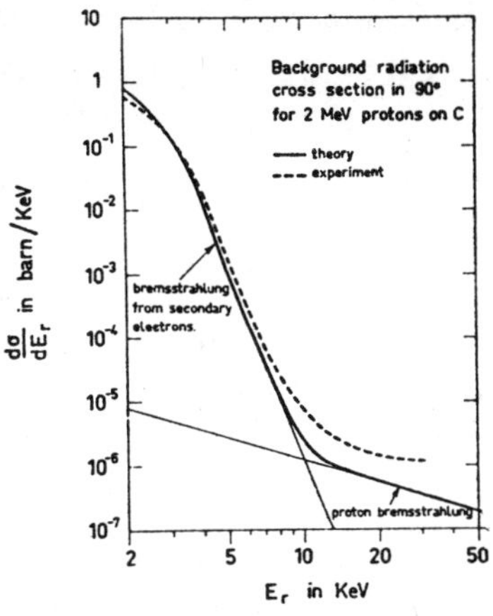

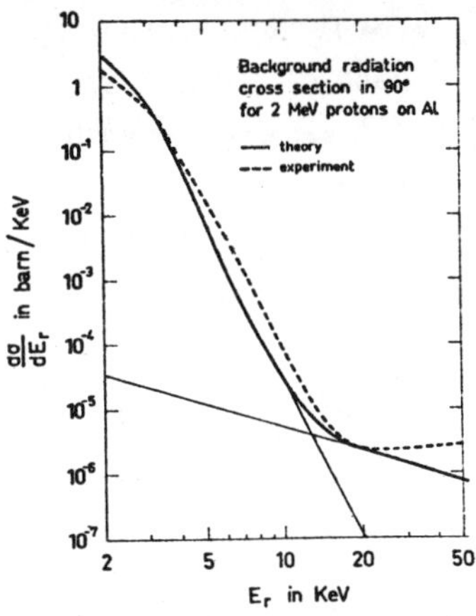

Fig. 2.8 - Experimental and theoretical cross sections for the background radiations from 2-MeV proton impact on thin C and Aℓ targets. (From [Fo74]).

Most applications of PIXE have been made with light ion beams and therefore much of the quantitative aspects of the analysis technique developed are based on such excitations. With protons, detection sensitivities of 10^{-7}-10^{-6} in concentration or 10^{-16}-10^{-9} g in mass quantity have been quoted [Fo76]. As the energy dependence of the x-ray production cross section is known and the stopping power is well characterized for light ions, elemental depth profile can be obtained by observing the change in the x-ray yield as a function of the trajectory of the projectile [Fe76a]. The depth sensitivity, however, does not approach those achieved in RBS or NRA techniques, for example. Finally, because of the steep decrease in the flourescence yield ω and the increase attenuation in the absorber for x-rays with decreasing Z, the detection of the Auger electrons rather than the x-rays will be more sensitive in the analysis of light elements.

2.4   Summary

The discussion here concerns primarily those ion beam analysis techniques which are easily implemented in laboratories equipped with small accelerators and standard nuclear instrumentations. As described, these techniques provide information on either impurity concentrations with high sensitivities or atomic composition as a function of depth with resolutions approaching tens of angstrom. Lateral profiling can also be achieved by collimating [Gr76, Fo79] or focussing [No77, Su79] the ion beam to a small spot size but the lateral resolutions attainable do not yet match those of electron microscopy. As currently applied, these ion beam techniques have nevertheless been shown to provide invaluable information on materials properties and they are usually simple, fast and non-destructive.

<u>References</u>

[Am62]G. Amsel and D. Samuel, J. Phys. Chem. Solids, $\underline{23}$, 1707 (**1962**).

[An77]H.H. Andersen and J.F. Ziegler, "Hydrogen Stopping Powers and Ranges in All Elements". Pergamon Press, N.Y., 1977.

[An80]H.H. Andersen, J. Bottiger, and K. Knudsen (ed.), "IVth Intl. Conf. on Ion Beam Analysis." North-Holland, Amsterdam, 1980.

[Ba72]M. Barat and W. Lichten, Phys. Rev. $\underline{A6}$, 211 (1972).

[Be72]W. Beezhold and E.P. EerNisse, Appl. Phys. Lett. $\underline{21}$, 592 (1972).

[Be72a]H.D. Betz, Rev. Mod. Phys. $\underline{44}$, 465 (**1972**).

[Bh76]C.P. Bhalla, in "Fourth Conf. on the Sci. and Ind. Appl. of Small Acc.," (J.L. Duggan and I.L. Morgan, ed.) p. 149, IEEE, N.J., 1976.

[Br05]W.H. Bragg and R. Kleeman, Phil. Mag. $\underline{10}$, 318 (1905).

[Br73]F. Brown and W.D. Mackintosh, J. Electrochem. Soc. $\underline{120}$,1096 (1973).

[Br73a]D.K. Brice, Thin Solid Films $\underline{19}$, 121 (1973).

[Ch74]W.K. Chu, H. Krautle, J.W. Mayer, H. Muller, M-A. Nicolet, and K.N. Tu, Appl. Phys. Lett. $\underline{25}$, 454 (1974).

[Ch75]W.K. Chu, S.S. Lau, J.W. Mayer, H. Muller, and K.N. Tu, Thin Solid Films $\underline{25}$, 393 (1975).

[Cu76]N. Cue, in "Fourth Conf. on the Sci and Ind. Appl. of Small Acc." (J.L. Duggan and I.L. Morgan, ed.) p. 299,IEEE, N.J., 1976.

[Fe76]A. Feuerstein, H. Grahmann, S. Kalbitzer, H.Oetzmann, in "Ion Beam Surface Layer Analysis," (O. Mayer, G. Linker, and F. Kappeler, ed.), Vol. 1, p. 471. Plenum, New York, 1976.

[Fe76a]L.C. Feldman and P.J. Silverman, in "Ion Beam Surface Layer Analysis" (O. Meyer, G. Linker, and F. Kappeler, ed.), Vol. 2, p. 735. Plenum, N.Y. 1976.

[Fi57]O.B. Firsov, Soviet Phys. JETP $\underline{5}$ 1192 (1957) and ibid $\underline{6}$, 534 (1958).

[Fo74]F. Folkmann, C. Gaarde, T. Huus, and K. Kemp, Nucl. Instr. Methods $\underline{116}$, 487 (1974).

[Fo76]F. Folkmann, in "Ion Beam Surface Layer Analysis" (O. Meyer, G. Linker, and F. Kappeler, ed.), Vol. 2, p. 695. Plenum, N.Y., 1976.

[Fo79]Cheng-Ming Fou, V.K. Rasmussen, C.P. Swann and D.M. VanPatter, IEEE Trans. on Nucl. Sci. $\underline{NS-26}$, 1378 (1979).

[Gr76]L. Grodzins, P. Horowitz and J. Ryan in "Proc. of the 4th Conf. on the Scientific and Industrial Appl. of Small Acc.", ed. by J.L. Duggan and I.L. Morgan, IEEE publ. 76CH 1175-9 NPS, 1976, p. 75.

[Gu76]G.J. van Gurp, D. Sigurd, and W.F. van de Weg, Appl. Phys. Lett. 29, 159 (1976).

[Ha73]A.M. Halpern and J. Law, Phys. Rev. Letters $\underline{31}$, 4 (1973).

[Hi76]J.K. Hirvonen and G.K. Hubler, in "Ion Beam Surface Layer Analysis" (O. Meyer, G. Linker, and F. Kappeler, ed.), Vol. 1, p. 457. Plenum, New York, 1976.

[Ho76]F. Hopkins, A. Little, N. Cue and V. Dutkiewicz, Phys. Rev. Letters $\underline{37}$, 1100 (1976).

[Hu76]M. Hufschmidt, W. Moller, V. Heintze, and D. Kamke, in "Ion Beam Surface Layer Analysis" (O. Meyer, G. Linker, and F. Kappeler, ed.) Vol. 2, p. 831. Plenum, NY, 1976.

[Ki42]L.O. Kirkendall, Trans. AIME $\underline{147}$, 104 (1942).

[La76]D.J. Land, D.G. Simons, J.G. Brennan and M.D. Brown, in "Ion Beam Surface Layer Analysis" (O. Meyer, G. Linker and F. Kappeler, ed., Vol. 2, p. 851. Plenum, N.Y., 1976.

[La76a]W.A. Lanford, H.P. Trautvetter, J.F. Ziegler and J. Keller, Appl. Phys. Lett. $\underline{28}$, 566 (1976).

[La77]W.A. Lanford, in "Third Intl. Conf. on Ion Beam Analysis" (E.A. Wolicki, J.W. Butler, and P.A. Treado, ed.). North-Holland, Amsterdam, 1978.

[Li63]J. Lindhard, M. Scharff and H.E. Schiøtt, Kgl. Dan. Vid. Selsk. Mat. Phys. Medd. $\underline{33}$, No. 14 (1963).

[Li74]W. Litchten, Phys. Rev. $\underline{A9}$, 1458 (1974).

[Lo74]See, e.g., A.S. Lodhi and D. Powers, Phys. Rev. $\underline{A10}$, 2131 (1974).

[Ma68]J.B. Marion and F.C. Young, "Nuclear Reaction Analysis," p. 163. North-Holland, Amsterdam, 1968.

[Ma74]"Proceedings of the Conference on Ion Beam Surface Layer Analysis, Yorktown Heights, N.Y., June 1973,"Thin Solid Films $\underline{19}$, 1 (1973).

[Ma74a]J.H. Macek and J.S. Briggs, J. Phys. B $\underline{7}$, 1312 (1974).

[Ma75]D.H. Madison and E. Merzbacher, in "Atomic Inner Shell Process" (B. Crasemann, ed.), Vol. 1, p. 1. Academic Press, N.Y. 1975.

[Ma77]"Ion Beam Handbook for Material Analysis", ed. by J.W. Mayer and E. Rimini, Academic Press, N.Y., 1977.

[Me76]O. Meyer, G. Linker, and F. Kappeler (ed.), "Ion Beam Surface Layer Analysis", Vol. 1 & 2. Plenum, New York, 1976.

[Me76a]G. Mezey, J. Gyulai, T. Nagy and E. Kotai, in "Ion Beam Surface Layer Analysis" (O. Meyer, G. Linker & F. Kappeler, ed.), Vol. 1, p. 303. Plenum, N.Y. 1976.

[No77]R. Nobeling, K. Traxel, F. Bosch, Y. Civelekoglu, B. Martin, B. Povh and D. Schwalm, Nucl. Instr. and Methods $\underline{142}$, 49 (1977).

[Pa65]D.W. Palmer, Nucl. Instr. & Meth. $\underline{38}$, 187 (1965).

[Pa76]G.M. Padawer, in "Fourth Conference on Application of Small
Accelerators, Denton, Texas, October 25-27, 1976" (J.L. Duggan and
I.L. Morgan, ed.). IEEE Publ. No. 76Ch1175-9 NTS, IEEE, N.Y.

[Po76]J.M. Poate, in "Ion Beam Surface Layer Analysis" (O. Meyer, G.
Linker, and F. Kappeler, ed.), Vol. 1, p. 317. Plenum, New York, 1976.

[Pr76]R. Pretorius, Z.L. Liau, S.S. Lau, and M-A. Nicolet, Appl. Phys.
Lett. 29, 598 (1976).

[Pr77]"Proceedings of the International Conference on Particle Induced
X-Ray Emission and its Analytical Applications, Lund, Sweden, 23-26
Aug. 1976." Nucl. Instr. Methods (1977).

[Ri67]E. Ricci and R.L. Hahn, Anal. Chem. 39, 794 (1967).

[Ru11]E. Rutherford, Phil. Mag. 21, 669 (1911).

[Sc76]B.M.U. Scherzer, P. Børgesen, M.A. Nicolet, and J.M. Mayer, in
"Ion Beam Surface Layer Analysis," (O. Meyer, G. Linker, and F.
Kappeler, ed.), Vol. 1, p. 33. Plenum, New York, 1976.

[Si74]P. Sigmund, in "Radiation Damage Processes in Materials," (C.H.
S. Dupuy, ed.), p. 3, Noordhoff, Leyden, 1975.

[Su79]M. Suter, G. Bonani, H. Jung, Ch. Stoller and W. Wolfli, IEEE
Trans. on Nucl. Sci. NS-26, 1373 (1979).

[Wa75]R.L. Watson, T. Chiao and F.E. Jenson, Phys. Rev. Letters 35,
254 (1975).

[Wi75a]J.S. Williams, Nucl. Instr. Meth. 126, 205 (1975).

[Wo75]E.A. Wolicki, in "New Uses of Ion Accelerators" ed. by J.F.
Ziegler, p. 159, Plenum Press, N.Y. 1975.

[Wo77]E.A. Wolicki, J.W. Butler, and P.A. Treado (ed.), "Proceedings of
the Third International Conference on Ion Beam Analysis." North-
Holland, Amsterdam, 1978.

[Zi74]J.F. Ziegler and W.K. Chu, At. Data and Nucl. Data Tables 13,
463 (1974).

[Zi76]J.F. Ziegler, R.F. Lever, and J.K. Hirvonen, in "Ion Beam Surface
Layer Analysis," (O. Meyer, G. Linker, and F. Kappeler, ed.), Vol. 1,
p. 163. Plenum, New York, 1976.

[Zi76a]J.F. Ziegler, W.K. Chu and J.S.Y. Feng, in "Ion Beam Surface
Layer Analysis," (O. Meyer, G. Linker, and F. Kappeler, ed.), Vol. 1,
p. 15. Plenum, New York, 1976.

[Zi77]J.F. Ziegler, "Helium: Stopping Powers and Ranges in All Elemental Matter". Pergamon Press, N.Y., 1977.

[Zi77a] J.F. Ziegler, et al.,Profiling Hydrogen in Materials Using Ion
Beams, in"Third Intl. Conf. on Ion Beam Analysis," (E.A. Wolicki, J.W.
Butler, and P.A. Treado, ed.). North-Holland, Amsterdam, 1978.

3. Channeling of Heavy Charged Particles

In directing a beam of charged particles towards a single crystal in which the atoms are arranged in a regular order, the distribution of impact parameters which governs the cross sections of various collisional processes may be expected, and indeed found to be target orientation dependent. This effect is commonly called "channeling" effect. In order to see how these crystal orientation effects are used in specific applications, a review of the salient features of channeling phenomena will be helpful. Our discussion here will be brief, however, since excellent expositions on the subject are available elsewhere [Mo73, Ge74].

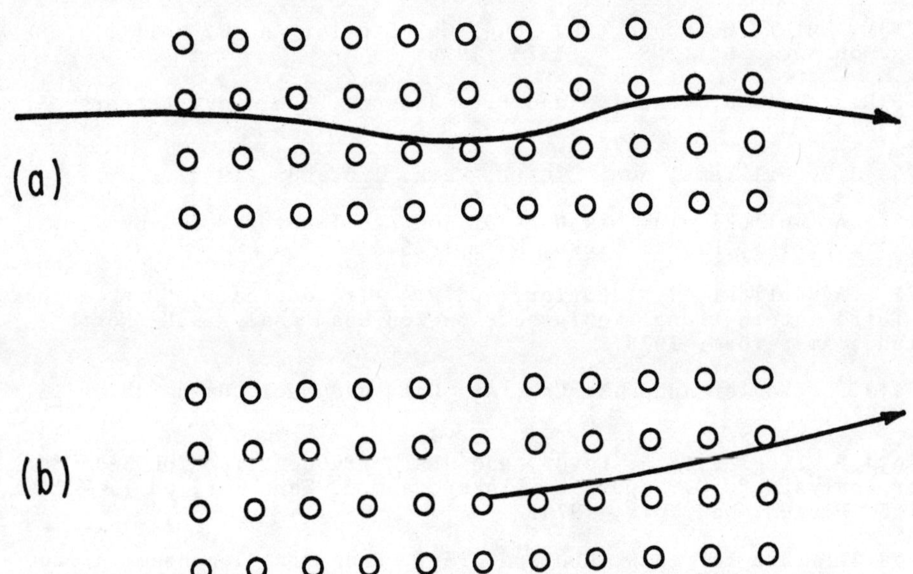

Fig. 3.1-Schematic of particle trajectories for a)channeling and b)blocking.

A schematic diagram of channeling motion is illustrated in Fig. 3.1(a) where the trajectory of a positively charged particle moving between two atomic rows is pictured. The steering motion arises because the particle, when directed close to a row, undergoes successive small angle scattering corresponding to strong correlations of impact parameters. One immediate consequence is the dramatic reduction of close collision events such as nuclear multiple scattering, nuclear reactions and atomic inner shell ionizations. The relative ease with which the variation of close-collision yield with crystal orientation can be measured is the main reason why channeling is extensively applied to studies of atom location in the bulk and on the surface, and of defects in crystals.

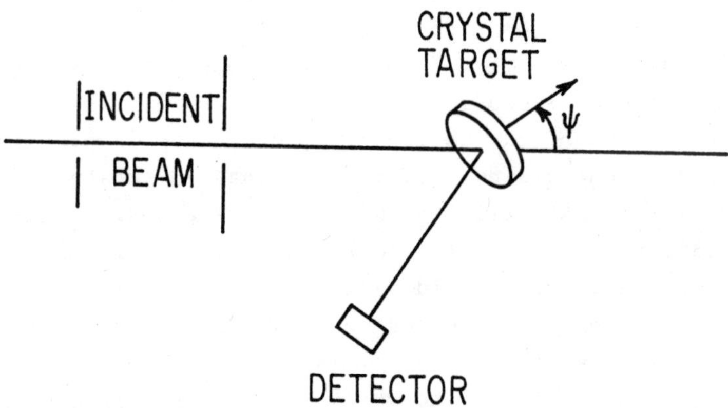

Fig. 3.2-Schematic of a typical arrangement for channeling measurement.

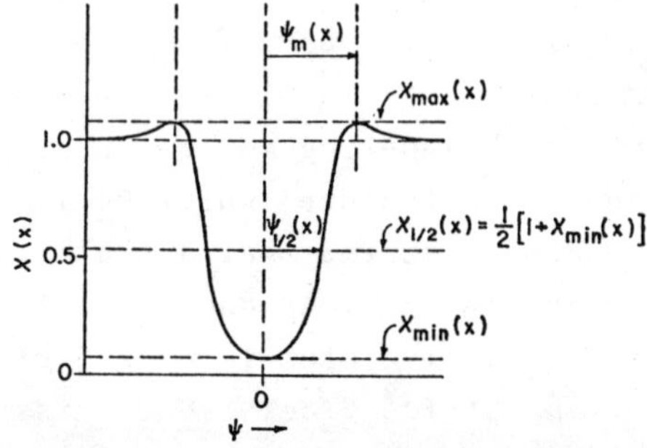

Fig. 3.3-Schematic diagram depicting the variation of a close-collision yield, corresponding to a depth x in the crystal, with the crystal tilt angle ψ (From Ref.[Ge74]).

A typical arrangement for channeling experiment is shown in Fig. 3.2. The angle between the incident beam direction and the crystal axis or plane of interest is called the tilt angle ψ. If detector #1 records the close-collision yield corresponding to a depth x in the crystal, the variation of this yield with ψ will be in a manner depicted in Fig. 3.3. The normalized yield of $\chi=1$ would be equivalent to that from a corresponding amorphous target. At $\psi=0$, the channeled fraction is maximum and thus χ will be a minimum χ_{min}. As ψ increases the yield rises to a maximum value χ_{max} before leveling off to $\chi=1$. This shoulder is seen in experiments and attributable to compensation effects [Li65].

Positively charged particles originating from lattice sites such as α-particles emitted from radioactive lattice atoms may be shadowed by neighboring atoms (see Fig. 3.1b). Their emergence from the crystal along directions close to a row or plane will therefore exhibit a blocking pattern much like that of Fig. 3.3. From a practical standpoint, this situation does not differ from that for channeled particles undergoing close-impact collisions with lattice atoms. According to Lindhard [Li65], not only should blocking and channeling dips (see Fig. 3.3) be similar, they should be identical if energy loss phenomena are ignored since the two processes are related by a rule of reversibility in a statistical treatment. This rule has been tested experimentally [An68] and the degree to which it is obeyed is excellent as can be seen in Fig. 3.4. In that case the elastic scattering yield of 400-keV H^+ on W was monitored. For the channeling curve, the <100> axis was varied relative to the beam but with the detector fixed at an angle intercepting emergence in random direction. For the blocking curve, it is the detector which was moved and this recorded emergent particles relative to the <100> axis for incident beam in a random direction. Over the small range of emergent angles examined, the scattering cross section can be considered to be uniform.

The close connection between channeling and blocking has been exploited in the detailed treatment of channeling motion [Li65]. It also permits qualitative understanding of features observed in channeling. For example, the shoulders observed near a channeling "dip" which correspond to larger than random yield can be explained by considering two identical sets of radioactive atoms, one set of which is in random order while the other in lattice order. The emission rate per unit solid angle will be the same in all directions for the random set. In contrast, this rate will be suppressed along atomic rows leading to "dips" for the lattice set. Since the total emission

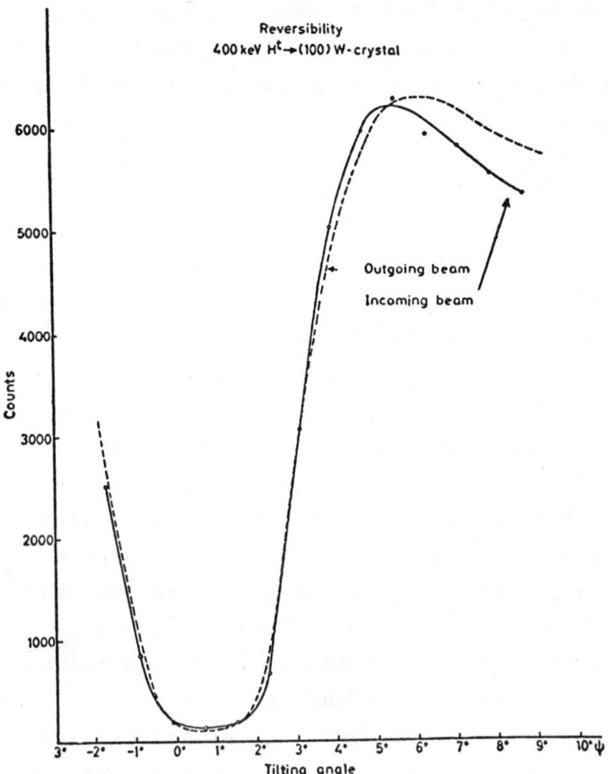

Fig. 3.4-Test of reversibility using the RBS of
400-keV H$^+$ on W. The solid and dashed curves
correspond to channeling and blocking, respectively
(From Ref. [An68]).

rate is independent of the underlying solid structure, the "dips" must
be compensated for by larger than random yield in some other directions.
These directions should not be far off the rows because of steering
motion, and thus the shoulders near the "dips".

In the following sections, more detailed aspects of channeling
are first discussed in the context of the continuum model. These will
then provide the basis for describing atom location, defect and
blocking lifetime studies.

3.1   The Continuum Model

When a moving particle enters a crystal at a small angle relative
to an atomic row or plane and the energy loss processes are neglected,
its velocity component along the row or plane will not change and thus
a description of its motion needs to address only the transverse
component. The continuum model [Le63, Li64a, Li65, Er65] asserts that

to a good approximation the motion of channeled particles is determined by a continuum potential U obtained by replacing the actual periodic feature of the crystal by one averaged over a direction parallel to the row or plane.

For the present cases, a classical treatment of directional effects is adequate and the accuracy of the continuum description in the axial case had been assessed by the halfway-plane treatment [Li65, Le67]. The planar case has not been assessed to the same degree, however. Nevertheless the model does provide reasonable descriptions of many of the planar measurements as well.

The procedure for obtaining the continuum potential U usually starts with a single isolated static row (axial) or plane (planar) [Ge74]. Averaging over the ion-atom potentials V along the row (z-axis) gives rise to a continuum potential V_{RS} which is a function only of the coordinate $r=(x^2+y^2)^{1/2}$ transverse to the row. A similar averaging over the two dimensions of the plane leaves a continuum potential $V_{ps}(y)$ which has the coordinate y transverse to the plane as the sole variable.

Inside the crystal the particle experiences the cummulative effects of rows or planes. The summation of these contributions [Ge74] then lead to the final static forms $U_{RS}(\bar{r})$ and $U_{PS}(y)$. Of course lattice atoms undergo thermal vibrations. The time of vibration is of order 10^{-13} s or more and, in all cases of interest, this is very long compared to the collision time d/v, where d is the atomic spacing and v is the projectile's velocity. Thus a projectile is deflected by rows and planes where atoms are still static but displaced from their ideal positions. Thermal effects can then be incorporated by convoluting into the static potential the displacement probability of lattice atoms obtained from the Debye theory of thermal vibrations [Ap67, Ba71].

The ion-atom potential V to be used in obtaining the continuum potential should be specific to the collision partners. However, in view of the averaging effects and the fact that important features of channeling are remarkably independent of the details of V, analytic approximations for V are usually employed in practice. Commonly used are the Moliere's analytic approximation to the Thomas-Fermi potential [Mo47] and the so-called standard potential due to Lindhard [Li65]. An example of axial continuum potential constructed from the standard potential and with thermal vibration effects incorporated is shown in Fig. 3.5 in the form of a contour plot for deuterons incident along the <110> axes of Si. A planar one using the Moliere's potential is

shown in Fig. 3.6 for protons channeled in the (110) planes of Si.
The difference between the static case and that with thermal vibrations
included can also be seen in the figure.

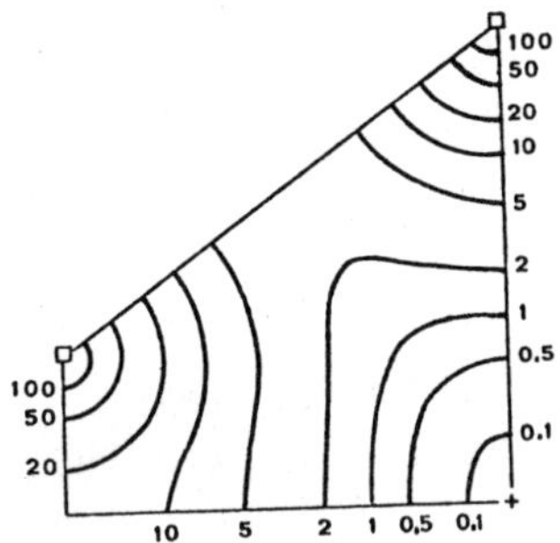

Fig. 3.5-Continuum potential energy contour
(labeled in eV) for deuterons incident along
the <110> axes of Si. The calculations are based
on the standard ion-atom potential. The value at the
lattice sites marked by squares is ∿130 eV. Only a
quarter of the channel is shown. (From Ref. [De74]).

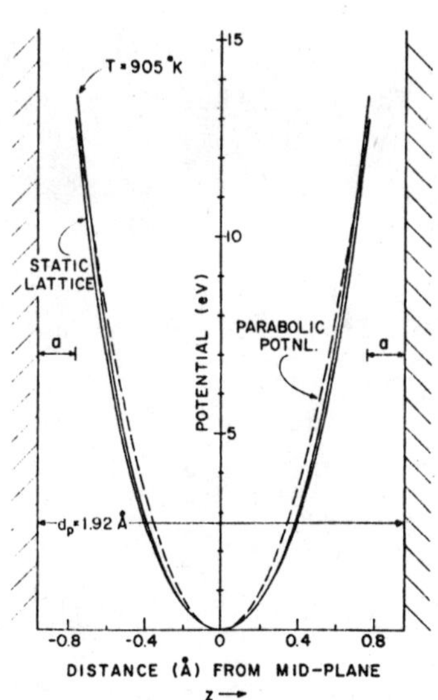

Fig. 3.6-Continuum potential energy for
protons channeled in the (110) planes of Si.
The calculations are based on the Moliere's
ion-atom potential. (From Ref. [Ge74]).

3.1.1 Estimates of $\psi_{1/2}$ and χ_{min}

When energy loss processes are neglected, the invariance of the velocity component along the row on plane leads to the conservation of transverse energy E_t for the channeled particles. In an isolated single row approximation, a particle directed at an incident angle ψ from a large distance to the row has a transverse energy given by

$$E_t = \frac{(p\psi)^2}{2M} = E\psi^2, \qquad\qquad 3.1$$

where p and E are the incident particle's momentum and kinetic energy, respectively. The distance of closest approach to the row r_{min} will be achieved when the particle trajectory and the row are coplanar and this is given by

$$E\psi^2 = U(r_{min}). \qquad\qquad 3.2$$

Violent collision will result if this r_{min} is smaller than the corresponding characteristic impact parameter which may be represented by either the Thomas-Fermi screening length [Li65]

$$a = 0.8853\, a_o\, [Z_1^{2/3} + Z_2^{2/3}]^{-1/2} \qquad\qquad 3.3$$

with a_o being the Bohr radius, or the rms thermal vibration amplitude ρ, whichever is the larger. There is thus a critical angle of incidence

$$\psi_c = [U(r_{min})/E]^{1/2} \qquad\qquad 3.4$$

beyond which stable channeling trajectories can not be sustained. This ψ_c may be identified with the $\psi_{1/2}$ in Fig. 3.3.

More accurate estimates for the $\psi_{1/2}$ based on the Lindhard's Standard potential [Ma75] can be written as (ψ in degrees, E in MeV, lengths in $\overset{o}{A}$)

$$\text{Axial } \psi_{1/2} = 0.25\, F_{ax}(\rho_2/a)\, [Z_1 Z_2/E_d]^{1/2} \qquad\qquad 3.5a$$

$$\text{Planar } \psi_{1/2} = 0.40\, F_{p\ell}(\rho_1/a, 1/na)\, [Z_1 Z_2\, na/E]^{1/2}, \qquad 3.5b$$

where d is the atomic spacing along the row, n is the atomic density (atoms/$\overset{o}{A}{}^2$) in the plane, F_{ax} and $F_{p\ell}$ are weakly varying functions of the appropriate ρ/a ratio; with typical F values for most lattices being in the 0.6-0.8 range.

The stability criterion that $r_{min} \gtrsim$ the larger of a and ρ also leads to simple estimates for the minimum yield χ_{min} at $\psi=0$ (see Fig. 3.3), since this χ_{min} is just the relative area of the forbidden zone for channeling. The explicit forms are:

$$\text{Axial } \chi_{min} = Ndr_{min}^2 \qquad\qquad 3.6a$$

$$\text{Planar } \chi_{min} = 2r_{min}/d_p, \qquad\qquad 3.6b$$

where N is the number of atoms per unit volume in the crystal and d_p is the spacing between planes. More accurate estimates have been suggested [Ba71] as

$$\text{Axial } \chi_{min} = Nd\pi(3\rho_2^2 + 0.5a^2) \qquad\qquad 3.7a$$

$$\text{Planar } \chi_{min} = 2(\rho_1^2 + a^2)^{1/2}/d_p. \qquad\qquad 3.7b$$

Typical values for $\psi_{1/2}$ range from $\sim 0.01^o$ to $\sim 1^o$, while those for χ_{min} are about 0.2-0.4 for planar channeling and about 0.01-0.05 for axial channeling. Both $\psi_{1/2}$ and χ_{min} are significantly affected by ρ and hence channeling effects are strongly enhanced at low temperature.

3.1.2 Flux Distribution

As mentioned previously, most applications of channeling rely on the strong directional effects of close collision yields. The yields due to crystal imperfections such as lattice defects and non-substitional foreign atoms would give rise to patterns different from that for the host atoms. The interpretation of these patterns requires a knowledge of the spatial distribution of channeled particles. Since axial channeling exhibits more pronounced effects than planar case and therefore more extensively used in applications, the discussion which follows will be concerned mainly with axial channeling.

Within the continuum model, the assumptions of transverse energy conservation and statistical equilibrium for the flux lead to a compact expression for the spatial distribution [Li65]. This is seen by noting that a particle entering the crystal at the transverse position $\bar{r}_i$, with incident angle ψ_i, will be confined to an area $A(E_t)$ limited by a contour line of the potential given by

$$U(\bar{r}) = E_t(\bar{r}_i, \psi_i) = E\psi_i^2 + U(\bar{r}_i), \qquad\qquad 3.8$$

where a thermal averaged form for U is implied. With statistical
equilibrium, there is an equal probability of finding an ion anywhere
within the accessible area $A(E_t)$. If the distance of closest approach
to the row is r_m and the critical distance for violent collision with
a lattice atom is a, the probability for a particle to be at a position $\bar{r}$ in axial channeling is

$$
P(E_t,\bar{r}) = \begin{cases} 1/A(E_t), & U(\bar{r}) \leq E_t \text{ and } |\bar{r}_j - \bar{r}| \geq r_m, \ r_m > a \\ 0 & , \ U(\bar{r}) > E_t \text{ and } |\bar{r}_j - \bar{r}| < r_m, \ r_m > a \\ 1/A_o & , \ r_m \leq a. \end{cases} \qquad 3.9
$$

Here the positions $\bar{r}$ and $\bar{r}_j$ are relative to the open channel axis and
$\bar{r}_j$ is that of the closest row, and A_o is the area of one channel.
The normalized flux along the equipotential contour $U(\bar{r})$ is thus

$$
F(\bar{r}) = \int_{A_c}^{A_o} \frac{dA}{A} = \ln \frac{A_o}{A_c} , \qquad 3.10
$$

where A_c is the area enclosed by the contour. A logarithmic divergence is seen for the mid-channel flux because $A_c \to 0$ as $r \to 0$ since
$U(\bar{r}) \to 0$. However, for such small values of U, fluctuations in E_t can
no longer be neglected. Instead, factors such as beam divergence,
scattering by electrons and surface disorder introduce a finite
spread in E_t, and this causes the flux to level off to some limiting
value.

Examples of flux distribution calculated using the analytical
model just described are shown in Fig. 3.7. These are for 3.5-MeV
^{14}N ions incident along the <100> axis of Fe and reproduced from
Alexander et al [A174]. The left panel shows the cross section of
the <100> channel with distinct sites marked as 0, A, B, and S, and
rows of Fe atoms represented by filled circles. The normalized flux
$F(\bar{r})$ along the line SOS' is shown in the center panel for three
incident angles ψ_i whose corresponding values can be read off from
the right panel which shows the flux at the different sites as a
function of the incident angle. Note that ψ_3 is $\psi_{1/2}$ for this case.

The simplicity of the analytical approach is attractive but one
needs to examine the extent to which the assumptions of statistical
equilibrium and conservation of transverse energy for channeled ions
are valid. In this connection, dechanneling effects due to multiple
scattering by electrons as well as by lattice atoms and to other
experimental factors such as lattice imperfections become important
considerations. Experimentally, RBS yield curves as a function of

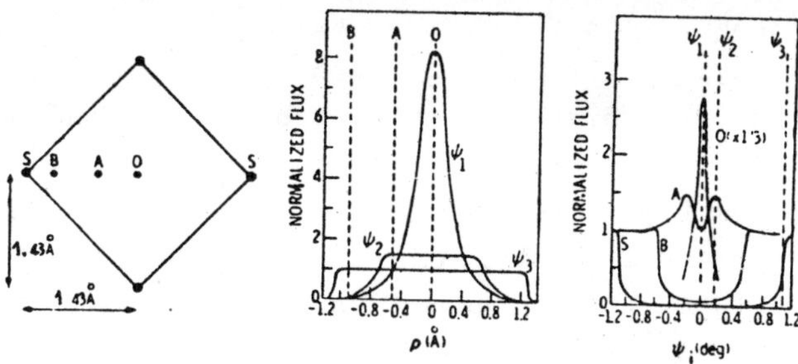

Fig. 3.7-Calculated normalized flux distribution of 3.5-MeV ^{14}N ions channeled along the <100> axis of bcc Fe lattice, based on the analytical model. (From Ref. [A174]).

depth provide important information on dechanneling. This can be seen in the work of Picraux et al. [Pi69] which is shown in Fig. 3.8. Note that the small high energy peak in the aligned spectrum is due to the unavoidable scattering from surface atoms. Comparison of the orientation dependences for depths of 0.1 and 0.6 μm clearly indicates the effects of dechanneling at the greater depth, as manifested by the absence of shoulders, a decreased $\psi_{1/2}$ and an increased χ_{min}. Such data, however, do not give direct information on the flux in the open channel. For this, theoretical approaches must still be relied upon. Analytical treatment of dechanneling by multiple scattering of electrons, through the Fokker-Planck diffusion equation, has been given by Kumakhov [Ku75]. The results show features much like those obtained in the more comprehensive Monte Carlo computer simulations which will now be discussed.

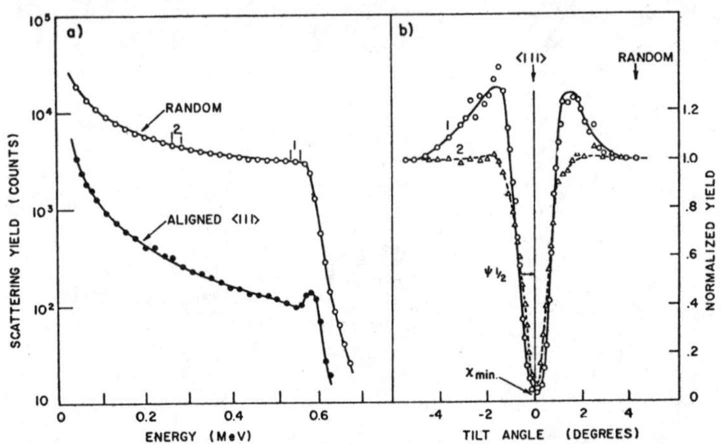

Fig. 3.8-a) Backscattered energy spectra for 1-MeV He ions incident on a Si crystal for aligned (<111>) and random direction of incidence. b) Orientation dependence of the normalized yield from scattered energy region 1 and 2 (see part a)) corresponding to depths of 0.1 and 0.6μm, respectively. (From Ref.[Pi69]).

Computer simulation has the advantage that multiple scattering
by lattice atoms is built in while such factors as electronic multiple scattering, beam divergence, surface disorder and lattice damage can be easily incorporated. Such computations, however, require
a large-scale computer. For the present cases of interest, the
binary collision model is applicable [Ja75]. In this model, the
interactions between the energetic ions and the lattice atoms are
treated as a series of independent two-body collisions described by
classical mechanics. Such calculations provide extensive checks on
the validity of the continuum model and assessments of the effects
of various experimental factors [Ba71, Vl71, Ry72]. They also show
the flux to exhibit significant oscillatory dependence with depth.
The amplitude of these oscillations is strongly damped by various
multiple scattering effects [Vl71]. Figure 3.9 displays the depth
dependence of the mid-channel flux calculated by Van Vliet [Vl71] for
1-MeV He ions along the <100> āxis of Cu, illustrating the flux
oscillations and the degree of damping when electronic multiple
scattering and finite beam divergence are taken into account.

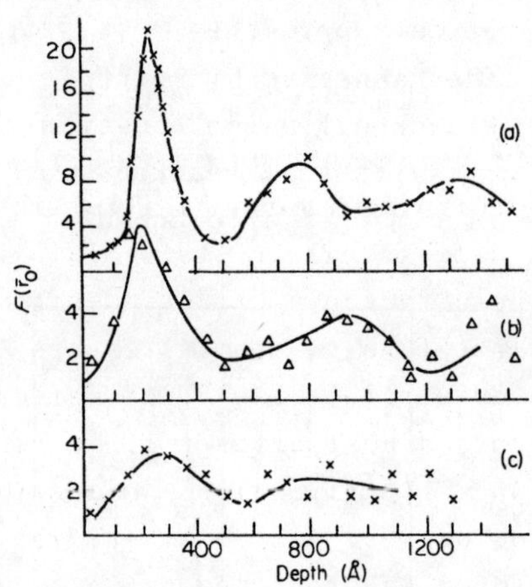

Fig. 3.9-The variation of mid-channel flux
$F(r=0)$ with depth for 1-MeV He ions in <100>Cu:
(a)no multiple scattering; (b)including multiple
scattering, thermal vibrations at $0^{o}C$ and a beam
collimation of $\pm 0.06^{o}$; (c)the same as (b) but with
a beam collimation of $\pm 0.23^{o}$. The solid curves are
best fits to the computed results. (From Ref. [Vl71]).

It would seem from the brief discussion here that, in the applications of channeling the foreign atom location studies, detailed
computer simulations for each individual system would be required if
accurate information is to be extracted. This will be true when the
atoms of interest are at depths of $\lesssim 1000\text{Å}$, as is usually the case for

ion implanted species. However, for depths $\gtrsim 1000$ Å, the flux oscillations will be largely damped out and statistical equilibrium sets in. The simpler analytical treatment of flux distribution may be adequate in these cases. Moreover, in many practical cases of interest, the question is in which one of the few known number of sites is the atom located. Such questions are often answered by simple triangulations using the various high-symmetry axes and/or planes, and with little demand on the detailed knowledge of the flux distributions.

3.2 Atom Location and Defect Studies

The strong directional effects of close collision yields in single crystals suggest a number of applications. Some of the more established of these techniques [Mo73, Th78] will be reviewed briefly here. In applications where the detection of foreign atoms in the presence of a large excess of host lattice atoms, any one of the close-impact processes discussed in Sec. 2 can be utilized. The choice will depend on the specific system under investigation. If the atomic concentration of foreign atoms exceeds approximately 10^{-3}-10^{-4}, a satisfactory one can usually be found.

3.2.1 Foreign Atom Locations in the Bulk

The basic idea behind the determination of a particular crystallographic site location of foreign atoms can be illustrated by considering an idealized two dimensional crystal. This is shown schematically in Fig. 3.10 where the open circles designate the host lattice atoms and three distinct foreign atom sites are marked. For the substitutional site (closed circle), the "dip" patterns of close-impact yields from the foreign and host atoms will be virtually identical along both <01> and <11> directions because both types of atoms are shadowed to the same degree. In contrast, for the most symmetric interstitial site (open square), shadowing occurs only along the <11> but not the <01>direction. Thus signals from the foreign atoms will exhibit a "dip" pattern for the <11> direction while in the <01> direction a "flux peaking" one. For the bridge sites marked by stars, shadowing of foreign atoms occurs along either the <01> or <10> direction but only one of the two sites will be shadowed. The "dip" pattern in this case will be shallower. From these simple considerations of triangulation, the determination of foreign atom locations is seen to be straignt forward provided that the foreign atoms occupy one particular site.

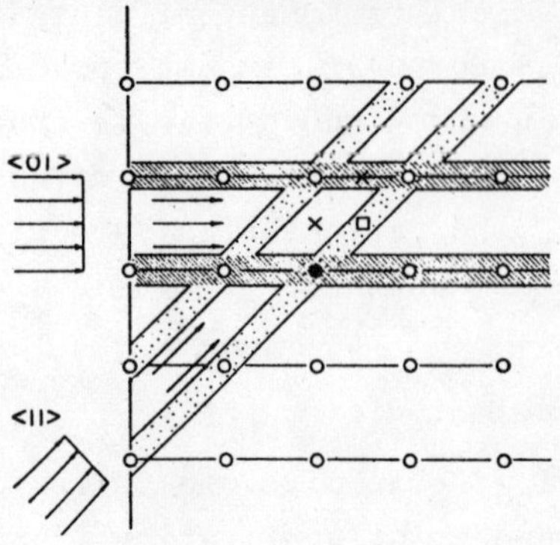

Fig. 3.10-Schematic of a two dimensional lattice illustrating how the channeling effect may be used to determine a foreign-atom site (From Ref.[Da73]).

One example of the technique just described is the work of Picraux and Vook [Pi74] on the determination of deuterium lattice location in Cr and W which helped clarify an apparent anomaly concerning hydrogen solubility in Cr. Interstitial foreign atoms are often located in well defined sites. In Cr and W which have a bcc structure, these are the octahedral and tetrahedral sites shown in Fig. 3.11. Also shown are their projections on the planes perpendicular to the three principal axial channels. In their study, the deuterons were implanted into Cr and W single crystals in a non-channeling direction with energies of 15- and 30-keV, respectively, corresponding to projected ranges of 1140 Å and 1270 Å. The deuteron dose were 3×10^{15} atoms/cm^2.

In Fig. 3.12, the sharp "flux peaking" pattern with no significant "dip" component for D in W clearly suggests a tetrahedral site for the D since the pattern implies no shadowing of the D at all. For D in Cr, a "flux peaking" feature is also observed but the amplitude is smaller, the width wider, and it sits on a "dip." Moreover, this D "dip" can be inferred to have a $\psi_{1/2}$ comparable to that for Cr($\psi_{1/2}=$ 1.38°) and a $\chi_{min}\sim 0.73$ which is closed to the value of 0.67 when 1/3 of the allowed sites are shadowed. An octahedral site for D in Cr is thus suggested. This interpretation is reinforced by the planar scans shown in Fig. 3.13. Here 2/3 of the tetrahedral sites lie in (100) planes and 1/3 lie between the planes, whereas all the octahedral sites lie in the (100) planes. The data indeed show a small "flux peaking" feature for D in W which is absent for D in Cr. Moreover

the "dip" in the latter case has a smaller χ_{min} which is consistent
with no exposed D for this case.

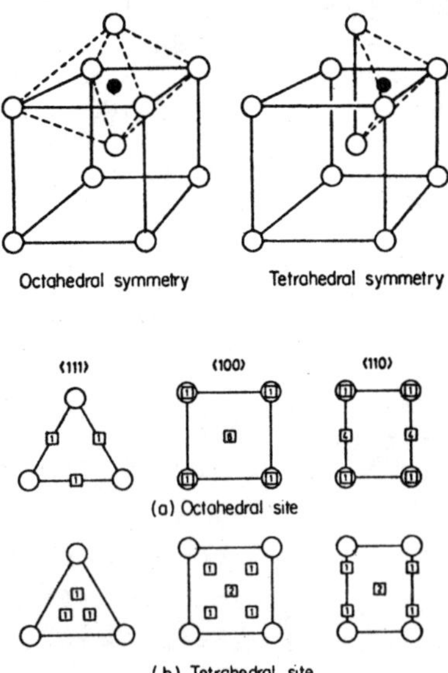

Fig. 3.11-Interstitial positions in bcc lattice
and their projections unto planes perpendicular
to the three principal axial channels. The number
inside the squares indicate the relative probability
of the interstitial site with the corresponding
projection. (From Ref. [Da73]).

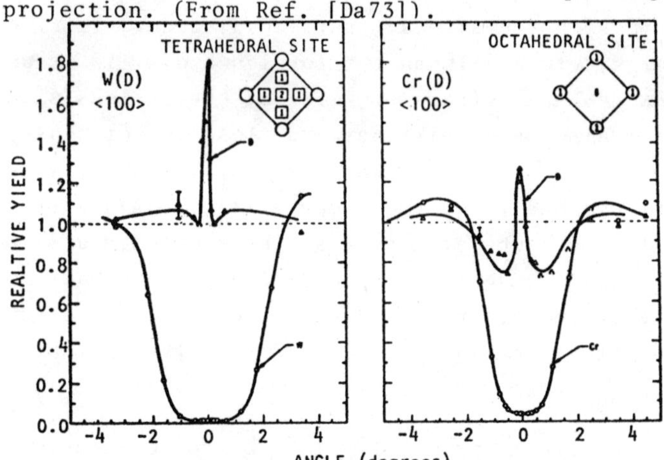

Fig. 3.12-Angular scans through the ⟨100⟩
axis for W(left) and Cr(right) implanted with
$3 \times 10^{15}/cm^2$ of deuterons at 30 and 15 keV,
respectively. A 750-keV ^{3}He beam was used in
which the RBS yields from W and Cr and the $D(^3He,p)^4He$
yield from deuteriums were recorded. (From Ref. [Pi74]).

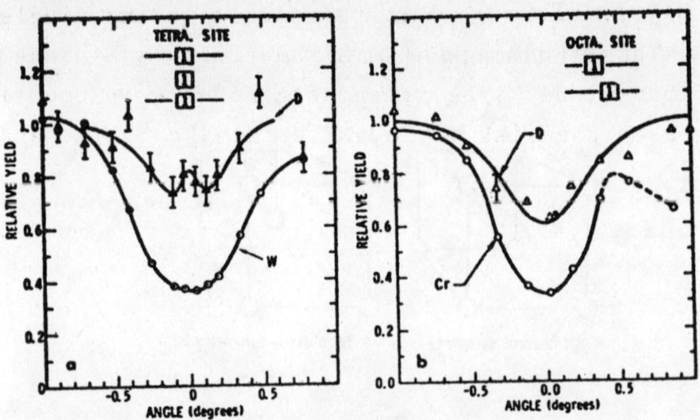

Fig. 3.13-Angular scans across the (100)
planes of W(left) and Cr(right) for the same
conditions as those of Fig. 3.12. (From Ref.
[Pi74]).

In cases where the foreign atoms occupy a less distinctive site
or multiple sites in the crystal, the interpretation of the observed
yields will require comparisons with detailed yield calculations for
the specific system. Furthermore serious complications may arise in
a number of situations: 1)when the foreign atom fraction is $\lesssim 10^{-4}$,
the signals become difficult to detect; 2)when crystal imperfection
become significant as reflected, for example, by a host atom χ_{min}
which is considerably greater than the theoretical value, because a
number of factors can be at play (surface disorder, lattice defects,
radiation damage, etc.); 3)when the foreign atoms lie at depths for
which the probing beam suffers significant energy loss since multiple
scattering will also be significant and close-impact cross sections
generally depend on energy; 4)when the crystal has a low symmetry
such as monoclinic and triclinic because triangulation procedures are
more difficult; and 5)when the analysis beam becomes a significant
source of radiation damage as will be the case in organic or ionic
crystals. These limitations, however, have not restricted the
applications of the technique to a large variety of problems.

3.2.2 Defect Studies

The concern here is primarily with regard to structural damage
or disorder. Thus RBS yield curves (see Fig. 3.8a) are particularly
useful since they provide information on dechanneling as a function
of depth and lattice defects are manifested as dechanneling effects.
Magnification of disorder effects in RBS spectra can be achieved by
the use of the so-called double alignment geometry [Bø67] illustrated
in Fig. 3.14. In the perfectly aligned condition, the minimum yield

will be $\chi_{2min} \approx \chi_{min}^2$, where χ_{min} is the **corresponding single** alignment value [Ap70]. Thus the range of minimum yield is expanded by roughly two orders of magnitude. The attendant increase in counting time, however, increases **the risk** of radiation damage by the analysis beam itself.

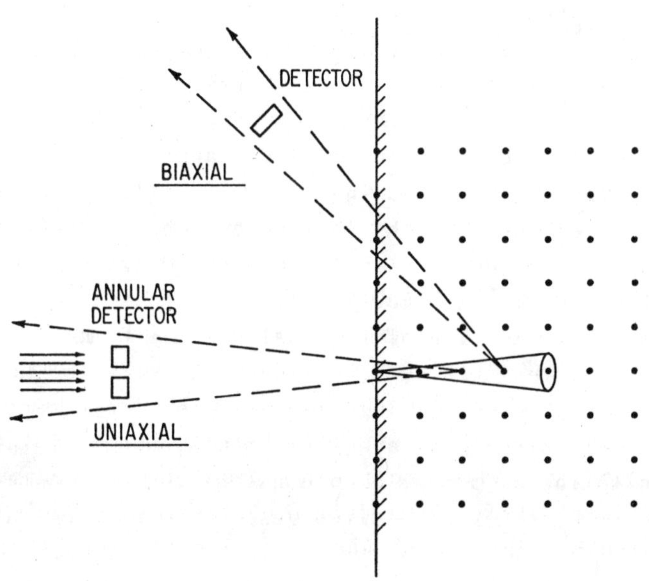

Fig. 3.14-Schematic diagram of double alignment
geometries which utilize shadowing effects in the
inward (channeling) and outward (blocking) paths.

In principle, the presence of disorder can be inferred from the RBS spectra by comparison with that of a crystal with no disorder. Even if such perfect crystal spectra are available, quantitative interpretation of disorder is difficult except for a particular type of disorder. Elaboration will not be attempted here. Instead the reader is referred to the article by Rimini [Ri78] for details. There it is concluded that only for disorder corresponding to randomly displaced atoms is the depth distribution directly reflected in the RBS spectra. Analysis of other disorder such as small atom displacements, dislocations, stacking faults, mosaic spread and etc. are

much more involved and contains some degree of ambiguity. Variation
of experimental parameters can help distinguish the types of disorder involved. Specifically, increasing the beam energy decreases
dechanneling caused by a random distribution of displaced atoms but
the opposite is true for dislocations and mosaic spread, while
stacking faults reflect no energy dependence.

3.2.3 Surface Studies

With improving high vacuum technology, the acquisition of clean
surface equipment adaptable to accelerator beam lines is no longer
prohibitively expensive. As a consequence the use of ion channeling
for surface investigations is increasingly being exploited. The
target chamber with the goniometer attached is usually maintained at
a pressure in the 10^{-10}-10^{11} Torr range. Surface cleaning accessories
are desirable if not necessary. These commonly consist of an argon
sputtering gun and a target heating stage for annealing out the
damage induced by the sputter cleaning process. It is also important
to ascertain the orderliness of the surface and this can be achieved
by the inspection of spot pattern in low or medium energy electron
diffraction(LEED or MEED). Surface impurities may still be present,
usually due to the condensation of residual gases. Commonly encountered are light impurities such as C and 0 and these can be monitored
by Auger electron spectroscopy using, for example, a cylindrical
mirror analyzer. A residual gas analyzer is also useful in allowing
not only the monitoring of probable condensates but also a means of
controlling the introduction of desired gaseous impurities in specific
experiments.

The basis for surface studies lies in the unavoidable scattering
from the surface atoms. For a clean surface, such scattering in an
axially aligned RBS spectrum are manifested as a peak in the high
energy end of the continuum spectrum. This so called surface peak
can be seen, for example, in Fig. 3.8a. A better illustration of
such a peak is displayed in Fig. 3.15. The area under the peak is
directly proportional to the number of atoms per unit area (surface
coverage), or atoms per row, exposed to the beam. If the analysis
beam damage is not significant, the use of a double alignment geometry can suppress considerably the background under the surface peak
as can be seen in Fig. 3.16.

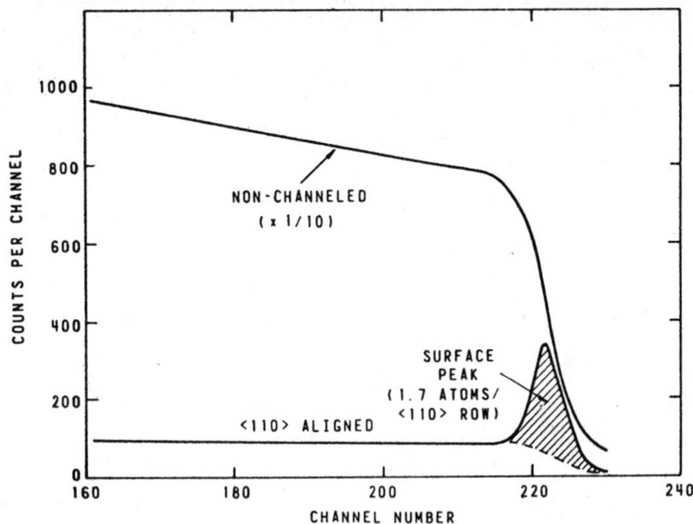

Fig. 3.15-Single-alignment RBS spectrum for 1.0 MeV He$^+$ on a (111) and Pt crystal, showing the well-resolved surface peak for <110> incidence. (From Ref. [Da78]).

492   J. A. DAVIES

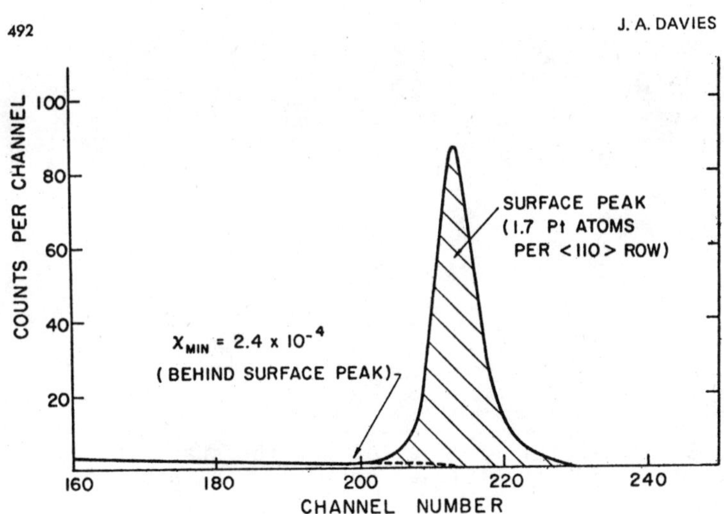

Fig. 3.16-Uniaxial double-alignment RBS spectrum for 1.0 MeV He$^+$ on a (111) cut Pt crystal along the <110> direction. (From Ref.[Da78]).

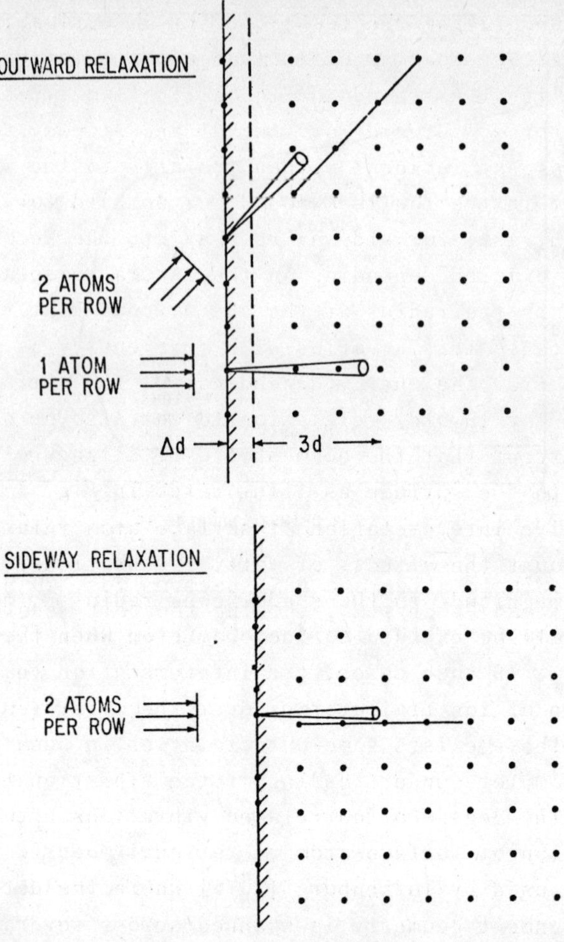

Fig. 3.17-Schematic illustration of how the shadowing technique can be used to determine the relaxation of surface atoms on a clean surface.

The concepts underlying the technique for investigating changes in structure and lattice spacing of surface atoms relative to the underlying lattice is illustrated in Fig. 3.17. Two types of surface relaxation are pictured: outward (or inward) and sideway relative to the surface. For a beam incident perpendicularly to the surface, only the first atom in the row is exposed for outward (or inward) relaxation. In contrast, for sideway relaxation, the second atom in a row will also be exposed depending on the lateral displacement Δd_L of the first atom and the radius of the shadow cone R at the second atom position. Because this R varies with beam energy as $E^{-1/2}$, Δd_L can be deduced from the energy dependence of the surface peak area. The outward (or inward) relaxation Δd can also be deduced in a similar manner except that the beam should be aligned with an axis not perpendicular to the surface as illustrated in Fig. 3.17.

The quantitative interpretation of surface atom relaxation data must take into account the effects of lattice vibrations. If these are comparable in magnitude to the shadow cone radius R, more than one atom per row will be exposed to the beam even when there is no surface relaxation. In such cases, the interpretation requires computer simulation of ion trajectories over the first few atomic planes. However, there exists some uncertainties in such simulations because of the incomplete understanding of the vibrational modes of surface atoms and the degree of correlated vibrations between adjacent atoms. Clarification of surface-atom vibrational modes is possible in the arrangement used by Turkenburg [Tu76] where the detector in a biaxial double alignment geometry is scanned across several blocking axes. Comparisons of the "dip" patterns for surface and "bulk" atoms provide more detailed information on the relaxation and vibrations.

Specific cases of surface relaxation studies in clean surfaces can be found in the review by Davies [Da78]. The technique is also applicable when surface impurites are present. Indeed such cases are of more practical interest. Thus the technique has been used to study, for example, the interface structure of thin layers of SiO_2 on Si [St78], the effects of terminating the surface dangling bonds of Si with H_2 [Fe80] and the initial stages of Au epitaxial growth on Ag [Fe81].

3.3 Blocking Lifetime Studies

As discussed previously, channeling and blocking "dip" patterns are very similar because particles originating from lattice sites have trajectories which are the time reversed of those for channeled particles ending in violent collisions. If the particles are emitted from atoms dislodged from the lattice sites, the "dip" will be narrower and more shallow due to the less effective blocking. The degree of changes in the "dip" pattern thus depends on the mean distance r_1 perpendicular to the row or plane from which the emission took place. This then provides a basis for nuclear lifetime measurements since recoiling excited nuclei produced in nuclear reactions or inelastic scattering can be displaced from their lattice sites during the time of their de-excitations. The situation is illustrated in Fig. 3.18 where the displacement is characterized by $r_1 = v_1 \tau$, with τ being the mean lifetime for decay. The corresponding blocking "dips" expected for prompt $p_0 (\tau=0)$ and delayed $p_1 (\tau>0)$ charged particle emission are sketched in the inset.

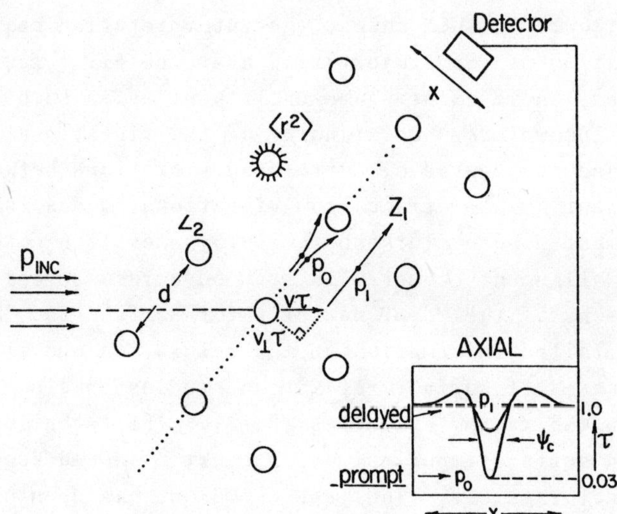

Fig. 3.18-Schematic illustration of blocking lifetime measurement with recoiling excited nuclei. (From Ref.[Gi 75]).

The range of lifetimes which can be measured is seen to be governed by the mean recoil distance r_1 which is bracketed on the low end $(r_1 \sim 0.1)$ by the requirement of measurable changes in the blocking distributions, and on the high end by the large deflection which may ensue when r_1 extends to the adjacent row or plane. For recoil velocities encountered in nuclear physics experiments, this translates to 10^{-18}-10^{-14} s range.

Due to the many factors affecting the blocking distribution, the lifetime is reflected most directly in the difference between the delayed and prompt (reference) distributions measured under the same conditions. Thus the experiments should be designed to make such simultaneous measurements possible. If the recoiling nuclei are produced in a preferred direction such as in the formation of compound nuclei by the capture of beam particles, the set up used by Sharma et al. [Sh73] in their study of ^{32}S lifetimes with the ^{31}P(p,α)^{28}Si reaction may be followed. As reproduced in Fig. 3.19, the GaP crystal is positioned such that the blocking distribution along the <111> axis can be measured in two directions. Thus the patterns at 10° and 81^{0} correspond to the prompt and delayed distributions, respectively, since all the ^{32}S recoil in a direction nearly parallel to the < 111> axis in the 10° case and nearly perpendicular in the 81° case. Figure 3.20 shows the α-particle distributions recorded for the resonance condition at 642 keV corresponding to ^{32}S recoiling near the surface (up to $\sim$1000 Å).

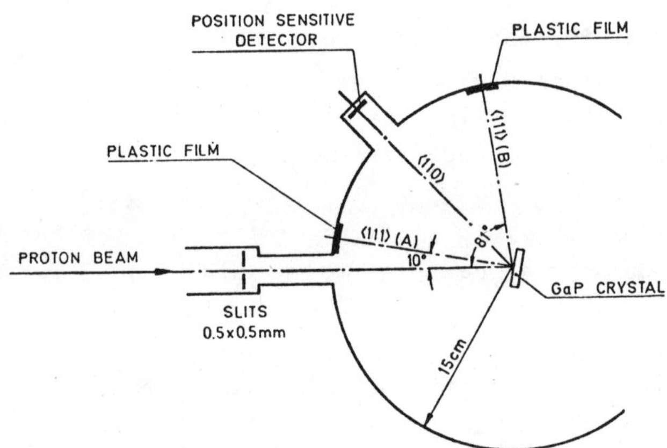

Fig. 3.19-Experimental arrangement used for lifetime measurements of ^{32}S formed in the ^{31}P(p,α)^{28}Si sections. (From Ref. [Sh73]).

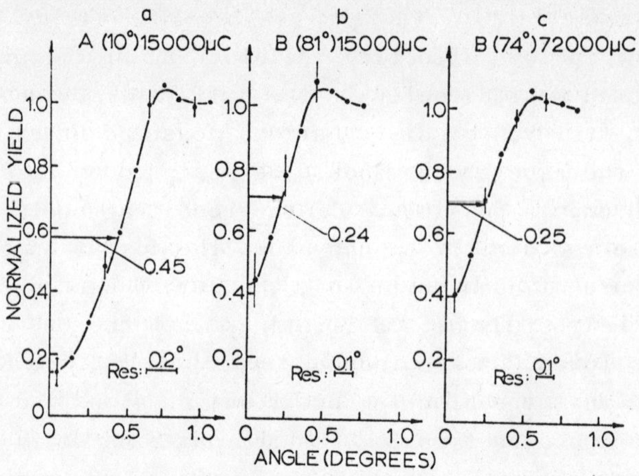

Fig. 3.20-Blocking patterns for α-particles
from ^{31}P(p,α)^{28}Si resonance reaction at 642
keV obtained using the arrangement shown in
Fig. 3.19. (From Ref.[Sh73]).

Another technique [Gi72] for measuring prompt and delayed patterns
simultaneously relies on the use of a thin crystal and a detector
which can distinguish the elastic and non-elastic peaks. For reactions
in which the non-elastic channels are governed by compound nuclear
mechanisms but shape elastic or potential scattering still dominates
the elastic channel, the blocking distribution for the elastic peak
then becomes the prompt reference for the delayed events in the non-
elastic channels.

The different analysis techniques for extracting nuclear lifetime
from the blocking data have been reviewed by Gibson [Gi75]. The
simplest one is based on the single-string continuum approximation
where the lifetime is contained in the expression for the minimum
yield $\chi = \chi_{min}$. With the assumption of statistical equilibrium for
the channeled flux distribution $F(r_1)$ and an exponential form for the
displacement probability of the recoiling emitters, $g(r_1) \propto \exp(-r_1/v_1\tau)$,
the χ calculated from the convolution of F and g leads to a difference
$\Delta\chi$ between the delayed and prompt cases which can be expressed as [Gi75]:

$$\text{Axial } \Delta\chi = 2\pi DNd(v_1\tau)^2$$

$$\text{Planar } \Delta\chi = Cv_1\tau/d_p,$$

where $D \overset{\sim}{-} 1.3 \pm 0.2$ and $C \overset{\sim}{-} 1.17 \pm 0.36$.

The unbounded increase of $\Delta\chi$ with increasing $v_1\tau$ is clearly not physical
and is a result, in the axial case, of $F \propto r_1^2$ for large r_1 in the

single string approximation. Nevertheless these $\Delta\chi$ may be applicable for $v_\perp \tau \lesssim 0.3$[Gi75].

A multi-string aprroach which correctly reflects the potential contours would be an improvement. But even in here the neglect of damping effects lead to a divergence in the mid-channel flux (see Sec. 3.2.3). Indeed all the complications entering in the calculations of channeled flux distribution discussed earlier in Sec. 3.2.3 are encountered here as well. In addition there are the effects of recoil emitters scattered in different transverse directions and a complex decay function in some cases. It would seem that lengthy Monte Carlo calculations would be needed in the analysis of individual result, but sufficient progress has been made in delineating the conditions under which the simpler analytic approximations may be used with reasonably accurate results.

[Al74] R.B. Alexander, P.T. Callaghan, and J.M. Poate, Phys. Rev. B9, 3022 (1974).

[An68] J.U. Andersen and E. Uggerhoej, Can. J. Phys. 46 517 (1968).

[Ap67] B.R. Appleton, C. Erginsoy and W.M. Gibson, Phys. Rev. 161, 330 (1967).

[Ap70] B.R. Appleton and L.C. Feldman, in "Proc. Sussex Conf. on Atomic Collision Phenomena in Solids," Amsterdam, North Holland, p. 417 (1970).

[Ba71] J.H. Barrett, Phys. Rev. B3, 1527 (1971).

[Bø67] E. Bøgh, in "Proc. Int. Conf. on Solid State Physics Research with Accelerators." BNL publication No. BNL-50083, p. 76 (1967).

[Da73] J.A. Davies, in p. 391 of Ref. [Mo73].

[Da78] J.A. Davies, in p. 483 of Ref. [Th78].

[De74] G. Della Mea, A.V. Drigo, S. Lo Russo, P. Mazzoldi, S. Yamguchi, G.G. Bentini, A. De Salvo, and R. Rosa, Phys. Rev. B10 1836 (1974).

[Er65] C. Erginsoy, Phys. Rev. Lett. 15, 360 (1965).

[Fe80] L.C. Feldman, P.J. Silverman, and I. Stensgaard, Nucl. Instr. Meth. 168, 589 (1980).

[Fe81] L.C. Feldman, private communication.

[Ge74] D.S. Gemmell, Rev. Mod. Phys. 46, 129 (1974).

[Gi72] W.M. Gibson, Y. Hashimoto, R.J. Kelly, M. Maruyama, and G.M. Temmer, Phys. Rev. Lett. 29, 74 (1972).

[Gi75] W.M. Gibson, Ann. Rev. Nucl. Sci. 25, 465 (1975).

[Ja75] D.P. Jackson, Atomic Collisions in Solids, Vol. I, ed. by S. Datz, B.R. Appleton, and C.D. Moak, (Plenum Press, N.Y., 1975), p. 185.

[Ku75] M.A. Kumakhov, Rad. Effects 36, 43 (1975).

[Le63] C. Lehmann and G. Leibfried, J. Appl. Phys. 34, 2821 (1963).

[Le67] P. Lervig, J. Lindhard and V. Nielsen, Nucl. Phys. A96, 481 (1967).

[Li64a] J. Lindhard, Phys. Lett. 12, 126 (1964).

[Li65] J. Lindhard, K. Dan. Vidensk. Selsk. Mat.-Fys. Medd. 34, No. 14 (1965).

[Ma75] J.W. Mayer and E. Rimini (ed.), Ion Beam Handbook for Material Analysis, Academic Press, N.Y., 1977.

[Mo47] G. Moliere, Z. Naturforsch. A2, 133 (1947).

[Mo73] D.V. Morgan (ed.), "Channeling-Theory, Observation and Applications." J. Wiley and Sons, London (1973).

[Pi69] S.T. Picraux, J.A. Davies, L. Eriksson, N.G.E. Johansson, and J.W. Mayer, Phys. Rev. 180, 873 (1969).

[Pi74] S.T. Picraux and F.L. Vook, Phys. Rev. Lett. 33, 1216 (1974).

[Ri78] E. Rimini, in p. 455 of Ref. [Th78].

[Ry72] V.A. Ryabou, Phys. Status Solidi B49, 467 (1972).

[Sh73] R.P. Sharma, J.U. Andersen and K.O. Nielsen, Nucl. Phys. A 204, 371 (1973).

[St78] I. Stensgaard, L.C. Feldman and P.J. Silverman, Surface Sci. 77 513 (1978).

[Th78] J.P. Thomas.and A. Cachard (ed.), "Material Characterization Using Ion Beams." Plenum, N.Y. 1978.

[Tu76] W.C. Turkenburg, W. Soszka, F.W. Saris, H.H. Kersten and B.G. Colenbrander, Nucl. Inst. Meth. 132, 587 (1976).

[Vl71] D. Van Vliet, Rad. Effects 10, 137 (1971).

4. Characteristic Radiation of Channeled Relativistic Electrons

Channeling motion in single crystals applies equally to negatively charged particles since the transverse potential they encounter differs from that for the corresponding positive ones only by a change of sign. Taking e^- and e^+ as examples in a classical description of planar channeling, Fig. 4.1 illustrates this inverted potential for e^-. The spatial regions forbidden to e^+ are precisely the allowed ones for e^-, and conversely. Although bound states of transverse motion can be expected for the electron case, the e^- are steered toward the planes of lattice atoms in contrast to the positron case. As a consequence, close-impact collision yields and dechanneling rate due to scattering from lattice atoms and electrons are enhanced for electron channeling.

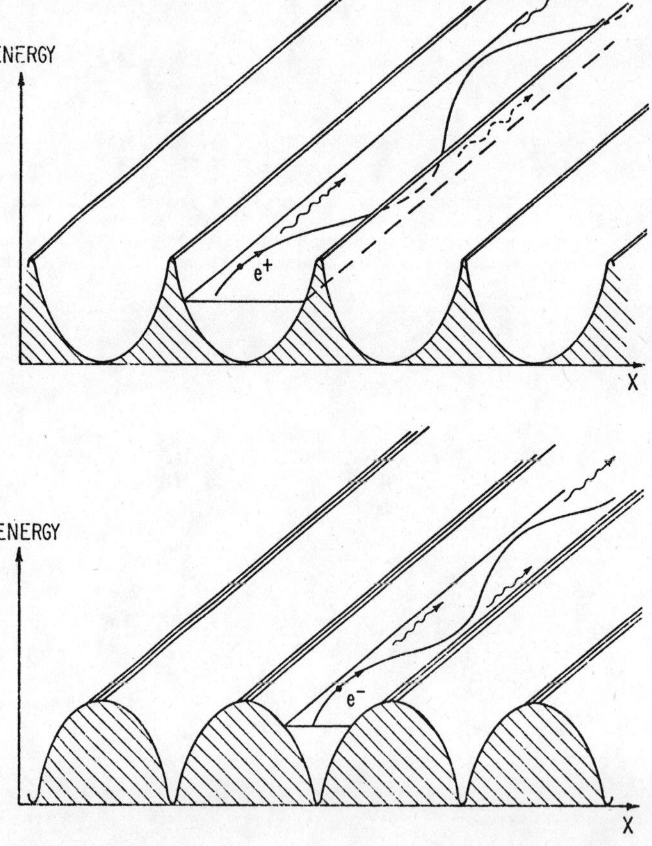

Fig. 4.1 - A schematic illustration of classical trajectories in the planar channeling of positrons(e^+) and electrons (e^-).

Based on classical considerations, oscillatory motions of the type shown in Fig. 4.1 should give rise to radiation because the charged particles undergo periodic accelerations. It is thus somewhat surprising that the interesting features of channeling radiation were called into attention only as recently as 1976 [Ku76, Te77] and their experimental confirmations followed a little later [Al79, Mi79, Ag79, Sw79, An80, Cu80]. The subject has been reviewed recently by Wedell [We80].

In Sec. 4.1, the general features of electron channeling radiation are described. Comparisons of experimental results with theoretical predictions then follow in Sec. 4.2 for the axial case and in Sec. 4.3 for the planar case. Comments in Sec. 4.4 on the prospects for applications serve to conclude the coverage here.

4.1 General Features

When relativistic electrons of total energy $E \gtrsim 1$ MeV are directed close to a high symmetry axis of a single crystal, they can be attracted into a spiral motion around the individual atomic rows. For such cases, a treatment of a single row in isolation (single string approximation) should be reasonable and Fig. 4.2 depicts this situation in the laboratory frame of reference (LAB frame). Because of the high speed ($v_z \simeq c$) along the row, the electron sees a line charge of density Z/d and the symmetry reduces the considerations to motion in a plane transverse to the row governed by a transverse potential $V(r) = -(Z/d) f(r)$. This is in fact the continuum approximation described in Sec. 3.1.1 but now applied to a negatively charged particle.

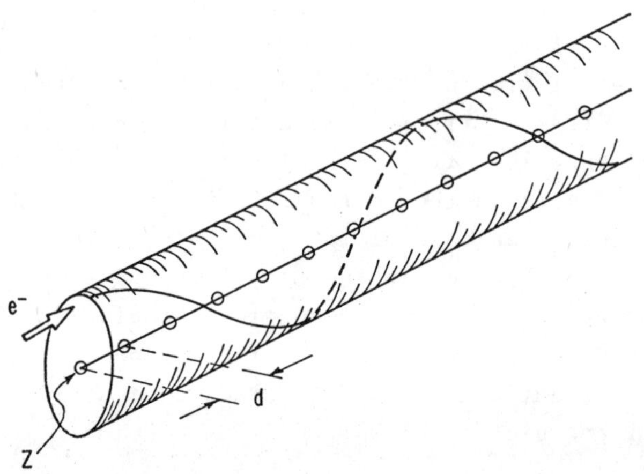

Fig. 4.2-Illustration of a classical spiral motion of a fast moving electron around a row of atoms.

The most general description of the motion should of course take into account quantum effects. Both the relativistic Dirac equation [Ku77] and the Klein-Gordon equation (appropriate when there is no spin interaction) [Le67] have been shown to lead, in this case, to a Schrödinger equation for transverse motion when terms of order $V/E \ll 1$ and smaller are neglected. The explicit form in the LAB frame can be written as

$$\text{LAB}: \ [\ \frac{P_t^2}{2\gamma m_o} + V(r)]\psi(r) = \varepsilon\psi(r), \qquad\qquad 4.1$$

where P_t is the transverse momentum of the particle with a rest mass of m_o, $\gamma = E/m_o c^2$ the usual relativistic factor, and ε the transverse energy. Looking at the problem from a frame of reference loosely called the 'Rest' frame in which $v_z = 0$, Eq. 4.1 becomes

$$\text{'Rest'}: \ [\frac{P_t^2}{2m_o} + \gamma V(r)]\psi(r) = \varepsilon'\psi(r) \qquad\qquad 4.2$$

since the transformation leaves the transverse momentum P_t and coordinate r unchanged, reduces γm_o to m_o, and increases the potential from V to γV because the electron now sees a contracted atomic spacing. In effect Eq. 4.2 can be obtained from Eq. 4.1 by a simple multiplication of a γ factor. Thus the corresponding solutions in the two frames are simply related: the eigenfunctions $\psi_n(r)$ are identical and the eigenvalues scale according to

$$\varepsilon_n' = \gamma\varepsilon_n. \qquad\qquad 4.3$$

Spontaneous transitions between eigenstates give rise to photons of discrete energy

$$\text{'Rest'}: \ \hbar\omega_R = \varepsilon_i' - \varepsilon_f'. \qquad\qquad 4.4$$

The advantages of viewing the radiation process from the 'Rest' frame now becomes clear. Firstly, the detailed radiation properties can be directly transcribed from the well studied atomic case and, secondly, recoil effects in the photon emission are more tractable. For $\hbar\omega_R \ll m_o c^2$ which is the case of interest here, recoil effects will not be important. Observables in the LAB frame can then be obtained by the use of the appropriate transformation formulas [Ha63]. In particular, the Doppler effects will confine the photons to the forward direction to essentially within a narrow cone angle of $1/\gamma$ rad, and amplify their energies at 0° to

$$\hbar\omega_L = \gamma(1+v/c)\hbar\omega_R \overset{\sim}{-} 2\gamma\hbar\omega_R. \qquad\qquad 4.5$$

The 0° intensity in the laboratory, according to the dipole approximation, is given by

$$\frac{dN}{d\Omega} = \frac{\alpha^4}{8\pi\gamma^2} \left(\frac{\hbar\omega_L}{e^2/a_o} \right) \frac{|<r>_{if}|^2}{a_o^3} L, \qquad\qquad 4.6$$

where $\alpha=1/137$ is the fine structure constant, a_o the Bohr radius, $<r>_{if}$ the dipole matrix element and L the population $P_i(\Theta,z)$ of the emitting state integrated over the entire length T of the target, i.e.,

$$L = \int_o^T dzP_i(\Theta,z). \qquad\qquad 4.7$$

For $z=0$, the initial population $P_i(\Theta,0)$ is determined by the overlap of incident plane wave with the emitting state wavefunction. This over-lap in turn depends on the tilt angle Θ of the row relative to the incident beam direction. The evaluation of L, in general, requires some knowledge of how P_i changes with z. It should be emphasized that one distinguishing characteristic of channeling radiation is that the line energies are independent of Θ, in contrast to the linear dependence expected for coherent bremmstrahlung [Ub56].

Within the simple continuum model, the descriptions above applies equally to electrons in planar channeling since a plane essentially consists of a superposition of rows, and the two-dimensional problem then reduces to a one-dimensional one. It also applies to positrons as long as the motion is confined to one open channel (cyclindrical or planar) and the transverse coordinate is measured relative to the potential minimum. These different cases differ only in the dimension-ality of the transverse space and the detailed shape of the transverse potential. The case of e^+ planar channeling is interesting because the approximately harmonic oscillator shape of the potential should lead to nearly equally-spaced eigenstates and thus to almost mono-chromatic radiation. This is indeed observed to be the case [Al79]. As will be seen later, the single row or plane approximation does provide reasonable descriptions of channeling radiation from the more tightly bound states.

From the above discussion, the study of channeling radiation is clearly seen to provide a sensitive technique for examining the details of the transverse potential. Moreover, the radiation constitutes a potentially versatile photon source for various applications. Versa-tile because the radiation is 1)highly directional (with a cone angle of $1/\gamma$ rad), 2) linearly polarized for planar channeling due to symmetry, 3) tunable by merely changing the beam energy since each line emission behaves as

$$\hbar\omega_L \propto (K\gamma)^b \gamma, \qquad\qquad 4.8$$

and 5)intense because the individual lines follow

$$\frac{d^2N}{d\Omega dz} \propto (K\gamma)^{3b}\gamma. \qquad\qquad 4.9$$

Here K is Z/d and $Z^{2/3}N_v d_p$, respectively, for the axial and planar case with N_v being the atomic volume density. The value of b depends on the potential shape; for example, b=0.5 for a harmonic oscillator and b=2 for a 1/r potential. The few measurements reported indicate that b≈0.5 for e^+ planar channeling [A179], b≈0.64-0.9 for e^- planar channeling [Pa79] and b≈1 for e^- axial channeling [Cu80].

The simple continuum description is attractive because it provides a convenient generalization for the various channeling geometries and crystal species. Comparisons of its predictions with experimental observations detailed in the following subsections will serve to point out the extend of their applicability, as well as their limitations.

4.2 Axial Case

Bound states in the axial channeling of electrons have previously been examined in the context of transmission [Ku72, Ta76] and back-scattering [An77] yields. Estimates there [An77] show a few number of bound states for γ≈3 and this number increases linearly with γ. Small accelerators thus seem tailored for investigating the details of channeling radiation since the limited number of bound states suggests well resolved lines.

At Albany, 2-4.5 MeV electrons from the Dynamitron accelerator were used to investigate the radiation along the <110> axis of a 3400-Å thick Si[Cu80]. The primary experimental considerations were directed toward obtaining the desired beam characteristics but minimizing as much as possible the background radiation falling on the Si(Li) semiconductor photon detector. Thus the beam was collimated by a pair of apertures and magnetically deflected towards the target without further collimation. All elements upstream of the deflection magnet were shielded from the target by a 45-cm thick concrete wall and the Si(Li) detector was placed inside a lead housing. The beam at the target position had a total divergence of $\lesssim 0.04°$ compared to the critical angle at γ=9 of $2\psi_{1/2}≈0.4°$, and a spot size of 3mm diameter compared to the 6mm diameter useful area of the Si crystal which was mounted on a two-axis goniometer.

A typical spectrum of photons along the incident beam direction for a Si<110> tilt angle of $\Theta=0.06°$ is shown in Fig. 4.3. The beam transmitted through the crystal was magnetically deflected so that the Si(Li) detector recorded only photon signals. Except for the Si-K line, all peaks can be identified with transitions between bound states of transverse motion. This resolved line feature should be contrasted to the broad hump feature observed at 28 and 56 MeV [Sw79]. Presumably the large number of bound states at the higher beam energies plus significant line broadening effects give rise to many overlapping lines which can not be resolved.

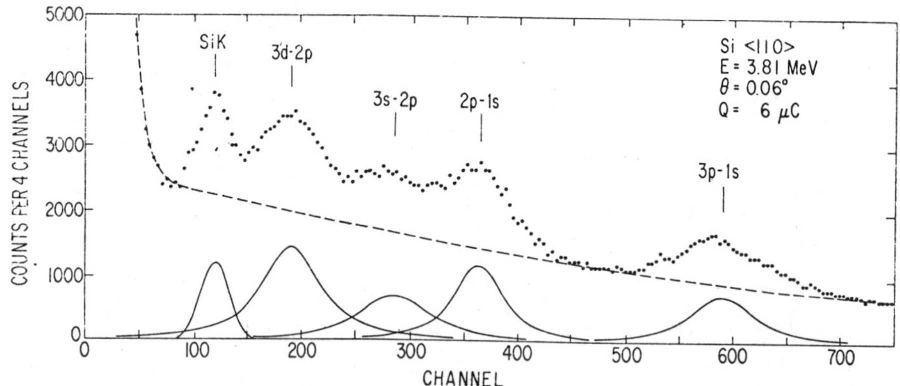

Fig. 4.3 - Typical radiation spectrum in the forward direction observed for 3.81 MeV electrons incident at 0.06° to a <110> axis of a 3400 Å Si crystal. The dashed curve is the assumed background while the solid curves are the result of peak fitting using a least-squares criterion.

The position, width and area for each line were extracted by a least squares procedure. A smooth background, with a shape obtained for off-axis cases, was scaled to match the high energy region. A gaussian was used for the Si-K peak and, for the others, a lorentzian convoluted with the instrumental response represented by a rectangular function with width equal to the FWHM of the Si-K peak. For a given beam energy, the peak positions do not vary with the crystal tilt angle θ while the peak areas diminish as θ exceeds the critical angle $\psi_{1/2}$, as expected for channeling radiation.

The observed line energies converted to the 'Rest' frame values via Eq. 4.5 are compared with predictions based on the single string approximation in Fig. 4.4b. Energies and wave functions for bound states were calculated in the 'Rest' frame using the static Moliere potential [Ap67]. The effect of thermal displacements of target atoms on the potential was taken into account by replacing the radial coordinate r by $(r^2 + \rho^2/2)^{1/2}$, where $\rho = 0.106\overset{o}{A}$ is the rms thermal amplitude in two dimensions. The calculated energy levels for 3.81 MeV electrons $(\gamma/d = 2.2 \overset{o}{A}{}^{-1})$ are shown in Fig. 4.4a with the arrows indicating the allowed dipole transitions.

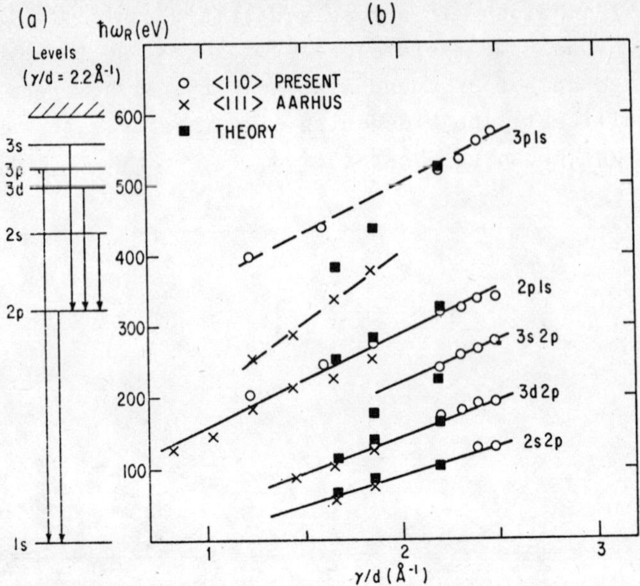

Fig. 4.4 - (a)Energy level diagram for transverse motion in the 'Rest' frame of 3.81 MeV electrons channeled along a Si<110> axis, and (b)the dipole transition energies connecting these states at two additional beam energies together with the corresponding experimental values showing systematic trends with γ/d. The lines drawn merely serve as guides to the eye.

According to Eq. 4.8, the variable $Z\gamma/d$ should serve to correlate the photon energies for the various beam energies, crystal axes and crystal species. The Si<111> case has been investigated by Andersen and Laegsgaard [An80] and their results are included in Fig. 4.4b where only γ/d is used since only one crystal species is involved. Both sets of data agree well with the predicted values and the γ/d scaling, except for the 3p-1s and 3s-2p transitions. In the 3p-1s case, a nearby transition, e.g., 4p-1s, may not be resolved in the measurement. In any event, the loosely bound 3p and 3s states extend into neighboring strings, one of which lies as near as 1.35Å in the Si<110> case. For such states the isolated string approximation may be inadequate.

Comparisons of the observed and predicted characteristic photon yields in the forward direction are shown in Fig. 4.5 for three prominent transitions at 3.81 MeV beam energy ($\gamma/d=2.2$ Å^{-1}). The Si-K yields are included for comparison. Displayed as a function of the crystal tilt angle Θ are the yields $d^2N/d\Omega dT$ per incoming electron per unit solid angle per unit crystal thickness. The relative

experimental uncertainties are typically ±10% but the absolute values
may be underestimated by a factor as large as 2 due to possible
detector misalignment and overestimates of background. The solid
curves are the predictions in which the beam angular divergence has
been folded in. These are based on the computed wavefunctions and Eq.
4.6 together with the assumption that the initial population of emitting state remains unaltered as the beam penetrates into the crystal,
i.e., $L=P_i(\Theta,0)T$. Calculated $P_i(\Theta,0)$ for the present cases are shown
in Fig. 4.6. The scaling factors required to bring the predictions
into agreement with data are indicated in Fig. 4.5. Agreement with
the tilt angle dependence of the data is good even for the 3p-1s case.
The magnitudes differ even if the experimental values are raised by a
factor of 2 corresponding to the underestimates mentioned earlier.

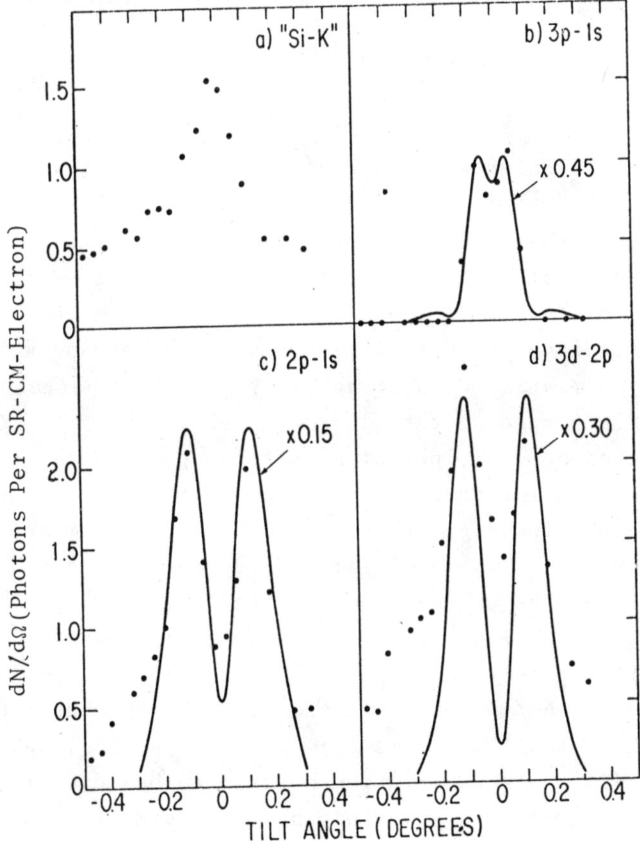

Fig. 4.5 - Tilt angle dependence of characteristic
photon yields observed in the laboratory along the
forward direction for 3.81 MeV electrons. The
theoretical predictions are shown as curves, scaled
down by the indicated factors.

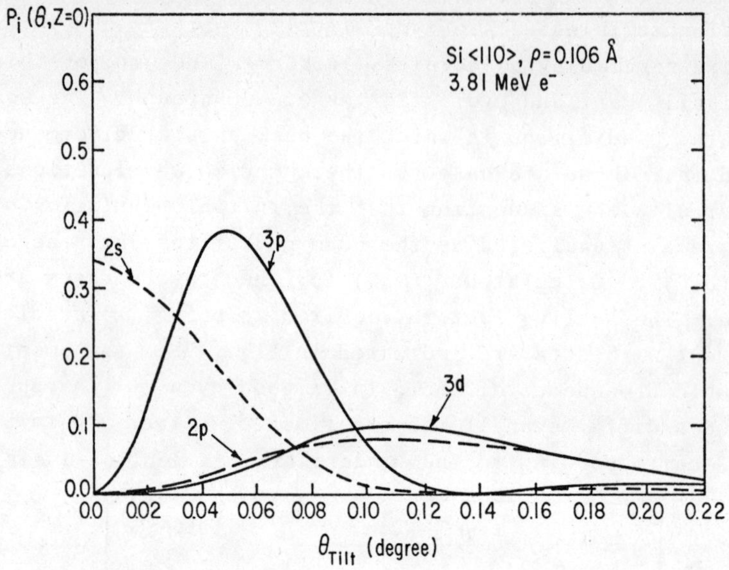

Fig. 4.6 - Calculated initial population of emitting
states as a function of crystal tilt angle.

The assumption that there is no redistribution of electrons among
states as they penetrate into the crystal is clearly unrealistic
since scattering by target electrons and lattice atoms must surely
occur. Such scattering effects can lead to a decrease as well as an
increase in the population of a given state. The good account of
the tilt angle dependence of the observed yields using predicted
initial populations however indicates that repopulation is not significant. Thus the smaller observed yields are attributable to depopulation effects which, for a given state, may be characterized by
$P_i(\Theta,z)=P_i(\Theta,0)\exp(-z/\ell)$ where ℓ is the depopulation length. Some
measure of these lengths can be found in the linewidths provided band
structure due to neighboring strings can be ignored, and this is
reasonable for the 2p-1s transition.

The length $\ell_{2p}=\gamma\hbar c/\Gamma_{2p}$ may be obtained from the measured 2p-1s
linewidth, which in the 'Rest' frame is $\Gamma_{1s}+\Gamma_{2p}=(47.2\pm5.3)$eV for $\gamma=8.5$,
if the ratio Γ_{1s}/Γ_{2p} is known. A rough estimate based on the probability of the bound electron overlapping the thermally displaced atoms
yields $\Gamma_{1s}/\Gamma_{2p}\approx8$, leading to $\ell_{2p}\approx3200$ Å. This is roughly 3 times
larger than that required to explain the observed 2p-1s magnitude.

Conclusions as to absolute yields are uncertain, but the isolated
string approximation is seen to provide a good quantitative account

of the observed line energies for tightly bound states as well as the tilt angle dependence of the corresponding intensities.

4.3 Planar Case

Planar potentials are generally weaker than the corresponding axial ones and thus, for a given beam energy, sustain correspondingly fewer number of bound states. With 1-4 MeV electrons available from small accelerators, characteristic planar channeling radiation even when not excluded by an insufficient number of bound states would generally involve emission from states which are susceptible to band structure effects.

Recently Andersen, Eriksen and Laegsgaard [An81] have undertook a comprehensive study along the principal planes of Si with 1.5-4 MeV electrons and crystal thicknesses of 0.3, 0.5 and 1μm. Their data show a number of new interesting features. Three types of photon lines were observed, with line energies independnet, quadratically dependent and linearly dependent upon the crystal tilt angle Θ. Taking into account explicitly the periodicity of the transverse potential, they have performed the so called N-beam calculations [An77]. The solutions in terms of transverse Bloch waves show energy eigenvalues ε periodic in the wavenumber space (k_x space) with period equal to the basic reciprocal-lattice vector perpendicular to the plane ($g=2\pi/d_p$), where the k_x relates to the incident momentum p as $k_x=\Theta(p/\hbar)$ and g to the Bragg angle Θ_B as $g=2\Theta_B(p/\hbar)$. The ε are also symmetric under inversion of k_x. Thus bound states (B) are characterized by $\varepsilon_B\tilde{}$ constant, and free states (F) by sections of parabola $\varepsilon\tilde{}\hbar^2k_x^2/2m_o+V_o$. The three types of photons were thus identified as bound-to-bound (BB), free-to-bound (FB) and free-to-free (FF) transitions since the corresponding transition energies are constant, proportional to k_x^2 and k_x, respectively, with respect to variation in $\Theta=\hbar k_x/p$.

Their explicit calculations for 4 MeV electrons in the Si (110) planes, which show only two bound states, reproduced well not only the observed photon energies corresponding to the various transitions but also the intensities as a function of Θ, based on the dipole approximation and computed initial populations of states. No scaling down factor was required to match the measured absolute intensities except for the sole BB case in which a value of 0.6 was needed. Presumably the redistribution of electrons among states due to scattering is more sensitively reflected in the bound state populations.

In the BB case the emitting state is also predicted to be loosely bound and with its energy decreasing by 9% in going from $\Theta=0$ to $\Theta=\Theta_B$. Since the critical angle for channeling is $\psi_{1/2}\tilde{}2\Theta_B$ here, the photon energy should oscillate with Θ of period Θ_B and $\pm 4.5\%$ excusion. This behavior was indeed confirmed by a careful examination of the experimental line energy. Their measurements also show the linewidth to broaden significantly when the crystal thickness is reduced from 1μ to $0.3\mu m$ indicating the increasing contribution of the finite target thickness effect. In the language of the first order time-dependent perturbation theory, this effect is the frequency broadening caused by the finite time duration in which the potential is active. For photons observed at 0°, the contribution is $\Delta(\hbar\omega_L) \simeq 2\gamma^2 h/T$, where T is the crystal thickness.

The FF transitions were both predicted and observed to have intensities only at $\Theta \gtrsim \psi_{1/2}$ and falling off as $I \propto 1/\Theta$. Also since the laboratory transition energies vary as $\hbar\omega_L \propto \Theta$, their intensity fall off may be reexpressed as $I \propto 1/\omega_L$. Both proportionalities are recognized as the markings of coherent bremmstrahlung radiation [Ub56]. The identification would be completed by assigning the higher energy FF transitions of decreasing intensities to the harmonics of the fundamental mode. Coherent bremmstrahlung is usually treated as transitions between plane-wave states, induced by the crystal potential which is treated as a small perturbation. Andersen et al. also showed that the perturbation treatment is equivalent to treating the radiation as a consequence of spontaneous transitions between eigenstates (Bloch waves) of the multi-plane transverse Hamiltonian.

Studies with MeV electrons from the more accessible small accelerators are seen to provide a clear understanding of radiation at or near channeling conditions. Not only have three distinct types of radiation lines been identified and categorized, a single theory has also been shown to successfully account for the features of all three types.

Investigations of electron channeling radiation have also been undertaken at the higher energies of $\gtrsim 28$ MeV. Although accelerators capable of such energies are not normally classified as small, the findings in the planar case are nevertheless directly relevant to the discussion here. With increasing beam energies, the transverse potential in the 'Rest' frame becomes deeper and the number of bound states increases correspondingly. Indeed, in the axial case, this number becomes so large for 28 MeV and higher that the individual BB transitions are no longer resolved and only a broad hump feature is observed [Sw79]. On the other hand, resolved line features are still prominent in the planar case, and such lines have been successfully interpreted as $\Delta n=1$ dipole allowed transitions in the single-plane continuum approximation [Pa79, Cu81].   More recently, much less intense lines in the Si (110) [Pa80] and diamond (110) [Cu81] spectra have also been identified as $\Delta n=3$ BB transitions, while coherent bremmstrahlung radiation near the (110) planes of a 1-μm Si crystal was traced out with 54 meV electrons [Cu81].

In their analysis of the Si(110) data at 28 and 56 MeV, Pantell and Swent [Pa79] used an empirical potential with three independent parameters, the values of which were chosen to produce transitions matching those observed. They also deduced the 0° photon energies to scale according to Eq. 4.8 with b=0.64, 0.82, 0.90, respectively, for the $2\rightarrow1$, $3\rightarrow2$, $4\rightarrow3$ transitions in which n=1 refers to the ground state. Measurements at Saclay [Cu81] of the same transitions at a higher energy of 80.3 MeV show the line energies to indeed follow these scalings. Actually a better way to correlate the various data is a plot analogous to Fig. 4.4b in which the observed line energies are first converted to their 'Rest' frame equivalents via Eq. 4.5, and then display these as a function of $K\gamma$. In such a plot, the use of a uniform value of b=1 for all observed transitions in Si (110) over the 28-80 MeV range is seen to provide an acceptable description of the trends (to within ±4%). Moreover the same transitions for diamond at 54 MeV [Cu81] can also be included in the plot. All observed lines from diamond indeed fall into the trends corresponding b=1, except for the $2\rightarrow1$ transition which lies roughly 14% higher. This exception is presumably due to the relatively tighter binding of the ground state because the cusp shape of the potential minimum is not filled in as much since the thermal vibration amplitude in diamond is only about half that in Si.

The effects of redistribution of electrons among states due to scattering appear to be relatively more significant at the higher energies. Just as in the MeV studies, most measurements have been made with Si, and for comparable beam angular divergence-to-$\psi_{1/2}$ ratios and 'Rest'-frame crystal lengths ($\sim$1μm for $\gamma\tilde{=}$10 and $\sim$10μm for $\gamma\approx$100). Yet crystal tilt angle dependence of line intensities which reflects the initial population of emitting state, e.g., a dip at Θ=0, has not been observed, in contrast to data obtained with MeV electrons. Additionally the peak-to-valley ratios of the observed line intensities fall far short of the expected increases based on Eq. 4.9 and γ^2 scaling for the background normal bremmstrahlung radiation. Indeed the distinctive peaks observed at 54 MeV seem to have dissolved into a broad hump when the beam energy is raised to 110 MeV [Cu81]. Electron scattering processes should increase with increasing Z of the lattice atoms. Thus no line was seen with a Ge crystal [Pa81] while much sharper and more intense lines were observed with a diamond crystal [Cu81]. With the many investigations still in progress, a quantitative description of scattering effects on the channeling radiation spectra may soon emerge.

4.4 Prospects for Applications

Among the suggested applications of channeling radiation, the most widely quoted has been its potential as a general versatile source of x- and γ-rays. Such a prospect must of course be assessed in the light of its actual features and be weighted against already existing sources such the rotating anode x-ray generator and synchrotron radiation. Thus the attractive monochromatic radiation from e^+ planar channeling would not be competitive because of the great difficulties associated with the production of intense e^+ beam. With electrons, the limitation is shifted to the consideration of radiation damage induced in the crystal. Although the question of crystal damage has not been addressed in details, Pantell et al. [Pa80] had compared the intensities expected for 123-keV photons (10% bandwidth) from the Stanford synchrotron source and the channeling of 100μA 54-MeV electrons in the (110) planes of a 18μm Si crystal ($2{\to}1$ transition), and found the latter to be more intense by ~6. The use of a diamond crystal under the same conditions would raise the factor by an order of magnitude. Higher photon energy from channeling may be obtained by simply raising the beam energy correspondingly. Although the resolved line features may eventually merged into a broad hump, the γ^4 scaling of the intensity makes channeling radiation attractive even when a monochromator is used. On the other hand, the γ^4 intensity scaling also implies that channeling radiation would not be competitive with the synchrotron source for the production of lower energy photons.

REFERENCES

[Ag79]A.O. Aganyants, Yu. A. Vartanov, G.A. Vartapetyan, M.A. Kumakhov, Chr. Trikalinos, V. Ya. Yarolov, Zh. Eksper. Teor. Fiz., Pisma 29, 554 (1979).

[A179]M.J. Alguard, R.L. Swent, R.H. Pantell,.B.L. Berman and S.D. Bloom, Phys. Rev. Lett. 42, 1148 (1979).

[Ap67]B.R. Appleton, C. Erginsoy and W.M. Gibson, Phys. Rev. 161, 330 (1967).

[An77]J.U. Andersen, S.K. Andersen and W.M. Augustyniak, Mat.-Fys. Medd. Dan. Vid. Selsk. 39, No. 10 (1977).

[An80]J.U. Andersen and E. Laegsgaard, Phys. Rev. Lett. 44, 1079 (1980).

[An81]J.U. Andersen, K.R. Eriksen, and E. Laegsgaard (to be published).

[Cu80]N. Cue, E. Bonderup, B.B. Marsh, H. Bakhru, R.E. Benenson, R. Haight, K. Inglis, and G.O. Williams, Phys. Lett. 80A 26(1980).

[Cu81]Lyon-Annecy-Saclay collaboration: N. Cue, M.J. Gaillard, R. Kirsch, J.C. Poizat, J. Remillieux, M. Gouanere, P. Sillou, M. Spighel, B.L. Berman, Ph. Cotillon, L. Roussel, and G. Temmer (to be published).

[Ha63]See, for example, R. Hagedorn, "Relativistic Kinematics", Benjamin, N.Y. (1963).

[Ku72]H. Kumm, F. Bell, R. Sizmann, and J. Kreiner, Rad. Effects 12, 53 (1972).

[Ku76]M.A. Kumakhov, Phys. Lett. 57A, 17 (1976).

[Ku77]M.A. Kumakhov, Sov. Phys. JETP 45, 781 (1977) [Translated].

[Le67]P. Lervig, J. Lindhard, and V. Nielsen, Nucl. Phys. A96, 481 (1967).

[Mi79]I.I. Miroshnishenko, J. Murray, R.O. Avakian, and T. Ch. Fieguth, Zh. Eksper. Teor. Fiz., Pisma 29, 786 (1979).

[Pa79]R.H. Pantell and R.L. Swent, Appl. Phys. Lett. 35, 910 (1979).

[Pa80]R.H. Pantell, R.L. Swent, S. Datz, M.J. Alguard, B.L. Barman and S.D. Bloom, in the "Proc. of 1980 Conf. on the Appl. of Small Acc. in Research and Industry, Denton, Texas," ed. by J.L. Duggan and I.L. Morgan, IEEE Trans. Nucl. Science (to be published).

[Pa81]R.H. Pantell, private communication.

[Sw79]R.L. Swent, R.H. Pantell, M.J. Alguard, B.L. Berman, S.D. Bloom and S. Datz, Phys. Rev. Lett. 43, 1723 (1979).

[Ta76]A. Tamura and T. Kawamura, Phys. Stat. Sol. (b) 73, 391 (1976).

[Te77]R.W. Terhune and R.H. Pantell, Appl. Phys. Lett. 30, 265 (1977).

[Ub56]H. Überall, Phys. Rev. 103, 1055 (1956).

[We80]R. Wedell, Phys. Stat. Sol. (b) 99, 11 (1980).

Chapter XII

NUCLEI FAR FROM STABILITY

Professor J.K.P. Lee
Foster Radiation Laboratory
McGill University
Montreal, Quebec, Canada

1. INTRODUCTION

The first international conference on the "Nuclides Far Off the Stability Line" was held in 1966 at Lysekil, Sweden[1]. At that time, study of the delayed proton emitter had just begun. Some on-line isotope separators (ISOL) had just started to operate and some were being planned. Many experimental techniques were being developed. The conference's main theme was "Why and How to Study Nuclei Far Off Stability". In the following years, many of these ideas discussed at the conference were put into practice. The two following international conferences[2,3] held at Leysin, Switzerland (1970) and Cargése, France (1976) summarized the progress at that time. Since then, much progress has been made[4-6]. The Interacting Boson Model[7] proved to be extremely successful in describing the collective states in nuclei over a large mass region. The level structure of the doubly major ^{132}Sn was studied. New radioactivity of delayed two-neutron emission was discovered. Mass-energy surface was extended further away from the stability. New generation of experiments making use of the intense beam of mass separated radioactive nuclides are being conducted and undoubtedly will have a major impact on our understanding of nuclear properties. In the next few years, several major heavy ion accelerators will become operational. Together with the new experimental facilities provided, it is safe to say that the field of "Nuclei Far Off Stability" will remain in the frontier of nuclear physics.

These notes are prepared as an introduction into the field. An experimenatlist's approach was adopted, and hopefully will compliment the other material to be covered in the Winter School. Three topics are treated with some detail - radioisotope production and transport methods, nuclear masses[8-10] and delayed particle emitters[11-12]. Results from γ-spectroscopy and in-beam γ-ray studies are complimentary to each other and have contributed substantially to our knowledge of nuclear structure. They can be treated as an extension from the nuclear properties near stability and are largely omitted with regret. Fortunately, the topic of nuclear structure will be covered extensively elsewhere during the Winter School. Another neglected topic is the heavy and superheavy elements. The experimental methods used are similar to those discussed here, but the topic is very specialized and is not easily linked to the others.

2. PRODUCTION METHODS

2.1 General Features

The reaction mechanisms leading to the production of various radioactive nuclei are generally complex. Although, in many cases, the mechanisms are not well understood, there exists sufficient data to guide the experimentalists to choose the appropriate ways for their particular studies.

For the neutron rich isotopes, nuclear fission remains to be one of the most important productive mechanisms. The other commonly used methods are fragmentation induced by GeV range protons and high energy heavy ions. One of the common features of these mechanisms is that the production is not selective: many nuclei over a wide element and mass range are produced simultaneously. Some kind of "sorting" method has to be used for the detailed studies of their properties.

The neutron deficient nuclei are readily produced when a target is bombarded by almost any accelerated projectile. Quite often, a particular radioactive nuclide can be selectively produced by proper choice of projectiles, bombarding energy and target nuclei. In these cases, "conventional" nuclear spectroscopic method can be used for the detailed studies of their properties. For producing more exotic nuclei lying far off the β stability valley, high energy projectiles must be used, resulting in the non-selective process in the production of many nuclides. In these cases, again some "sorting" will be necessary.

2.2 Nuclear Fission

The most commonly used reaction is the thermal neutron induced fission of ^{235}U. The yield of various mass chains has been extensively studied, exhibiting a double-hump feature centered around A=95 and 140. The cumulative and independent yields of many isotopes have also been determined. Fig. 2.1 shows the production cross-section yield of various isotopes from this reaction. Over 400 isotopes, extending from the stable ones to those over ten neutrons in excess, are produced in sufficient abundance such that detailed spectroscopic studies can be attempted. This non-selective production method necessitates a certain sorting process before detailed spectroscopic studies can be attempted. This is usually accomplished by chemical and/or mass separation process. The production cross-sections are high for many isotopes and reactors of moderate flux density can produce sufficient quantities for their detailed studies. Since the direct cost of production can be low, it remains the most popular method to study neutron rich nuclei in the medium mass region.

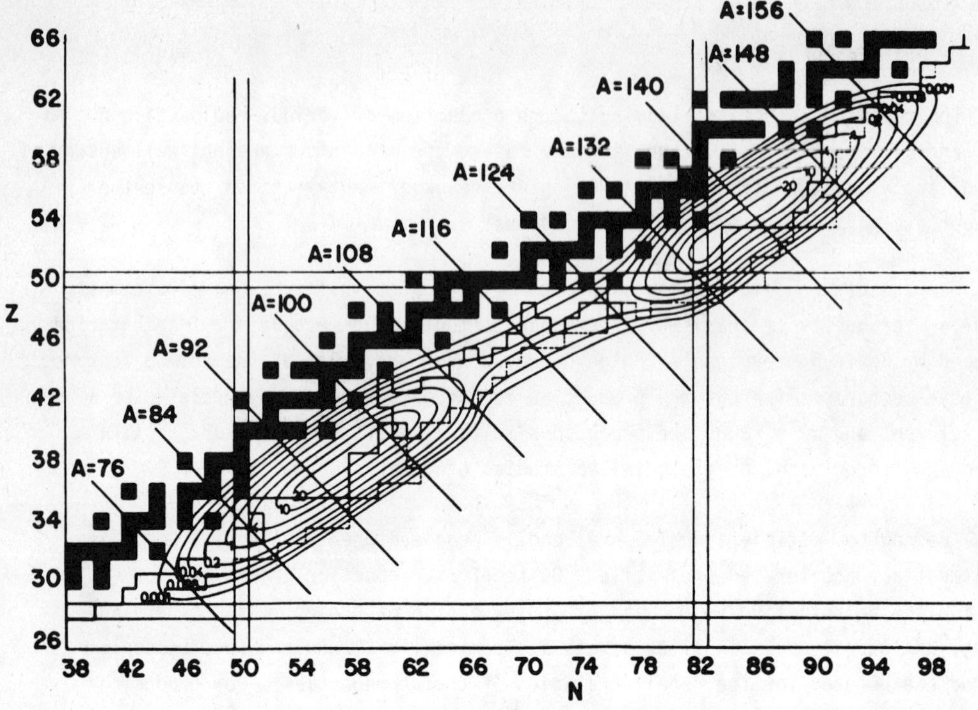

<u>Fig. 2.1</u> - Contour of Independent Cross-Sections (barns) for ^{235}U(n,f)

One of the drawbacks of the thermal neutron induced fission is its relatively low yield near the symmetric fission region. This can be remedied by the use of fission induced by low or medium energy projectiles. In these reactions, the yields of mass chains over a wide region are essentially constant. In addition, natural uranium or thorium targets can be used, with the added advantage that since the target is more neutron rich, the fission product in general has higher N/Z value, resulting in high proportion of neutron rich isotopes. Undoubtedly, accelerator-based production methods will be more costly to operate. However, the results of the ISOLDE operation at CERN indicates that the production rate can be much higher than any reactor based facility so far.

The other fission process is the spontaneous fission of heavy nuclei such

as ^{252}Cf. The mass yield curve is similar to that of ^{235}U(n,f) reaction, but generally it will not produce enough activities for detailed spectroscopic studies. However, it has the unique advantage that the fission source is portable, and the experiment can be set up in a very compact form such that the spectroscopy of the
prompt fission product can be studied.

2.3   Proton Induced Reactions

The reaction between the incident proton and the target nucleus is a complex
process[13]. Various models and approximations have been applied in an effort to
estimate the production rate under various conditions. Qualitatively, the reaction
can be considered to proceed in two stages: the pre-equilibrium stage where the
incident proton interacts with a few nucleons of the target nucleus, leading
eventually to the formation of the compound nucleus (CN); and followed by the de-
excitation of the compound nucleus. For low incident energies ($\lesssim 30$ MeV), the formation of compound nucleus (N,Z+1) with excitation energy of (E_p+B_p) is dominant;
here N and Z are respectively the neutron and atomic number of the target nucleus,
E_p the proton center of mass (c.m.) energy and B_p the proton binding energy. The
de-excitation of the CN is predominantly accomplished by neutron evaporation.
Therefore (p,xn) products are the principal products.

When the bombardment energy increases, the process during the pre-equilibrium
stage becomes more complicated and often leads to the emission of nucleons or
cluster of nucleons before the formation of CN. For example, when bombarded by 100
MeV protons, the heavy elements may emit an average of 2.5 nucleons (including 0.3
protons) in the pre-equilibrium stage. The effect will be more severe when targets
of lighter elements are used. The resultant CN will have higher excitation energies
and can spread over several elements. The competition among various de-excitation
modes also becomes more severe. The probability for the emission of a particle x
can be considered as being proportional to:

$$P_x \propto \exp\left[-\frac{B_x^*}{T}\right] \qquad\qquad 2.1$$

where B_x^* is the effective binding energy of particle x in the CN and T is the
nuclear temperature. For neutron emission, B_n^* is the neutron binding energy. For
charge particles (p or α) $B_p^* = B_p + B_{cp}$, where B_{cp} is the effective proton Coulomb
barrier height. The ratio of P_n to P_p is therefore:

$$\frac{P_n}{P_p} = \exp\left[-\frac{B_n-B_p-B_{cp}}{T}\right] \qquad\qquad 2.2$$

When (B_n-B_p) is small compared to the proton effective Coulomb barrier, $P_n/P_p >> 1$.

This usually is the situation for nuclei near the beta stability line and predominant de-excitation mode of CN is neutron emission. With higher excitation energies, successive neutron evaporation will lead to the region where (B_n-B_p) is of the order of B_{cp}. In this case, proton emission becomes competitive and may even be favoured. Therefore, when this point is reached, the neutron emission probability will be hindered and for more neutron-deficient isotopes, the production cross-sections will decrease very rapidly. In general, one can therefore visualize an effective (B_n-B_p) boundary in the nuclide chart. Beyond that boundary, the production via CN formation will be very low. Since (B_n-B_p) is $(Q_{ec} + 0.78$ MeV$)$, this boundary basic-ally follows the curve.

$$Q_{ec} + 0.78 \text{ MeV} > B_{cp} \qquad\qquad 2.3$$

Usually, the curve is reached about 5-10 neutrons from the stability line for medium weight masses and much higher for heavier masses. However, exceptional cases do occur. One such example is the case for the production of technicium isotopes. ^{93}Tc has N=50 and the (B_n-B_p) is 8.8 MeV, already greater than B_{cp}. The subsequent more neutron deficient isotopes all have (B_n-B_p) values comparable to or larger than B_{cp}. As a consequence, the ^{92}Mo(p,xn) reaction cross-sections decrease very rapidly - (p,3n) cross-section for ^{90}Tc is already low and the isotope ^{89}Tc produced via (p,4n) reaction is only identified recently.

Further increase in proton energy leads to very complex pre-equilibrium process, where increasingly more nucleons and clusters of nucleons are emitted: the spallation reaction. Large numbers of nuclides are produced, and the concept of CN plays little role, and with it the (B_n-B_p) effect discussed above. Usually,

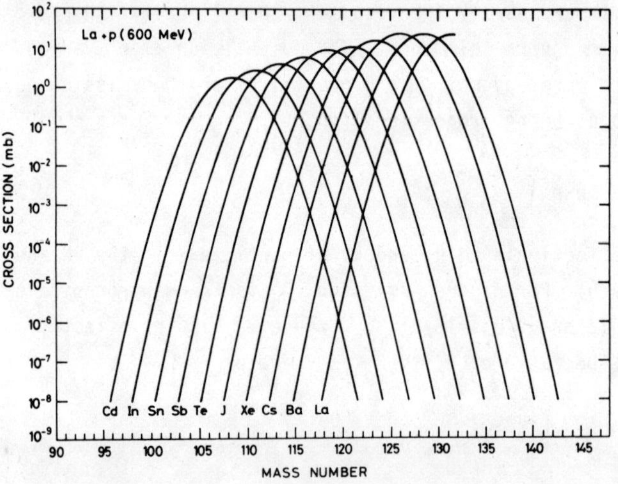

<u>Fig. 2.2</u> - Estimated Isotopic Distributions from Spallation of La Target by 600 MeV Protons (From Reference 3)

empirical methods were used to estimate the production cross-section of various isotopes. Fig. 2.2 shows an estimated isotope yield distribution for 600 MeV protons on La target, and clearly demonstrates the non-selective nature of this process. Although, the production yield is spread over many nuclides, with efficient sorting procedure, it will be an effective way to produce very neutron deficient isotopes.

The cluster of nucleons emitted from such reaction can have a wide range of N and Z combinations. That is, they can form many varieties of nuclides. Fig. 2.3 shows the identification of various nuclides from 800 MeV proton bombardment of a thin uranium target. With further increase in bombarding energy, these fragmentation products tend to spread over even wider nuclides. Recent works using the 24 GeV proton beam led to the study of many neutron rich sodium and potassium isotopes.

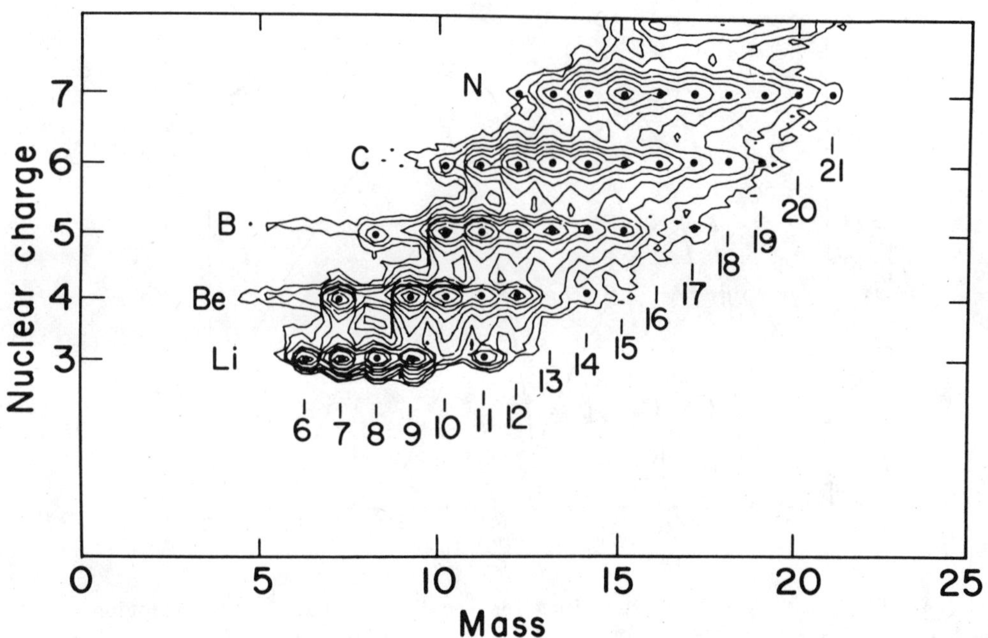

Fig. 2.3 - Contour of Nuclear Charge vs Mass. Solid points indicate the positions of peaks and contour levels are logrithmically spaced (From Reference 3).

2.4 Heavy Ion Induced Reactions

When the incident projectiles are complex light or heavy ions, the reaction process with the target nuclei can still be percieved to proceed along the two stages discussed above. At a bombarding energy below the effective Coulomb barrier, the predominant reaction processes are Coulomb excitation and scattering. With increasing incident energy, the complete fusion of the colliding nuclei at high excitation energy is most likely. The CN thus formed will then de-excite as discussed in the previous section. Therefore, with incident energies not too much higher than the Coulomb barrier, the heavy ion induced reaction is very selective: the fusion of incident at target nuclei is dominant, the excitation energy is well defined and often only a few isotopes are produced in appreciable amounts. This situtation is demonstrated in Fig. 2.4 where typical yields were estimated for ^{181}Ta(^{16}O,xn)$^{197-x}$Tℓ reaction at different bombarding energies. Since there are many possible combinations of projectiles, incident energies and target nuclei, it is quite possible to choose the right one to maximize the yield of a particular isotope of interest. It should be recalled that this selectivity is still limited by the competition among the evaporation of various particles: when $(B_n-B_p-B_{cp})$ or $(B_n-B_\alpha-B_{c\alpha})$ becomes zero, the production of more neutron deficient isotopes will decrease rapidly.

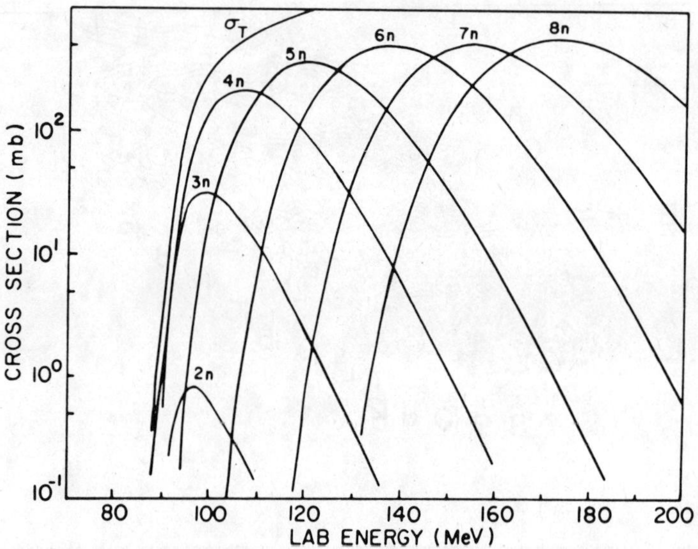

<u>Fig. 2.4</u> - Estimated Excitation functions for the ^{181}Ta(^{16}O,xn)$^{197-x}$Tℓ Reaction. (From Reference 13).

The use of heavy ions for isotope production has some noticeable advantages over proton bombardment, particularly in the medium and heavy mass region. In this region, the target has N>Z while the projectiles can have N≈Z. The CN formed is usually a few neutrons more deficient than those that can be obtained by proton bombardment. Therefore, for the production of a particular isotope, the excitation energy required is smaller. This usually means higher selectivity in the production yields. Besides this higher selectivity in isotope production, heavy ion bombardment has the unique feature for forming CN of high angular momentum. The centrifugal barrier retards the neutron emission process, and often these CN de-excite predominantly via γ-rays, providing a unique chance to study the high-spin states in these nuclei.

Besides the fusion process, there exists another important reaction unique to the collision between complex nuclei, the deep inelastic transfer (DIT) reaction[14]. It has the peculiar feature of exhibiting characteristics of both the direct interaction (DI) and CN formation. It is found that the incident projectile can transfer many nucleons with the target nuclei. The emerging particles are forward peaked, and these distributions show a maximum in the vicinity of the incident particle - a DI characteristic. On the other hand, the kinetic energy of these DIT products is close to the exit Coulomb barrier and is largely independent of the bombarding energy - a CN characteristic. Angular distributions of various isotopes were studied and some systematics can be obtained. Fig. 2.5 shows the differential cross-sections at 40° for various isotopes shown in a Q_{gg} plot here:

$$Q_{gg} = (M_1+M_2) - (M_3+M_4) \qquad\qquad 2.4$$

where M_1 and M_2 are masses in the incident channel, M_3 and M_4 are masses in exit channel. Q_{gg} therefore is the Q value for such reaction. The cross-section is not very small and can be used to produce nuclei far off stability. For example, if a neutron rich projectile such as ^{48}Ca is used, very neutron rich neighbouring nuclei may be produced. Also, DIT involving a large number of nucleons may offer an opportunity to produce superheavy elements.

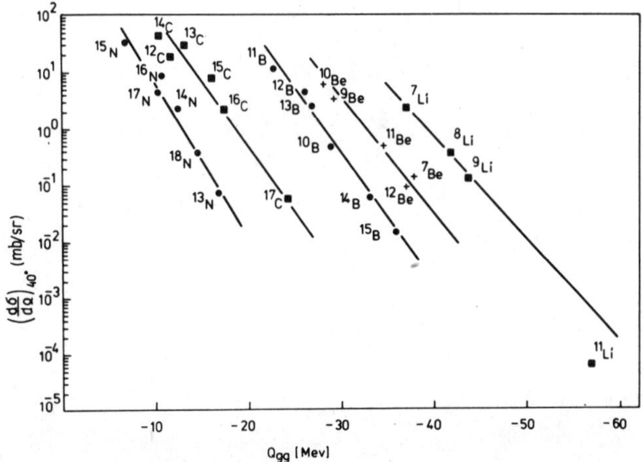

Fig. 2.5 - Q_{gg} Systematics for the Reaction ^{16}O+^{232}Th at a Bombarding Energy of 137 MeV (from Reference 14).

With further increases in heavy ion energies the selectivity aspect will gradually deteriorate. At relativistic bombarding energies (~hundred of MeV per nucleon), many hadrons are emitted and situations will be similar to those encountered with high energy protons. Some form of spallation and fragmentation process will become important.

3.   EXPERIMENTAL METHODS

3.1   Introduction

Many different experimental techniques have been developed for the study of various aspects of nuclei far from beta stability. In this section, we shall not be concerned with specific measuring apparatus but rather with general methods to "prepare" the desired source for various studies. The particular choice of method is affected by many factors such as the degree of contamination of the source, the decay properties of the nucleus of interest, etc. In general, the methods can be divided into two categories: in some, "sorting" process is implemented to enhance the relative abundance of the desired nuclei before actual measurements are to be performed, and in the other, the source has the basic composition as originally produced. In the following sections, some of the general methods are described.

3.2   Direct In-Beam Measurements

In this method, the detector system is monitoring particles or γ-rays directly emitted from the target during the time of bombardment. Under these circumstances, the background level is usually high and often, very sophisticated shielding and particle discrimination methods are necessary. For the detection of exotic heavy particles, often some kind of electromagnetic deflecting system is used to reduce the general counting rate of the detectors. The identification of these particles is necessary and often their energies are also measured. Very often, these direct in-beam measurements are used first to identify the existence of certain isotopes. Fig. 2.4 is an example showing the isotope yields from complex reactions.

The other type of in-beam experiment involves the detection of prompt γ-rays. In these experiments, it is very desirable that these γ-rays originate from few nuclei. Heavy ion induced reaction is very much preferred since it is often possible to choose the appropriate projectile, bombarding energies and targets such that only one particular final nucleus is dominantly populated. In this situation, the prompt γ-rays from the target are predominantly originated from that nucleus. Much information about the nuclear structure has been obtained through these experiments.

3.3   Pulsed Beam Technique

In the accelerated beam induced reactions, the activity of interest often decays with life-times of nanoseconds or longer. In these cases, it will be desirable to study the activities under beam-off conditions where the background level will be much reduced. Repetitive pulsed beam technique is very useful and often is the only viable method for such studies. Various methods have been developed ranging from mechanical beam chopper, electromagnetic deflectors to internal accelerator pulsing and beam burst of sub-nanosecond durations are now available for most types of accelerators. This method is particularly effective in the studies of delayed protons and alphas. Efficient particle discrimination methods are available to suppress the positron and γ-ray background. In fact, most of the early delayed proton and alpha emitters in the light mass regions were studied in this manner.

3.4   Rabbit System

In this system, targets are mounted on a carrier, often called the rabbit. It can be transported rapidly between a low background counting area and the irradiation position, usually via some pneumatic transfer system. The system is designed to deliver the target to a counting area with the shortest possible delay. Typically, the transport time is < 0.2 sec over a distance of a few meters or a few seconds over say 100 m. They are usually designed to suit particular facility layout and type of experiment being performed.

Rabbit systems have been in use for a long time. Its main drawback is the speed of transportation and the necessity to use target material in certain forms. However, they are still useful in particular applications. Over the years, many sophisticated additions have been made such as automatic target change, repeated operations, delivery of target to and from vacuum systems, etc. In some cases, the rabbit becomes an integral part of the facility such as automated fast chemical separation system.

3.5   Gas-Jet Recoil Transport System

Although the rabbit can transport the irradiated source to a low background area, it has some fundamental drawbacks. It is basically very cumbersome to operate, particularly if repetitive bombardment is necessary, the transport time has a practical lower limit and the method is not suitable for charged-particle studies. The gas-jet recoil transport system is versatile, and is capable of transporting rapidly most elements with high efficiency over a long distance. The basic idea of this method is quite straightforward and is shown in Fig. 3.1. The reaction products recoil out of the target material, and are thermalized in the surrounding gas. These reaction products have a tendency to stick to large molecules. When the gas is pumped out of the chamber, these large molecules are swept along and emerge from the capillary and can be collected on drums or tapes. Many different types of large molecules have been used as the carrier with good (>80%) transport efficiency. The thermalized gas is usually helium (hence the term helium-jet) but other types of gas are also common. The transport time is a function of transport length, tube size and throughput of pumping speed,   typically only a fraction of a second, and isotopes with half-lives of milliseconds have been studied using this method. A

comprehensive collection of articles on the gas transport system can be found in
Ref. 15.

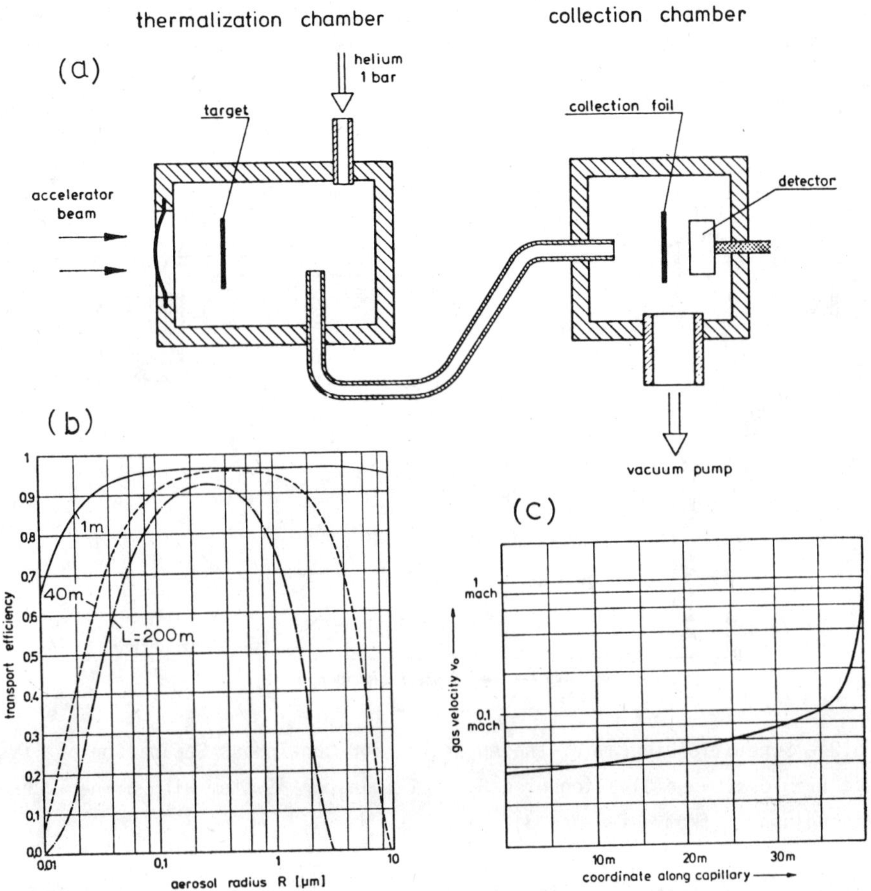

Fig. 3.1 - General Characteristics of a Helium-Jet Transport System (from Reference
15). (a) Principle of a He-jet transport system. An accelerator beam hits a target.
The thermalized reaction products are swept to a collection chamber by a gas stream.
In the collection chamber they are collected on some foil so that their decay
properties can be investigated; (b) The transmission probability for an aerosol of
of radius R through capillaries of 4 mm diameter and of different lengths; (c) The
velocity of the sweep gas along capillary of a gas jet.

3.6   Rapid Chemical Separation

The methods discussed so far are basically non-sorting processes and the
activities entering the detecting system have essentially the same constituents as
that from the target. For non-selective production reactions such as fission, the

contamination would be too severe to make detailed studies. Chemical separations
are often used to select particular element(s). For isotopes far off the stability,
the problem is the speed of separation and the continuous or repetitive supply of
these sources. Much progress has been made in the past few years. Fig. 3.2 shows
an example of separation of bromides by thermochromatographic column[3]. A gas jet
system delivers the reaction product to a reaction chamber where bromination takes
place. The volatile bromides are then swept through the thermochromatographic

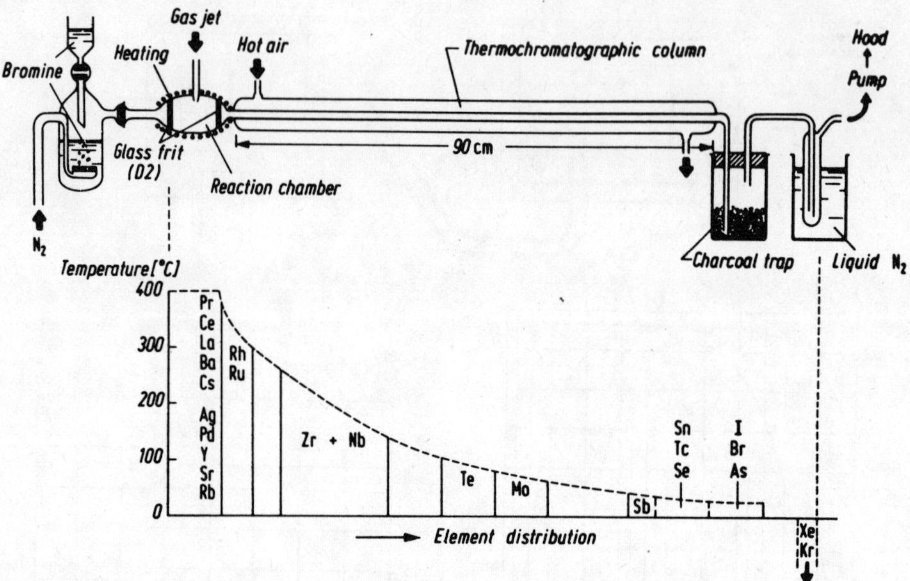

Fig. 3.2 - Schematic Diagram of the Apparatus for Continuous Separation of Fission
Product Bromides by Combination of a Gas-Jet Transport System with a Thermochromatographic Column. (from Reference 3).

column where a temperature gradient is maintained. Different bromides then will
deposit at different sections along the column. This system is simple to
operate and the gas jet will provide a continuous supply of activities. The drawback certainly is the buildup of long-lived isotopes. Another example is shown in
Fig. 3.3, where a rabbit is used to deliver the reaction products to the apparatus
where fully automated chemical processes are carried out. The separated element(s)
is again delivered by a rabbit to the counting area. The complete process takes
slightly over 2 sec.

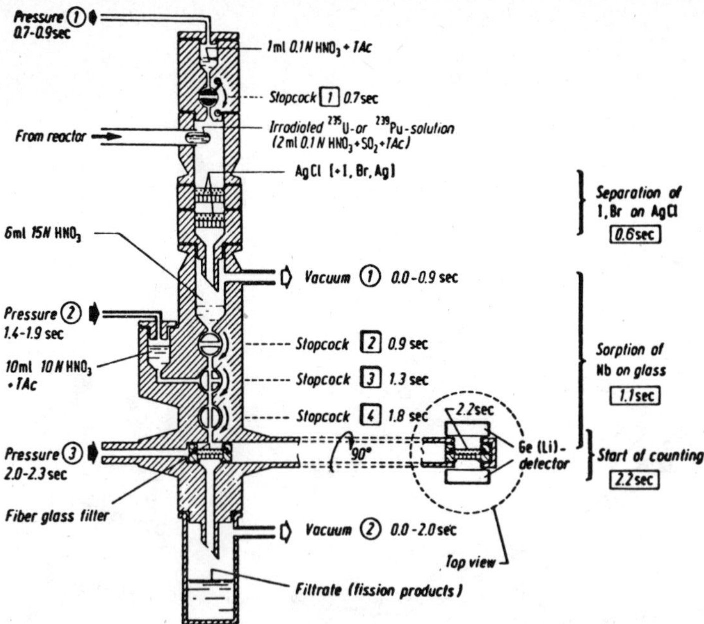

Fig. 3.3 - Apparatus and Time Schedule for the Rapid Separation of Niobium from Fission Products. At the right-hand side the mean-time for each separation step is indicated. (from Reference 3).

3.7   Electromagnetic Separation of Fast Reaction Products

With the exception of the direct on-line measurement as described in Section 3.2, all the other methods involve the transportation of the activity to a low background area. These methods are inherently slow due to the limited speed of the carrier (rabbit or gas). It is therefore very tempting to make use of the fast recoil velocity of the reaction products for the transportation. If this can be done, then the time lapsed before actual counting can be performed is minimized.

The characteristic of these prompt products may differ for different reactions. Generally, they can spread over a wide mass region, with large energy and velocity dispersions. Their ionic charges, when leaving the target, also spread over a few charge units. Therefore, the electromagnetic system used to transport these fast products are designed according to their specific purpose.

3.7.1   The Parabola Mass Spectrograph

When a particle of mass M and charge q moves with velocity v in an electric field E and magnetic field B, the corresponding radius of curvature

(ρ_e and ρ_m, respectively) are given by:

$$\rho_e = \frac{M_v^2}{qE}$$

3.1

and

$$\rho_m = \frac{M_v}{Bq}$$

3.2

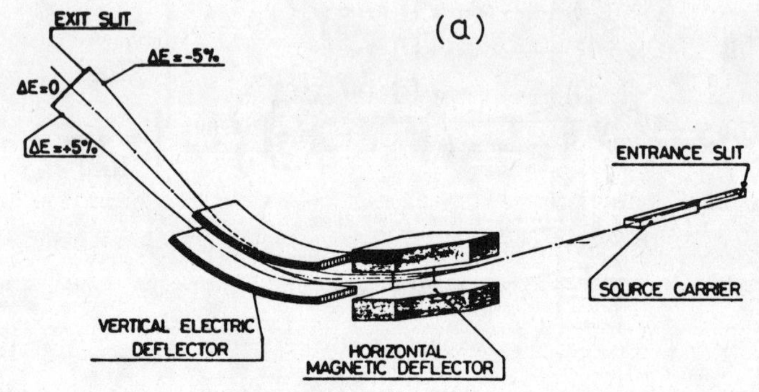

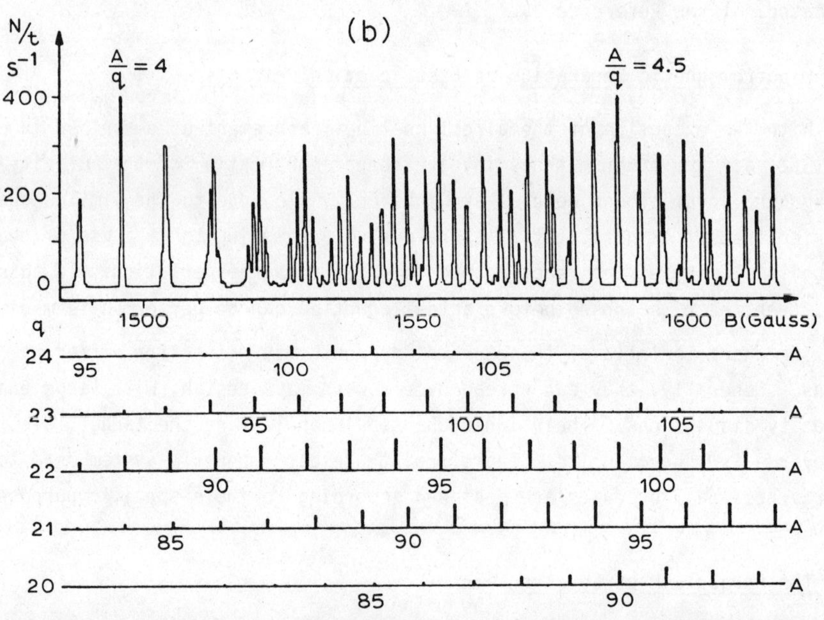

Fig. 3.4 - (a) Perspective View of the Arrangement of Fields at Lohengrin;
(b) Typical A/q Spectrum with Constant Electric Field (from Reference 15).

Consider a system with parallel and successive magnetic and electrostatic fields. The magnetic deflection spread out the entrance particles according to their momentum per charge magnitude, while for the electric deflector, spread them out according to their kinetic energy per charge. Also, since E and B are parallel, these deflections take place in perpendicular planes. For particles of definite M/q value, the locus for different kinetic energy will then follow a parabolic path on a plane perpendicular to the medium path, thereby the name parabola mass spectrograph.

Figure 3.4 shows one such device[15] (LOHENGRIN) installed at the high flux reactor of ILL, Grenoble, France. It has a total beam path length of over 20 m and accepts an energy spread of 10% with a 60 cm parabolic arc length. With a 0.4 mg/cm^2 thick ^{235}U target, a total yield of 10^4 atoms/sec for products of 2% independent yield is obtained. This production rate is adequate for some spectroscopic studies. However, the extended spatial distribution of the product means some activity transport system such as tape or gas transport methods will be necessary, resulting in additional delays.

It should be noted that the apparatus was primarily designed as a direct on-line spectrograph to study the mass, isotopic, ionic charge, and kinetic energy distributions of the prompt fission products. Much work along that line has been performed as well as those associated with nuclear spectroscopy.

3.7.2 Gas-Filled Spectrometer

A heavy ion with atomic mass A and atomic number Z, moving in a gas media at velocity v, will have an equilibrium ionic charge distribution and the most probable ionic charge $\bar{q}$ is approximately given by[15]

$$\bar{q} = \frac{v}{v_0} f(Z) \qquad\qquad 3.3$$

where v_0 is the electron velocity in a Bohr orbit and $f(Z)$ is a certain function dependent on Z. Consider a magnetic field B over this media, then the radius of curvature ρ of the ion is:

$$\rho = \frac{Mv}{Bq} = \frac{v_0 A}{B\, f(Z)} \qquad\qquad 3.4$$

It is therefore possible to separate the particles according to their A and Z values. The situation is particularly advantageous for the case of fission fragments. Here, the kinematic energies of the fragments occur with an energy range where $f(Z)$ is particularly simple, namely:

$$f(Z) = Z^\alpha \qquad\qquad 3.5$$

Here, α is a constant and its value is dependent on the type of gas and its pressure in the media. Therefore, for a particular element, the Bρ value is linearly dependent on A.

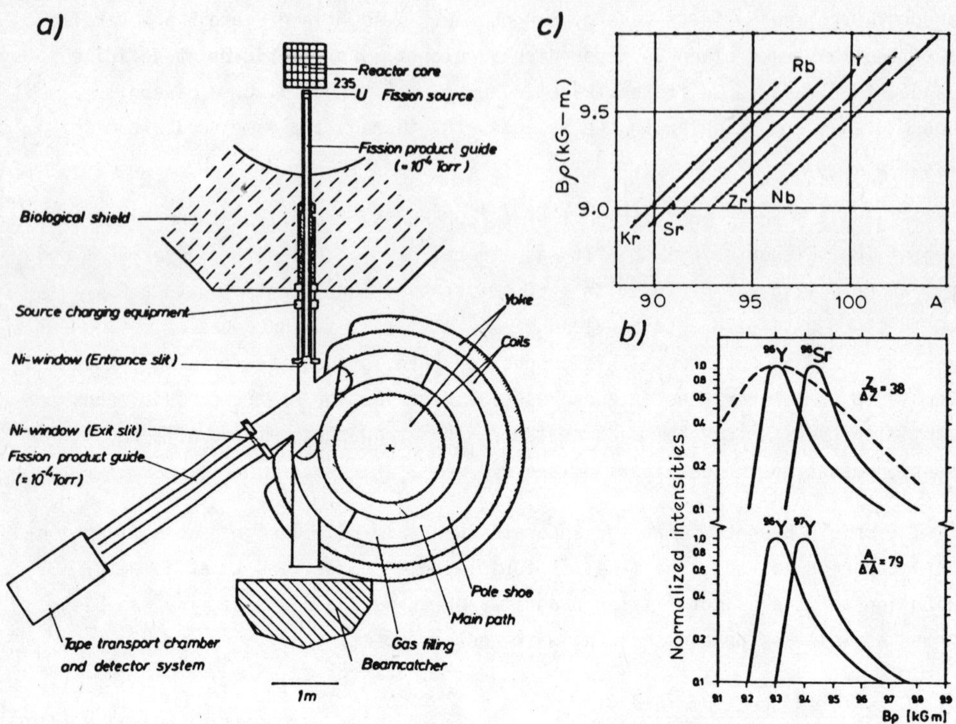

<u>Fig. 3.5</u> - JOSEF. (a) Schematic Presentation; (b) Intensity Distributions of Selected Isotopes as a Function of B_ρ Demonstrating the Mass and Atomic Number Resolution; (c) B_ρ-Values of Primary Fission Products as a Function of Mass for He of 4 Torr as a Filling (from Ref. 15).

Fig. 3.5a shows the schematic diagram of the gas-filled separator JOSEF located at the Julich reactor[15) in Germany. It has a mass resolution of about 79 and is capable of providing >10^6 atoms/sec for a fission yield of 2%. Also shown in Fig. 3.5b is the mass and charge resolution obtained. The selection of the type of gas and its pressure will affect the type and degree of overlapping isotopes. As can be seen, contamination of neighbouring isotopes are tolerated.

One advantage of this spectrometer is the higher yield of separated istopes. This is due to the fact that the present selection includes a wide range of energy and ionic charge state of the primary fragment. Also, the focal spot is smaller (~12 cm diameter), allowing more efficient use of the selected activity.

Like the Lohegrin, this spectrometer has also been extensively used for the fission studies.

3.7.3 <u>Velocity Filter</u>

From equations 3.1 and 3.2, one obtains:

$$v = \frac{B\rho_m}{E\rho_e} \qquad\qquad 3.5$$

which means that for a combination of electric and magnetic deflections, particles of certain velocity will follow specific trajectory and can therefore be separated according to their velocity. This kind of velocity filter is particularly useful in heavy ion induced reactions. In these experiments, the primary beam and the recoil nuclei all emerge from the target in a narrow cone. For the fusion reaction, which is the dominant process for much of the non-relativistic heavy ion reactions, the recoil velocity of nuclei is:

$$v = \frac{v_1 A_1}{A_1 + A_2} \qquad\qquad 3.6$$

where A_1 and A_2 are mass numbers for the projectile and target nuclei, respectively, and v_1 is the incident projectile velocity. A velocity filter will therefore separate the recoil nuclei from the primary beam.

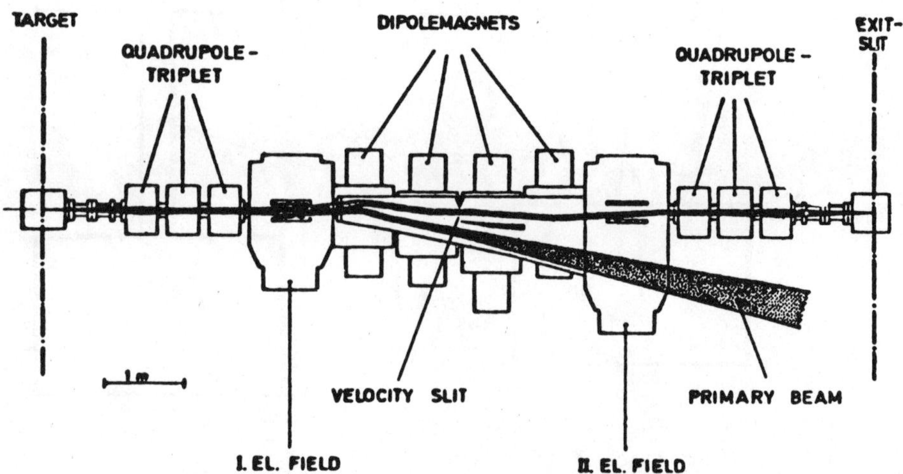

<u>Fig. 3.6</u> - Schematic View of the Velocity Filter Arrangement. (from Reference 15).

Fig. 3.6 shows a schematic view of one such device SHIP[15] installed at GSI in Darmstadt, Germany. It consists of two filtering stages with a total flight path of 11 m. The transmission efficiency can be 70% with 10% velocity window and 20% charge window. A suppression of about 10^{11} for the projectile particles has been achieved.

3.8   On-Line Isotope Separator (ISOL)

The various methods discussed above are all very useful, and have contributed much towards the study of nuclei far off stability. However, they all have their limitations. An ideal system will require a sorting process that can select only one pure isotope and deliver the activity to a low background area with minimum time delay and high efficiency. The ISOL system is designed aiming to meet as much of these optimal conditions as possible. The inherent limitation is the transport time and the challenge is element selection.

The basic components of an ISOL system consist of the production chamber, ionization chamber, mass selection device, and detecting system with transportation sections between the sequential components as shown in Fig. 3.7. The Z selection can be applied at various stages as indicated in the diagram. A recent review[16], listed some forty elements that can be isolated in fairly or extremely pure form. They were achieved by specific design of individual component or a combination of

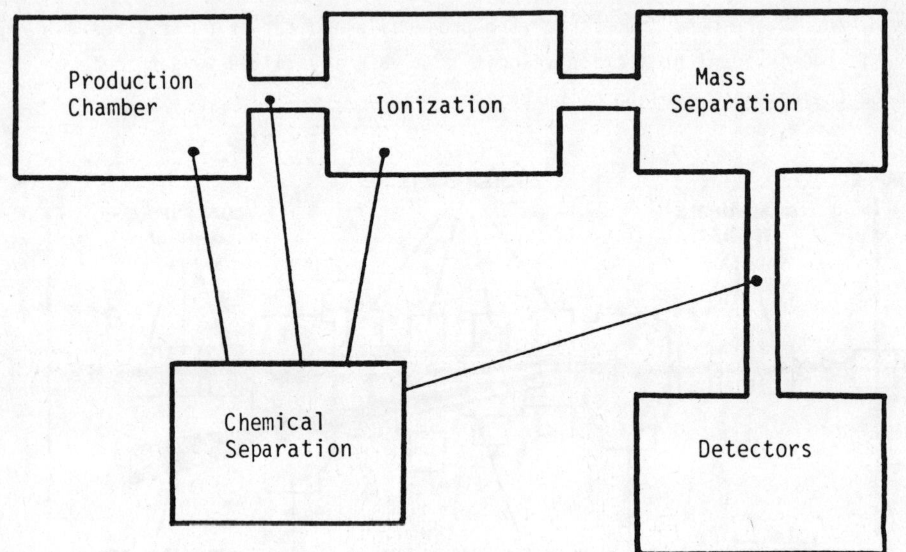

<u>Fig. 3.7</u> - Basic Components of an On-Line Isotope Separator. Chemical separation can be applied at different stages of the process.

them. For example: emanating targets can release noble gases at room temperature; volatile elements such as zinc, cadmium and mercury can emerge from molten germanium, tin and lead targets readily. Using these target designs, particular element selection can be achieved. Specific chemicals can be introduced either in the production chamber or in the transmission line to form specific molecules. For example, alkaline earth elements will react with fluoride vapour to form stable monofluorides, and

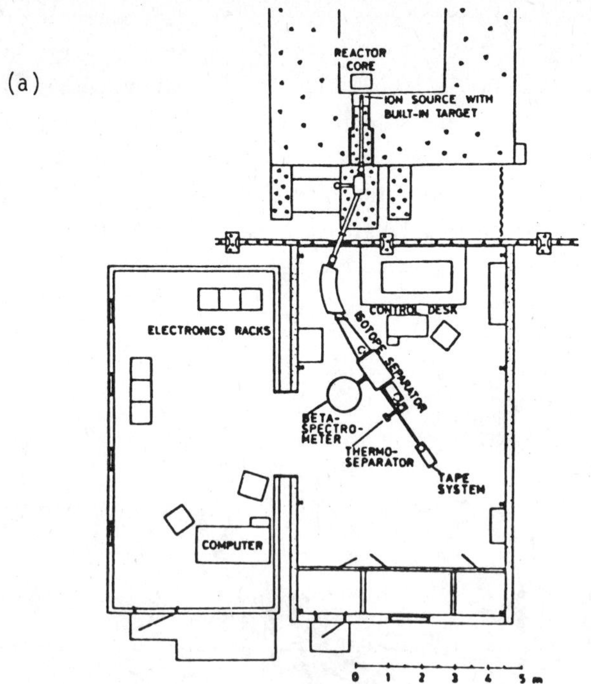

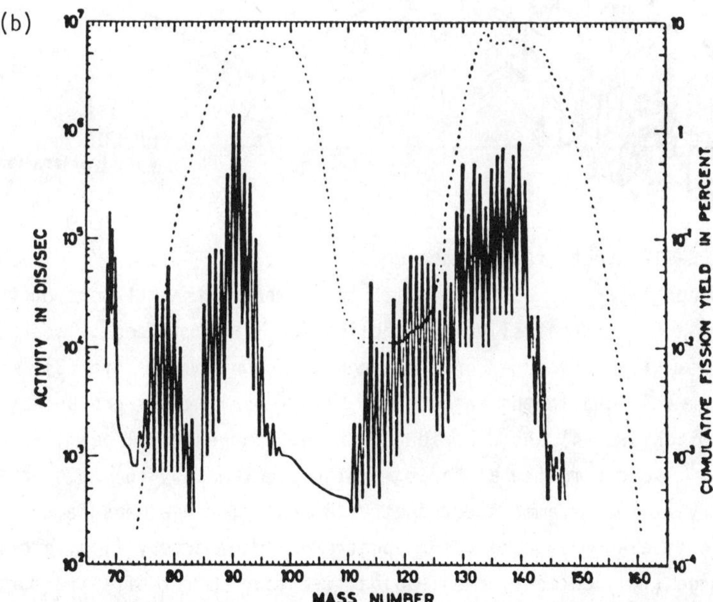

Fig. 3.8 - (a) Layout of OSIRIS Separator; (b) Counting Rates of Different Isobars, Broken curve shown is mass yield from ^{235}U(n,f). (from Reference 15).

ionized molecules will shift the element mass by 19 and can be separated from the alkalis. Another form of element selection is the use of surface ionization. By choosing suitable surface material, elements with low ionization potentials are favourably ionized. Alkali metals are selected this way, and in principle, Ga, In and Th can also be isolated in a similar fashion. Halogens can form negative ions when in contact with surface of low work function elements or compounds, and can thus be selected. Finally, chemical selection process can still be carried out after mass selection. As can be seen, much has been achieved but still, there are many possibilities for improvements.

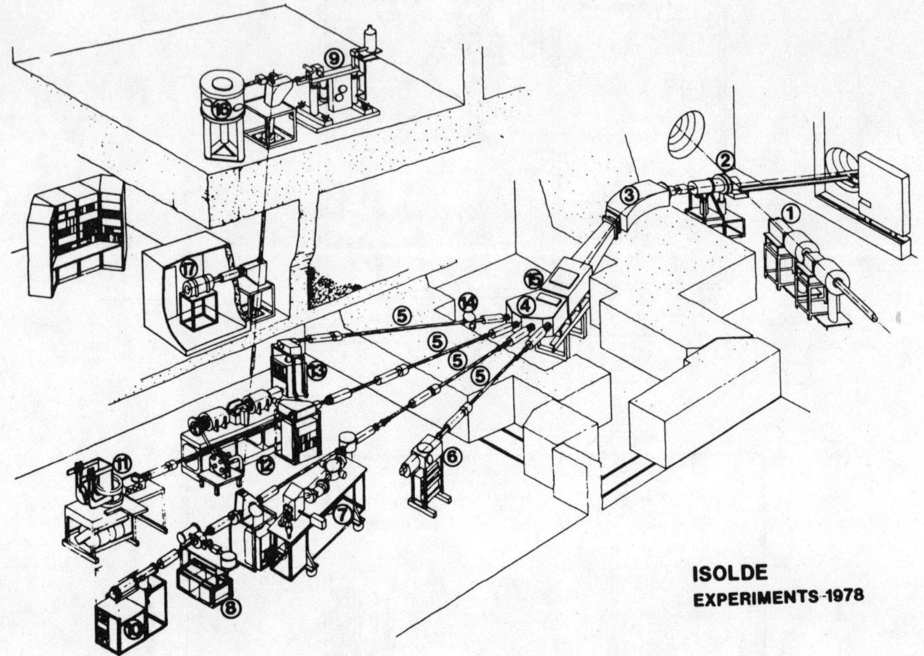

<u>Fig. 3.9</u> - ISOLDE-II Layout (from Reference 6).
The on-line isotope separator ISOLDE-II with the experiments that were on the floor in 1978. Not shown are electronics, power supplies, on-line computers, lasers, etc. The 600-MeV proton beam (1) is focused on the target and ion-source unit (2), and the 60-keV ions are mass-analysed in the magnet (3). Individual masses are selected in the electrostatic switchyard (4) and distributed through the external beam lines (5) to the experiments. These comprise alpha and proton spectroscopy (6), high-resolution mass spectrometry(7), beta-gamma spectrometry (8 and 9), range measurements of ions in gases (10), optical-pumping and laser spectroscopy on mercury (11), atomic beam magnetic resonance (12), collection of radioactive sources for off-line work [hyperfine interactions in solids, determination of shifts in the energies of K X-rays, targets for nuclear reaction studies (13,14,15)], beta-decay Q values measured by coincidences with a magnetic "orange spectrometer (16), and spectroscopy of beta-delayed neutrons (17).

Besides the surface ionization mentioned above, the most commonly used method is to create a discharge plasma. In this case, any neutral atom can be ionized and unless some element selection is provided elsewhere, only mass separation is provided. These ion sources are usually called universal ion-source and are the only effective alternative method for ionization so far.

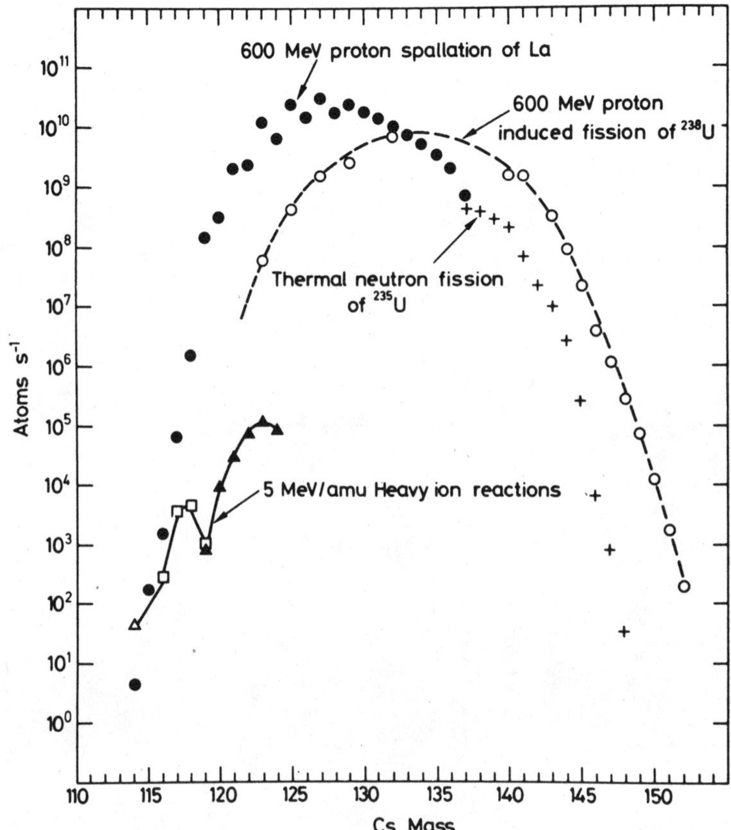

Fig. 3.10 - Production (in atom s^{-1} arriving at the collector plate) of cesium isotopes by various techniques: (a) spallation of molten lanthanum with 600-MeV protons (b) fission of uranium carbide with 600-MeV protons, (c) heavy-ion reactions with the GSI Unilac, and (d) with reactor neutrons at the TRIGA reactor in Mainz (from Reference 16).

Many ISOL systems are operational at various accelerators and reactors. Fig. 3.8 shows the OSIRIS in Sweden. It is a reactor based ISOL system employing a universal ion source. The isobaric yield is also shown. A thermoseparator based on the principle of thermochromatographic method is being developed for chemical separation at the receiving end of the separator. The most elaborate ISOL facility is the ISOLDE using the 600 MeV proton beam from the synchrocyclotron at CERN, Switzerland.

Fig. 3.9 shows the facility layout. Many different types of target design and ion sources have been used. Fig. 3.10 shows the yield of Cs isotopes from fission and spallation reaction. The intense separated ion beams made possible many exotic experiments.

4. NUCLEAR MASS

4.1 Introduction

The mass of a nucleus is one of its fundamental properties. Mass differences between appropriate pairs of nuclei yield various binding energies. It determines the proton and neutron drip lines and gives the limit for stability against alpha emission and spontaneous fission. The systematic trend of the binding energies is a very effective way to evaluate the importance of shell effects and the onset of deformation. Accurate knowledge of the masses are important in the testing of such fundamental principles of the charge symmetry of nuclear forces and the weak coupling constants in beta decay. Fig. 4.1 shows the chart of nuclides. The exterior contour

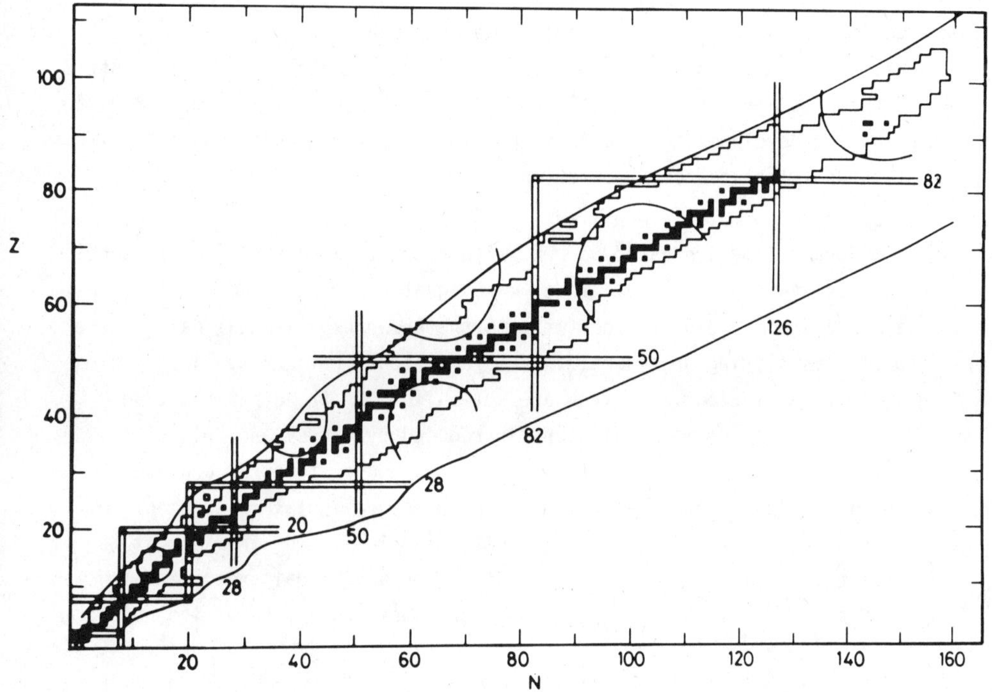

Fig. 4.1 - Chart of the nuclides, showing the limits of our present knowledge, particle drip-line predictions, magic nucleon numbers, and regions of deformation.

shown here indicate the approximate limit of proton and neutron stability - the so-

called proton and neutron drip lines where the corresponding nucleon binding energy
becomes zero. Nuclides that are known to be stable against nucleon emission are enclosed by the histogram line. At present, the limit for stability against proton
emission has been verified in the light nuclides, and may be reached in some medium
and heavy elements. On the neutron rich side, the stability limit is only reached
in the very light nuclei.

4.2 Mass Formulae

Ever since the advent of nuclear physics, attempts have been made to produce
a mass formula that could predict the general behaviour of nuclear mass over a wide
mass region. The standard text book Weizsacker semi-empirical mass formula based on
Liquid Drop Model (LDM) of nuclei was very successful in explaining many of the
general trends of nuclear stability. When more extensive and accurate experimental
results become available, substantial modification is necessary. Various efforts
were made to improve the formula and innovative approaches were also attempted. The
international conferences on nuclear masses[8,9] reported the progress made. After
the 1975 conference, a review article[11] brought together some of the mass formulae
and summarized the development of various approaches, their foundations and perform-
ance, and is a useful source of information.

Most of these formulae are primarily concerned with the mass prediction over
large mass regions. Some are based on LDM, with sophisticated shell corrections,
while some others start with a shell model mass equation. The parameters of these
equations are adjusted to give the best fit to the known experimental data. As a
consequence of this fitting procedure, their predictions in the very light nuclei
usually give large deviations and often are not valid. On the other hand, some
other formulae are based on mass relations introduced by Garvey and Kelson[17].
These relations link the masses of some neighbouring nuclei together and are therefore capable of predicting an unknown mass from the experimental measurements of
its neighbours. In this way, the mass prediction is "localized" and is most
effective in the very light nuclei region. Often one additional experimental
measurement of a mass can substantially change the mass prediction of its neighbouring nuclei.

4.3 Garvey-Kelson Mass Relations

These mass relations can be derived using an independent-particle model.
The aim is to find a relation in the form:

$$\sum_{i=1}^{n} C_i \, M_i(N_i, Z_i) = 0 \qquad\qquad 4.1$$

where N_i and Z_i are the neutron and proton numbers for the ith nucleus, and the C_i

can have values ± 1. To satisfy this relation, all the single particle energies as well as the residual interactions must cancel out. That is:

$$\sum_{i=1}^{n} C_i N_i = 0 \qquad\qquad 4.2$$

$$\sum_{i=1}^{n} C_i Z_i = 0 \qquad\qquad 4.3$$

and

$$\sum_{i=1}^{n} C_i N_i Z_i = 0 \qquad\qquad 4.4$$

To satisfy these conditions, non-trivial solutions are possible for $n \geq 6$. For n=6, one can get:

$$M(N_1,Z_1) + M(N_2,Z_2) + N_3,Z_3)$$
$$-M(N_1,Z_2) - M(N_2,Z_3) - M(N_3,Z_1) = 0 \qquad\qquad 4.5$$

Here, the equations 4.2 and 4.3 are satisfied, and eq. 4.4 can also be satisfied if:

$$N_1 Z_1 + N_2 Z_2 + N_3 Z_3 - N_1 Z_2 - N_2 Z_3 - N_3 Z_1 = 0 \qquad\qquad 4.6$$

or: $\qquad (Z_1-Z_2)(N_1-N_2) = (Z_1-Z_3)(N_3-N_2)$

To avoid generating identities, it is necessary that all the Z's and N's are different. The simplest solutions are:

$$\left.\begin{array}{ll} Z_2 = Z_1+1, & N_1 = N_2-1 \\ Z_3 = Z_1-1, & N_3 = N_2+1 \end{array}\right\} \qquad 4.7$$

or:

$$\left.\begin{array}{ll} Z_2 = Z_1-1, & N_1 = N_2-1 \\ Z_3 = Z_1+1, & N_3 = N_2+1 \end{array}\right\} \qquad 4.8$$

For the solution given by eqs. 4.7, the mass relation 4.5 becomes:

$$M(N+2,Z-2) - M(N,Z)$$
$$+M(N+1,Z) - M(N+2,Z-1)$$
$$+M(N,Z-1) - M(N+1,Z-2) = 0 \qquad\qquad 4.9$$

here: $\qquad Z_2 = Z$ and $N_1 = N$. Similarly the solution represented by eq. 4.8 yields:

$$M(N+2,Z+2) - M(N,Z)$$
$$+M(N+1,Z) - M(N+2,Z+1)$$
$$+M(N,Z+1) - M(N+1,Z+2) = 0 \qquad\qquad 4.10$$

These two mass relations can be represented graphically in N vs Z plot as shown in Fig. 4.2. Eq. 4.9, which is represented by Fig. 4.1a, is a mass relation among mass differences of three isobar pairs. It tends to relate masses with larger (N-Z) differences and is called the traverse mass relation. Eq. 4.10, on the other hand, relates mass differences of pairs along constant N-Z line (or constant T_z pairs), and is called the longitudinal mass relation.

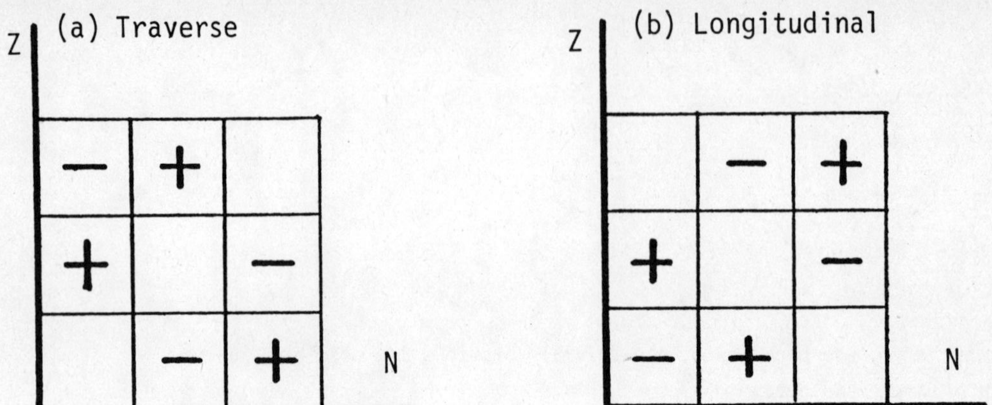

Fig. 4.2 - Schematic representation of the traverse and longitudinal mass relations. The presence of a plus or minus sign in box indicates that the mass value of respective nucleus is to be added or subtracted.

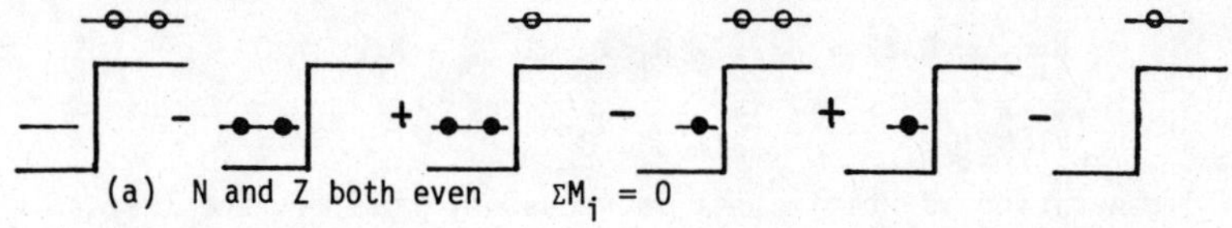

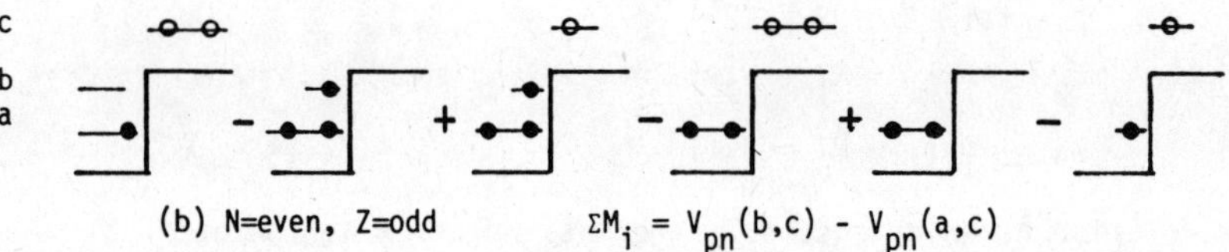

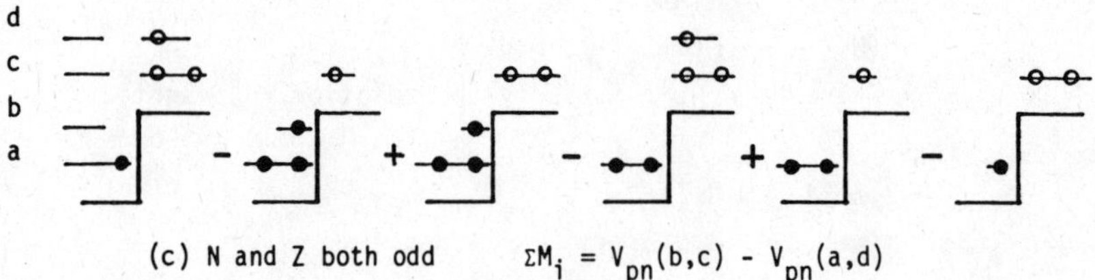

Fig. 4.3 - Representation of the traverse mass relation (eq. 4.9) based on four-fold degenerate levels.

The validity of these mass relations is hinged on the assumptions implicated in eqs. 4.2 and 4.3. Generally, the single particle energies and residual inter-

actions in different nuclei will not remain constant. However, over a narrow mass
region, these variations may be expected to be small. One exception to this slow
variation is the neutron-proton interaction. The effective interaction between the
neutron and proton is much stronger when they share the same spatial orbitals than
for different orbitals. Therefore, if any member of the mass relation is an odd-odd
self-conjugate nucleus, the mass relations would not hold. (It should be noted that
in the j-j coupling shell model calculation with seniority scheme, the relations 4.9
and 4.10 still hold except for N=Z = odd case).

Consider a model with four-fold degenerate Hartree-Fock or Nilsson-like
single particle levels, the equation 4.9 can then be represented graphically in Fig.
4.3. Here, the single particle energies in the sum all cancel out, so do the two-
body interactions between the like particles, regardless whether N or Z is even or
odd. As for the interaction between protons and neutrons, the situation is more
complicated. If both N and Z are even, then the interactions cancel each other. If
either N or Z or both are odd, the cancellation may not be complete, that is, the
mass relation is reduced to the difference of the two-body interactions between
proton and neutron in their respective orbits.

$$\Sigma M_i = V_{pn}(a,c) - V_{pn}(b,d) \qquad 4.11$$

where V_{pn} indicates p-n interaction while a,b,c,d are orbital designations. This
difference will be small as long as the protons and neutrons are in different orbits.
On the other hand, if one of these terms represent protons and neutrons in the same
orbit (i.e. a=c or b=d), then the two terms will have vastly different magnitudes,
resulting in the failure of the mass region. Similar conclusions can be reached for
the longitudinal mass relation, eq. 4.10.

These mass relations have some interesting implications when they are
applied across the N=Z line. These relations are still valid as long as the N=Z =
odd nucleus is not included in the relation. For eq. 4.9, this relation involves
the mass differences of three pairs of mirror nuclei. To change the notation, let
$M(A,T_z) = M(N,Z)$ where A=N+Z is the mass number and $T_z = \frac{1}{2}(N-Z)$ is the z-component
of isospin T. The eq. 4.9 is then equivalent to:

$$M(A,+1) - M(A,-1) + M(A+1,-\tfrac{1}{2}) - M(A+1,+\tfrac{1}{2})$$
$$+ M(A-1,-\tfrac{1}{2}) - M(A-1,+\tfrac{1}{2}) = 0 \qquad 4.12$$

Note that the mass difference between mirror nuclei is the Coulomb displacement
energy ΔE_C corrected for the mass difference of neutron and proton. That is:

$$\Delta E_C(A,T) = M(A,-T) - M(A,+T) + 2T(M_n - M_H) \qquad 4.13$$

Eq. 4.12 then becomes a relation of ΔE_C,

$$\Delta E_C(A,T=1) = \Delta E_C(A+1,T=\tfrac{1}{2}) + \Delta E_C(A-1,T=\tfrac{1}{2}) \qquad 4.14$$

Such equations can also be derived using a simple assumption on the linear coefficients in the isobaric multiplet mass equation:

$$M(A,T,T_z) = a(A,T) + b(A,T)T_z + c(A,T)T_z^2 \qquad\qquad 4.15$$

These two mass relations can be treated as recurring relations. Starting near the stability valley, the unknown mass of the nucleus can be estimated from the known masses of five adjacent nuclei. This step can be repeated and eventually, predicted mass tables can be obtained, and stability of nuclei far from β-stability can be estimated. Naturally, when such repeated processes are applied, deviations from experimental values are likely to increase. Methods to minimize the deviation of mass prediction have been suggested. These are also discussed in the review articles[10].

The mass relations presented here are only the simplest solutions to eq. 4.6. It is likely to be more accurate since the mass region covered is small. Many other relations can be derived. For example, if one assumes that the mass relation still only involves six nuclei (i.e. using eq. 4.6), but adapts different values for the N's and Z's, such that eq. 4.6 is still valid, quite different mass relations can be obtained. Some of these examples are presented graphically in Figs. 4.4. The relations represented by Fig. 4.4c can be obtained either by assuming ten members in the mass relation (n=10) or by repeatedly applying n=6 mass relations. Such mass relations are useful for particular purposes. For example, from Fig. 4a, the neutron binding energy of a nucleus far from stability can be estimated from the neutron and proton binding energies of nuclei near the stability.

$$S_n(N+4,Z-4) = S_n(N+4,Z) + S_p(N+3,Z) - S_p(N,Z) \qquad\qquad 4.16$$

here $\qquad S_n(N,Z) = M(N,Z) - M(N-1,Z) - M_n$ and

$$S_p(N,Z) = M(N,Z) - M(N,Z-1) - M_H$$

Similarly, the Q_β for nucleus far from stability can be estimated from difference of neutron and proton binding energies of nuclei near the stability as shown in Fig. 4.3b. That is,

$$Q_\beta(N,+4,Z-4) = M(N+4,Z-4) - M(N+3,Z-3)$$
$$= S_n(N+4,Z) - S_p(N,Z-3) + (M_n-M_H) \qquad\qquad 4.17$$

The relation represented by Fig. 4.4c is particularly interesting if the diagonal line is the N=Z line. Then, the mass difference of mirror nuclei with large isospin is expressed as the sum of a set of $T=\frac{1}{2}$, mirror nuclei mass differences.

The mass relations presented here are simple examples of what could be related. They are necessarily only approximate, and the amazing point is that they worked well in many cases. In fact, when the prediction is extended to the entire chart of nuclides, the standard deviation from the experimental data is about 100 keV. Systematic deviation from prediction will point towards possible reason for their failure. For example, the over-estimation of neutron deficient nuclei masses near or beyond the proton drip line in the very light mass region led

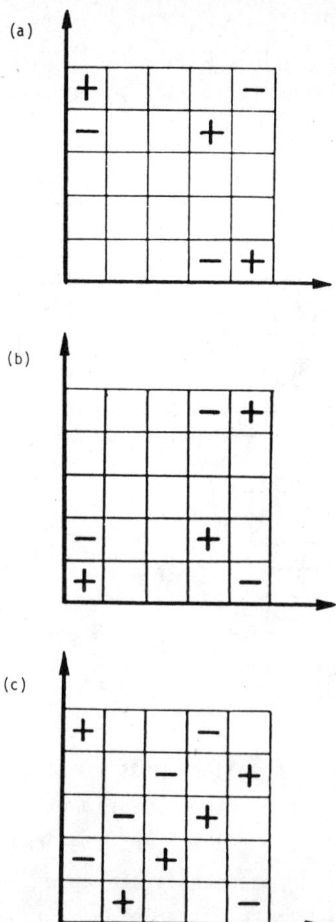

<u>Fig. 4.4</u> - Examples of possible mass relations.

to the identification of Thomas-Ehrman shift. Detailed comparison in the other
regions off the stability line will be very useful.

4.4   <u>Mass Measurements</u>

Nuclear masses are generally determined by reaction and decay Q-value
measurements or via mass spectrometric method. For nuclei far off stability, the
same principles will apply but different experimental technique will be necessary.
In the following, some typical methods are presented and their results discussed.

4.4.1   <u>Exotic Nuclear Reactions</u>

The limit of nucleon-emission stability can be reached in the light-mass
region, and it would be extremely important to verify its position. Fig. 4.5 shows
the present status in this region. For the neutron deficient side, some of the
proton-unbound nuclei masses have been determined. As for the neutron rich side,

stability limit is only reached for the very light nuclei and many of the nuclei have been identified via reaction production studies but their masses and decay properties are not yet known.

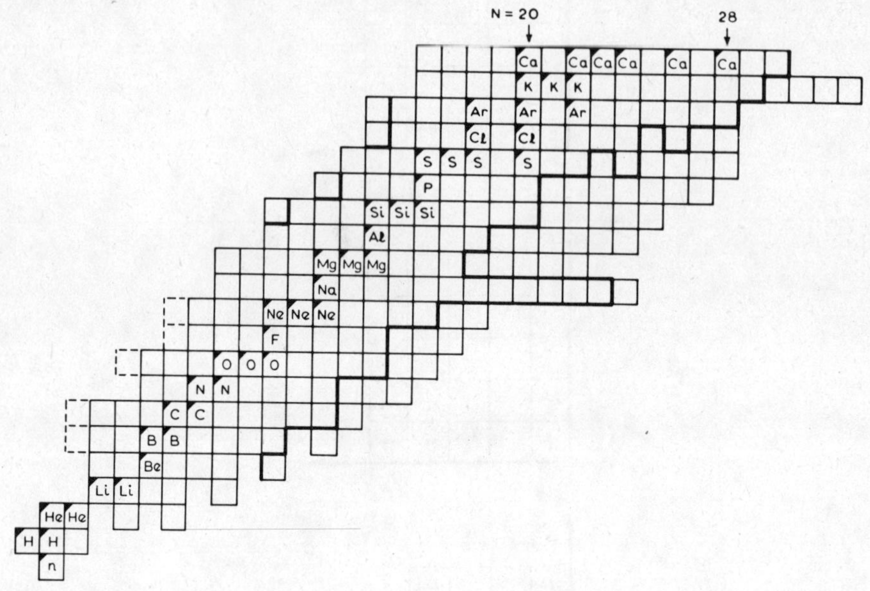

Fig. 4.5 - Chart of Nuclides in the light mass region. Squares with solid line sides are nuclei whose existence has been experimentally verified. Squares with dark upper left corners are stable nuclei, and the heavy-lined histograph represent the region where reasonably accurate masses are known. The dotted squares are nuclei experimentally verified to be unstable against nucleon emission.

EXOTIC NUCLEAR REACTIONS EMPLOYED FOR MASS MEASUREMENTS OF NUCLEI FAR FROM STABILITY

Description		Reaction	Typical Cross Section $(nb/sr)_{lab}$
Three Nucleon	+3p	$(^{7}Li, {}^{10}C)$	1500
Transfer	+3n	$(^{3}He, {}^{6}He)$ & $(^{18}O, {}^{21}O)$	300 - 1600
	−3n	$(^{11}B, {}^{8}B)$	160 - 300
	+2p − 1n	$(^{7}Li, {}^{8}B)$	700
	+1p − 2n	$(^{9}Be, {}^{8}B)$	100 - 400
	−1p + 2n	$(^{18}O, {}^{19}N)$	200
Four Nucleon	+4n	$(^{4}He, {}^{8}He)$	4 - 60
Transfer	−4n	$(^{18}O, {}^{14}O)$	300
	+2p − 2n	$(^{18}O, {}^{18}Ne)$	100 - 500
Five Nucleon	+5n	$(^{3}He, {}^{8}He)$	0.2
Transfer	+1p + 4n	$(p, {}^{6}He)$ & $(^{3}He, {}^{8}Li)$	200 - 400
	−1p − 4n	$(^{7}Li, {}^{2}He)$ & $(^{7}Li, 2p)$	10 - 2500
Six Nucleon Transfer	+1p + 5n	$(^{3}He, {}^{9}Li)$	5

Fig. 4.6 - Examples of multi-nucleon transfer reactions used for mass measurements (from Reference 3).

One of the useful methods is via multinucleon transfer reaction. Fig. 4.6 shows a table of reactions for such studies with typical cross-sections. Certainly, this does not exhaust the possibility and recently, eight-nucleon transfer reactions have been reported[8,17]. Since the reaction cross-sections are low and precise energy determination is required, the detector system used must be able to handle high counting rate, and make clear particle discrimination and yet preserve the necessary good energy determination such that meaningful test of the mass formula can be made. On such example is shown in Fig. 4.7 used in the (^{4}He,^{8}He) and

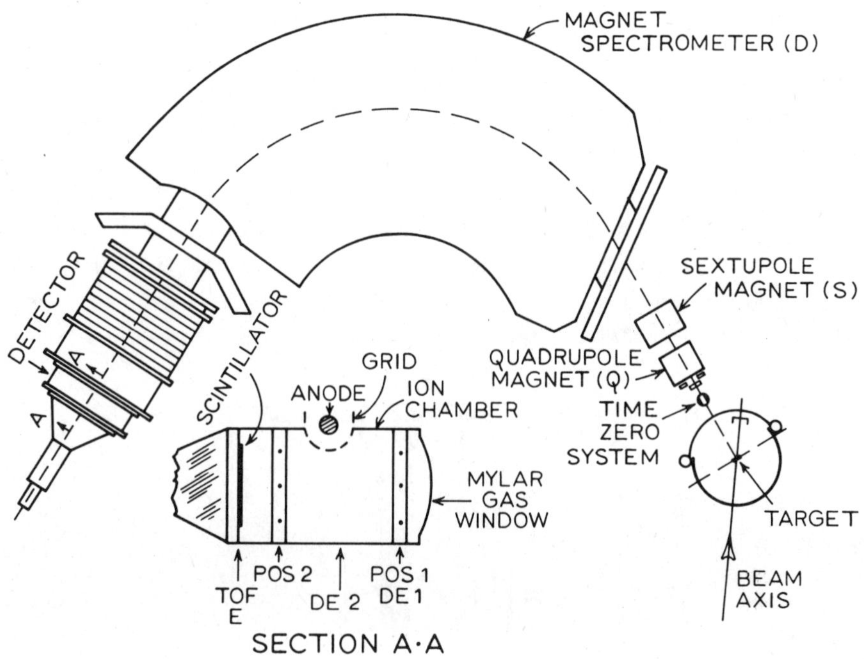

Fig. 4.7 - The Quadrupole-sextupole-dipole spectrometer used for identification of exotic nuclei. See text for further explanation.

^{3}He,^{8}Li) reactions. Here, the scattered particles pass through a time zero detector before entering a quadrupole-sextupole-dipole (QSD) spectrometer. Particles with a certain range of magnetic rigidity (B_ρ) are focused on the focal plane where the detector system as shown in the diagram is used. Signals representing positions (POS 1 and POS 2), differential energy losses (DE 1 and DE 2), total energy (E) and time-of-flight (TOF) were obtained. Particles of mass M, atomic number Z entering the QSD spectrometer with ionic charge q, and definite B_ρ (determined by POS 1 and POS 2) will have DE 1 and DE 2 proportioned to $(MZ/q)^2$, E proportinal to q^2/M and TOF proportional to M/q. The overdetermination of parameters helps to improve particle discrimination and rejection of accidental coincidence. In the particular reaction reported, the system can reliably pick out one ^{8}He particle among 10^7

incident charged particles.

Another interesting reaction is the double charge exchange (DCX) reaction (π^-,π^+). Only recently, it was thought that DCX reaction between analog states such as $^{18}O(\pi^+,\pi^-)^{18}Ne$ would have much higher cross-sections than the other DCX between non-analog states. Recent experimental results[18], showed that (π^-,π^+) reaction has comparable cross-sections and can lead to discrete nuclear states and since this reaction has $\Delta T_z = +2$, it is very useful in studying the light neutron rich nuclei.

The accurate mass measurements are important to test the rigidity of charge symmetry of nuclear force and to evaluate the weak coupling constants in beta decay. It is useful to study systematically the Coulomb displacement energies and the effect is reflected by the isobaric mass multiplet equation (IMME).

$$M(A,T,T_z) = a + bT_z + cT_z^2 + dT_z^3 \qquad\qquad 4.18$$

Here $M(A,T,T_z)$ is the mass of a member of an isospin multiplet. If charge symmetry is valid, then d is identically zero. To test this formula, at least four members of a isotopic multiplet will be needed, that is $T > 3/2$. Much effort was made to measure these multiplet masses accurately. Fig. 4.8 shows the summary of results (in 1975) of the known $T=3/2$ quartets. Since then, some new data were added,

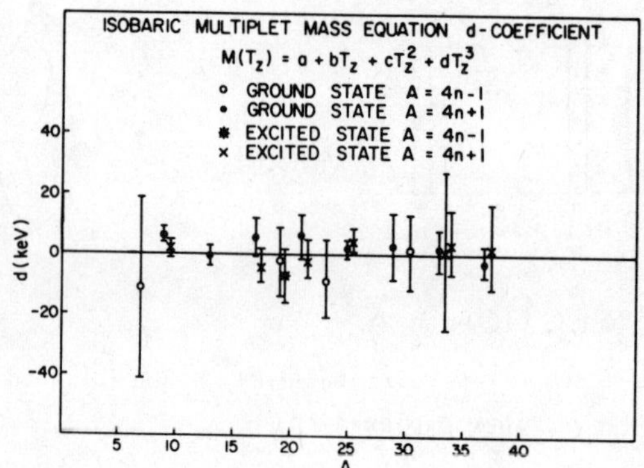

Fig. 4.8 - d-coefficient of the isobaric multiplet equation versus A. Excited states are displaced slightly to the right of the appropriate A (from Reference 3).

particularly those from $T=2$ and $T=5/2$ multiplets, and more accurate measurements were performed. The results show that the IMME is almost valid. However, there are definitive non-zero values for d-coefficient such as the $A=9$ case shown and many other higher T multiplets. Similar situations apply also for beta decay studies. In this mass region, there are many superallowed pure Fermi transitions between $J=0$, $T=1$ analog states. The ft-value is then a measure of the nuclear overlap matrix and the vector coupling constant. If the overlap of analog state is complete, then the ft-values should all be identical. Since ft-value is dependent

on mass difference and life-time, accurate measurements of masses contributed to reduce the difficulties encountered in the early days.

4.4.2   Beta Decay Q-Values

The beta decay of far unstable nuclei is characterized by their short half-
lives, high Q_β values and often complex decay schemes. The determination of their
end-point energies becomes quite difficult and some knowledge of their decay
properties is necessary. For relatively long-lived activities, the most accurate
Q_β values are usually obtained via magnetic spectrometer. However, this method is
not practical for short-lived nuclei. Scintillation detectors are handicapped by
their inherent poor resolution resulting in large uncertainties in the determination
of the end-point energies. For solid state detectors, the available detector size
is still small and detector efficiency for full energy loss of electrons is low, and
is dependent on the particular experimental set-up used. This made the analysis of
the continuous beta energy spectrum difficult. Recently, a beta spectrometer using
a thick high purity germanium detector placed inside the bore of a superconducting
solenoid has been developed. Fig. 4.9 shows a schematic diagram of the layout. In

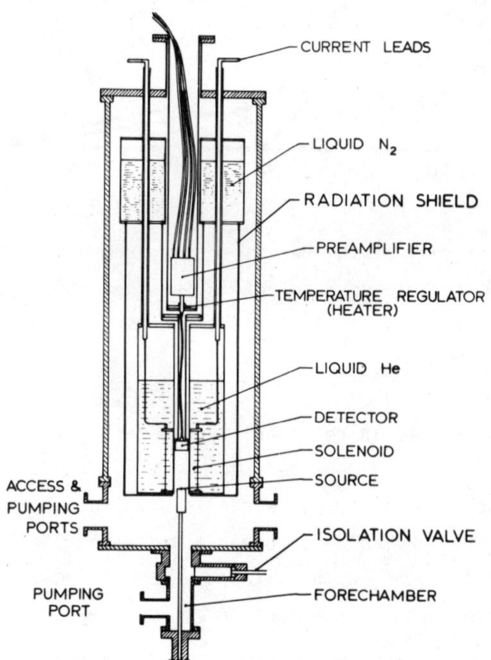

Fig. 4.9 - Sectional view of the beta spectrometer using a superconducting solenoid
and a high-purity Ge detector.

this system, the emitted beta particles from the source are confined to a small
cylindrical volume and the effective solid angle of the detector for β particles is

almost 2π. This sufficiently suppresses the γ-ray efficiency such that energy spectrum in singles are relatively free of γ-ray contamination. With its good energy resolutions, and a reasonable choice of beta response function for the detector, detailed analysis of the beta spectrum is possible. Fig. 4.10 shows the fit to a beta singles spectrum from a separated [89]Rb source. The branching ratios obtained

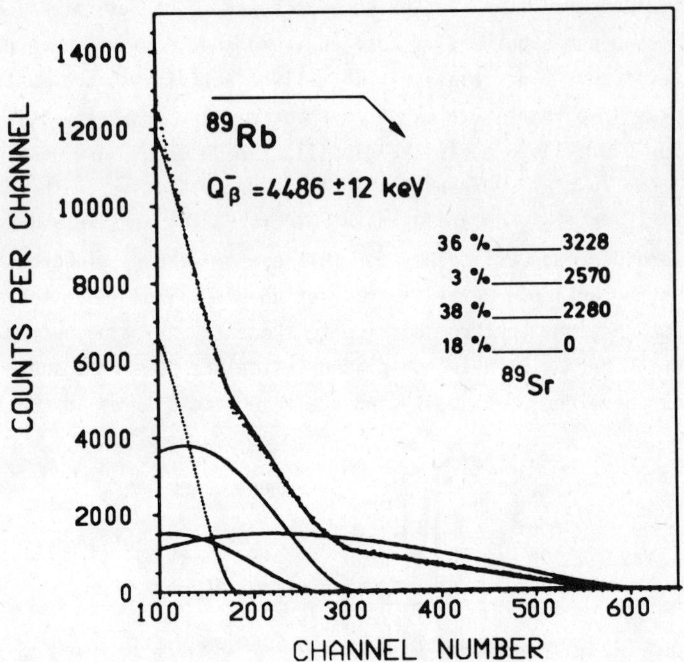

Fig. 4.10 - Detailed analysis of the beta spectrum from decay of [89]Rb. Branching ratio as shown in the insert is used to obtain the theoretical curves shown. After folding in the detector response, calculated spectrum is compared to experimental data.

from beta spectrum analysis is in reasonable agreement with those from decay scheme studies. With a reliable analysis procedure, it will then be possible to determine precisely the end-point energies from the single spectrum if the decay scheme is favourable. However, it is not always possible and often it is necessary to obtain the beta spectrum in coincidence with suitable γ-rays. These are usually lengthy experiments and sometimes results are open to ambiguities.

For neutron deficient nuclei where Q_{EC} is larger than B_p(or B_α), the binding energy of proton (or alpha) of the daughter nuclei, delayed proton (or alpha) may occur. Fig. 4.11 shows the delayed alpha spectrum from the [118]Cs precursor. Assuming a slow variation of beta strength function, the alpha end-point energy $(E_{\alpha max})$ can be obtained and thus Q_{EC} (equals $E_{\alpha max} + B_x$). Another alternative is to determine the EC/B$^+$ ratio as a function of the particle energy. Since this ratio is strongly dependent on the beta energy, $(Q_\beta - B_x)$ can be obtained.

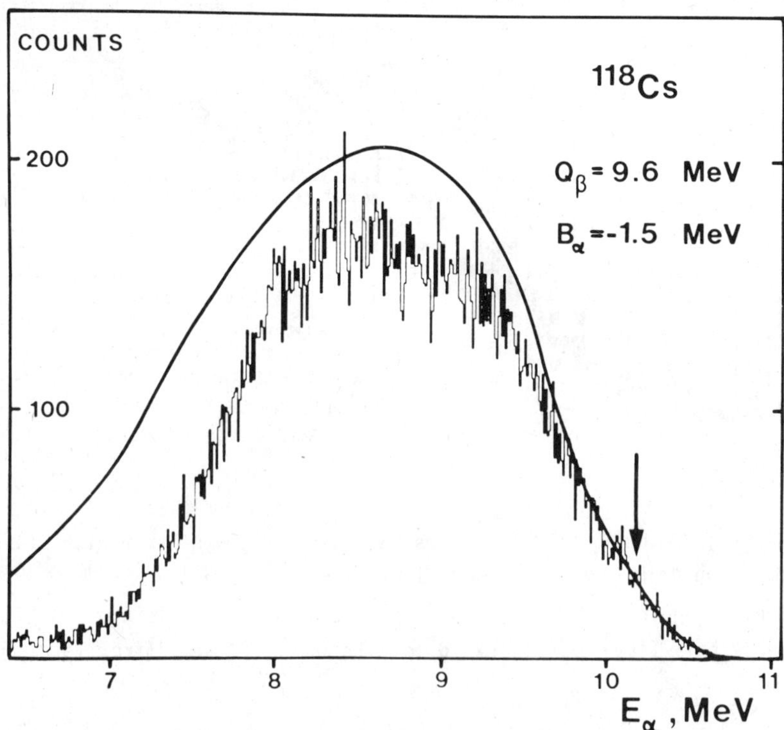

Fig. 4.11 - Delayed alpha spectrum from ^{118}Cs measured with a 100 µm, 150 mm^2 singles surface barrier detector. The experimental resolution was 20 keV FWHM. The end-point of the spectrum is fitted to a calculation assuming constant beta and alpha strength functions. The arrow indicates the normalization point (from Reference 3).

4.4.3   Alpha Decay Q-Values

In the medium and heavy mass region, many nuclei close to the stability valley are actually unstable against alpha emission. However, their transition probability are hindered by the Coulomb barrier and often have bery small branching ratios as compared to the beta decay. However, when farther away from stability, this decay process can compete effectively with beta decay, thereby offering another way to determine the nuclear masses. In fact, for an alpha decay, due to the strong energy dependence of penetrating factor, the predominant transitions often consist of a few peaks. Therefore, the mass difference of the corresponding pair of nuclei is much easier to determine than in the case of beta decay.

Fig. 4.12 is part of the chart of nuclides showing the neutron deficient alpha emitters in the medium and heavy mass range. These emitters tend to cluster

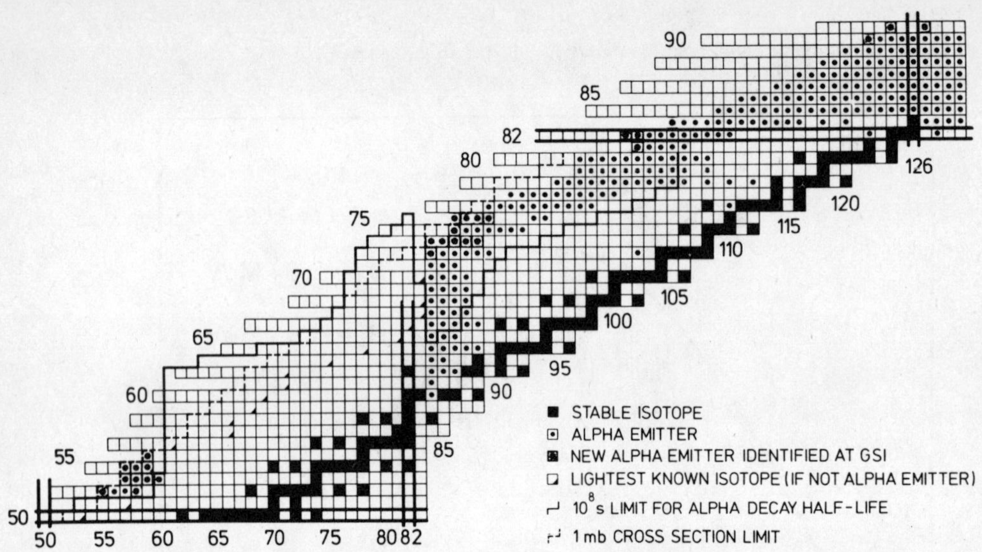

<u>Fig. 4.12</u> - Section of the Chart of Nuclides extending on the neutron rich side to the predicted proton drip line. To the left of the dash line, the estimated cross-sections for heavy ion induced reaction fall below 1 mb. To the left of solid line, the α-decay half-lives are estimated to fall below 10^8 sec. (from Reference 5).

together and their decay Q-values yield the relative masses of a chain of nuclides and eventually, their masses can be determined by linking them to some known ones. This vast mass-energy surface far from β-stability is very useful in testing various mass formulaie, and to estimate the strength of shell closure. For example, the Q_α values in the Te, I and Cs region give mass excesses in general agreement to the droplet-model and indicate the existence of double-shell closure[4] at N-Z-50. Also, the systematics of the Q_α values in the $62 < Z < 82$ region indicate large energy gaps above Gd (Z=64) and Os (Z=76). These gaps are likely due to subshell closure below and above the $h_{11/2}$ orbitals.

4.4.4   Direct Mass Measurement

The methods discussed above give the mass differences, and for nuclei far from stability, the mass excesses can only be determined by the measurement of successive decay Q-value. These measurements are often complicated and may be subject to many uncertainties, particularly if the decay schemes are not adequately studied. The direct mass measurement method eliminates these uncertainties and can thus be used as known masses for other Q-value measurements.

The experimental set-up used at ISOLDE for the mass determination of Rb and Cs isotopes is shown in Fig. 4.13. The separated isotope beam from ISOLDE was first stopped, re-ionized and then mass analysed by a subsequent mass spectrometer of higher resolving power. During the measurement, the magnetic field was kept constant and the accelerating voltage was varied. The accelerating voltage (V_i) required to send the isotope with mass M_i through a specific trajectory was determined. To improve the precision, one unknown and two reference masses were scanned sequentially and they are related by:

$$M_a(V_a+\delta) = M_b(V_b+\delta) = M_c(V_c+\delta)$$

where δ is a small value to account for deviation from constant M_iV_i requirement. The masses for the isotopes were determined and the precision obtained varied from about 25 keV to 400 keV. The deduced two-neutron separation energies for Rb isotopes

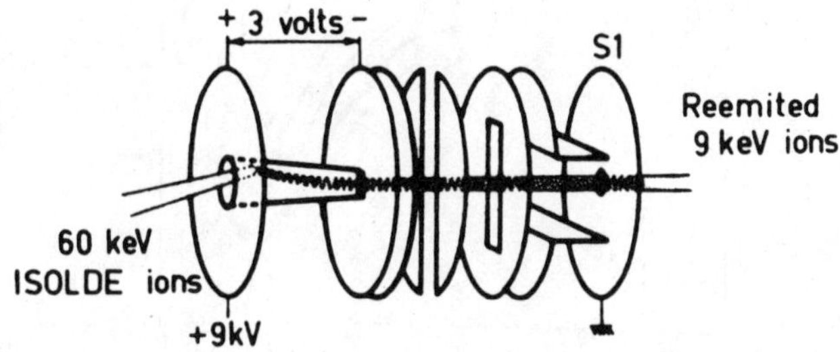

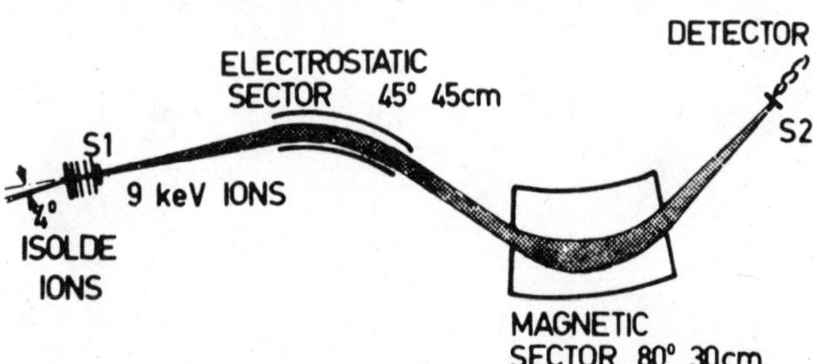

Fig. 4.13 - Schematic view of the set-up for direct on-line mass measurement. The upper diagram shows the 60 keV ISOLDE ions stopped in the tantalum tube. The atoms are re-ionized, and reemitted. Lower figure shows the mass spectrometer layout (from Phys. Rev. C19, 1504).

(A=76 to 99) are shown in Fig. 4.14. The systematic clearly show the closure of major shell at N=50, a smaller drop at N=56 presumably due to the closure of $d_{5/2}$ subshell,

and the on-set of deformation at N~60. An earlier similar direct mass measurement
was also carried out on the neutron rich Na isotopes produced by 24 GeV proton bombardment of uranium target. Results show that the heavy isotopes are more bound than
expected and are interpreted as the on-set of a new deformation near N~20 region.

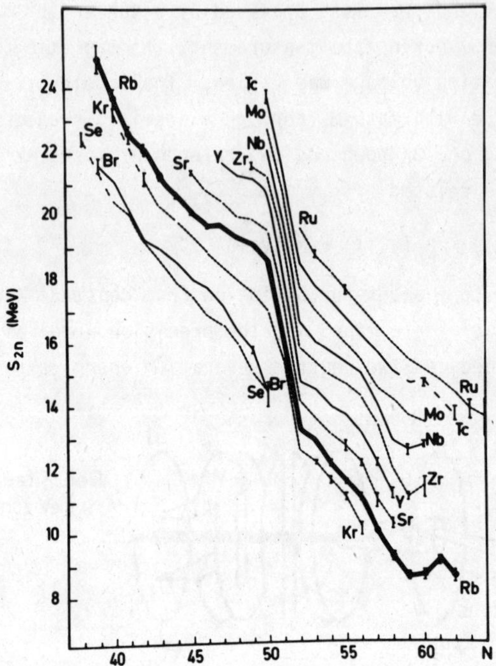

<u>Fig. 4.14</u> - Experimental two-neutron separation energies S_{2n} vs. neutron numbers in
Rb region (from Reference 6).

5.   DELAYED PARTICLE EMITTER

5.1   Introduction

When a radioactive nucleus has relatively low decay energy (Q_β or Q_{EC}), it cannot feed many daughter states. The resulting γ-ray spectrum usually consists of well resolved peaks. When the decay energy increases such that the daughter states may have excitation energy higher than the binding energy for the emission of certain particle x (B_x) with mass number A_x, delayed emission of particle x is then possible. Fig. 5.1 shows a typical decay scheme for a β^+-delayed particle -x precursor. Here, the precursor (N,Z) decays via β^+ (and/or EC) to state i in the emitter with excitation energy E_i. If E_i is larger than the binding energy of particle x (B_x), particle emission may occur, feeding a level f, with excitation energy E_f in

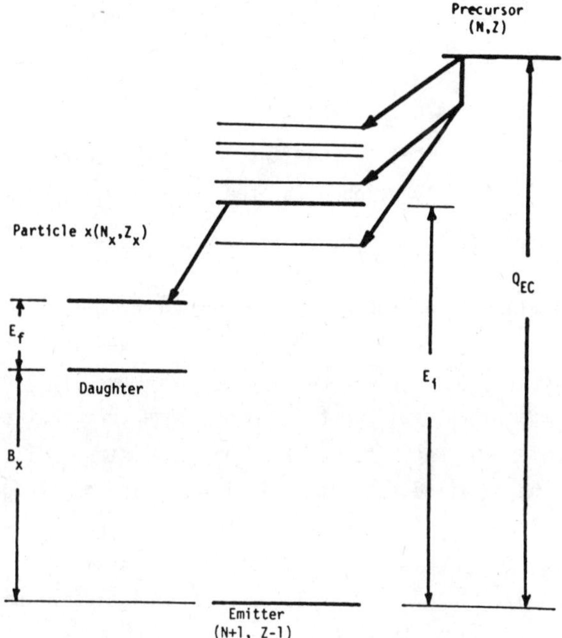

Fig. 5.1 - Typical decay of $\beta^{\pm}$delayed x-particle precursor, illustrating various terminology used.

the daughter nuclei. The energy of the emitting particle E_x is then:

$$E_x = (E_i - B_x - E_f)\frac{N+Z-A_x}{N+Z} \qquad 5.1$$

Similar decay schemes and terminologies can be obtained for the β^- decay precursor.

For delayed particle emission, the particle energy distribution is determined by two factors: (i) the beta intensity feeding the energy levels in the emitter, and (ii) the partial width of particle emission from state i in the emitter to level f in the daughter. The beta intensity distribution $I_\beta(E_i)$ can be expressed as:

$$I_\beta(E_i) = S_\beta(E_i) \cdot f(Z-1, Q_\beta-E_i) \qquad 5.2$$

within a normalization factor. Here $S_\beta(E_i)$ is the beta strength function and $f(Z,E_\beta)$ is the statistical rate function for beta decay. (For simplicity, the competing EC process is not explicitly shown, but should also be included in the total feeding to state i in the emitter). The total width of level i (Γ^i_{tot}) is the sum of all possible decay channels, including emission of competing particles x and y.

$$\Gamma^i_{tot} = \Gamma^i_\gamma + \sum_f \Gamma^{if}_x(E_x) + \sum_{f'} \Gamma^{if'}_y(E_y) \qquad 5.3$$

The probability for emission of particle x with energy E_x from level i to f is given by:

$$p^{if}_x(E_x) = \Gamma^{if}_x / \Gamma^i_{tot} \qquad 5.4$$

The particle energy spectrum $I_x(E_x)$ is then the sum of all transitions that would yield particle energy E_x properly weighted by the actual beta intensity.

$$I_x(E_x) = \sum_{i,f} W(J,J_i) \, I_\beta(E_i) \, p^{if}_x(E_x) \qquad 5.5$$

Here $W(J,J_i)$ is a weighing factor for decay from precursor with spin and parity J^π to the possible J_i of the emitter state. The summation over i and f states has the energy constraint given by Eq. 5.1. In cases where the level spacings in emitter and daughter are large compared to their widths and the detector resolution, the summation in Eq. 5.5 contains only one single term. The energy spectrum will then consist of individual peaks and the detailed spectroscopic information can be obtained for these levels. When the level density is high and average level spacing becomes smaller than the detector resolution, it will then become increasingly difficult to extract specific information about individual levels, and a statistical approach will therefore be needed.

To illustrate the above discussion in graphical form, a simplified picture for a delayed-precursor is shown in Fig. 5.2. The proton spectrum is the product of two factors (eq. 5.5). The beta intensity $(I_\beta(E_i))$ is dominated by the statistical rate function and decreases rapidly for higher E_i (and therefore low E_β). On the other hand, the proton partial width is dominated by the penetration of the Coulomb barrier and increases with E_i. The proton energy spectrum therefore should show a bell-shaped structure in general. In cases where the individual levels are involved, then the I_β curves are modified by delta functions representing transition matrices between precursor and individual levels in the emitter at appropriate energies. The spectrum will consist of individual peaks, but the overall bell shape should still prevail.

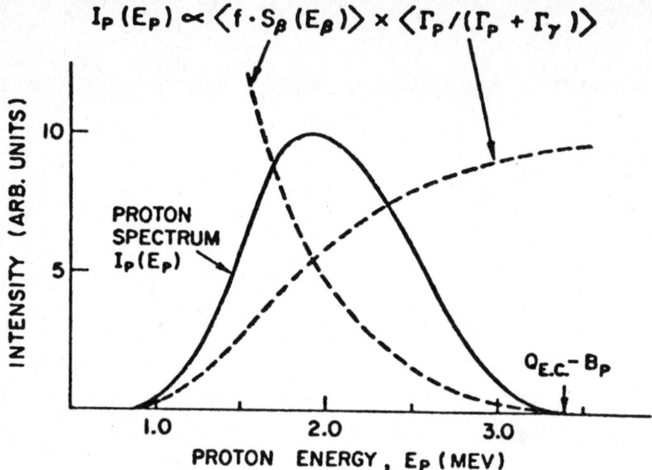

Fig. 5.2 - Typical proton spectrum calculated for a medium mass nucleus, illustrating the contributions from beta-decay and proton widths (from Reference 3).

In general, the level density is low for light nuclides particularly for low excitation energies and individual peaks in the delayed particle energy spectrum should be the dominant feature. On the other hand, for heavier elements, continuum spectrum is more likely. Some of these examples are shown in the following sections.

5.2 Delayed Particle Spectroscopy N<Z

In this group, the precursors are all neutron deficient nuclei. Their ground states usually have the lowest isospin allowed, that is $T=T_z$ where $T_z=\frac{1}{2}(N-Z)$. For N≮Z nuclei, the T_z values are negative and they decay by β^+ transition to emitters lying closer to the N=Z line (except the $T_z=-\frac{1}{2}$ precursors, where the emitters will be their mirror nuclei). Due to the Coulomb displacement energy, there exists in the emitter an isobaric analog state (IAS) of the precursor ground state lying at a lower energy. Therefore, the beta intensities are dominated by the superallowed transition to the IAS, and the excitation energy of the IAS plays an important role affecting the nature of the delayed particle energy spectrum.

5.2.1 $T_z=-\frac{1}{2}$ Precursors

In this case, the beta decay occurs between a mirror nuclei pair. The IAS is the ground state of the emitter, and the other beta transitions are allowed Gamow-Teller (GT) transitions. They are relatively weak and the beta intensities are dominated by the statistical rate function f. The delayed proton emission probability is strongly affected by B_p, the proton binding energy in the emitter.

For the Z=odd precursor, the B_p is usually higher, and there is little chance to observe the delayed protons. The best candidates are heavier Z=even precursors. Recently, one such precursor ^{59}Zn has been reported[17]. A group of 15 proton peaks with total branching of 3×10^{-3} was obtained. In such studies, the Q_β measurements give the Coulomb displacement energies. Branching ratio for transitions to IAS is useful for the accurate determination of ft-values, and thus the GT matrix element. Detailed comparisons to model calculations can then be made.

5.2.2 $\underline{T_z=1 \text{ Precursors}}$

The even-even members of this series has 0^+ ground state and they decay via pure Fermi superallowed decay to their analog states in the daughter. These IAS's have low excitation energies and often are the ground state. No delayed particle emission has been reported.

Precursor	$t_{1/2}$ (msec)	Particle	$Q_\beta^* - B_a$ (MeV)	Particle branch (per disintegration)	Major peaks (relative intensities) (E_L in MeV)
$^{8}_{5}$B$_{3}$	769	α	18.07	1	1.50(100) broad, 8.36(8)
		p	0.72	—	None
$^{12}_{7}$N$_{5}$	10.97	α	9.98	3.5×10^{-2}	Continuum
		p	1.39	—	None
$^{20}_{11}$Na$_{9}$	446	α	9.16	1.9×10^{-1}	2.15(100), 2.48(5), 3.80(2), 4.44(17), 4.90(1)
		p	1.04	—	None
$^{24}_{13}$Al$_{11}$	2066	α	4.57	7.7×10^{-5}	1.57(49), 1.98(100), 2.27(1), 2.33(4), 2.36(3)
		p	2.19	—	None observed
$^{24}_{13}$Al$_{11}^{m}$	130	α	5.01	$\lesssim 1 \times 10^{-2}$	1.40(100), 1.77(58), 1.83(12)
		p	2.63	—	None observed
$^{32}_{17}$Cl$_{15}$	298	α	5.74		1.61(65), 2.20(100)
		p	3.83	$\sim 5 \times 10^{-4}$	0.76(32), 1.02(52), 1.35(29)
$^{40}_{21}$Sc$_{19}$	182.4	α	7.29	—	None observed
		p	6.00	$\sim 5 \times 10^{-3}$	1.05(100), 1.24(10), 1.44(3), 1.83(4), 2.10(7), 2.38(5)
$^{44}_{23}$V$_{21}$	90	α	8.59^a	—	2.78(100)
		p	5.06^a	—	None observed

Fig. 5.3 - Summary for T_z=-1, Precursors (from Reference 13)

The odd-odd T_z=-1 nuclei are delayed-particle precursors. For $A \lesssim 20$, the analog states are situated above the particle emission threshold and therefore relatively higher delayed particle branch is possible. Fig. 5.3 shows a table listing the precursors in this series. In all the emitters, the proton binding energies are higher, therefore most of these are delayed-alpha precursors. ^{32}Cl has comparable delayed-alpha and delayed-proton branches, while ^{40}Sc is pure delayed-proton precursor.

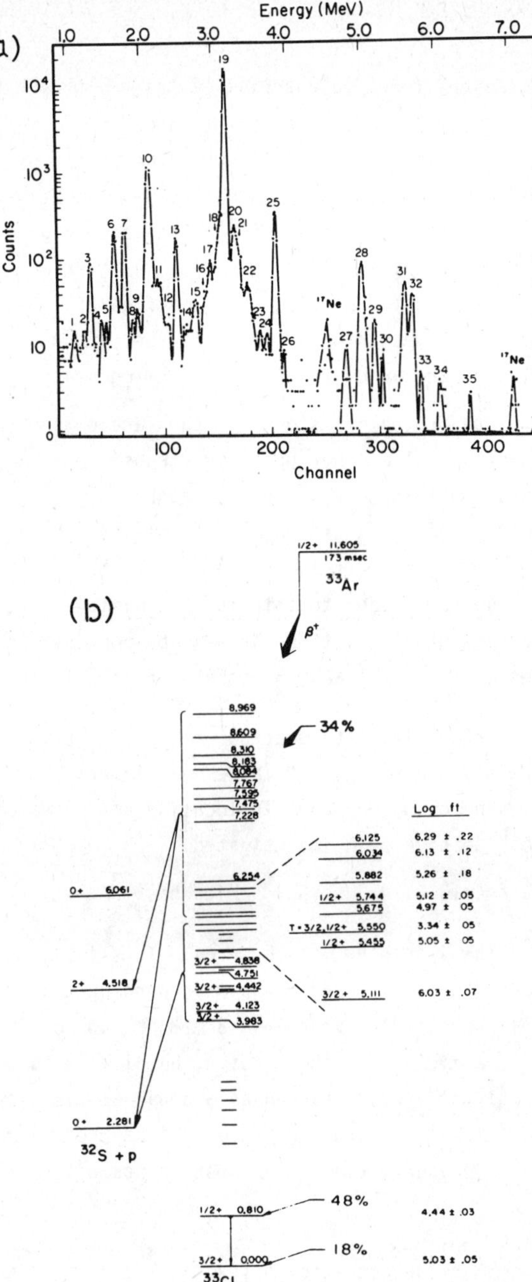

<u>Fig. 5.4</u> - Decay of ^{33}Ar - an example of $T_z = -\frac{3}{2}$, A=4n+1 delayed proton emitter (from Reference 13). (a) Energy spectrum; (b) Resultant decay scheme.

5.2.3 $T_z = -\frac{3}{2}$ Precursors

The $T_z=-\frac{3}{2}$, A=4n+1 (or N=odd) series of delayed-proton precursors are the first ones studied. The lighter members can be produced via (p,2n) and (^{3}He,2n) reactions and the heavier ones via (HI,xn) reactions. The proton emission branches are high and typically, the energy spectrum is dominated by a few strong transitions as represented by that obtained from the decay of ^{33}Ar shown in Fig. 5.4. These peaks represent transitions between individual levels in the emitter and daughter, and a decay scheme can then be constructed as shown in same figure. The main feature which is common to all the decay schemes in this series is the strong superallowed transition from the $(T,T_z) = (\frac{3}{2},-\frac{3}{2})$ precursor to the $(\frac{3}{2},-\frac{1}{2})$ analog state in the emitter. The dominant peak corresponds to proton emission from this analog state to the ground state in the daughter nucleus. The other peaks are transitions between individual levels in the emitter and low lying levels in the daughter. It is interesting to determine the ft-values for the β-transitions, particularly those feeding the IAS, which will then give an estimate of its isospin purity. (Note that the proton transition to low lying $T_z=0$ daughter is due to this isospin mixing in the analog state). The detailed level structure of the emitter can also be compared to model calculations. In principle, the masses of the $T_z=-\frac{1}{2}$ and $-\frac{3}{2}$ members of the T=$\frac{3}{2}$ quartet can be obtained to test the isobaric multiplet equation (IMME). In practice, these masses are obtained more accurately via reaction Q-value measurements but the knowledge obtained in these decay schemes were useful in locating these analog states. Fourteen members from ^{9}C to ^{61}Ge have been investigated.

The $T_z=-\frac{3}{2}$, A=4n+3 (i.e. N=even) series has ^{23}Aℓ as its lightest member, and delayed protons with low branching was observed. This is due to the fact that in these cases, the emitters have even Z and the proton binding energy is much higher than the cases for A=4n+1 series. Heavier A=4n+3 members up to ^{35}K have been identified and are stable against nucleon emission, but little is known about their decay properties. Recently, ^{35}K was obtained as a separated isotope and its decay property has been investigated via both γ-ray and delayed proton studies (first of such studies for all $T_z=-\frac{3}{2}$ precursors). Ten proton groups with a total branching of 3×10^{-3} were observed.

All the $T_z=-\frac{3}{2}$ precursors studied (except ^{35}K) are summarized in table presented in Fig. 5.5.

Table 1 Observed β^+-delayed proton precursors with $T_z = -\frac{3}{2}$

Precursor	Production reaction[a]	$t_{1/2}$ (msec)	$Q_\beta - B_p$ (MeV)[b]	Proton branching ratio
N odd				
$^{9}_{6}C_3$	$^{10}B(p,2n)$	127 ± 1	16.68	~1.0
$^{13}_{8}O_5$	$^{14}N(p,2n)$	8.9 ± 0.2	15.82	0.12
$^{17}_{10}Ne_7$	$^{16}O(^3He,2n)$	109 ± 1	13.93	0.99
$^{21}_{12}Mg_9$	$^{20}Ne(^3He,2n)$	123 ± 3	10.67	0.33
$^{25}_{14}Si_{11}$	$^{24}Mg(^3He,2n)$	221 ± 3	10.47	0.32
$^{29}_{16}S_{13}$	$^{28}Si(^3He,2n)$	188 ± 4	11.04	0.47
$^{33}_{18}Ar_{15}$	$^{32}S(^3He,2n)$	174 ± 2	9.34	0.34
$^{37}_{20}Ca_{17}$	$^{36}Ar(^3He,2n)$	175 ± 3	9.78	0.76
$^{41}_{22}Ti_{19}$	$^{40}Ca(^3He,2n)$	80 ± 2	11.77	1.0
$^{45}_{24}Cr_{21}$	$^{32}S(^{16}O,3n)$	50 ± 6	10.80	0.25
$^{49}_{26}Fe_{23}$	$^{40}Ca(^{12}C,3n)$	75 ± 10	11.00	0.60
$^{53}_{28}Ni_{25}$	$^{40}Ca(^{16}O,3n)$	45 ± 15	11.63	0.45
$^{57}_{30}Zn_{27}$	$^{40}Ca(^{20}Ne,3n)$	40 ± 10	13.99	0.65
$^{61}_{32}Ge_{29}$	$^{40}Ca(^{24}Mg,3n)$	~40	~12.2	~0.50
N even				
$^{23}_{13}Al_{10}$	$^{24}Mg(p,2n)$	470 ± 30	4.66	—

Fig. 5.5 - Table for observed $T_z=-\frac{3}{2}$, A=4n+1 delayed proton precursors (from Reference 11).   Recently another even N number (^{35}K) has been restudied using isotope separated source.

5.2.4 T_z=-2 Precursors

The A=4n, T_z=2 series are predicted to be delayed-particle precursors. For A<40, the analog states in the T=2, T_z=-1 emitter lie above the proton emission threshold, which should be strongly populated by the pure Fermi superallowed transition from the J=0 precursor. The proton energy spectrum is expected to show the dominant branch of the emission process. Recent (^{4}He,^{8}He) reactions indentified ^{20}Mg to be the lightest member of this series, and its decay property was also studied showing a single-peak proton spectrum. Similar spectra were obtained also for ^{24}Si, and ^{32}Ar and ^{36}Cℓ precursors.

The decay properties of the T_z=-2, A=4n nuclides are interesting. It offers an opportunity to test the pure Fermi superallowed beta transitions and to compare the ft-values to the decay of T=1, T_z=-1 cases. Also, the T=2 mass quintet allows further tests of the IMME. Accurate mass for T=2, T_z=-2 members can be determined by four neutron transfer reactions. The analog state in T_z=-1 members can be reached via $(^3\mathrm{He},^6\mathrm{He})$ reactions. However, such reactions are not selective and many T=1 states are also populated. The delayed-proton spectrum shows the location of the IAS and their precise mass excesses can then be obtained via these reactions.

The odd-odd A=4n+2, T_z=-2 members are poorly studied. Present mass equations suggest that $^{22}\mathrm{A\ell}$ is stable against proton emission. It would be interesting to establish its existence and to measure its mass excess.

5.2.5 $T_z \leqslant -\frac{5}{2}$ Precursors

Some mass equations[17] suggest the existence of three T_z=-$\frac{5}{2}$, A=4n+3 (even Z) precursors, namely $^{23}\mathrm{Si}$, $^{27}\mathrm{S}$ and $^{35}\mathrm{Ca}$. The decay properties of these members should be very similar to those of the A=4n+1, T_z=-$\frac{3}{2}$ series, namely, strong β^+-decay to analog states followed by isospin forbidden proton emission to low lying states in the daughter. Completion of T=$\frac{5}{2}$ sextet mass measurements will be very interesting.

No nuclide with $T_z < -3$ is predicted to be stable against particle emission. Experimentally, this has not yet been verified.

5.3   Delayed Particle Spectroscopy N>Z

For β^+ decay from N≮Z nuclei, the analog state lies at a lower energy and is the dominating feature in the β-transition process. For N>Z nuclei, the precursor, usually having the ground state $T=T_z$, decays to nuclides with isospin T+1. There is no dominating single branch transition and the β^+ decay strength is spread over many allowed transitions via Gamow-Teller (GT) interactions.

The β^- decay from N≮Z precursor (isospin T) will populate nuclei with isospin (T-1), but its analog state lies at a higher energy and GT β-decay is still the predominant process. This spreading of beta strength weakens the relative individual transition strength.

5.3.1 $T_z=\frac{1}{2}$ Delayed Proton Precursors

This series of delayed proton precursors started at $^{65}\mathrm{Ge}$ and delayed proton spectra for four of these members were obtained[3] and shown in Fig. 5.6. As expected, they are all bell-shaped, with fluctuations superimposed on them. The solid curves were obtained based on the approach outlined in Section 5.1. The beta intensities were obtained from the gross theory of β-decay[18], the proton reduced width r_p^{if} were estimated via optical potential and the γ-width was obtained from experimental dates on photoexcitation and (n,γ) reaction. The fit was obtained by varying the end-point energy of the spectrum $(Q_{EC}-B_p)$ and the overall normalization.

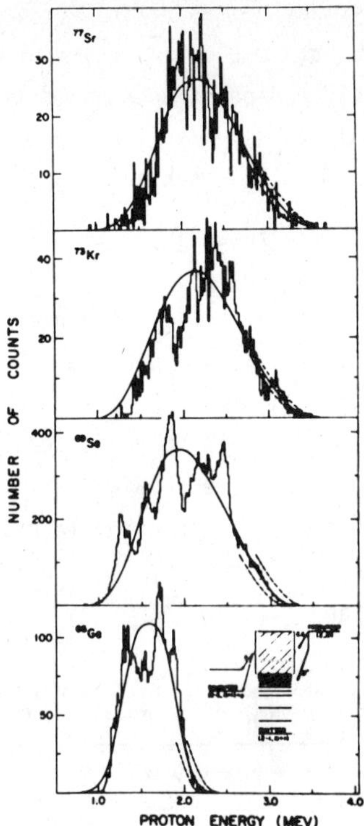

<u>Fig. 5.6</u> - Spectra of protons observed following the decays of ^{65}Ge,^{69}Se,^{73}Kr and ^{77}Sr. Smooth curves are the results of statistical model calculations with the dashed lines at high energy representing ± 100 keV changes in (Q_{EC}-B_p) values. A general decay scheme is also shown (from Reference 3).

As can be seen, the general bell-shape is well produced. The fluctuations could be due to single level effects, but they can also be caused by statistical fluctuations in the distribution of transition probabilities and level spacings.

5.3.2   <u>Other Heavy-Mass Neutron Deficient Precursors</u>

In the medium and heavy mass region, a large number of nuclides are potentially delayed proton and (or) alpha emitters. With the heavy-ion accelerators and on-line isotope separators, many of these precursors are now being studied in detail. The delayed particles usually have very smooth bell-shaped energy spectra as demonstrated in Fig. 4.11. Obviously, statistical approach for the analysis of such spectrum is necessary. In general, the upper part of the spectrum is more sensitive to the beta strength function and the lower energy part is more affected by the particle emission probability. Detailed analysis of the spectrum can yield information about the strength functions and other nuclear properties. Some examples are illustrated here.

Q_{EC} values - As already discussed in Section 4.42, and also demonstrated in the case for the $T_z = \frac{1}{2}$, the upper part of the delayed particle spectrum can be fitted by varying the particle end-point energy and thus the Q_{EC}-value. A note of precaution: reliability of analysis depends strongly on the beta strength function used. Good counting statistics is also essential.

Spins and Parities - As shown in eq. 5.5, the particle spectrum is dependent on the spins and parities of the precursor (J^π), the emitter level ($J_i^{\pi_i}$) and the final state in the daughter ($J_f^{\pi_f}$). If coincidence measurements of the delayed particles with the de-exciting γ-rays in the daughter nuclei can be made, much more information can be obtained. One such example is the deduction of J^π of the precursor from the relative intensities of proton branches feeding various levels in the daughter as illustrated by the Table shown in Fig. 5.7. Here, the delayed proton branches from the decay of ^{115}Xe precursor feeding levels in ^{114}Ba daughter were measured. By assuming a J^π for the precursor, these branches can be calculated. Results are shown in the Table and a J^π of $5/2^+$ for ^{115}Xe is thus deduced.

<u>Table 4</u>

Calculation[a] of the relative and absolute proton intensities from ^{115}Xe for various values of the initial spin and parity

Assumed I^π	Relative intensity (%)			Absolute intensity P_p
	0.0 MeV (0^+)	0.71 MeV (2^+)	1.48 MeV (4^+)	
$1/2^-$	92.3	7.7	0.0	9.6×10^{-3}
$3/2^-$	79.7	20.0	0.3	5.6×10^{-3}
$5/2^-$	62.4	36.0	1.6	2.8×10^{-3}
$7/2^-$	2.1	87.8	9.8	7.0×10^{-4}
$1/2^+$	83.5	16.4	0.0	9.6×10^{-3}
$3/2^+$	72.8	27.1	0.1	6.4×10^{-3}
$5/2^+$	41.1	55.7	3.2	2.3×10^{-3}
$7/2^+$	33.2	56.3	10.5	1.1×10^{-3}
Experiment [b]	~ 40	58±7	< 2	$(3.6 \pm 0.6) \times 10^{-3}$

<u>Fig. 5.7</u> - Result of calculation illustrating the determination of $J^\pi = \frac{5}{2}^+$ for ^{115}Xe (from Reference 3).

Strength Functions - The overall fit of the particle energy spectrum is the summation of products of the beta intensity and the particle partial width (eq. 5.5). Since the exact nature of these factors are not well understood, it is difficult to isolate one from the other. At present, the penetration probability of low energy charged particles cannot yet be estimated with certainty and the prediction of energy dependence of particle partial width is not very definite. These spectra may provide a good opportunity for such studies. As for the beta strength function, it can be tested via other experimental results such as the prediction of half-lives, the deduction of γ-ray and delayed neutron spectra, etc. However, no firm conclusion of the general behaviour of beta strength function can be drawn yet.

5.3.3   Delayed Neutron Precursors

Beta delayed neutron emissions should be a very common occurrence for neutron rich nuclides. Over a vast mass region, Q_β is larger than B_n, and delayed neutron emissions from the highly excited states are possible. In fact, due to the absence of Coulomb barrier, the neutron partial width for the states with $E_i > B_n$ should be the dominant ones. In the light mass region, these neutron rich nuclides can be produced via fragmentation or deep inelastic scattering processes. For the medium mass region, fission is obviously the most suitable mechanism. All these methods are non-selective production processes and some sorting will be necessary. Despite the abundance of these delayed neutron precursors, earlier work along that line is very scarce. This is mainly due to the lack of means for the production and isolation of these nuclides, and the general low efficiency and poor energy resolution of the neutron detectors. By 1972, only two neutron energy spectra from fission products were obtained, the ^{87}Br (56 sec half-life) and ^{137}I (24 s half-life) precursors.

The situation changed drastically in the past decade. First, a ^{3}He filled ionization chamber neutron detector with an energy resolution of < 20 keV was developed. Although, the absolute efficiency was still low ($\sim 10^{-4}$ for 1 MeV neutrons), neutron spectroscopy became a distinct possibility. More recently, automated fast chemical separation methods were developed, together with the installation of new on-line isotope separators at various facilities such as reactors and high energy proton accelerators, vastly improved yield of the neutron rich nuclides became available and much progress has been made since.

Early work in the light mass region was confined to nuclides that can be produced via simple reactions such as (t,p). A series of precursors with $T_z = \frac{3}{2}$, A=4n+1, (Z=odd), namely, ^{9}Li, ^{13}B and ^{17}N were identified. The neutron spectrum exhibits distinct peaks, which correspond to transition between individual states in the emitter and daughter. The other precursors lie further away from the stability and are most effectively produced via fragmentation reaction. Recent results identified many more delayed neutron precursors. In particular, a new radioactivity was discovered: delayed two-neutron emission was detected from decay of ^{11}Li, and $^{30-32}$Na. The proposed decay property and the detection method[4] in the study of ^{11}Li is illustrated in Fig. 5.8.

In the medium mass region, early delayed neutron spectra from the relatively long-lived precursors ^{87}Br and ^{137}I showed generally broad continuum. With improved energy resolution of the ^{3}He detector, the continuum is resolved into

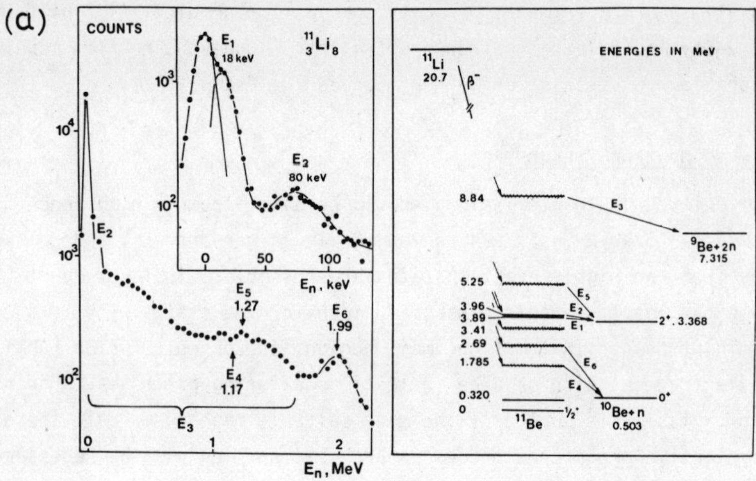

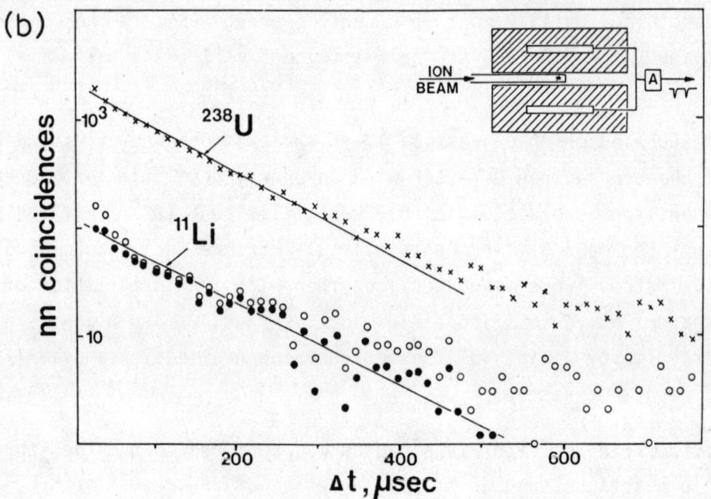

Fig. 5.8 - Delayed two -neutron emission from ^{11}Li (from Reference 5). (a) The pulse height spectrum of delayed neutrons from ^{11}Li, and the proposed origin of the various energy groups; (b) Time correlation of delayed neutrons from ^{11}Li as com-pared to the neutrons from ^{238}U fission. Neutrons from ^{9}Li show no time correlation effect.

many overlapping peaks. This situation seems to be common even for heavier mass pre-cursors. Many such spectra were obtained using sources obtained after fast chemical or on-line isotope separation. In favourable cases, neutron spectra in coincidence with γ-rays in the daughter nuclei were also obtained[11]. Fig. 5.9 shows a delayed neutron spectrum[3]) from the 1.7 sec ^{135}Sb decay. This is a particularly favourable case for the observation of individual energy groups: the precursor has N=84 and the emitter has just one excess neutron beyond the N=82 shell. The $(Q_\beta - B_n)$ of 4 MeV also allows the neutron emission to populate excited states in ^{134}Te. Detailed analysis

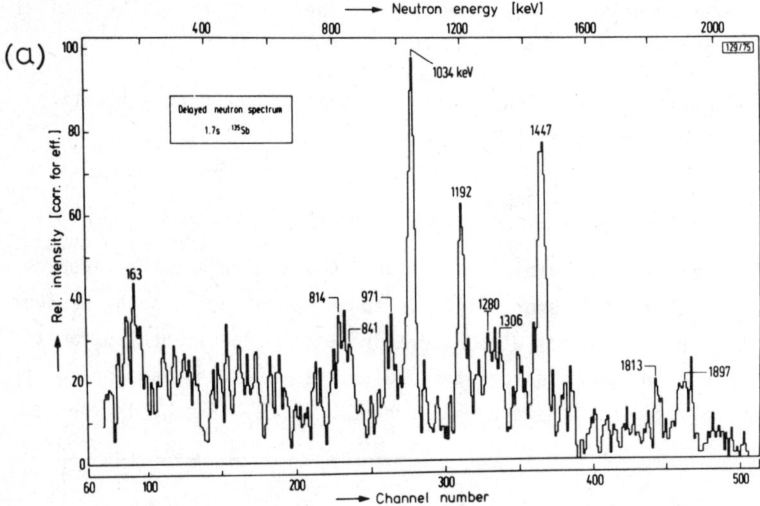

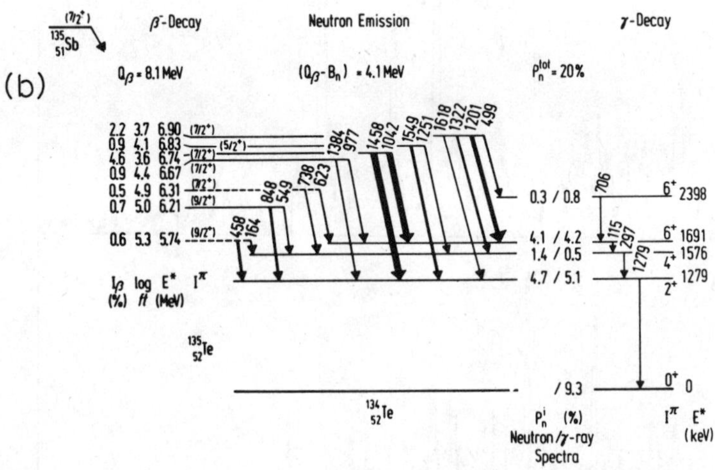

of the spectrum, together with the γ-ray measurement led to the identification of highly excited individual states in the emitter, and the suggestion of appreciable nuclear structure effect on the selectivity of β-decay and neutron emission.

From the statistical model point of view, fluctuation in the delayed particle energy spectrum can also occur due to the limited number of participating levels and transitions. This effect is mainly caused by the fluctuation in the transition probabilities and appropriate analysis will show the extent of statistical fluctuation and relfect on the level density of the emitter.

The fluctuation in the transition probability for a single reaction channel is governed by the Porter-Thomas (PT) law.

$$P(\Gamma_X) = \left(\frac{1}{2\pi < \Gamma_X > \Gamma_X}\right)^{\frac{1}{2}} \exp\left(-\frac{\Gamma_X}{2 < \Gamma_X >}\right) \qquad 5.6$$

where Γ_X is the transition probability for channel x and $< \Gamma_X >$ is the average width. The delayed particle emission probability is then proprotional to the product of two such PT distribution functions. With a certain set of experimental parameters, the statistical fluctuation can be calculated and expressed as the variance of the intensities $(\mathrm{Var}(I_n / < I_n >))$. The enrgy dependence of the variance is then compared

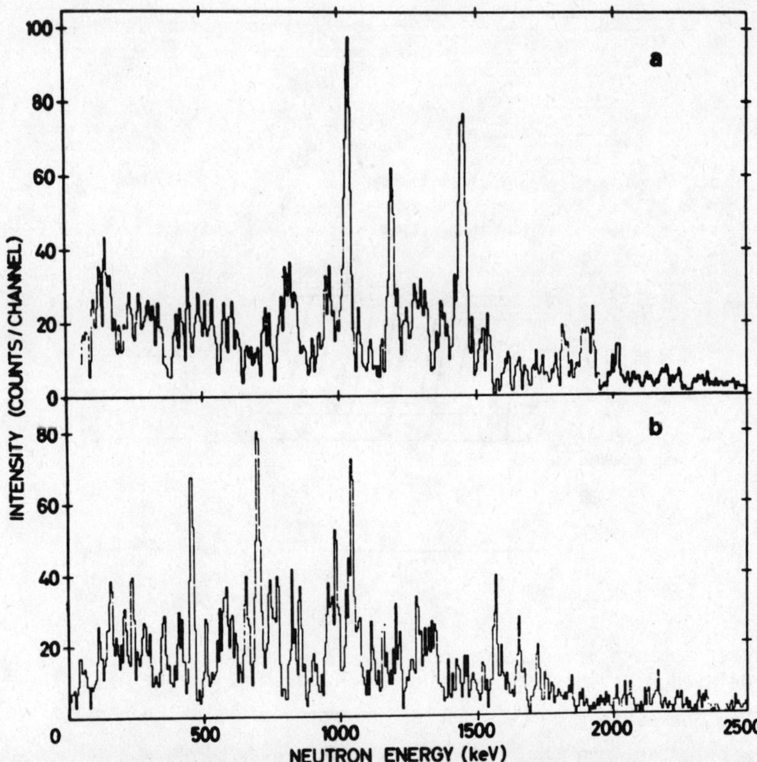

Fig. 5.10 - Beta-delayed neutron spectra (from reference 6). (a) From ^{135}Sb decay (same as Fig. 5.9); (b) From a "pandamonium" calculation.

to the actual experimental data. This calculation was carried out for a pseudonucle-
us named "pandemonium" using Monte-Carlo technique. In this calculation, pure
statistical approach was used. Delayed neutron spectrum was generated allowing the
population of different states in the daughter. Fig. 5.10 shows a comparison between
the pandemonium simulated for ^{135}Sb precursor and the actual experimental data. As
can be seen, large fluctuations were also obtained from statistical effects alone,
suggesting the necessity of thorough analysis of experimental data in order to isolate the nuclear structure effect.

5.4   Other Delayed Emission Precursors

In the very light mass regions, some neutron rich nuclides are delayed alpha
precursors. The resulting α spectra are usually broad continuum shaped. One interesting exception is the decay of ^{16}N(7.1 sec), where a weak branch of alpha emission
from the 8.87 MeV (2^-) level in ^{16}O to ground (0^+) in ^{12}C was detected; a parity-
violating transition.

In the heavy mass regions ($A > 200$), many more delayed particle and delayed
fission precursors exist. These are interesting areas but will not be pursued
further here.

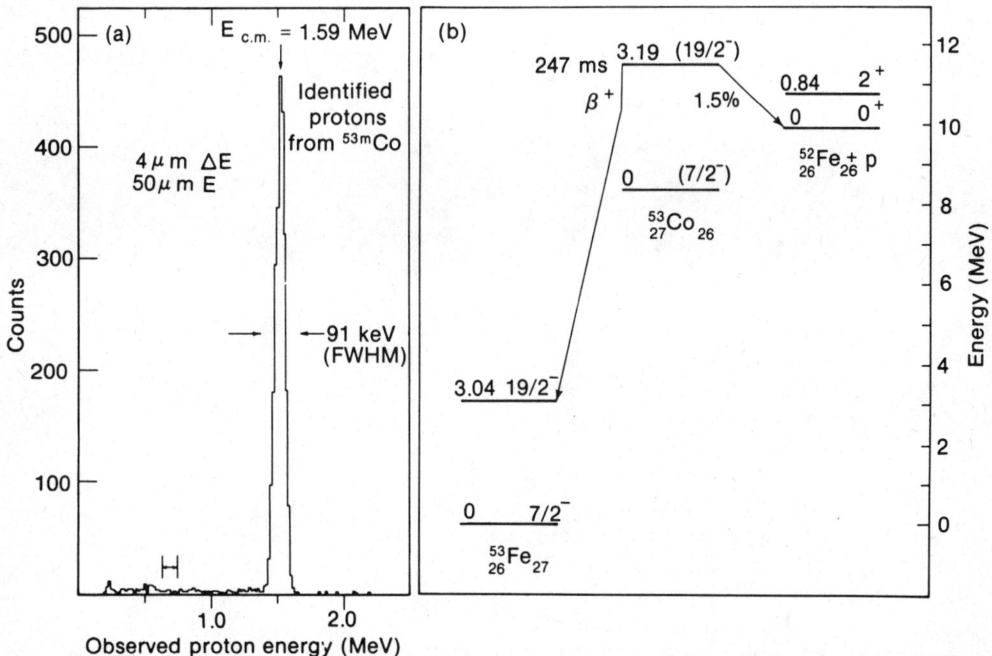

Fig. 5.11 - Observed proton energy (MeV) (from Reference 11).(a) An identified proton
energy spectrum from the decay of ^{53m}Co produced by the ^{54}Fe(p,2n) reaction induced by
35-MeV protons. The horizontal arrow indicates the location of any possible transitions to the ^{52}Fe* (0.84-MeV) state; (b) The decay scheme of ^{53m}Co.

5.5   Proton Radioactivity

At the proton drip line, B_p becomes negative, i.e. the nucleus is unstable against proton emission. In these cases, the nucleus will decay by one-step emission of a proton, similar to the situation for alpha emitters.   If the hindrance due to Coulomb and other barrier is severe, the life-time may be long enough for a positive identification of such radioactivity. Although, all the nuclei beyond the proton drip line are potential candidates, the requirement for the barrier effect is such that B_p can only have a narrow energy range. So far, one such activity has been identified[11], - a 1.59 MeV proton group from the isomer ^{53m}Co(J^{π}=19/2$^-$) to the ground state of ^{52}Fe as shown in Fig. 5.11. The partial life-time is very long and the retardation is due to a strong dependence of this transition rate on the residual interaction of the $(f_{7/2})^{-3}$ configuration of the isomeric state.

The possibility of two-proton radioactivity has also been speculated. This can occur if a nucleus is stable against single proton emission but not stable aginst di-proton emission. No evidence exists yet for such activity.

REFERENCES

1. Proceedings of First International Conference on "Nuclei Far Off Stability Line", Lysckil, Sweden, 1966, ed. W. Forsling, C.J. Herrlander and H. Ryde. Almgrist & Wiksell, Stockholm.

2. Proceedings of Second International Conference on "Properties of Nuclei Far From the Region of Beta Stability", Leysin, Switzerland (1979), CERN 70-30, Vol. I and II.

3. Proceedings of Third International Conference on "Nuclei Far From Stability", Cargése, France (1976), CERN 76-13.

4. Proceedings of the Isotope Separator On-Line Workshop, Brookhaven National Laboratory, U.S.A. (1977).

5. Proceedings of International Symposium on "Future Directions in Studies of Nuclei Far From Stability", Nashville, U.S.A. (1979), ed. J.H. Hamilton, E.H. Spejewski, C.R. Bingham and E.F. Zganjar, North-Holland.

6. P.G. Hansen, Ann. Rev., Nucl. and Part. Phys $\underline{29}$ (1979) 69.

7. Proceedings of First Symposium on "Interacting Bosons in Nuclear Physics, Sicily, Italy (1978), Plenum Press.

8. Proceedings of Fifth International Conference on "Atomic Masses and Fundamental Constants", Paris, France (1975) Plenum Press.

9. Proceedings of Sixth International Conference on "Atomic Masses and Fundamental Constants", East Lansing, U.S.A. (1979).

10. Atomic Data Nuc. Data Table 17 (1976).

11. J. Cerny and J.C. Hardy, Ann. Rev. Nucl. Sc. $\underline{27}$ (1977) 333.

12. K.L. Kratz, "Review of Delayed Neutron Energy Spectra", IAEA Consultant's Meeting on Delayed Neutron Properties".

13. "Nuclear Spectroscopy and Nuclear Reactions", ed. J. Cerny, Part A-D, Academic Press (1974).

14. V.V. Volkov, Phys. Rep. $\underline{44}$ (1978) 93.

15. Proceedings of the Nineth International Conference on "Electromagnetic Isotopte Separators and Related Ion Accelerators", Kiryat Anavim, Israel (1976), Nucl. Inst. & Meth. 139 (1976).

16. H.L. Ravn, Phys. Rep. $\underline{54}$ (1979) 201.

17. G.T. Garvey and I. Kelsen, Phys. Rev. Lett. $\underline{16}$ (1966) 197, and Phys. Rev. Lett. $\underline{23}$ (1966) 689.

18. Proceedings of International Conference on Nuclear Physics, Berkeley, U.S.A. 1980.

19. S.I. Koyama, K. Takahashi and M. Yamada, Progr. Theor. Phys. $\underline{44}$ (1970) 633 and: K. Takahashi, M. Yamada and T. Kondoh, Atomic Data and Nucl. Data Tables $\underline{12}$ (1973) 101.

Lecture Notes in Physics

Vol. 114: Stellar Turbulence. Proceedings, 1979. Edited by D. F. Gray and J. L. Linsky. IX, 308 pages. 1980.

Vol. 115: Modern Trends in the Theory of Condensed Matter. Proceedings, 1979. Edited by A. Pekalski and J. A. Przystawa. IX, 597 pages. 1980.

Vol. 116: Mathematical Problems in Theoretical Physics. Proceedings, 1979. Edited by K. Osterwalder. VIII, 412 pages. 1980.

Vol. 117: Deep-Inelastic and Fusion Reactions with Heavy Ions. Proceedings, 1979. Edited by W. von Oertzen. XIII, 394 pages. 1980.

Vol. 118: Quantum Chromodynamics. Proceedings, 1979. Edited by J. L. Alonso and R. Tarrach. IX, 424 pages. 1980.

Vol. 119: Nuclear Spectroscopy. Proceedings, 1979. Edited by G. F. Bertsch and D. Kurath. VII, 250 pages. 1980.

Vol. 120: Nonlinear Evolution Equations and Dynamical Systems. Proceedings, 1979. Edited by M. Boiti, F. Pempinelli and G. Soliani. VI, 368 pages. 1980.

Vol. 121: F. W. Wiegel, Fluid Flow Through Porous Macromolecular Systems. V, 102 pages. 1980.

Vol. 122: New Developments in Semiconductor Physics. Proceedings, 1979. Edited by F. Beleznay et al. V, 276 pages. 1980.

Vol. 123: D. H. Mayer, The Ruelle-Araki Transfer Operator in Classical Statistical Mechanics. VIII, 154 pages. 1980.

Vol. 124: Gravitational Radiation, Collapsed Objects and Exact Solutions. Proceedings, 1979. Edited by C. Edwards. VI, 487 pages. 1980.

Vol. 125: Nonradial and Nonlinear Stellar Pulsation. Proceedings, 1980. Edited by H. A. Hill and W. A. Dziembowski. VIII, 497 pages. 1980.

Vol. 126: Complex Analysis, Microlocal Calculus and Relativistic Quantum Theory. Proceedings, 1979. Edited by D. Iagolnitzer. VIII, 502 pages. 1980.

Vol. 127: E. Sanchez-Palencia, Non-Homogeneous Media and Vibration Theory. IX, 398 pages. 1980.

Vol. 128: Neutron Spin Echo. Proceedings, 1979. Edited by F. Mezei. VI, 253 pages. 1980.

Vol. 129: Geometrical and Topological Methods in Gauge Theories. Proceedings, 1979. Edited by J. Harnad and S. Shnider. VIII, 155 pages. 1980.

Vol. 130: Mathematical Methods and Applications of Scattering Theory. Proceedings, 1979. Edited by J. A. DeSanto, A. W. Sáenz and W. W. Zachary. XIII, 331 pages. 1980.

Vol. 131: H. C. Fogedby, Theoretical Aspects of Mainly Low Dimensional Magnetic Systems. XI, 163 pages. 1980.

Vol. 132: Systems Far from Equilibrium. Proceedings, 1980. Edited by L. Garrido. XV, 403 pages. 1980.

Vol. 133: Narrow Gap Semiconductors Physics and Applications. Proceedings, 1979. Edited by W. Zawadzki. X, 572 pages. 1980.

Vol. 134: $\gamma\gamma$ Collisions. Proceedings, 1980. Edited by G. Cochard and P. Kessler. XIII, 400 pages. 1980.

Vol. 135: Group Theoretical Methods in Physics. Proceedings, 1980. Edited by K. B. Wolf. XXVI, 629 pages. 1980.

Vol. 136: The Role of Coherent Structures in Modelling Turbulence and Mixing. Proceedings 1980. Edited by J. Jimenez. XIII, 393 pages. 1981.

Vol. 137: From Collective States to Quarks in Nuclei. Edited by H. Arenhövel and A. M. Saruis. VII, 414 pages. 1981.

Vol. 138: The Many-Body Problem. Proceedings 1980. Edited by R. Guardiola and J. Ros. V, 374 pages. 1981.

Vol. 139: H. D. Doebner, Differential Geometric Methods in Mathematical Physics. Proceedings 1981. VII, 329 pages. 1981.

Vol. 140: P. Kramer, M. Saraceno, Geometry of the Time-Dependent Variational Principle in Quantum Mechanics. IV, 98 pages. 1981.

Vol. 141: Seventh International Conference on Numerical Methods in Fluid Dynamics. Proceedings. Edited by W. C. Reynolds and R. W. MacCormack. VIII, 485 pages. 1981.

Vol. 142: Recent Progress in Many-Body Theories. Proceedings. Edited by J. G. Zabolitzky, M. de Llano, M. Fortes and J. W. Clark. VIII, 479 pages. 1981.

Vol. 143: Present Status and Aims of Quantum Electrodynamics. Proceedings, 1980. Edited by G. Gräff, E. Klempt and G. Werth. VI, 302 pages. 1981.

Vol. 144: Topics in Nuclear Physics I. A Comprehensive Review of Recent Developments. Edited by T.T.S. Kuo and S.S.M. Wong. XX, 567 pages. 1981.

Vol. 145: Topics in Nuclear Physics II. A Comprehensive Review of Recent Developments. Proceedings 1980/81. Edited by T. T. S. Kuo and S. S. M. Wong. VIII, 571-1.082 pages. 1981.